Logic in Computer Science

Hantao Zhang • Jian Zhang

Logic in Computer Science

 Springer

Hantao Zhang
Department of Computer Science
University of Iowa
Iowa City, IA, USA

Jian Zhang
Institute of Software
Chinese Academy of Sciences
Beijing, China

ISBN 978-981-97-9815-5   ISBN 978-981-97-9816-2   (eBook)
https://doi.org/10.1007/978-981-97-9816-2

© The Editor(s) (if applicable) and The Author(s), under exclusive license to Springer Nature Singapore Pte Ltd. 2025

This work is subject to copyright. All rights are solely and exclusively licensed by the Publisher, whether the whole or part of the material is concerned, specifically the rights of translation, reprinting, reuse of illustrations, recitation, broadcasting, reproduction on microfilms or in any other physical way, and transmission or information storage and retrieval, electronic adaptation, computer software, or by similar or dissimilar methodology now known or hereafter developed.
The use of general descriptive names, registered names, trademarks, service marks, etc. in this publication does not imply, even in the absence of a specific statement, that such names are exempt from the relevant protective laws and regulations and therefore free for general use.
The publisher, the authors and the editors are safe to assume that the advice and information in this book are believed to be true and accurate at the date of publication. Neither the publisher nor the authors or the editors give a warranty, expressed or implied, with respect to the material contained herein or for any errors or omissions that may have been made. The publisher remains neutral with regard to jurisdictional claims in published maps and institutional affiliations.

This Springer imprint is published by the registered company Springer Nature Singapore Pte Ltd.
The registered company address is: 152 Beach Road, #21-01/04 Gateway East, Singapore 189721, Singapore

If disposing of this product, please recycle the paper.

Preface

Mathematical logic is an important basis for mathematics, computer Science, and artificial intelligence. The application of logic as a problem-solving tool is an important aspect of education of computer scientists and engineers. This book provides a comprehensive introduction to various logics, including the classical propositional logic and first-order predicate logic, as well as equational logic, temporal logic, and Hoare logic. The focus of the book is to present algorithms in the form of proof procedures for the classical logics and decision procedures for checking the satisfiability of logical formulas. A large portion of the book is devoted to the introduction of software tools based on these algorithms and the practical problems which can be solved by these tools in constraint satisfaction, formal verification, and artificial intelligence.

The book assumes no background in logic. It is appropriate for (junior and senior) undergraduate and graduate students majoring in computer science or mathematics. Each chapter has about a dozen or so exercises to help the reader understand the materials.

The first chapter of the book is a general introduction of logic (with a bias toward mathematical logic) and the notions to be used through the book. The rest of the book is divided into four parts:

- Part I: Propositional logic, Chaps. 2–4
- Part II: First-order logic, Chaps. 5–7
- Part III: Logic in Programming, Chaps. 8–10
- Part IV: Logic of Computability, Chaps. 11 and 12

The topics of this book were chosen from the field of logic with a focus on the algorithmic aspect of logic. We have selected only most effective algorithms from the field. We did not list all the important references to these topics as they are rarely needed for the beginners and can be found easily on the internet. We thank all the original contributors of these contents, including research papers and software tools, and hope they will forgive us for not listing them all.

This book also contains some original materials:

- In Chap. 5, we show that the set definition by a first-order formula is an interesting application of first-order logic, where the defined set is a special model of the formula. If the set is a new entity (type or sort), the set itself is the extensional definition of the entity and the formula is its intentional definition.
- A new presentation of an almost-linear time unification algorithm in Chap. 6.
- In Chap. 8, we introduce the notions of *first-termination* and *last termination* of Prolog programs.
- An interpretation of Gödel's incompleteness theorems in terms of Turing machines is given in Chap. 11.
- Also in Chap. 11, a new concept of "computably countable" is shown to be equivalent to computable or recursively enumerable, and is the result of adding computability to countability.
- In Chap. 12, we pointed out that Deepak Kapur's ground congruence algorithm can be simplified by using only "splitting," that is, "flattening" is optional.

Without the help from many friends, colleagues, and our families, there would be no printing of this book. The authors thank everyone who helped one way or the other.

This book grew out of notes from a course titled "Logic in Computer Science" at the University of Iowa. The first author taught this course many times. Students in his course used these notes as textbooks and provided good feedback. We apologize for not listing their names here. We thank Cesare Tinelli for introducing this course to Iowa, Zubair Shafiq for using an early version of this book in teaching this course, and Alberto Segre for the support when the first author expressed the idea of converting the notes into a textbook. We also thank Clark Barrett, Pascal Fontaine, and Cesare Tinelli for permitting us to use a figure of SMT-LIB in this book.

We wish to thank the teachers and colleagues who enriched the authors' knowledge on the topics of the book. The first author thanks Jean-Luc Rémy, Pierre Lescanne, Jean-Pierre Jouannoud, Hélène Kirchner, Claude Kirchner, and Michael Rusinowitch for making the oversea life enjoyable for a new graduate student to study rewrite systems. The first author is grateful to his thesis advisor, Deepak Kapur, for supporting and directing his research on automated induction and ordered resolution within rewrite systems. The second author is grateful to his thesis advisors, Yunmei Dong and C.S. Tang, from whom he learned about formal specifications, rewrite systems, and temporal logic. We also thank the colleagues who worked with us on many topics of the book. They are Peter Baumgartner, Frank Bennett, Maria Paola Bonaciana, Jieh Hsiang, Emmanual Kounalis, William McCune, Mark Stickel, to name a few.

Many people read different versions of the draft, and provided us with useful comments and suggestions. We are grateful to all of them, including Shaowei Cai, Wenjian Chai, Yuhang Dong, Rui Han, Jiangang Huang, Fuqi Jia, Changwen Li, Tian Liu, Yongmei Liu, Kunhang Lv, Zongyan Qiu, Yidong Shen, Langchen Shi, Yuefei Sui, Sicheng Tan, Bohua Zhan, Wenhui Zhang, Yu Zhang, Chunlai Zhou, and Xueyang Zhu. Special thanks go to Tian Liu, who provided a very good technique for proving a language is unrecognizable and whose comments lead to the concept of "computably countable" in Chap. 11.

It has been a pleasure to work with the folks at Springer Nature in creating the final product. We mention Celine Lanlan Chang, Sudha Ramachandran, and Kamesh Senthilkumar because we have had the most contact with them, but we know that many others have been involved, too. We would also like to thank the National Science Foundation of China (NSFC) for its continuous support.

Finally, we would like to thank deeply our families for their patience, inspiration, and love. The first author is grateful in particular to his son, Roy, and daughter, Felice, for comments and proofreading of the book, and his wife, Ling Rao, for logistic of writing the book. This book is dedicated to Ling.

Iowa City, IA, USA	Hantao Zhang
Beijing, China	Jian Zhang
August 2024	

Contents

1	**Introduction to Logic**		1
	1.1 Logic Is Everywhere		2
		1.1.1 Statement or Proposition	3
		1.1.2 A Brief History of Logic	5
		1.1.3 Thinking or Writing Logically	6
	1.2 Logical Fallacies		8
		1.2.1 Formal Fallacies	8
		1.2.2 Informal Fallacies	9
	1.3 A Glance of Mathematical Logic		10
		1.3.1 Set Theory	11
		1.3.2 Computability Theory	20
		1.3.3 Model Theory	24
		1.3.4 Proof Theory	26
	Exercises		29
	References		32

Part I Propositional Logic

2	**Propositional Logic**		35
	2.1 Syntax		35
		2.1.1 Logical Operators	35
		2.1.2 Formulas	37
	2.2 Semantics		38
		2.2.1 Interpretations	39
		2.2.2 Models, Satisfiability, and Validity	41
		2.2.3 Equivalence	43
		2.2.4 Entailment	45
		2.2.5 Theorem Proving and the SAT Problem	47
	2.3 Normal Forms		48
		2.3.1 Negation Normal Form (NNF)	49
		2.3.2 Conjunctive Normal Form (CNF)	51

		2.3.3	Disjunctive Normal Form (DNF)	53
		2.3.4	Full DNF and Full CNF from Truth Table	54
		2.3.5	Binary Decision Diagram (BDD)	55
	2.4	Optimization Problems		59
		2.4.1	Minimum Set of Operators	59
		2.4.2	Logic Minimization	61
		2.4.3	Maximum Satisfiability	65
	2.5	Using Propositional Logic		66
		2.5.1	Bitwise Operators	67
		2.5.2	Decision Problems in Propositional Logic	68
	Exercises			73
	References			78
3	**Reasoning in Propositional Logic**			79
	3.1	Proof Procedures		79
		3.1.1	Types and Styles of Proof Procedures	80
		3.1.2	Semantic Tableau	81
		3.1.3	α-Rules and β-Rules	83
	3.2	Deductive Systems		85
		3.2.1	Inference Rules and Proofs	86
		3.2.2	Hilbert System	88
		3.2.3	Natural Deduction	89
		3.2.4	Inference Graphs	92
	3.3	Resolution		93
		3.3.1	Resolution Rule	93
		3.3.2	Resolution Strategies	96
		3.3.3	Preserving Satisfiability	97
		3.3.4	Completeness of Resolution	100
		3.3.5	A Resolution-Based Decision Procedure	102
		3.3.6	Simplification Strategies	103
	3.4	Boolean Constraint Propagation (BCP)		106
		3.4.1	*BCP*: A Simplification Procedure	106
		3.4.2	A Decision Procedure for Horn Clauses	108
		3.4.3	Unit Resolution Versus Input Resolution	109
		3.4.4	Head/Tail Literals for *BCP*	111
	Exercises			115
	References			117
4	**Propositional Satisfiability**			119
	4.1	The DPLL Algorithm		120
		4.1.1	Recursive Version of *DPLL*	120
		4.1.2	All-SAT and Incremental SAT Solvers	123
		4.1.3	*BCPw*: Use of Watch Literals in *DPLL*	124
		4.1.4	Iterative Implementation of *DPLL*	127
	4.2	Conflict-Driven Clause Learning (CDCL)		128
		4.2.1	Generating Clauses from Conflicting Clauses	129

	4.2.2	DPLL with CDCL	131
	4.2.3	Unsatisfiable Cores	133
	4.2.4	Random Restart	134
	4.2.5	Branching Heuristics for DPLL	135
4.3	Use of SAT Solvers		137
	4.3.1	Specify SAT Instances in DIMACS Format	138
	4.3.2	Sudoku Puzzle	139
	4.3.3	Latin Square Problems	140
	4.3.4	Graph Problems	141
4.4	Maximum Satisfiability		143
	4.4.1	2SAT Versus Max2SAT	143
	4.4.2	Weighted and Hybrid MaxSAT	145
	4.4.3	Local Search Methods	147
	4.4.4	The Branch-and-Bound Algorithm	149
	4.4.5	Use of Hybrid MaxSAT Solvers	154
Exercises			155
References			158

Part II First-Order Logic

5 First-Order Logic ... 161
- 5.1 Syntax of First-Order Languages ... 161
  - 5.1.1 Terms and Formulas ... 162
  - 5.1.2 Quantifiers ... 165
  - 5.1.3 Unsorted and Many-Sorted Logic ... 167
- 5.2 Semantics ... 169
  - 5.2.1 Interpretation ... 170
  - 5.2.2 Models, Satisfiability, and Validity ... 173
  - 5.2.3 Equivalence and Entailment ... 175
  - 5.2.4 Set Constructions and Many-Sorted Logic ... 179
  - 5.2.5 First-Order Logic Versus Higher-Order Logic ... 182
- 5.3 Proof Methods ... 184
  - 5.3.1 Semantic Tableau ... 185
  - 5.3.2 Natural Deduction ... 187
- 5.4 Conjunctive Normal Form (CNF) ... 188
  - 5.4.1 Prenex Normal Form ... 188
  - 5.4.2 Skolemization ... 190
  - 5.4.3 Clausal Form ... 194
- Exercises ... 198
- References ... 200

6 Unification and Resolution ... 201
- 6.1 Unification ... 201
  - 6.1.1 Substitutions and Unifiers ... 202
  - 6.1.2 Combining Substitutions ... 203
  - 6.1.3 Rule-Based Unification ... 204

		6.1.4 Practically Linear Time Unification Algorithm	207
	6.2	Resolution	213
		6.2.1 Formal Definition	213
		6.2.2 Factoring	215
		6.2.3 A Refutational Proof Procedure	215
	6.3	Simplification Orders and Ordered Resolution	217
		6.3.1 Well-Founded Partial Orders	217
		6.3.2 Simplification Orders	219
		6.3.3 Completeness of Ordered Resolution	224
	6.4	Prover9: A Resolution Theorem Prover	226
		6.4.1 Input Formulas to Prover9	226
		6.4.2 Inference Rules and Options	229
		6.4.3 Simplification Orders in Prover9	231
		6.4.4 The TPTP Library	232
	Exercises		233
	References		235
7	**First-Order Logic with Equality**		**237**
	7.1	Equality of Terms	237
		7.1.1 Axioms of Equality	238
		7.1.2 Semantics of "="	239
		7.1.3 Theory of Equations	241
	7.2	Rewrite Systems	243
		7.2.1 Rewrite Rules	244
		7.2.2 Termination of Rewriting	246
		7.2.3 Confluence of Rewriting	247
		7.2.4 The Knuth-Bendix Completion Procedure	248
		7.2.5 Special Rewrite Systems	255
	7.3	Inductive Theorem Proving	259
		7.3.1 Inductive Theorems	259
		7.3.2 Structural Induction	261
		7.3.3 Induction on Two Variables	263
		7.3.4 Many-Sorted Algebraic Specification	264
		7.3.5 Recursion, Induction, and Self-Reference	268
	7.4	Resolution with Equality	269
		7.4.1 Paramodulation	270
		7.4.2 Simplification Rules	271
		7.4.3 Equality in Prover9	273
	7.5	Finite Model Finding in First-Order Logic	275
		7.5.1 Mace4: A Finite Model Finding Tool	275
		7.5.2 Finite Model Finding by SAT Solvers	278
	Exercises		280
	References		283

Contents xiii

Part III Logic in Programming

8 Prolog: Programming in Logic 287
    8.1   Prolog's Working Principle 287
            8.1.1   Horn Clauses in Prolog 288
            8.1.2   Resolution Proof in Prolog 289
            8.1.3   A Goal-Reduction Procedure 291
    8.2   Prolog's Data Types 295
            8.2.1   Atoms, Numbers, and Variables 296
            8.2.2   Compound Terms and Lists 297
            8.2.3   Popular Predicates over Lists 298
            8.2.4   Sorting Algorithms in Prolog 300
    8.3   Recursion in Prolog 301
            8.3.1   Program Termination 302
            8.3.2   Focused Recursion 304
            8.3.3   Tail Recursion 305
            8.3.4   A Prolog Program for N-Queen Puzzle 306
    8.4   Beyond Clauses and Logic 307
            8.4.1   The Cut Operator "!" 307
            8.4.2   Negation as Failure 309
            8.4.3   Beyond Clauses 310
            8.4.4   Variations and Extensions 311
    Exercises 312
    References 313

9 Hoare Logic 315
    9.1   Formal Verification of Computer Systems 315
            9.1.1   Verification of Imperative Programs 316
            9.1.2   Verification of Functional Programs 317
    9.2   Hoare Triples 318
            9.2.1   Hoare Rules 320
            9.2.2   Examples of Formal Verification 325
            9.2.3   Partial and Total Correctness 328
    9.3   Automated Generation of Assertions 329
            9.3.1   Verification Conditions 330
            9.3.2   Proof of Theorem 9.3.4 332
            9.3.3   Implementing VC in Prolog 336
    9.4   Obtaining Good Loop Invariants 340
            9.4.1   Invariants from Generalizing Postconditions 341
            9.4.2   Program Synthesis from Invariants 343
            9.4.3   Choosing A Good Language for Assertions 345
            9.4.4   Verification Tools for Conventional Languages ... 347
    Exercises 348
    References 349

10 Temporal Logic 351
10.1 An Approach from Modal Logic 352
10.1.1 Modal Operators 352
10.1.2 Kripke Semantics 353
10.1.3 Restrictions and Limitations 357
10.2 Linear Temporal Logic 358
10.2.1 Timeline as Interpretation Sequence 359
10.2.2 Properties of LTL 362
10.2.3 Model Checking with Kripke Frames 364
10.3 Semantic Tableaux for LTL 366
10.3.1 Rules for Modal Operators 367
10.3.2 Deciding Satisfiability by Tableaux 370
10.4 Binary Temporal Operators 377
10.4.1 The *until* and *release* Operators 377
10.4.2 The *weak until* and *strong release* Operators 380
10.4.3 Extension of Semantic Tableau 381
10.5 Verification of Concurrent Programs 381
10.5.1 Model Checking with Finite State Machines 382
10.5.2 Properties of Concurrent Programs 384
10.5.3 The BAKERY(2) Program 384
Exercises 389
References 392

Part IV Logic of Computability

11 Decidable and Undecidable Problems 395
11.1 A Bit of History 395
11.1.1 Gödel's Incompleteness Theorems 395
11.1.2 Three Well-Known Computing Models 397
11.1.3 Halting Problem 398
11.2 Turing Machines 398
11.2.1 Formal Definition of Turing Machines 400
11.2.2 High-Level Description of Turing Machines 404
11.2.3 Recognizable Versus Decidable 405
11.3 Decidability of Problems 406
11.3.1 Encoding of Decision Problems 406
11.3.2 Decidable Problems 408
11.3.3 Undecidable Problems 410
11.3.4 Reduction 412
11.3.5 Rice's Theorem 416
11.4 Computability of Counting Bijections 418
11.4.1 Properties of Counting Bijections 419
11.4.2 Equivalence of Recognizable and Recursively Enumerable 420

		11.4.3	Equivalence of Recognizable and Computably Countable	421

        11.4.4 Identifying Unrecognizable Languages 424
  11.5 Turing Completeness... 426
        11.5.1 Turing Completeness of Prolog 427
        11.5.2 Turing Completeness of Rewrite Systems................. 429
  Exercises 431
  References.. 432

12 Decision Procedures 433
  12.1 DPLL(T): Extend DPLL with Theories T 435
        12.1.1 Propositional Abstraction 435
        12.1.2 Examples of DPLL(T) 438
        12.1.3 $DPLLT(X)$: An Algorithm of DPLL(T) 440
  12.2 Equality with Uninterpreted Functions............................ 442
        12.2.1 Uninterpreted Functions................................... 442
        12.2.2 The Congruence Closure Problem 443
        12.2.3 The NOS Algorithm 445
        12.2.4 Ground Congruence by Completion 448
  12.3 Linear Arithmetic 453
        12.3.1 Simplex Method by Example 454
        12.3.2 The Simplex Algorithm 458
        12.3.3 Linear Programming....................................... 460
        12.3.4 Integer Programming 461
        12.3.5 Difference Logic 463
  12.4 Making DPLL(T) Practical.. 465
        12.4.1 Making DPLL(T) Efficient 465
        12.4.2 Combination of Theories.................................. 468
        12.4.3 SMT-LIB: A Library of SMT Problems 471
        12.4.4 SMT-COMP: Competition of SMT Solvers.............. 472
  Exercises 473
  References.. 475

Index 477

About the Authors

Hantao Zhang is Professor of Computer Science with the University of Iowa, USA. His research interests include automated reasoning, constraint solving, and discrete mathematics. He is the recipient of numerous NSF awards, including the prestigious NSF Young Investigator Award. He has published more than 100 papers in these areas, including a book on automated mathematical induction and a book chapter for the *Handbook of Satisfiability*. He won the Skolem Award by CADE (International Conference on Automated Deduction) in 2015 as his paper on SATO has passed the test of time as one of the most influential papers in the field.

Jian Zhang is a researcher with the Institute of Software, Chinese Academy of Sciences, and a professor with the University of Chinese Academy of Sciences. His research interests include automated reasoning, constraint solving, program analysis, and software testing. Jian Zhang serves on the editorial boards of several journals including *IEEE Transactions on Reliability*, *Journal of Computer Science and Technology*, *Frontiers of Computer Science*, *Science China – Information Sciences*, and *Chinese Journal of Computers*. He is the author of *Deciding the Satisfiability of Logical Formulas—Methods, Tools and Applications* (in Chinese, Science Press, 2000) and a co-author of *Automatic Generation of Combinatorial Test Data* (Springer, 2014).

Chapter 1
Introduction to Logic

Do you like to solve Sudoku puzzles? If so, how long will it take on average to solve a Sudoku puzzle appearing in a newspaper, minutes or hours? An easy exercise in Chap. 1 of this book asks you to write a small program in your favorite language that will solve a Sudoku puzzle, no matter the difficulty, in less than a second on your laptop.

Did you ever play the *Tower of Hanoi*? This puzzle consists of three rods and a number of disks of different sizes, which can slide onto any rod. The puzzle starts with the disks stacked in ascending order of size on one rod, making a conical shape with the smallest at the top. The objective of the puzzle is to move the entire stack to another rod, following three rules:

- Only one disk can be moved at a time.
- Each move consists of taking the top disk from one of the stacks and placing it on top of another stack or on an empty rod.
- No larger disk may be placed on top of a smaller disk.

In the illustration figure, the three rods are named a, b, and c (Fig. 1.1). There are three disks initially on rod a, and each disk is named by its size. A solution of the puzzle of moving the disks from rod a to rod b is given as follows:

```
Move disk 1 from a to b
Move disk 2 from a to c
Move disk 1 from b to c
Move disk 3 from a to b
Move disk 1 from c to a
Move disk 2 from c to b
Move disk 1 from a to b
```

This solution was produced by a Prolog program of 3 lines and 168 characters (including commas and periods). The same program can produce solutions for various numbers of disks. Prolog is a programming language based on logic and is introduced in Chap. 8 of the book.

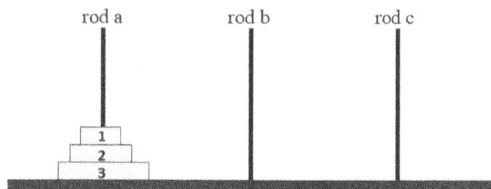

Fig. 1.1 Illustration of the Tower of Hanoi

The above two examples illustrate that logic is not only a theoretic foundation of computer science but also a problem-solving tool. The emphasis of this book is on how to make this tool efficient and practical. After a rigorous introduction to basic concepts, we will study various algorithms based on logic and introduce software tools which are implementations of these algorithms.

To solve a problem using a logic-based tool, you need to think about the problem in logic and specify the problem in logic. This is a skill that needs to be learned like any other, and it does take some training and practice to master this skill. The same skill can be applied to abstract situations such as those encountered in formal proofs.

Logic is innate to all of us—indeed, you probably use the laws of logic unconsciously in your everyday speech and in your own internal reasoning. Because logic is innate, the logic principles that we learn should make sense—if you find yourself having to memorize one of the principles in the book, without feeling a mental "click" or comprehending why that law should work, then you will probably not be able to use that principle correctly and effectively in practice.

In 1953 Albert Einstein wrote the following in a letter to J. E. Switzer:

> The development of Western Science has been based on the two great achievements, the invention of the formal logical system (in Euclidean geometry) by the Greek philosophers, and the discovery of the possibility of finding out causal relationships by systematic experiment (at the Renaissance).

This book demonstrates the application of these two great achievements of modern science in one setting: developing formal logical systems by systematic experiment. That is, our focus is to convert the algorithms of logical systems into logic tools, and both the algorithms and the tools need systematic experiments.

1.1 Logic Is Everywhere

Logic has been called the *calculus of computer science*, because logic is fundamental in computer science, similar to calculus in the physical and engineering sciences. Logic is used in almost every field of computer science: computer architecture (such as digital gates, hardware verification), software engineering (specification and verification), programming languages (semantics, type theory, abstract data types, object-oriented programming), databases (relational algebra), artificial intelligence

1.1 Logic Is Everywhere

(automated theorem proving, knowledge representation), algorithms and theory of computation (complexity, computability), etc.

Logic also plays important roles in other fields, such as mathematics, psychology, and philosophy. In mathematics, logic includes both the mathematical study of logic and the applications of formal logic to other areas of mathematics. The unifying themes in mathematical logic include the study of the expressive power of formal systems and the deductive power of formal proof systems. Mathematical logic is further divided into the fields of (*a*) set theory, (*b*) model theory, (*c*) proof theory, and (*d*) computability theory. These areas share basic results of logic.

Symbolic logic in the late nineteenth century and mathematical logic in the twentieth are also called *formal logic*. Topics traditionally treated by logic but not part of formal logic are often labeled *philosophical logic*, which deals with arguments in the natural language.

Example. Given the premises that (a) "all men are mortal" and (b) "Adam is a man," we may draw the conclusion that (c) "Adam is mortal," by the inference rule of *modus ponens* (Definition 5.3.1).

Philosophical logic is the investigation, critical analysis, and intellectual reflection on issues arising in logic. It is the branch of studying questions about reference, predication, identity, truth, quantification, existence, entailment, modality, and necessity. Many of these topics are shared by mathematical logic and discussed in this book. Hence, studying this book will help us be logical in thinking, writing, or conversation.

1.1.1 Statement or Proposition

Logic comes from natural languages as most sentences in natural languages are statements or propositions. A *proposition* is a sentence that expresses a judgment or opinion and has a truth value. We will use *statement* and *assertion* as synonyms of proposition. Here are some examples:

- The color of the apple is green.
- Today is either Monday or Tuesday.
- He is a 20-year-old sophomore.

Every statement can be either *true* or *false*. For instance, the first statement above can be *true* for one apple but *false* for another. *True* or *false* are called the *truth values* of a statement. Some sentences in natural languages are not statements, such as commands or exclamatory sentences. For example, *"Run fast!" Coach shouts.* The first part of the sentence, "Run fast!", is **not** a statement; the second part, "Coach shouts," is a statement. In fact, a sentence is a statement if and only if it has a truth value.

In natural languages, we can combine or relate statements with words such as "not" (negation), "and" (conjunction), "or" (disjunction), "if-then" (implication), etc. That is, a statement can be obtained from other statements by these words. In logic, these words are called *logical operators*, or equivalently, *logical connectives*. A statement is *composed* if it can be expressed as a composition of several simpler statements; otherwise, it is *simple*. In the above three examples of statements, the first statement is simple; the other two are composed. That is, "today is either Monday or Tuesday" is the composition of "today is Monday" and "today is Tuesday," using the logical operator "or." The second sentence, "He is a 20-year-old sophomore," is the composition of "he is a 20-year-old" and "he is a sophomore," using the implicit logical operator "and." We often regard "the color of that apple is not green" as a composed statement: It is the negation of a simple statement.

Logical operators are indispensable for expressing the relationship between statements. For example, the following statement is a composition of several simple statements.

(∗) If either taxes are not raised or expenditures rise, then the debt ceiling is increased.

To see this clearly, we need to introduce symbols, such as t, e, and d, to denote simple statements:

1. t: "taxes are raised"
2. e: "expenditures rise"
3. d: "the debt ceiling is increased"

To express that the statement (∗) is composed, let us use these symbols for logical operators: $\neg$ for negation, $\vee$ for "either-or," and $\rightarrow$ for the "if-then" relation (i.e., implication). Then (∗) becomes the following formula:

$$((\neg t) \vee e) \rightarrow d$$

Here each of the symbols t, e and d is called a *propositional variable*, which denotes a statement, either simple or composed. Naturally, these propositional variables, like statements, can take on only the truth values: *true* and *false*. Each propositional variable can be assigned an *interpretation* or *meaning*. For instance, the interpretation of t in the above example is "taxes are raised."

Propositional variables are also called *Boolean variables* after their inventor, the nineteenth century mathematician George Boole. Boolean logic includes any logic in which the considered truth values are *true* and *false*. This book studies exclusively Boolean logic. The study of propositional variables with logic operators is called *propositional logic*, which is the simplest one in the family of Boolean logic. This can be differentiated from probability logic, which is not a Boolean logic because probability values are used to represent various degrees of truth values. Statements which contain subjective adjectives, such as "that apple is delicious," are better expressed in probability logic, as "delicious" is subjective. Questions regarding the degrees of truth of subjective statements are ignored in Boolean logic. For example, the statement, "that apple is delicious," can be *true* in one's view and *false* in the other's. Despite this simplification, or indeed because of it, Boolean logic is scientifically successful. One does not even have to know exactly what the truth values *true* and *false* are.

1.1.2 A Brief History of Logic

Sophisticated theories of logic were developed in many cultures, including Greece, India, China, and the Islamic world [1]. Greek methods, particularly Aristotelian logic (or term logic), found wide application and acceptance in Western science and mathematics for millennia.

Aristotle (384–322 BC) was the first logician to attempt a systematic analysis of logical syntax, of nouns (or terms), and of verbs. He demonstrated the principles of reasoning by employing variables to show the underlying logical form of an argument. He sought relations of dependence which characterize necessary inference, and distinguished the validity of these relations, from the truth of the premises. He was the first to deal in a systematic way with the principles of *contradiction*, which states that a proposition cannot be both true and false, and *excluded the middle*, which states that for every proposition, either this proposition or its negation is true. Aristotle has had an enormous influence in Western thought. He developed the theory of *syllogism*, where three important principles are applied for the first time in history: the use of symbols, a purely formal treatment, and the use of an axiomatic system. Aristotle also developed the theory of fallacies, as a theory of non-formal logic.

Another famous ancient Greek logician is Euclid (325–265 BC), who established the foundations of geometry that largely dominated the field until the early nineteenth century. Euclidean geometry is an axiomatic system, in which all theorems are derived from a small number of simple axioms. Euclid's reasoning from assumptions to conclusions remains valid after the discovery of non-Euclidean geometry.

Christian and Islamic philosophers such as Boethius (died 524), Ibn Sina (Avicenna, died 1037), and William of Ockham (died 1347) further developed Aristotle's logic in the Middle Ages, reaching a high point in the mid-fourteenth century with Jean Buridan (1301–1358/62 AD).

In the nineteenth century, attempts to treat the operations of formal logic in a symbolic or algebraic way had been made by philosophical mathematicians, including Leibniz and Lambert, but their works remained isolated and little known.

In the beginning of the nineteenth century, logic was studied with rhetoric, through syllogism, and with philosophy. Mathematical logic emerged in the mid-nineteenth century as a field of mathematics independent of the traditional study of logic. The development of modern symbolic logic or mathematical logic during this period by the likes of Boole, Frege, Russell, and Peano is the most significant in the 2000-year history of logic and is arguably one of the most important and remarkable events in human intellectual history.

1.1.3 Thinking or Writing Logically

Language is a tool for thinking and writing. Being logical means that the reality of the world expressed by language is accurate. Hence, language and logic are inseparable, as logic is about truth. To think or write logically, we need to pay attention to the following points:

- **Matching words to facts or entities**
  A *fact* is something made or done. An *entity* is a thing with distinct and independent existence. We will use fact for both fact and entity. Fact has an objective status. A word can be used to denote a fact. The meaning of a word changes if the word is used to represent a different fact, just like the interpretation of a propositional variable changes in different applications. Being logical requires that the word unambiguously represents a fact in a given context. Many words have multiple meanings. For example, "apple" can be a fruit name or a company's name. Often, we cannot come up with the right word for a fact. We may need a novel word or to add an adjective to a word so that the word matches the fact. Mixing meanings of a word in the same context causes confusion. In a conversation, the two parties should attach the same meaning to a word.

- **Making true statements**
  Most sentences in a natural language are statements. Many statements can be expressed in propositional logic. Some of them cannot, but can be expressed in other logic, such as first-order logic. A true simple statement should be the representation of an objective fact. When a simple statement cannot be convincingly true, it gives us a distorted representation of the objective world. The truth value of complex statements can be decided by logical reasoning, which can be found later in the book. Speaking or writing incomplete sentences is a bad habit, as these sentences cannot be statements. We should be cautious when a statement contains subjective adjectives as the truth of this statement is subjective. We should try to be straightforward in writing or speaking to help the reader or audience to understand and accept the truth of your statements.

Violating the above two points leads to mistakes, many of which are so-called logical fallacies.

Example 1.1.1 The word "logic" has many meanings. In this section, it means "the study of correct reasoning," including both formal and informal logic. The focus of this book is on a small portion of formal logic used in computer science for building reasoning tools. Hence, the word "logic" in the title of this book means "a science that deals with the principles and criteria of validity of inference and demonstration." On the other hand, "logic" in "propositional logic" means "the formal principles of a branch of knowledge." This way of using the word "logic" may cause some confusion, but will never be an obstacle for learning logic in computer science. □

1.1 Logic Is Everywhere

If we match one word to two facts, confusion may arise as shown by the *Paradox of the Court*, also known as *Protagoras' paradox*, which originated in ancient Greece.

> **Paradox of the Court** The famous sophist Protagoras took on a promising pupil, Euathlus, on the understanding that the student would pay Protagoras for his instruction after he wins his first court case. After instruction, Euathlus decided to not enter the profession of law, and so Protagoras decided to sue Euathlus for the amount he is owed. Protagoras argued that if he won the case, he would be paid his money. If Euathlus won the case, Protagoras would still be paid according to the original contract, because Euathlus would have won his first case. Euathlus, however, claimed that if he won, then by the court's decision he would not have to pay Protagoras. If, on the other hand, Protagoras won, then Euathlus would still not have won a case and would therefore not be obliged to pay. The question is then, which of the two men is in the right?

The story continues by saying the jurors were embarrassed to make a decision and postponed the case indefinitely. The confusion of this paradox lies on the mix-up of "two types of payments": one by the contract and the other by the court order. If we distinguish these two types of payments, then the paradox no longer exists.

Let p denote "Euathlus won his first case" and $q(x)$ denote "Euathlus pays \$ x to Protagoras." Then the following two statements are true:

- **By the contract**: If p then $q(t)$ else $q(0)$, where \$$t$ is the tuition of the instruction.
- **By the court order**: If p then $q(x)$ else $q(y)$, where x can be negative if the court asks Protagoras to pay Euathlus.

Merging the above two statements, the statement "if p then $q(t+x)$ else $q(y)$" is true. That is, Euathlus needs to pay Protagoras \$ $(t+x)$ if he won the case, and \$$y$, otherwise. If $x = 0$ and $y = t$, Protagoras will get paid the same amount in both cases. Euathlus does not need to pay Protagoras if $x = -t$ (when he won) or $y = 0$ (when he lost), two unlikely court decisions.

The interesting number paradox claims that, if we classify every natural number as either "interesting" or "uninteresting," then every natural number is interesting. A "proof by contradiction" goes as follows: If there exists a non-empty set S of uninteresting natural numbers, then the smallest number of S is interesting by being the smallest uninteresting number, thus producing a contradiction. The paradox is due to lack of a formal definition of "interesting numbers," so that a number can be both interesting and uninteresting.

In logic, paradox serves as an acute example of a theory's inconsistency and is often a motivation for the development of a new theory. In natural language, finding a paradox in the opponent's arguments is a good way to win a debate. Sometimes, paradox can be a useful expression. The phrase "indescribable feeling" is such an example. If the feeling is indescribable, then this phrase conveys nothing. But if the word "indescribable" communicates something about the feeling, then it does convey something: This is self-contradictory in logic but useful in natural language. As another example, "'Impossible' is not in our vocabulary!" Though a false statement in logic, it is a powerful expression in a motivation speech. Plato's "I know that I know nothing" is also a meaningful false statement.

1.2 Logical Fallacies

Logical fallacies are logical errors one makes when writing or speaking. In this section, we will give a brief introduction to logical fallacies, which are the subjects of philosophical logic. Knowing these fallacies will help us to avoid them. Writing logically is related to, but not the same as, writing clearly, or efficiently, or convincingly, or informatively; ideally, one would want to do all of these at once, but one often does not have the skill to achieve them all. Though with practice, you will be able to achieve more of your writing objectives concurrently. The big advantage of writing logically is that you can be absolutely sure that your conclusion will be correct, as long as all your hypotheses are correct, and your steps are logical.

In philosophical logic [2], an *argument* has similar meaning to a proof in formal logic, as the argument consists of premises and conclusions. A *fallacy* is the use of invalid or otherwise faulty reasoning in the construction of an argument. A fallacious argument may be deceptive by appearing to be better than it really is. Some fallacies are committed intentionally to manipulate or persuade by deception, while others are committed unintentionally due to carelessness or ignorance.

Fallacies are commonly divided into "formal" and "informal." A formal fallacy can be expressed neatly in formal logic, such as propositional logic, while an informal fallacy cannot be expressed in formal logic.

1.2.1 Formal Fallacies

In philosophical logic, a formal fallacy is also called *deductive fallacy*, *logical fallacy*, or *non sequitur* (Latin for "it does not follow"). This is a pattern of reasoning rendered invalid by a flaw in its logical structure.

> **Example.** Given the premises that (a) my car is a car, and (b) some cars are red, we draw the conclusion that (c) my car is red.

This is a typical example of a conclusion that does not follow logically from premises or is based on irrelevant data. Here are some common logical fallacies.

- **Affirming the consequent**: Any argument with the invalid structure of: If A then B. B, therefore, A.

    > **Example.** If I get a B on the test, then I will get the degree. I got the degree, so it follows that I must have received a B. In fact, I got an A.

- **Denying the antecedent**: Any argument with the invalid structure of: If A then B. Not A, therefore not B.

1.2 Logical Fallacies

Example. If it's a dog, then it's a mammal. It's not a dog, so it must not be a mammal. In fact, it's a cat.

- **Affirming a disjunct**: Any argument with the invalid structure of: It is the case that A or B. A, therefore, not B.

  Example. I am working or I am at home. I am working, so I must not be at home. In fact, I am working at home.

- **Denying a conjunct**: Any argument with the invalid structure of: It is not the case that both A and B. Not A, therefore B.

  Example. I cannot be both at work and at home. I am not at work, so I must be at home. In fact, I am at a park.

- **Undistributed middle**: Any argument with the invalid structure of: Every A has B. C has B, so C is A.

  Example. Every bird has a beak. That creature has a beak, so that creature must be a bird. In fact, the creature is a dinosaur.

A formal fallacy occurs when the structure of the argument is incorrect, despite the truth of the premises. A valid argument always has a correct formal structure, and if the premises are true, the conclusion must be true. When we use false premises, the formal fallacies disappear, but the argument may be regarded as a fallacy as the conclusion is invalid.

As an application of modus ponens, the following example contains no formal fallacies:

If you took that course on CD player repair right out of high school, you would be doing well and vacationing on the moon right now.

Even though there is no logic error in the argument, the conclusion is invalid because the premise is contrary to the fact. With a false premise, you can make any conclusion, so that the composed statement is always true. However, an always true statement has no value in reasoning.

By contrast, an argument with a formal fallacy could still contain all true premises:

(a) If someone owns the world's largest diamond, then he is rich. (b) King Solomon was rich. Therefore, (c) King Solomon owned the world's largest diamond.

Although (a) and (b) are true statements, (c) does not follow from (a) and (b) because the argument commits the formal fallacy of "affirming the consequent."

1.2.2 Informal Fallacies

There are numerous kinds of informal fallacies that use an incorrect relation between premises and conclusion. These fallacies can be grouped into four groups, and each group contains several types of fallacies:

- Fallacies of improper premise
- Fallacies of faulty generalizations
- Fallacies of questionable cause
- Relevance fallacies

For instance, *circular reasoning*, where the reasoner begins with what he or she is trying to end up with, belongs to the group of improper premises. An example of circular reasoning is given below.

> You must obey the law, because it's illegal to break the law.

Some informal fallacies do not belong to the above four groups. For example, the *false dilemma* fallacy presents a choice between two mutually exclusive options, implying that there are no other options. One option is clearly worse than the other, making the choice seem obvious. Also known as the *either/or* fallacy, false dilemmas are a type of informal logical fallacy in which a faulty argument is used to persuade an audience to agree. False dilemmas are everywhere.

- Vote for me or live through 4 more years of higher taxes.
- America: Love it or leave it.
- Subscribe to our streaming service or be stuck with cable.

For a complete list of informal fallacies, the reader is recommended to read *Wikipedia*'s page on the same topic for details. Understanding these logical fallacies can help us more confidently parse the arguments we participate in on a daily basis, separating fact from dressed fiction.

1.3 A Glance of Mathematical Logic

Mathematical logic is divided into four areas [3]:

- Set theory
- Computability theory
- Model theory
- Proof theory

Each area has a distinct focus, although many techniques and results are shared between multiple areas. The borders between these areas, and the lines between mathematical logic and other fields of mathematics, are not always distinct. Each area contains rich materials which can be studied in graduate courses of multiple levels. As an introductory book on logic, we will not give an overview of mathematical logic, but pick some basic concepts and results, which are closely related to this book.

1.3 A Glance of Mathematical Logic

1.3.1 Set Theory

A *set* is a mathematical model for a collection of zero or more distinct objects. All objects in a set are unequal (distinct) and unordered. Set theory is a branch of mathematical logic that studies sets and deals with operations between, relations among, and statements about sets. Although any type of object can be collected into a set, set theory is applied most often to objects that are relevant to mathematics. The language of set theory can be used to define nearly all mathematical objects. Set theory is an area of major interest for logicians and philosophers of mathematics and has many applications in computer science.

German mathematician Georg Cantor is commonly considered the founder of set theory, which often goes under the name of *naive set theory*. After the discovery of paradoxes within naive set theory, various axiomatic systems were proposed in the early twentieth century, of which Zermelo–Fraenkel set theory is the best-known. Zermelo–Fraenkel set theory ensures that the sets formulated by a set of axioms in first-order logic are general enough to represent all entities in the universe of discourse, but also exclude the sets which are their own members, thus, formulating a theory of sets free of paradoxes.

Basic Concepts and Notations

The main purpose of this presentation is to show the notations used throughout the book.

People often use $\mathcal{N}$ to denote the set of natural numbers, i.e.,

$$\mathcal{N} = \{0, 1, 2, 3, \ldots\}$$

$\mathcal{Z}$ to denote the set of integers,

$$\mathcal{Z} = \{\ldots, -3, -2, -1, 0, 1, 2, 3, \ldots\}$$

and $\mathcal{R}$ to denote the set of rational numbers.

$$\mathcal{R} = \{m/n \mid m, n \in \mathcal{N}, n \neq 0\}$$

$\mathcal{N}$, $\mathcal{Z}$, and $\mathcal{R}$ are naturally infinite sets.

Let U be the collection of all objects under consideration. U is often called the *universal set*, or the *universe of discourse*, or simply the *universe*. For example, U can be the set $\mathcal{N}$ of all natural numbers or the population of the world. Any object x under consideration is a *member* of U, written as $x \in U$. A *set* is any collection of objects from U. The *empty set* is often denoted by $\emptyset$ or simply $\{\}$. Given two sets A and B, the following are common set operations:

- $x \in A$: it is *true* iff (if and only if) x is a member of A.

- $A \cup B$: the *union* of A and B. For any object x, $x \in A \cup B$ iff x is in A or B or both.
- $A \cap B$: the *intersection* of A and B. For any $x \in U$, $x \in A \cap B$ iff $x \in A$ and $x \in B$.
- $A - B$: the *set difference* of A and B. For any $x \in U$, $x \in A - B$ iff $x \in A$ but not $x \in B$.
- $\overline{A}$: the *complement* of A (with respect to U). $\overline{A} = U - A$. $A - B$ is also called the *complement* of B with respect to A: $A - B = A \cap \overline{B}$.
- $A \subseteq B$: A is a *subset* of B. $A \subseteq B$ is *true* iff every member of A is a member of B. Naturally, $A \subseteq U$ and $B \subseteq U$ for the universal set U.
- $A \subset B$: A is a *proper subset* of B. $A \subset B$ is *true* iff $A \subseteq B$ and $A \neq B$.
- $A = B$: $A = B$ iff $A \subseteq B$ and $B \subseteq A$, that is, A and B contain exactly the same elements.
- $A \times B$: the *Cartesian product* of A and B. An ordered pair (a, b) is in $A \times B$ iff $a \in A$ and $b \in B$. That is, $A \times B = \{\langle a, b \rangle \mid a \in A, b \in B\}$.
- A^i: the *Cartesian product* of i copies of A, where $i \in \mathcal{N}$, a natural number. When $i = 0$, A^0 denotes the *empty sequence*, not the empty set; $A^1 = A$; $A^2 = A \times A$, $A^3 = A \times A \times A$, etc.
- $|A|$: the *cardinality* or *size* of a set. When A is finite, $|A|$ is the number of members in A. $|A|$ is called a *cardinal number* if A is infinite.
- $\mathcal{P}(A)$: the *power set* of A. $\mathcal{P}(A)$ is the set whose members are subsets of A. Formally, $\mathcal{P}(A) = \{X \mid X \subseteq A\}$. That is, for any set X, $X \in \mathcal{P}(A)$ iff $X \subseteq A$. When $|A| = n$, $|\mathcal{P}(A)| = 2^n$.

Elementary set theory can be studied informally and intuitively and can be taught in primary schools using Venn diagrams. Many mathematical concepts can be defined precisely using only set theoretic concepts, such as relations and functions.

Relations

Given two sets A and B, a *relation* between A and B is a subset of $A \times B$. If $A = B$, that happens in many applications, we say the relation is on set A. For example, the less-than relation, $<$, on $\mathcal{N}$ (the set of natural numbers), is

$$< \ = \{\langle 0, 1 \rangle, \langle 0, 2 \rangle, \langle 0, 3 \rangle, \langle 1, 2 \rangle, \langle 0, 4 \rangle, \langle 1, 3 \rangle, \langle 0, 5 \rangle, \langle 1, 4 \rangle, \langle 2, 3 \rangle, \ldots\}$$

which is a subset of $\mathcal{N} \times \mathcal{N}$. In general, given a relation R on set S, i.e., $R \subseteq S \times S$, we assume that $R(a, b)$ is *true* if $\langle a, b \rangle \in R$ and $R(a, b)$ is *false* if $\langle a, b \rangle \notin R$. Of course, we are used to write $<$ as an infix operator, such as $0 < 1$, instead of a prefix operator such as $<(0, 1)$.

Let $\succeq$ be a relation on the set S and a, b, c denote arbitrary elements of S.

- The *converse* of $\succeq$ is $\preceq$: $a \preceq b$ iff $b \succeq a$.
- $\succeq$ is *reflexive* if $a \succeq a$ is true.
- $\succeq$ is *irreflexive* if $a \succeq a$ is false. $\succeq$ can be neither reflexive nor irreflexive.

1.3 A Glance of Mathematical Logic

- $\succeq$ is *antisymmetric* if $a \succeq b$ and $b \succeq a$ imply $a = b$.
- $\succeq$ is *asymmetric* if $b \not\succeq a$ when $a \succeq b$. $\succeq$ is asymmetric iff $\succeq$ is both antisymmetric and irreflexive.
- $\succeq$ is *transitive* if $a \succeq b$ and $b \succeq c$ imply $a \succeq c$.
- $\succeq$ is *comparable* if at least one of the following is true: $a = b$, $a \succeq b$, or $b \succeq a$.
- $\succeq$ is a *partial order* if $\succeq$ is antisymmetric and transitive. In literature, a "partial order" is often referred to a reflexive partial order.
- An irreflexive partial order, often denoted by $\succ$, is called *strict partial order* and can be defined from a partial order $\succeq$: $a \succ b$ iff $a \succeq b$ and $a \neq b$. On the other hand, a reflexive $\succeq$ can be defined from $\succ$: $a \succeq b$ iff $a \succ b$ or $a = b$. If a partial order is required to be reflexive, then $\succ$ cannot be a partial order.
- $\succeq$ is a *total order* if $\succeq$ is a comparable partial order.
- Given a partial order $\succeq$, a *minimal* element of $X \subseteq S$ is an element $m \in X$ such that if $x \in X$ and $m \succeq x$, then $x = m$.
- A partial order $\succeq$ is *well-founded* if there is a minimal element $m \in X$ for every non-empty $X \subseteq S$. Equivalently, $\succeq$ is well-founded iff there exists no infinite sequence of distinct elements $x_1, x_2, \ldots, x_i, \ldots$, such that

$$x_1 \succeq x_2 \succeq \cdots \succeq x_i \succeq \cdots$$

- $\succeq$ is a *well-order* if $\succeq$ is a well-founded total order.

For example, the common relation $\geq$ on integers is well-founded on the set $\mathcal{N}$ of natural numbers but not well-founded on the set $\mathcal{Z}$ of integers.

Functions

A *function* f is a relation $R \subset A \times B$ with the restriction that for every $a \in A$, there exists at most one $\langle a, b \rangle \in R$ for some $b \in B$, written $f(a) = b$. Equivalently, the restriction can be expressed as: For every $a \in A$ and $b, c \in B$, if $\langle a, b \rangle \in R$ and $\langle a, c \rangle \in R$, then $b = c$.

- Function f is often denoted by $f : A \mapsto B$ to show that f is a relation of $A \times B$, and we say f is a function from A to B.
- Given $f : A \mapsto B$, if $f(a) = b$, we say $f(a)$ is *defined* and b is the *output* of f on *input* a. Obviously, A is the set of possible input values to f and B is the set of possible output values of f.
- *Mapping* is a synonym of function. If $f(a) = b$, we say a is *mapped* to b and b is the *image* of a under f.
- Given $f : A \mapsto B$, $dom(f) = \{x \mid f(x) \text{ is defined }\}$ is called the *domain* of f and $ran(f) = \{f(x) \mid x \in dom(f)\}$ is the *range* of f. Obviously, $dom(f) \subseteq A$ and $ran(f) \subseteq B$.
- If a function g takes two values as input, say $g(a, a') = b$, we denote g by $g : A \times A' \mapsto B$ to show that $g \subset A \times A' \times B$. By convention, we write

$g(a, a') = b$ instead of $g(\langle a, a' \rangle) = b$, where $a \in A, a' \in A'$, and $b \in B$. This practice is generalized to functions with more than two inputs.

- Function $f : A \mapsto B$ is said to be *total* if $f(a)$ is defined for every $a \in A$. Otherwise, f is *partial*. Obviously, f is total iff $dom(f) = A$. For instance, the division function on the integers is not total as the quotient is not defined when the divisor is zero. People generally assume that a function is total unless mentioned explicitly.
- When both A and B are finite, the set of all functions from A to B is finite. Let $F = \{f \mid f : A \mapsto B, f \text{ is total }\}$, then $|F| = n^m$ if $|A| = m$ and $|B| = n$. This is because there are m choices for $a \in A$ and there are n choices for each $f(a)$. For example, if $B = \{0, 1\}$ and $A = B \times B$, then $|B| = 2, |A| = 4$ and the number of total functions from A to B is $2^4 = 16$. In other words, there are 16 distinct total binary Boolean functions. The number of all functions from A to B is equal to the number of all total functions from A to $B \cup \{u\}$, where $u \notin B$ and $f(a) = u$ means "$f(a)$ is undefined for $f : A \mapsto B$" (see Example 11.4.1).
- Function $f : A \mapsto B$ is said to be *surjective* if f is total and $ran(f) = B$. That is, for every $b \in B$, there exists $a \in A$ such that $f(a) = b$. A surjective function is also called *onto*.
- Function $f : A \mapsto B$ is said to be *injective*, also called *one-to-one*, if for any $x, y \in A$ and $w, z \in B$, if $x \neq y$, $f(x) = w$, and $f(y) = z$, then $w \neq z$. If f is injective, then the *inverse* of f, denoted by f^{-1}, is an injective function from B to A: $f^{-1}(y) = x$ iff $f(x) = y$.
- Function $f : A \mapsto B$ is said to be *bijective* if f is both injective and surjective. A bijective function is also called a *one-to-one correspondence* or *bijection* between A and B. In this case, we say A and B have the *same size* and write $|A| = |B|$.

In the above definitions, a surjective function is required to be total. If this requirement is dropped, then a bijection needs to be total, since if a bijection is a partial function, its inverse cannot be a bijection and such bijections are not useful. For instance, define $f : \mathcal{P}(\mathcal{N}) \mapsto \mathcal{N}$ by $f(\{x\}) = x$ for all $x \in \mathcal{N}$, then f is injective and surjective (if not required to be total), and the inverse of f is not surjective.

Proposition 1.3.1 $f : A \mapsto B$ *is bijective iff both f and the inverse of f are total functions.*

The proof is left as an exercise.

Countable Sets

Let S be any set. If S is finite, $|S|$ is a number in $\mathcal{N}$, and we can compare these numbers easily. How to compare the sizes of infinite sets? Intuitively, if X is a proper subset of S, then $|X| < |S|$. However, this intuition is only correct for finite sets. For instance, let $\mathcal{E}$ be the set of all even natural numbers, then $\mathcal{E} \subset \mathcal{N}$. However, it

1.3 A Glance of Mathematical Logic

is wrong to say $|\mathcal{E}| < |\mathcal{N}|$. Let $f : \mathcal{N} \mapsto \mathcal{E}$ be defined by $f(i) = 2i$. Then f is bijective; $\mathcal{N}$ and $\mathcal{E}$ have the same size, i.e., $|\mathcal{N}| = |\mathcal{E}|$ even though $\mathcal{E} \subset \mathcal{N}$.

Hilbert's Hotel Paradox David Hilbert imagines a hotel with rooms numbered 0, 1, 2, and so on with no upper limit and every room is occupied. There are yet an infinite number of new visitors coming and each expecting their own room. The manager then orders all the guests in room i move simultaneously to room $2i$. Thus, all the odd-numbered rooms become available for the new visitors who can each have their own room in this hotel.

This paradox shows that our thinking is quite different when dealing with infinite sets. A proper subset can have the same size as its superset.

Cantor introduced the following concept for the comparison of infinite sets.

Definition 1.3.2 A set S is *countable* if either S is finite or $|S| = |\mathcal{N}|$. When $|S| = |\mathcal{N}|$, the bijection $r : S \mapsto \mathcal{N}$ is called the *rank* function of S and $r(x)$ is the *rank* of $x \in S$. S is *uncountable* if S is not countable. S is *countably infinite* if S is both countable and infinite.

Proposition 1.3.3

(a) *All countably infinite sets have the same size, and each countably infinite set has a rank function.*
(b) *If r is the rank function of a countably infinite set S, then r induces a well-founded total order $\succeq$ on S: $x \succeq y$ iff $r(x) \geq r(y)$.*

The proof is left as an exercise.

Given a set S, we can write $S = \{a_0, a_1, a_2, \ldots\}$ if S is countably infinite. In this notation, a rank function $r : S \mapsto \mathcal{N}$ is assumed such that $r(a_i) = i$ for any $i \in \mathcal{N}$. Since $\mathcal{N} - \{0\}$ is also countably infinite, we can write $S = \{a_1, a_2, a_3, \ldots\}$, too. So, the set of natural numbers can contain 0 or not, and the rank can start with 0 or 1. If S is uncountable, it is wrong to write $S = \{a_0, a_1, a_2, \ldots\}$.

Today, countable sets form the foundation of *discrete mathematics*. We first check what sets are countable.

Proposition 1.3.4 *The following sets are countable:*

1. *Any finite set*
2. $\mathcal{N} - \{k\}$, *where $k \in \mathcal{N}$*
3. $A \times B$, *if both A and B are countable sets*
4. $\mathcal{Z}$, *the set of all integers*
5. $\{0, 1\}^*$, *the set of all binary strings of finite length*
6. Σ^*, *the set of all strings of finite length built on the symbols from Σ, a finite alphabet*

Proof Some of the proofs are sketches.

1. This comes from the definition.

2. The bijection $f : \mathcal{N} \mapsto (\mathcal{N} - \{k\})$ can be defined as follows: $f(i) = i$ if $i < k$ else $i + 1$. Then $f^{-1}(x) = x$ if $x < k$ else $x - 1$. Both f and f^{-1} are total functions.
3. $A \times B$ is countable when either A or B is finite, and its proof is left as an exercise. When both A and B are countably infinite, without loss of generality, we assume $A = B = \mathcal{N}$. Let $g(k) = k(k + 1)/2$, which is the sum of the first $k + 1$ natural numbers, and $f : \mathcal{N} \times \mathcal{N} \mapsto \mathcal{N}$ be the function:

$$f(i, j) = g(i + j) + j$$

The inverse of f exists: For any $k \in \mathcal{N}$, let $m = h(k)$ be the integer such that $g(m) \leq k < g(m + 1)$. Then $f^{-1}(k) = \langle i, j \rangle$, where $i = g(m) + m - k = g(h(k)) + h(k) - k$ and $j = k - g(m) = k - g(h(k))$. It is easy to check that $i + j = m$ and $f(i, j) = k$. For example, if $k = 12$, then $m = 4$, and $i = j = 2$. Table 1.1 shows the first few values of $k = f(i, j)$. Since both f and f^{-1} are total functions, f is bijective. So, $\mathcal{N} \times \mathcal{N}$ and $\mathcal{N}$ have the same size.
4. To show $\mathcal{Z}$ is countable, we define a bijection f from $\mathcal{Z}$ to $\mathcal{N}$: $f(x) = $ if $x \geq 0$ then $2x$ else $-2x - 1$. This function maps 0 to 0, positive integers to even natural numbers, and negative integers to odd natural numbers. f is bijective because both f and f^{-1} are total functions: For any $n \in \mathcal{N}$, $f^{-1}(n) = $ if n is even then $n/2$ else $(-n - 1)/2$. Note that $\geq$ on $\mathcal{Z}$ is not well-founded and cannot be used to list $\mathcal{Z}$ from the smallest to a larger one. Using f as the rank function of $\mathcal{Z}$, we list $\mathcal{Z}$ by the rank of each element as follows: $0, -1, 1, -2, 2, -3, 3, \ldots$.
5. Let $S = \{0, 1\}^*$ be the set of all finite-length binary strings. To show S is countable, let us find a bijection from S to $\mathcal{N}$. For any binary string $s \in S$, let the length of s be n, i.e., $|s| = n$. There are $1 + 2 + 2^2 + \ldots + 2^{n-1} = 2^n - 1$ strings in S whose lengths are less than $|s|$. Let $v(s)$ be the decimal value of s: $v(\epsilon) = 0$, where ϵ is the empty string, $v(x0) = 2v(x)$ and $v(x1) = 2v(x) + 1$, then there are $v(s)$ strings of length n whose decimal values are less than $v(s)$. Define $\gamma : S \mapsto \mathcal{N}$ by $\gamma(s) = 2^{|s|} + v(s) - 1$, then γ must be a bijection, because both γ and $\gamma^{-1} : \mathcal{N} \mapsto \{0, 1\}^*$ are total functions: For any $x \in \mathcal{N}$, let $n = \lfloor log_2(x + 1) \rfloor$, then $\gamma^{-1}(x) = s$ such that $|s| = n$ and $v(s) = x - 2^n + 1$.
6. Let $\Sigma = \{a_0, a_1, \ldots, a_{k-1}\}$. If $k = 1$, then $\Sigma^* = \{a_0^i \mid i \in \mathcal{N}\}$ is countable. If $k \geq 2$, we generalize the proof in the previous problem from $\{0, 1\}^*$ to Σ^*. Let $\theta : \Sigma \mapsto \mathcal{N}$ be $\theta(a_i) = i$ for $0 \leq i \leq k - 1$. Every string $s \in \Sigma^*$ is either ϵ or xa, where $x \in \Sigma^*$ and $a \in \Sigma$. The decimal value of $s \in \Sigma^*$, $v(s)$, is

Table 1.1 $k = f(i, j) = (i + j)(i + j + 1)/2 + j$

$i \backslash j$	0	1	2	3	4
0	0	2	5	9	14
1	1	4	8	13	19
2	3	7	12	18	25
3	6	11	17	24	32
4	10	16	23	31	40

1.3 A Glance of Mathematical Logic

computed as follows: $v(\epsilon) = 0$ and $v(xa) = v(x) * k + \theta(a)$. If $|s| = n$, there are $(k^n - 1)/(k - 1)$ strings of Σ^* whose lengths are less than n, and there are $v(s)$ strings of length n whose decimal values are less than $v(s)$. Define $\gamma : \Sigma^* \mapsto \mathcal{N}$ by $\gamma(s) = (k^{|s|} - 1)/(k - 1) + v(s)$, then it is easy to check that both γ and γ^{-1} (an exercise) are total functions. □

The last three examples are instances of countably infinite sets. Σ^* as well as $\{0, 1\}^*$ in the last two examples plays an important role in theory of computation. If we order the strings of Σ^* by γ, i.e., listing them as $\gamma^{-1}(0), \gamma^{-1}(1), \gamma^{-1}(2)$, and so on, this order will choose (a) shorter length first; (b) for strings of the same length, smaller decimal value first. In this order, $s \in \Sigma^*$ is preceded by $(k^{|s|} - 1)/(k - 1)$ strings whose lengths are less than $|s|$, plus $v(s)$ strings of length $|s|$ and whose values are less than $v(s)$. Hence, the rank of s in this order is exactly $\gamma(s) = (k^{|s|} - 1)/(k - 1) + v(s)$.

Definition 1.3.5 Let $\gamma : \Sigma^* \mapsto \mathcal{N}$ be the rank function defined in Proposition 1.3.4(6). The order induced by γ is called the *canonical order* of Σ^*.

For instance, $\{0, 1\}^*$ is sorted by the canonical order as follows:

$$\{\epsilon, 0, 1, 00, 01, 10, 11, 000, 001, 010, 011, 100, 101, 110, 111, 0000, \ldots\}$$

The canonical order of Σ^* is uniquely defined by the rank function γ. We may speak indifferently about a natural number and a string because $w \in \Sigma^*$ can be viewed as $\gamma(w) \in \mathcal{N}$ and $n \in \mathcal{N}$ as $\gamma^{-1}(n) \in \Sigma^*$.

Example 1.3.6 Let $\Sigma = \{a, b\}$, then the first few strings of Σ^* are given below with γ, where a^i denotes i copies of a:

$x \in \Sigma^*$	ϵ	a	b	a^2	ab	ba	b^2	a^3	a^2b	aba	ab^2	ba^2	bab	b^2a	b^3	a^4	a^3b	...
$\gamma(x)$	0	1	2	3	4	5	6	7	8	9	10	11	12	13	14	15	16	...

Σ^* becomes binary strings if we replace a by 0 and b by 1, □

Uncountable Sets

In 1874, in his first set theory article, Georg Cantor introduced the term "countable set" and proved that the set of real numbers is not countable, thus showing that not all infinite sets are countable. In 1878, he used bijective proof to identify infinite sets of the same size, distinguishing sets that are countable from those that are uncountable by the well-known diagonal method.

Proposition 1.3.7 *The following sets are uncountable:*

1. $\mathcal{B}$: *the set of binary strings of infinite length*
2. $\mathcal{R}_1$: *the set of non-negative real numbers less than 1*

3. $\mathcal{P}(\mathcal{N})$: *the power set of the natural numbers*
4. $D_F = \{f \mid f : \mathcal{N} \mapsto \{0, 1\}\}$ *is total, the set of total functions from the natural numbers to Boolean values, called* decision functions *or* characteristic functions

Proof For the set $\mathcal{B}$ of infinite-length binary strings, we will use Cantor's diagonal method to show that $\mathcal{B}$ is uncountable by a refutational proof. Assume that $\mathcal{B}$ is countable, then there exists a bijection $f : \mathcal{N} \mapsto \mathcal{B}$ and we write $B = \{s_0, s_1, s_2, \ldots\}$, where $s_i = f(i)$. Let the jth symbol of s_i be s_{ij}. Now, we construct a binary string x of infinite length as follows: The jth symbol of x, i.e., x_j, is the complement of s_{jj}. That is, if $s_{jj} = 0$, then $x_j = 1$; if $s_{jj} = 1$, then $x_j = 0$. Clearly, $x \in \mathcal{B}$. Let $x = s_k$ for some $k \in \mathcal{N}$. However, x differs from s_k on the kth symbol, i.e., $x_k = \overline{s_{kk}}$, the complement of s_{kk}, where $\overline{0} = 1$ and $\overline{1} = 0$. The contradiction comes from the assumption that $\mathcal{B}$ is countable. Table 1.2 illustrates the idea of Cantor's diagonal method.

Once we know $\mathcal{B}$ is uncountable, if we have a bijection from $\mathcal{B}$ to R_1, then R_1 must be uncountable. The function $f : \mathcal{B} \mapsto R_1$ is defined by $f(s) = 0.s$, as each number of R_1 can be represented by a binary number, where $s \in \mathcal{B}$. It is obvious that $f^{-1}(0.s) = s$, assuming an infinite sequence of 0 is appended to the right end of s if s is a finite sequence. Since f and f^{-1} are total functions, f is indeed a bijection.

To show $\mathcal{P}(\mathcal{N})$ is uncountable, let $g : \mathcal{B} \mapsto \mathcal{P}(\mathcal{N})$ be $g(s) = \{j \mid (j + 1)^{th}$ symbol of s is 1 }. For example, if $s = 1101000\ldots$ (the rest of s are zeros), then $g(s) = \{0, 1, 3\}$. If $g(s) = A \in \mathcal{P}(\mathcal{N})$, then $g^{-1}(A) = \lambda(0)\lambda(1)\lambda(2) \cdots \in \mathcal{B}$, where, for every $i \in \mathcal{N}$, $\lambda(i) = 1$ if $i \in A$ and 0 if $i \notin A$. Obviously, g is bijective because both g and g^{-1} are total functions. In literature, $g^{-1}(A)$ is called the *characteristic sequence* of A.

Similarly, to show the set D_F of decision functions is uncountable, let $h : \mathcal{B} \mapsto D_F$ be $h(s) = f$, where, for every $j \in \mathcal{N}$, $f(j) =$ the $(j + 1)^{th}$ symbol of s. Since s is a binary string of infinite length, $f : \mathcal{N} \mapsto \{0, 1\}$ is well-defined. For instance, if s is a string of all zeros, then $f(j) = 0$ for all j. If $h(s) = f$, then

Table 1.2 Illustration of Cantor's diagonal method: If $\mathcal{B}$ can be enumerated as $s_0, s_1, s_2, \ldots$, then $x = \overline{s_{00}}\,\overline{s_{11}}\,\overline{s_{22}}\ldots$ is a binary string but cannot be in $\mathcal{B}$, where $\overline{0} = 1$ and $\overline{1} = 0$

$s_0 =$	**0**	0	0	0	0	0	0	0	0	...
$s_1 =$	0	**0**	0	0	0	0	0	1	0	...
$s_2 =$	0	0	**0**	0	0	1	0	0	0	...
$s_3 =$	0	0	0	**1**	0	0	0	0	0	...
$s_7 =$	0	0	0	1	**0**	1	0	0	0	...
$s_6 =$	0	0	0	0	1	**0**	0	0	0	...
$s_4 =$	0	0	0	1	0	1	**0**	1	0	...
$s_5 =$	0	0	1	0	0	0	0	**1**	0	...
$s_8 =$	0	1	0	0	1	0	0	1	**0**	...
$s_9 =$	0	1	0	0	1	1	0	1	0	**1**	...
					......					
$x =$	**1**	**1**	**1**	**0**	**1**	**1**	**1**	**0**	**1**	**0**	...

1.3 A Glance of Mathematical Logic

$h^{-1}(f) = f(0)f(1)f(2)\cdots \in \mathcal{B}$. Obviously, h is bijective because both h and h^{-1} are total functions. □

A total decision function $f = h(s)$, $s \in \mathcal{B}$, is also called *characteristic function*, because $s = g^{-1}(A)$, $A \in \mathcal{P}(\mathcal{N})$, is a characteristic sequence of $A \subseteq \mathcal{N}$, where h and g are defined in the above proof.

The result that $\mathcal{P}(\mathcal{N})$ is uncountable while $\mathcal{N}$ is countable can be generalized: For every set S, the power set of S, i.e., $\mathcal{P}(S)$, has a larger size than S. This result implies that the notion of the *set of all sets* is an inconsistent notion. If S were the set of all sets, then $\mathcal{P}(S)$ would at the same time be bigger than S and a subset of S.

Cantor chose the symbol $\aleph_0$ for the size of $\mathcal{N}$, $|\mathcal{N}|$, a cardinal number. Symbol $\aleph_0$ is read aleph-null, where $\aleph$ is the first letter of the Hebrew alphabet. The size of the reals is often denoted by $\aleph_1$, or **c** for the continuum of real numbers, because the set of real numbers is larger than $\mathcal{N}$. Cantor showed that there are infinitely many infinite cardinal numbers, and there is no largest cardinal number.

$$\begin{aligned}
\aleph_0 &= |\mathcal{N}| \\
\aleph_1 &= |\mathcal{P}(\mathcal{N})| &&= 2^{\aleph_0} > \aleph_0 \\
\aleph_2 &= |\mathcal{P}(\mathcal{P}(\mathcal{N}))| &&= 2^{\aleph_1} > \aleph_1 \\
\aleph_3 &= |\mathcal{P}(\mathcal{P}(\mathcal{P}(\mathcal{N})))| &&= 2^{\aleph_2} > \aleph_2 \\
&\cdots
\end{aligned}$$

The famous inconsistent example of naive set theory is the so-called Russell's paradox, discovered by Bertrand Russell in 1901:

Russell's Paradox Let $T = \{S \mid S \notin S\}$, then $T \in T \equiv T \notin T$.

That is, if "$T \in T$" is *false*, then "$T \in T$" is *true* by the definition of T because condition "$S \notin S$" is true and $T = S \in T$. If "$T \in T$" is *true*, then "$T \notin T$" is also *true* because every member of T satisfies $S \notin S$. This paradox showed that some attempted formalization of the naive set theory created by Cantor et al. led to a contradiction. Russell's paradox shows that the concept of "a set contains itself" is invalid. If "$S \in S$" is always *false*, then $S \notin S$ *is always true* and $T = \{S \mid S \notin S\}$ is the "set of all sets," which does not exist, either.

A layman's version of Russell's paradox is called the *Barber's paradox*.

Barber's Paradox In a village, there is only one barber who shaves all those, and those only, who do not shave themselves. The question is, does the barber shave himself?

Answering this question results in a contradiction. The barber cannot shave himself as he only shaves those who do not shave themselves. Conversely, if the barber does not shave himself, then he fits into the group of people who would be shaved by the barber, and thus, as the barber, he must shave himself.

The discovery of paradoxes in naive set theory caused mathematicians to wonder whether mathematics itself is inconsistent, and to look for proofs of consistency. Ernst Zermelo (1904) gave proof that every set could be well-ordered, using the *axiom of choice*, which drew heated debate and research among mathematicians. In

1908, Zermelo provided the first set of axioms for set theory. These axioms, together with the *axiom of replacement* proposed by Abraham Fraenkel, are now called *Zermelo–Fraenkel set theory* (ZF). Besides ZF, many set theories are proposed since then, to rid paradoxes from naive set theory.

1.3.2 Computability Theory

Computability theory, used to be called *recursion theory*, is a branch of mathematical logic, of computer science, and of computation theory that originated in the 1930s with the study of computable functions. The field has since expanded to include the study of generalized computability. In these areas, computability theory overlaps with set theory and proof theory.

Recursion

The main reason for the name of "recursion theory" is that recursion is used to construct objects and functions.

Example 1.3.8 Given a constant symbol 0 of set T and a function symbol $suc : T \mapsto T$, the objects of set T can be recursively constructed as follows:

1. 0 is an object of T.
2. If n is an object of T, so is $suc(n)$.
3. Nothing else will be an object of T.

If we let $suc^0(0) = 0$, $suc^1(0) = suc(0)$, $suc^2(0) = suc(suc(0))$, etc., then T can be expressed as

$$T = \{0, suc(0), suc^2(0), suc^3(0), \ldots, suc^i(0), \ldots\}$$

Obviously, there exists a bijection f between T and $\mathcal{N}$: $f(suc^i(0)) = i$, where $\mathcal{N}$ is the set of natural numbers. That is, for any $i \in N$, $f^{-1}(i) = suc^i(0)$. By giving suc the meaning of *successor* of a natural number, mathematicians use T as a definition of $\mathcal{N}$.

In ZF (Zermelo–Fraenkel set theory), $\mathcal{N}$ can be defined by the *axioms of pairing* and *infinity*:

$$\begin{aligned}
0 &= \{\} &&= \emptyset \\
1 &= \{0\} &&= \{\emptyset\} \\
2 &= \{0, 1\} &&= \{\emptyset, \{\emptyset\}\} \\
3 &= \{0, 1, 2\} &&= \{\emptyset, \{\emptyset\}, \{\emptyset, \{\emptyset\}\}\} \\
&\ldots &&\ldots
\end{aligned}$$

Members of $\mathcal{N}$ are called *von Neumann ordinals* in ZF. □

1.3 A Glance of Mathematical Logic

Functions can be recursively defined in an analogous way. Let pre, add, sub, and mul be the predecessor, addition, subtraction, and multiplication functions on the set of natural numbers:

$$pre(0) = 0$$
$$pre(s(x)) = x$$
$$add(0, y) = y$$
$$add(s(x), y) = s(add(x, y))$$
$$sub(x, 0) = x$$
$$sub(x, s(y)) = sub(pre(x), y)$$
$$mul(0, y) = 0$$
$$mul(s(x), y) = add(mul(x, y), y)$$

Note that the subtraction defined above is different from the subtraction on the integers: We have $sub(s^i(0), s^j(0)) = 0$ when $i \le j$. This function holds the property that $sub(x, y) = 0$ iff $x \le y$.

The above functions are examples of so-called *primitive recursive functions*, which are total functions and members of a set of so-called *general recursive functions*. A primitive recursive function can be computed by a computer program that uses an upper bound (determined before entering the loop) for the number of iterations of every loop. The set of general recursive functions (also called partial recursive functions) is equivalent to those functions that can be computed by Turing machines.

Backus–Naur Form (BNF)

In computer science, *Backus–Naur form* (BNF) is a notation technique for *context-free grammars*, often used to describe the syntax of languages used in computing, such as computer programming languages, document formats, instruction sets, and communication protocols. John Backus and Peter Naur are the first to use this notation to describe Algol, an early programming language. BNF is applied wherever formal descriptions of languages are needed. We will use BNF to describe the syntax of logic as a language.

BNF can be extended to define recursively constructed objects in a succinct style. For example, the set of natural numbers is expressed in BNF as follows, where suc denotes the *successor* function:

$$\langle N \rangle ::= 0 \mid suc(\langle N \rangle)$$

The symbol ::= can be understood as "is defined by," and "|" separates alternative parts of a definition. The above formula can be read as follows: the member of N is either 0, or suc applying to (another) member of N; nothing else will be in N. From this example, we see that to define a set N, we use multiple items on the right of

::=. For a recursive definition, we have a basic case (such as 0) and a recursive case (such as $suc(\langle N \rangle)$, where $\langle N \rangle$ denotes a member of N).

Symbols like $\langle N \rangle$ are called *variables* in a context-free grammar and symbols like 0, suc, (, and) are called *terminals*. A context-free grammar specifies what set of strings of terminals can be derived for each variable. For the above example, the set of strings derived from BNF is

$$N = \{0, suc(0), suc^2(0), suc^3(0), \ldots, suc^i(0), \ldots\}$$

This set is the same as the objects defined in Example 1.3.8, but BNF gives us a formal and compact definition.

Example 1.3.9 To define the set Σ^* of all strings (of finite length) over the alphabet Σ, we may use BNF. For instance, if $\Sigma = \{a, b, c\}$, then Σ^* is the set of all strings of finite length:

$$\langle symbol \rangle ::= a \mid b \mid c$$
$$\langle \Sigma^* \rangle ::= \epsilon \mid \langle symbol \rangle \langle \Sigma^* \rangle$$

That is, the empty string $\epsilon \in \Sigma^*$; if $w \in \Sigma^*$, for every symbol $s \in \Sigma$, $sw \in \Sigma^*$; nothing else will be in Σ^*. Thus, Σ^* contains exactly all the strings built on Σ:

$$\Sigma^* = \{\epsilon, a, b, c, aa, ab, ac, ba, bb, bc, ca, cb, cc, aaa, \ldots\}.$$

This way, we may construct Σ^* for any finite alphabet Σ. □

Since there is a one-to-one corresponding between a string and a sequence, sequences can be also recursively defined in an analogous way. In fact, all recursively constructed objects can be defined this way.

Computable Functions

One of the basic questions addressed by computability theory is the following:

What does it mean for a function on the natural numbers to be computable?

There are two different answers to this simple question:

- A function is *computable* if there is a *computing model* (also called *computing device*, or *machine*), to compute it.
- A function is *computable* if there is an algorithm which computes the function.

The distinguishing between the two meanings of "computable functions" is very important in computability theory. To avoid confusion, we will stick to the first meaning of "computable functions" and call the functions of the second meaning "total computable functions."

1.3 A Glance of Mathematical Logic

Logicians and computer scientists introduced different computing models. These computing models can be Gödel's general recursive functions, Turing machines, Church's λ-calculus, Kleene's recursion theory, Post's canonical systems, Chomsky's grammars, Minsky's counter machines, or von Neumann model. By the *Church–Turing thesis* (also known as *computability thesis*, *Church–Turing conjecture*), if a function on the natural numbers can be calculated by a computing model, it can be computed by a Turing machine. Assuming the Church–Turing thesis is true, we will use Turing machines as our computing model. A Turing machine may loop forever on some input and the computed function is thus partial because of these loops. Thus, by "computable functions," we mean those total or partial functions that are computed by Turing machines.

Historically, an *algorithm* was defined as a finite sequence of step-by-step instructions and has been discussed by many within different computing models. An algorithm could be a program, a procedure, or a Turing machine. In today's study of algorithms, the scope of *algorithm* is reduced to those programs that must halt on any input. That is, by *algorithm*, we mean that the computing process will halt in a finite number of steps. For old meaning of "algorithm," we will use "procedure" instead.

Thus, "total computable functions" are more restrictive than "computable functions" of the first meaning: It requires that there is a computing model for the function and the computing model must **halt** on every input of the function. In this case, the computed function is total. That is, "total computable function" and "total and computable function" have the same meaning, so does "computable total function." Assuming the Church–Turing thesis is true, every computable function can be computed by a Turing machine. Every total computable function can be computed by a Turing machine that halts on every input and these Turing machines will be called "algorithms."

In theory of recursive functions, "computable functions" are equivalent to general recursive functions, and "total computable functions" are a superset of all primitive recursive functions. Note that the division function "/" on natural numbers may be total by allowing $x/0 = 0$.

Definition 1.3.10 A *decision problem* is a function that takes any input of the problem and returns 0 (false or no) or 1 (true or yes) as output.

A decision problem is *computable* if there exists a Turing machine to compute its function.

A decision problem is *decidable* if there is a Turing machine which computes it and halts on every input. This Turing machine is called *decider*.

Now, functions are divided into three disjoint groups: (*a*) total computable, (*b*) computable but not total computable, and (*c*) uncomputable. Decision problems are also divided into three disjoint groups: (*a*) decidable, (*b*) computable but undecidable, and (*c*) uncomputable. By the Church–Turing thesis, no computing models are available for uncomputable functions or uncomputable decision problems.

In Sect. 1.3.1, we saw that the set of decision functions on natural numbers is uncountable. These functions belong to the intersection of decision problems and general functions over natural numbers. Since the set of finite-length binary strings is countable and there exists a bijection between binary strings and Turing machines, the set of Turing machines is countable. In other words, we have only a countable set of Turing machines but decision problems are uncountable. As a result, there exist massive number of decision problems which are uncomputable.

One notable undecidable problem is to decide if a Turing machine M will halt on an input w. If we encode M and w as a single positive integer denoted by $\langle M, w \rangle$, and define the function $f(\langle M, w \rangle) = 1$ if M halts on w, and 0, otherwise, then f is an undecidable decision function.

As the counterpart of Church–Turing thesis, the concept of *Turing completeness* concerns whether a computing model can simulate a Turing machine. A computing model is *Turing complete* if it can simulate any Turing machine. By the Church–Turing thesis, a Turing machine is theoretically capable of doing all tasks done by computers; on the other hand, nearly all digital computers are Turing complete if the limitation of finite memory is ignored. Some logics are also Turing complete as they can be used to simulate any Turing machine. As a result, some problems such as the halting problem for these logics are undecidable. Computability theory will help us to decide if there exist or not decision procedures for some logic, and we will see some examples in Chap. 11.

1.3.3 Model Theory

Model theory is the study of the relationship between formal theories and their models. A formal theory is a collection of sentences in a formal language expressing statements. Models are mathematical structures (e.g., groups, fields, algebras, graphs, logics) in which the statements of the theory hold. The aspects investigated include the construction, the number and size of models of a theory, the relationship of different models to each other, and their interaction with the formal language itself. In particular, model theorists also investigate the sets that can be defined in a model of a theory, and the relationship of such definable sets to each other. Every formal language has its syntax and semantics. Models are a semantic structure associated with syntactic structures of a formal language. Following this approach, every formal logic is defined inside of a formal language.

Syntax and Semantics

The syntax of a formal language specifies how various components of the language, such as symbols, words, and sentences, are defined. For example, the language for propositional logic uses only propositional variables and logic operators as its symbols, and well-formed formulas built on these symbols as its sentences. In model theory, a set of sentences in a formal language is one of the components that form a theory. The language for logic often contains the two constant symbols, either

1.3 A Glance of Mathematical Logic

1 and 0, or $\top$ and $\bot$, which are interpreted as *true* and *false*, respectively. They are usually considered to be special logical operators which take no arguments, not propositional variables.

The semantics of a language specifies the meaning of various components of the language. For example, if we use symbol q to stand for the statement "Today is Monday," then the meaning of q is "Today is Monday." q can be used to denote a thousand different statements, just like a thousand Hamlets in a thousand people's eyes. On the other hand, a formal meaning of the formula $q \vee r$ is the truth value of $q \vee r$, which can be decided uniquely when the truth values of q and r are given. In model theory, the formal meaning of a sentence is explored: It examines semantic elements (meaning and truth) by means of syntactical elements (formulas and proofs) of a corresponding language.

In model theory, semantics and model are synonyms. A *model* of a theory is an interpretation that satisfies the sentences of that theory. Universal algebras are often used as models. In a summary definition, dating from 1973,

$$\text{model theory} = \text{universal algebra} + \text{logic} \tag{1.1}$$

Universal algebra, also called *categorical algebra*, is the field of mathematics that studies algebraic structures and their models or algebras.

Boolean Algebra

In universal algebra, the most relevant algebra related to the logic discussed in this book is *Boolean algebra*. Boolean algebra is not necessarily a topic of model theory, but is often quoted as an example in model theory.

Many syntactic concepts of Boolean algebra carry over to propositional logic with only minor changes in notation and terminology, while the semantics of propositional logic are defined via Boolean algebras in a way that the tautologies (theorems) of propositional logic correspond to equational theorems of Boolean algebra.

In Boolean algebra, the values of the variables are the truth values *true* and *false*, usually denoted by 1 and 0, respectively. The main operations of Boolean algebra are the multiplication "·" (conjunction), the addition "+" (disjunction), and the inverse i (negation). It is thus a formalism for describing logical operations in the same way that elementary algebra describes numerical operations, such as addition and multiplication. In fact, *Boolean algebra* is any set with binary operations + and · and a unary operation i thereon satisfying the Boolean laws (equations), which define the properties of the logical operations.

Boolean algebra was introduced by George Boole in his first book *The Mathematical Analysis of Logic* (1847). Sentences that can be expressed in propositional logic have an equivalent expression in Boolean algebra. Thus, Boolean algebra is sometimes used to denote propositional logic performed in this way. However,

Boolean algebra is not sufficient to capture logic formulas using quantifiers, like those from first-order logic.

1.3.4 Proof Theory

In an equivalent way to model theory, proof theory is situated in an interdisciplinary area among mathematics, philosophy, and computer science. In proof theory, proofs are studied as formal mathematical objects. Proofs are typically recursively defined as data structures such as lists (or trees) of formulas. In a logical system, axioms and inference rules are provided to guide the construction of proofs. Consequently, proof theory is syntactic in nature, in contrast to model theory, which is semantic in nature.

Axioms and Theorems

In formal logic, the *axioms* are a set of sentences which are assumed to be *true*. Typically, the axioms contain by default the definitions of all logical operators. For example, the negation operator $\neg$ is defined by $\neg(\top) = \bot$ and $\neg(\bot) = \top$.

Intuitively, the set of *theorems* are the formulas which can be proved to be true from the axioms by the given proof system. For example, for any propositional variable p, $\neg(\neg(p)) = p$ is a theorem. Using the case analysis method, if the truth value of p is $\top$, then $\neg(\neg(\top)) = \neg(\bot) = \top$; if the truth value of p is $\bot$, then $\neg(\neg(\bot)) = \neg(\top) = \bot$. It is easy to see that different sets of axioms or different proof systems would lead to different set of theorems.

A better approach to define *theorems* is to use the concept of models. Given a set of axioms, a *theorem* is any formula that will accept any model of the axioms as its model. According to this definition, if a set of axioms has no models, then every formula will be a theorem. Later, we will see that this definition of theorems leads to two important properties of a proof system: *completeness* (every theorem can be proved) and *soundness* (everything proved is a theorem).

There are two important properties concerning a set of axioms:

- **Consistency** The axiom set is *consistent* if every formula in the axiom set can be *true* at the same time.
- **Independence** The axiom set is *independent* if no axiom is a theorem of the other axioms. An independent set of axioms is also called a *minimum* set of axioms.

1.3 A Glance of Mathematical Logic

Proof Procedures

Given a set A of axioms and a formula B, we need a procedure P that can answer the question "if B is a theorem of A or not": If $P(A, B)$ returns "yes," we say B is proved; if $P(A, B)$ returns "no," we say B is disproved; if it does not return a thing, we do not know if B is a theorem or not. P is called a *proof procedure*. This procedure can be carried out by hand or executed on a computer.

There are four important properties concerning a proof procedure P:

- **Consistency** $P(A, B)$ is *consistent* iff either A is inconsistent, or it is impossible for $P(A, B)$ to return both *true* and *false* for any B. This concept assumes that a proof procedure may return more than one output on the same input.
- **Soundness** $P(A, B)$ is *sound* iff whenever $P(A, B)$ returns *true*, B is a theorem of A; whenever $P(A, B)$ returns *false*, B is not a theorem of A.
- **Completeness** $P(A, B)$ is *complete* iff for any theorem B of A, $P(A, B)$ returns *true*.
- **Termination** $P(A, B)$ is *terminating* iff $P(A, B)$ halts for any axiom set A and any formula B.

Soundness is a stronger property than consistency as a proof procedure may be consistent but not sound. For example, in propositional logic, if $P(\{p \vee q\}, p)$ returns *true* but $P(\{p \vee q\}, \neg p)$ does not return *true*, then P is consistent. However, P is not sound, since p is not a theorem of $p \vee q$ in propositional logic.

Proposition 1.3.11

(a) *If a proof procedure $P(A, B)$ is sound, then $P(A, B)$ is consistent.*
(b) *If a proof procedure $P(A, B)$ is sound and terminating, then $P(A, B)$ is complete.*

Proof

(a) If A is inconsistent, then $P(A, B)$ is consistent by definition. Now assume A is consistent. If $P(A, B)$ is not consistent, then there exists a formula B such that $P(A, B)$ returns both *true* and *false*. In this case, $P(A, B)$ cannot be sound, because B cannot be a theorem and a non-theorem at the same time when A is consistent.
(b) For any theorem B of the axioms A, $P(A, B)$ must return *true* because if $P(A, B)$ halts with *false*, that is a contradiction to the soundness of $P(A, B)$. □

The proof of the above proposition shows that, under the condition of soundness, the termination of P implies the completeness of P. In this case, P is called a *decision procedure* for the logic if P is both sound and terminating. By the above result, every decision procedure is sound and complete. In general, a *decision procedure* usually refers to an algorithm which always halts with a correct answer to a decision problem, such as "A is valid or not." Some algorithms always halt with one of the outcomes: "yes," "no," or "unknown"; they cannot be called "decision procedures," because they may return "unknown" on some theorems, thus not complete.

Under the condition of soundness, the completeness of a proof procedure P does not imply the termination of P, because a sound and complete proof procedure may loop on a formula that is not a theorem and thus cannot be a decision procedure by definition. In literature, people call a sound and complete proof procedure as a *semi-decision procedure* for the logic, because it will halt on any formula which is a theorem.

For any logic, we always look for its decision procedure that is both sound and terminating. For example, propositional logic has different decision procedures for different decision problems. If a decision procedure does not exist, we look for a semi-decision procedure. For some logic, even a semi-decision procedure cannot exist. These comments will be more meaningful once we have formally defined the logic in question.

Inference Rules

In structural proof theory, which is a major area of proof theory, inference rules are used to construct a proof. In logic, an *inference rule* is a pattern which takes formulas as premises and returns a formula as conclusion. For example, the rule of *modus ponens* (MP) takes two premises, one in the form "If p then q" (i.e., $p \to q$) and another in the form p, and returns the conclusion q. A rule is *sound* if whenever the premises are *true* under any interpretation, so is the conclusion.

An inference system S consists of a set A of axioms and a set R of inference rules. The soundness of S comes from the soundness of every inference rule in R. A *proof* of formula F_n in $S = (A, R)$ is a sequence of formulas $F_1, F_2, \ldots, F_n$ such that each F_i is either in A or can be generated by an inference rule of R, using the formulas before F_i in the sequence as the premises.

Example 1.3.12 Let $S = (A, R)$, $A = \{p \to (q \to r), p, q\}$, and $R = \{MP\}$. A proof of r from A using R is given below:

$$
\begin{array}{lll}
1. & p \to (q \to r) & axiom \\
2. & p & axiom \\
3. & q \to r & MP, 1, 2 \\
4. & q & axiom \\
5. & r & MP, 3, 4 \\
\end{array}
$$

where "$MP, 1, 2$" means the formula $q \to r$ is derived by MP (modus ponens) from formulas 1 and 2 as premises. □

Thus, a *formal proof* is a sequence of formulas, starting with axioms and continuing with theorems derived from earlier members of the sequence by rules of inference.

One obvious advantage of such a proof is that these proofs can be checked either by hand or automatically by computer. Checking formal proofs is usually simple, whereas finding proofs (automated theorem proving) is generally hard. Checking an informal proof in the mathematics literature, by contrast, requires weeks of peer review and may still contain errors. Informal proofs of everyday mathematical practice are unlike the proofs of proof theory. They are like high-level sketches that would allow an expert to reconstruct a formal proof at least in principle, given enough time and patience. For most mathematicians, writing a fully formal proof is too pedantic and long-winded to be in common use.

In proof theory, proofs are typically presented as recursively defined data structures such as lists (as shown here), or trees, which are constructed according to the axioms and inference rules of the logical system. As such, proof theory is syntactic in nature, in contrast to model theory, which is semantic in nature.

Every inference system can be converted to a proof procedure if a *fair strategy* is adapted to control the use of inference rules. Fair strategy ensures that if an inference rule is applicable at some point to produce a new formula, this new formula will be derived eventually. If every inference rule is sound, the corresponding proof procedure is sound. Extra effort is needed in general to show that the proof procedure is complete or halting.

Besides structural proof theory, some of the major areas of proof theory include ordinal analysis, provability logic, reverse mathematics, proof mining, automated theorem proving, and proof complexity.

One of the main aims of this book is to present algorithms invented for logic and the software tools built up based on these algorithms. That is why we will invest a disproportionate amount of effort (by the standard of conventional logic textbooks) in the algorithmic aspect of logic. Set theory and model theory help us to understand and specify the problems; proof theory helps us to create algorithms; and computability theory helps us to know what problems have algorithms for them, and what problems do not have algorithms at all.

Exercises

1. Decide if the following sentences are statements or not. If they are, use propositional variables (i.e., symbols) for simple statements and logic connectives to express them.

    (a) If I win the lottery, I'll be poor.
    (b) The man asked, "shut the door behind you."
    (c) Today is the first day of school.
    (d) Today is either sunny or cloudy.

2. Pinocchio, an animated puppet, is punished for each lie that he tells by undergoing further growth of his nose. The Pinocchio paradox arises when Pinocchio says "My nose grows now." Please explain why this is a paradox.

3. Yablo's paradox is given as an ordered infinite sequence of statements, each of which says that "the next statement is false." Please explain why this is a paradox. Decide whether this paradox relies on self-reference or not.
4. Answer the questions in the following examples regarding logical fallacies.

   (a) During their argument, Anthony tells Marie that he hates her and wished that she would get hit by a car. Later that evening, Anthony receives a call from a friend who tells him that Marie is in the hospital because she was struck by a car. Anthony immediately blames himself and reasons that if he hadn't made that comment during their fight, Marie would not have been hit. What logical fallacy has Anthony committed?

   (b) Louise is running for class president. In her campaign speech she says, "My opponent does not deserve to win. She is a smoker and she cheated on her boyfriend last year." What fallacy has Louise committed?

   (c) Bill: "I think capital punishment is wrong." Adam: "No it isn't. If it was wrong, it wouldn't be legal." Of which fallacy is this an example?

   (d) Ricky is watching television when he sees a commercial for foot cream. The commercial announcer says, "This is the best foot cream on the market because no other foot cream makers have been able to prove otherwise!" What fallacy has the announcer committed?

   (e) Maria has been working at her current job for more than 30 years at the same wage. She desperately wants a raise, so she approaches her boss to ask for one. She says, "You are one of the kindest people I know. You are smart and good looking, and I really love your shoes." What type of fallacy is this?

   (f) Jeff's mom is concerned when she finds out that he skipped class one day. She tells him that she is concerned that since he skipped one class, he will start skipping more frequently. Then he will drop out altogether, never graduate or get into college, and end up unemployed and living at home for the rest of his life. What type of fallacy has Jeff's mom committed?

   (g) Dana is trying to raise money for her university's library. In her address to the board of trustees, she says, "We must raise tuition to cover the cost of new books. Otherwise, the library will be forced to close." Of what fallacy is this an example?

   (h) Jeff is preparing to create a commercial for a new energy drink. He visits a local high school and surveys students in an English class about their beverage preferences. Most of the class says they prefer grape-flavored drinks, so Jeff tells his superiors that grape is the flavor favored most by high school students. What error in reasoning has Jeff made?

5. In the last three problems of the previous exercise, please identify which sentences are statements and which are not. For the simple statements in these problems, please introduce propositional variables (i.e., symbols) to denote them. Use these variables and logical operators to denote the composed statements.
6. If $\succeq$ is a partial order, show that the reverse of $\succeq$ is a partial order.

7. Let $A = \{0, 1\}$; please provide the solution of $\mathcal{P}(A)$ and $\mathcal{P}(\mathcal{P}(A))$.
8. Show that all countably infinite sets have the same size, and each countably infinite set has a rank function.
9. Show that if r is the rank function of a countably infinite set S, then r induces a well-order $\succeq$ on S: $x \succeq y$ iff $r(x) \geq r(y)$.
10. Let a, b be real numbers and $a < b$. Define a function $f : [0, 1) \mapsto [a, b)$ by $f(x) = a + x(b - a)$, where $[a, b)$ denotes the interval $\{x \mid a \leq x < b\}$. Show that f is bijective.
11. Answer the following questions with justification:

    (a) Is the set of all odd integers countable?
    (b) Is the set of real numbers in the interval $[0, 0.1)$ countable?
    (c) Is the set of angles in the interval $[0^o, 90^o)$ countable?
    (d) Is the set of all points in the plane with rational coordinates countable?
    (e) Is the set of all Java programs countable?
    (f) Is the set of all words using an English alphabet countable?
    (g) Is the set of sands on the earth countable?
    (h) Is the set of atoms in the solar system countable?

12. Prove the following properties:

    (a) Function $f : A \mapsto B$ is injective iff its inverse is well-defined and injective.
    (b) An injective function $f : A \mapsto B$ is total iff its inverse is surjective (assuming a surjective function is not required to be total).

13. Prove Proposition 1.3.1: $f : A \mapsto B$ is bijective iff both f and f^{-1} are total functions.
14. Prove that the set $\mathcal{R}$ of rational numbers is countable.
15. Prove that the set $\mathcal{N} \times \{a, b, c\}$ is countable, where $\mathcal{N}$ is the set of natural numbers. Your proof should not use the property that countable sets are closed under Cartesian product.
16. Decide with proof if the set $\mathcal{N}^k$ is countable or not, where $\mathcal{N}$ is the set of natural numbers and $k \in \mathcal{N}$.
17. Decide with proof if the set $S_k = \{A \mid A \subset \mathcal{N}, |A| = k\}$ is countable or not, where $\mathcal{N}$ is the set of natural numbers and $k \in \mathcal{N}$.
18. Prove that if both A and B are countable and $A \cap B = \emptyset$, then $A \cup B$ is countable.
19. Let each of $A_0, A_1, \ldots, A_i, \ldots$ be countably infinite and for all $i, j \in \mathcal{N}$, $A_i \cap A_j = \emptyset$ for $i \neq j$. Prove that $S = \bigcup_{i=0}^{\infty} A_i$ is countably infinite.
20. Decide with proof if the set of all total functions $f : \{0, 1\} \mapsto \mathcal{N}$ is countable or not.
21. Please provide a bijection $f : \mathcal{N} \mapsto \{0, 1\}^*$ and show how to compute $f(x)$ for any $x \in \mathcal{N}$. What is the time complexity of your algorithm for $f(x)$?
22. For the function γ in Definition 1.3.5, please define γ^{-1} explicitly as a total function from $\mathcal{N}$ to Σ^*.
23. Prove that the set $\{a, b, c\}^*$ of all strings (of finite length) on $\{a, b, c\}$ is countable.

24. Prove that the set of all formal languages is uncountable, where a *formal language* is any subset of Σ^* and $\Sigma \neq \emptyset$.
25. Prove that the following statements are logically equivalent:

    (a) Any subset of a countable set is countable.
    (b) Any subset of $\mathcal{N}$ is countable.
    (c) If there is an injective function from set S to $\mathcal{N}$, then S is countable.
    (d) If there is a surjective function from $\mathcal{N}$ to S, then S is countable.

26. **Barber's paradox**: In a village, there is only one barber who shaves all those, and those only, who do not shave themselves. The question is, does the barber shave himself? Let the relation $s(x, y)$ be that "x shaves y" and b denote the barber.

    (a) Define A to be the set of the people shaved by the barber using $s(x, y)$ and b.
    (b) Define B to be the set of the people who does not shave himself.
    (c) Use A and B to show that the answer to the question "does the barber shave himself?" leads to a contradiction.

27. Suppose the set of natural numbers is constructed by the constant 0 and the successor function s. Provide a definition of $<$ and $\leq$ on the natural numbers. No functions other than 0 and s are allowed to use when defining $<$ or $\leq$.
28. Provide a definition of power function, $pow(x, y)$, which computes x^y, on the natural numbers built up by 0 and the successor function s. You are only allowed to use the functions *add* (addition) and *mul* (multiplication).
29. Provide a BNF (Backus–Naur form) for the set of binary trees, where each non-empty node contains a natural number. You may use *null: Tree* for the empty tree and *node: Nat, Tree, Tree $\mapsto$ Tree* for a node of the tree.

References

1. Paul Vincent Spade and Jaakko J. Hintikka, *History of logic*. Encyclopedia Britannica, Dec. 17, 2020, www.britannica.com/topic/history-of-logic, retrieved Nov. 1, 2023
2. Daniel Cohnitz and Luis Estrada-Gonzalez, *An Introduction to the Philosophy of Logic*, Cambridge University Press, Jun. 27, 2019
3. Joseph Mileti, *Modern Mathematical Logic* (Cambridge Mathematical Textbooks), Cambridge University Press, Sep. 22, 2022

Part I
Propositional Logic

Chapter 2
Propositional Logic

Following model theory presented in Chap. 1, we will present propositional logic as a language. We introduce first the syntax of this language and then its semantics [1, 2].

2.1 Syntax

The syntax of propositional logic specifies what symbols are allowed and in what forms these symbols are used. Propositional logic uses *propositional variables* and *logical operators* as the only symbols. In the first chapter, we said that propositional variables are symbols for denoting statements and take only true and false as truth values. For convenience, we will use 1 for true and 0 for false. We will treat $\top$ and $\bot$ as the two nullary logical operators such that the truth value of $\top$ is 1 and the truth value of $\bot$ is 0. Note that $\top$ and $\bot$ are syntactic symbols while 1 and 0 are semantic symbols.

2.1.1 Logical Operators

Mathematicians use the words "not," "and," "or," etc., for operators that change or combine propositions. The meaning of these logical operators can be specified as a function which takes Boolean values and returns a Boolean value. These functions are called *Boolean functions*. For example, if p is a proposition, then so is $\neg p$ and the truth value of the proposition $\neg p$ is determined by the truth value of p according to the meaning of $\neg$: $\neg(1) = 0$ and $\neg(0) = 1$. That is, if the value of p is 1, then the value of $\neg(p)$ is 0; if the value of p is 0, then the value of $\neg(p)$ is 1. Note that the same symbol $\neg$ is used both as a Boolean operator (syntax side) and a Boolean function (semantic side).

Table 2.1 Meanings of ∧, ∨, →, ⊕, and ↔

p	q	p∧q	p∨q	p→q	q→p	(q→p)∧(p→q)	p⊕q	p↔q
1	1	1	1	1	1	1	0	1
1	0	0	1	0	1	0	1	0
0	1	0	1	1	0	0	1	0
0	0	0	0	1	1	1	0	1

In general, the definition of a Boolean function is displayed by a table, called *truth table*, where the output of the Boolean function is given for each possible assignment of truth values to all the input variables.

For example, the meanings of ∧, ∨, and → are defined by Table 2.1. The table has four rows, since there are four pairs of Boolean values for the two variables. According to the table, the truth value of $p \wedge q$ is 1 when both p and q are 1. The truth value of $p \vee q$ is 1 when either p or q, or both are true. This is not always the intended meaning of "or" in everyday dialog, but this is the standard definition in logic. So, if a logician says, "You may have cake, or you may have ice cream," he means that you could have both. If you want to exclude the possibility of having both cake and ice cream, you should combine them with the exclusive-or operator, ⊕, which is also defined in Table 2.1. The exclusive disjunction ⊕ corresponds to addition modulo 2 and is therefore given the symbol +, where the values are 0 and 1.

The truth table for implication is given in Table 2.1, too. This operator is worth remembering, because a large fraction of all mathematical statements is of the if-then form. Now let's figure out the truth of the following example: If elephants fly, I'll be on Mars. What is the truth value of this statement? This statement can be represented by $p \rightarrow q$, where p is "elephants can fly" and q is "I'll be on Mars." Elephants do not fly, so the truth value of p is 0 and we fall to the last two rows of Table 2.1: In both rows, $p \rightarrow q$ is 1. In Sect. 1.2.2 on informal fallacies, we pointed out that with a false premise, you can make any conclusion, so that the composed statement is always true. However, an always true statement has no value in reasoning, and it does not imply the causal connection between the premise and the conclusion of an implication. In logic, it is important to accept the fact that logical implications ignore causal connections.

In Table 2.1, we also give the definition of ↔ (if and only if). If p denotes "Tom wears a blue jean" and q denotes "Sam wears a blue jean," then the formula $p \leftrightarrow q$ asserts that "Tom and Sam always wear a blue jean at the same time." This is because the sentence implies that "Tom wears a blue jean when Sam does" and "Sam wears a blue jean when Tom does." So, we have $(p \rightarrow q) \wedge (q \rightarrow p)$, which has the same truth value as $p \leftrightarrow q$ does in every interpretation. In fact, $p \leftrightarrow q$ is true iff p and q have the same truth value. Note that $p \oplus q$ and $p \leftrightarrow q$ always have opposite values, that is, $\neg(p \oplus q)$ and $p \leftrightarrow q$ always have the same value.

Let $B = \{0, 1\}$; the meaning of any logical operator is a Boolean function $f : B^n \mapsto B$. There are two nullary operators (when $n = 0$): ⊤ denotes 1 and ⊥ denotes 0. There are four unary operators (when $n = 1$): $f_1(x) = 0$; $f_2(x) = 1$; $f_3(x) = x$; and $f_4(x) = \neg x$. There are 16 binary operators (when $n = 2$), and ∧, ∨, →, ⊕,

2.1 Syntax

and ↔ are some of them. We may use operators when $n > 2$. For example, let $ite : B^3 \mapsto B$ be the if-then-else operator, if we use a truth table for the meaning of $ite(p, q, r)$, then the truth table will have eight rows. In general, there are 2^{2^n} n-ary Boolean functions, because the number of functions $f : A \mapsto B$ is $|B|^{|A|}$ and for n-ary Boolean function, $|B| = 2$ and $|A| = |B^n| = |B|^n = 2^n$.

2.1.2 Formulas

Not every combination of propositional variables and logic operators makes sense. For instance, p () $q \vee \neg$) is a meaningless string of symbols. The *propositional formulas* are the set of well-formed strings built up by propositional variables and logical operators by a rigorous set of rules.

Let *op* be the binary operators and V be the propositional variables used in an application, then the *Formulas* for this application is defined by the following BNF grammar:

$\langle op \rangle ::= \wedge \mid \vee \mid \rightarrow \mid \oplus \mid \leftrightarrow$
$\langle V \rangle ::= p \mid q \mid r \mid s \mid t$
$\langle Formulas \rangle ::= \top \mid \bot \mid \langle V \rangle \mid \neg \langle Formulas \rangle \mid (\langle Formulas \rangle \, \langle op \rangle \, \langle Formulas \rangle)$

We may add/delete operators or propositional variables at will in the above definition. According to the definition, every member of *Formulas* is either the constants $\top$ or $\bot$, a propositional variable of V_P, or the negation of a formula, or a binary operator applying to two formulas. Here are some examples:

$$\top, \; p, \; \neg q, \; (p \wedge \neg q), \; (r \rightarrow (p \wedge \neg q)), \; ((p \wedge \neg q) \vee r)$$

Throughout this chapter, we will use $p, q, \ldots$, for propositional variables, and $A, B, \ldots$, for formulas.

For every binary operator, we introduce a pair of parentheses, and some of them may be unnecessary. We will use a precedence, which is a total order over all the operators, to remove them. The typical precedence is given as the following list, where operators with higher precedence go first. Note that we assign $\oplus$ and $\leftrightarrow$ the same precedence.

$$\neg, \; \wedge, \; \vee, \; \rightarrow, \; \{\oplus, \leftrightarrow\}$$

Thus $(r \rightarrow (p \wedge \neg q))$ can be simply written as $r \rightarrow p \wedge \neg q$, and we will not take it as $(r \rightarrow p) \wedge \neg q$, because $\wedge$ has higher precedence than $\rightarrow$. Similarly, $((p \wedge \neg q) \vee r)$ is written as $p \wedge \neg q \vee r$ without ambiguity.

When a well-formed expression, like the formulas defined above, is represented by a string (a sequence of symbols), we use parentheses and commas to identify the structures of the expression. Such a structure can be easily represented by a tree, where we do not need parentheses and commas.

Definition 2.1.1 A *tree* for formula A is defined recursively as follows:

1. If A is p, $\top$, or $\bot$, the tree is a single node with A as its label.

Fig. 2.1 The formula tree for $(\neg p \wedge q) \rightarrow (p \wedge (q \vee \neg r))$

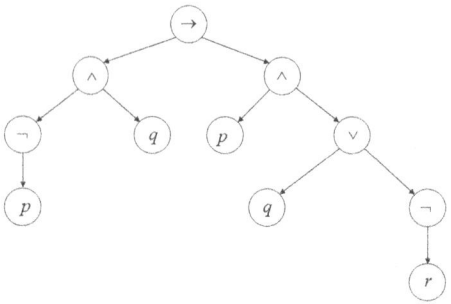

2. If A is $\neg B$, then the root node is labeled with $\neg$ and has one branch pointing to the formula tree of B.
3. If A is $B \ op \ C$, then the root node is labeled with op and has two branches pointing to the formula trees of B and C, respectively.

For example, the formula tree of $(\neg p \wedge q) \rightarrow (p \wedge (q \vee \neg r))$ is displayed in Fig. 2.1. The tree representation of a formula does not need parentheses.

A formula and its formula tree are the two representations of the same object. We are free to choose one of the representations for the convenience of discussion.

Definition 2.1.2 Given a formula A, we say B is a *subformula* A if either $B = A$ or $A = (A_1 \ op \ A_2)$, where op is a binary operator, or $A = \neg A_1$ and B is a subformula of A_1 or A_2. If B is a subformula of A and $B \neq A$, then B is a proper subformula of A.

For example, $\neg p \vee \neg(q \vee r)$ contains $p, q, r, \neg p, q \vee r$, and $\neg(q \vee r)$ as proper subformulas. Intuitively, a subformula is just a formula represented by any subtree in the formula tree.

Definition 2.1.3 Given a subformula B of A and another formula C, we use $A[B \leftarrow C]$ to denote the formula obtained by substituting all occurrences of B by C in A. If B is a propositional variable, then $A[B \leftarrow C]$ is called an *instance* of A.

For example, if A is $\neg p \vee \neg(q \vee r)$, B is $\neg(q \vee r)$, and C is $\neg q \wedge \neg r$, then $A[B \leftarrow C]$ is $\neg p \vee (\neg q \wedge \neg r)$. $A[p \leftarrow p \wedge q] = \neg(p \wedge q) \vee \neg(q \vee r)$ is an instance of A.

2.2 Semantics

The title of the previous section is syntax, though we talked about Boolean values, truth tables, and the meaning of logical operators, which are, strictly speaking, semantic concepts of the logic. Now, let us focus on the meaning of propositional variables and formulas.

2.2 Semantics

The informal meaning of a propositional variable is the statement that symbol represents. In different applications, the same symbol can represent different statements and thus has different meanings. However, each statement can have only two truth values, i.e., true or false. We will use these truth values of propositional variables, extending to the formulas which contain them, as the formal semantics of propositional logic.

2.2.1 Interpretations

A propositional variable may be assigned a truth value, either true or false. This truth value assignment is considered to be the *semantics* of the variable. For a formula, an assignment of truth values to every propositional variable is said be an *interpretation* of the formula. If A is a formula built on the set V_P of propositional variables, then an interpretation of A is the function $\sigma : V_P \mapsto \{1, 0\}$. It is easy to check that there are $2^{|V_P|}$ distinct interpretations.

Suppose σ is an interpretation over $V_P = \{p, q\}$ such that $\sigma(p) = 1$ and $\sigma(q) = 0$. We may write $\sigma = \{p \mapsto 1, q \mapsto 0\}$. We may also use $(1, 0)$ for the same σ, assuming V_P is a list (p, q) and $\sigma(p, q) = (1, 0)$.

An alternative representation of an interpretation σ is to use a subset of V_P. Given σ, let

$$X_\sigma = \{x \in V_P \mid \sigma(x) = 1\}$$

Then there is one-to-one relation between σ and X_σ. So, we may use X_σ to represent σ. For example, if $V_P = \{p, q\}$, then the four interpretations over V_P are the power set of V_P:

$$\mathcal{P}(V_P) = \{\emptyset, \{p\}, \{q\}, \{p, q\}\}$$

Another alternative representation of σ is to use a set of literals. A *literal* is either a propositional variable or the negation of a propositional variable. Let

$$Y_\sigma = \{x \in V_P \mid \sigma(x) = 1\} \cup \{\neg y \mid y \in V_P, \sigma(y) = 0\}$$

then Y_σ is a set of literals such that every variable of V_P appears in Y_σ exactly once. For instance, let $V_P = \{p, q\}$ and $X_\sigma = \emptyset$, then $Y_\sigma = \{\neg p, \neg q\}$; if $X_\sigma = \{p\}$, $Y_\sigma = \{p, \neg q\}$. That is, by adding the negations of the missing variables in X_σ, we obtain such a representation Y_σ for σ. Using a set of literals to represent an interpretation has an additional advantage: It can represent a *partial* interpretation where some propositional variables do not have a truth value.

For a compact representation of an interpretation, we may write a set of literals as a conjunction of these literals, often omitting the conjunction operator $\wedge$. For example, $\{p, \neg q\}$ can be written as $p \wedge \neg q$, or simply $p\overline{q}$.

Given σ, we can check the truth value of the formula under σ, denoted by $\sigma(A)$. This notation means that we treat σ as a function from the formulas to $\{0, 1\}$, as the unique extension of $\sigma : V_P \mapsto \{0, 1\}$. On the other hand, the notation $A\sigma$ is used to denote the formula obtained by substituting each propositional variable p by $\sigma(p)$ if we regard 1 as $\top$ and 0 as $\bot$.

Example 2.2.1 Given the formula $A = (p \wedge \neg q) \vee r$, and an interpretation $\sigma = \{p \mapsto 1, q \mapsto 1, r \mapsto 0\}$, we can rewrite σ as $\{p, q, \neg r\}$. Replacing p by 1, q by 1, and r by 0 in A, we have $A\sigma = 1 \wedge \neg 1 \vee 0$. Applying the meanings of $\neg$, $\wedge$, and $\vee$, we obtain 0. In this case, we say the formula is *evaluated* to 0 under σ, denoted by $\sigma(A) = 0$. Given another interpretation $\sigma' = \{p, q, r\}$, the same formula will be evaluated to 1, i.e., $\sigma'(A) = 1$. □

Recall that we use $A[B \leftarrow C]$ to denote an instance of A where every occurrence of propositional variable B is replaced by C. An interpretation σ is a special case of the substitution of formulas where B is a variable and C is either 1 or 0. For example, if $\sigma = \{p, \neg q\}$, then $A\sigma = A[p \leftarrow 1][q \leftarrow 0]$. Strictly speaking, $A\sigma$ is not a propositional formula: It is the meaning of A under σ. Hence, we assume $\top \sigma = 1$ and $\bot \sigma = 0$ for any σ, where $\top$ and $\bot$ are propositional formulas, and 1 and 0 are their truth values, respectively.

To obtain $\sigma(A)$, we may use the algorithm *eval*.

Algorithm 2.2.2 The algorithm *eval* will take a formula A and an interpretation σ and returns a Boolean value.

proc $eval(A, \sigma)$
    if $A = \top$ **return** 1
    if $A = \bot$ **return** 0
    if $A \in V_P$ **return** $\sigma(A)$
    if A is $\neg B$ **return** $\neg eval(B, \sigma)$
    else A is $(B\ op\ C)$ **return** $eval(B, \sigma)\ op\ eval(C, \sigma)$

Example 2.2.3 Let $\sigma = \{p, \neg q\}$. Then the execution of $eval(p \rightarrow p \wedge \neg q, \sigma)$ will return 1:

$eval(p \rightarrow p \wedge \neg q, \sigma)$ calls
    $eval(p, \sigma)$, which returns 1; and
    $eval(p \wedge \neg q, \sigma)$, which calls
        $eval(p, \sigma)$, which returns 1; and
        $eval(\neg q, \sigma)$, which calls
            $eval(q, \sigma)$, which returns 0; and
            returns $\neg 0$, i.e., 1;
        returns $1 \wedge 1$, i.e., 1;
    returns $1 \rightarrow 1$, i.e., 1.

□

What *eval* does is to travel the formula tree of A bottom-up: If the node is labeled with a variable, use σ to get the truth value; otherwise, compute the *truth value of*

2.2 Semantics

this node under σ using the operator at that node with the truth values from its children. The process of running *eval* is exactly what we do when constructing a truth table for $p \to p \land \neg q$. The truth values under p and q in the truth table give us all the interpretations σ and the truth values of $A = \neg q$, $p \land \neg q$, or $p \to p \land \neg q$ are the values of $eval(A, \sigma)$.

p	q	$\neg q$	$p \land \neg q$	$p \to p \land \neg q$
0	0	1	0	1
0	1	0	0	1
1	0	1	1	1
1	1	0	0	0

Theorem 2.2.4 *Algorithm* $eval(A, \sigma)$ *returns* 1 *iff* $\sigma(A) = 1$, *and runs in* $O(|A|)$ *time, where* $|A|$ *denotes the number of symbols, excluding parentheses, in A.*

$|A|$ is also the number of nodes in the formula tree of A. The proof is left as an exercise.

Corollary 2.2.5 *For any formulas A and B, and any interpretation σ, the following equations hold.*

$$\sigma(A \lor B) = \sigma(A) \lor \sigma(B)$$
$$\sigma(A \land B) = \sigma(A) \land \sigma(B)$$
$$\sigma(\neg A) = \neg \sigma(A)$$

The truth value of a formula in propositional logic reflects the two foundational principles of Boolean logic: the *principle of bivalence*, which allows only two truth values, and the *principle of extensibility* that the truth value of a general formula depends only on the truth values of its parts, not on their informal meaning.

2.2.2 Models, Satisfiability, and Validity

Interpretations play a very important role in propositional logic and introduce many important concepts.

Definition 2.2.6 *Given a formula A and an interpretation σ, σ is a model of A if* $eval(A, \sigma) = 1$. *If A has a model, A is satisfiable.*

Given a set V_P of propositional variables, we will use All_P to denote the set of all interpretations over V_P. For example, if $V_P = \{p, q\}$, then $All_P = \{\{p, q\}, \{p, \neg q\}, \{\neg p, q\}\}, \{\neg p, \neg q\}\}$, or abbreviated as $All_P = \{pq, p\overline{q}, \overline{p}q, \overline{pq}\}$. In general, $|All_P| = 2^{|V_P|}$. We may use *eval* to look for a model of A by examining every interpretation in All_P.

Definition 2.2.7 Given a formula A, let $\mathcal{M}(A)$ be the set of all models of A. If $\mathcal{M}(A) = \emptyset$, A is *unsatisfiable*. If $\mathcal{M}(A) = All_P$, i.e., every interpretation is a model of A, then A is *valid*, or *tautology*. We write $\models A$ if A is valid.

Being valid is different from being useful or efficient. For instance, the statement $p \vee \neg p$ is valid but neither useful nor efficient (the formula $\top$ is more precise).

If there is a truth table for a formula A, $\mathcal{M}(A)$ can be easily obtained from the truth table, as the truth values under the variables of A are all the interpretations and the models of A are given in those rows of the table where the truth value of A is 1.

Example 2.2.8 Let $V_P = \{p, q\}$. If A is $\neg p \to \neg q$, then $\mathcal{M}(A) = \{pq, p\overline{q}, \overline{pq}\}$. If B is $p \vee (\neg q \to \neg p)$, then $\mathcal{M}(B) = All_P$. From $\mathcal{M}(A)$ and $\mathcal{M}(B)$, we know that both A and B are satisfiable, and B is also valid, i.e., $\models B$ is true. □

A valid formula is one which is always true in every interpretation, no matter what truth values its variables may take, that is, every interpretation is its model. The simplest example is $\top$, or $p \vee \neg p$. There are many formulas that we want to know if they are valid, and this is done by so-called theorem proving, either by hand or automatically.

We may think about valid formulas as capturing fundamental logical truths. For example, the transitivity property of implication states that if one statement implies a second one, and the second one implies a third, then the first implies the third. In logic, the transitivity can be expressed as the following formula:

$$(p \to q) \wedge (q \to r) \to (p \to r)$$

The validity of the above formula confirms the truth of this property of implication. There are many properties like this for other logical operators, such as the commutativity and associativity of $\wedge$ and $\vee$, and they all can be stated by a tautology.

Theorem 2.2.9 (Substitution) *If A is valid, so is any instance of A. That is, if p is a propositional variable in A and B is any formula, then $A[p \leftarrow B]$ is valid.*

Proof For any interpretation σ of $A[p \leftarrow B]$, let $\sigma(p) = \sigma(B)$. Applying σ to the formula trees of A and $A[p \leftarrow B]$, then the truth values of all the nodes of A must be identical to those of the corresponding nodes of $A[p \leftarrow B]$. Since A is valid, the root node of A must be 1 under σ, so the root node of $A[p \leftarrow B]$ must have the same truth value, i.e., 1. Since σ is arbitrary, $A[p \leftarrow B]$ must be valid. □

The significance of the above theorem is that if we have a tautology for one variable, the tautology holds when the variable is substituted by any formula. For example, from $p \vee \neg p$, we have $A \vee \neg A$ for any A. On the other hand, when we try to prove a tautology involving symbols $A, B, \ldots$, we may treat each of these symbols as a propositional variable. For example, to prove $(A \wedge B) \leftrightarrow (B \wedge A)$, we may prove instead $(p \wedge q) \leftrightarrow (q \wedge p)$.

An unsatisfiable formula is one which does not have any model, that is, no interpretation is its model. The simplest example is $\bot$, or $p \wedge \neg p$. Validity and unsatisfiability are dual concepts, as stated by the following proposition.

2.2 Semantics

Proposition 2.2.10 *A formula A is valid iff $\neg A$ is unsatisfiable.*

From the view of theorem proving, proving the validity of a formula is as hard (or as easy) as proving its unsatisfiability.

The following theorem shows the relationship between logic and set theory.

Theorem 2.2.11 *For any formulas A and B over V_P,*

$$(a) \ \mathcal{M}(A \vee B) = \mathcal{M}(A) \cup \mathcal{M}(B).$$
$$(b) \ \mathcal{M}(A \wedge B) = \mathcal{M}(A) \cap \mathcal{M}(B).$$
$$(c) \ \mathcal{M}(\neg A) = All_P - \mathcal{M}(A).$$

Proof To show $\mathcal{M}(A \vee B) = \mathcal{M}(A) \cup \mathcal{M}(B)$, we need to show that, for any interpretation σ, $\sigma \in \mathcal{M}(A \vee B)$ iff $\sigma \in \mathcal{M}(A) \cup \mathcal{M}(B)$. By definition, $\sigma \in \mathcal{M}(A \vee B)$ iff $\sigma(A \vee B) = 1$. By Corollary 2.2.5, $\sigma(A \vee B) = \sigma(A) \vee \sigma(B)$. So $\sigma(A \vee B) = 1$ iff $\sigma(A) = 1$ or $\sigma(B) = 1$. $\sigma(A) = 1$ or $\sigma(B) = 1$ iff $\sigma(A) \vee \sigma(B) = 1$. $\sigma(A) \vee \sigma(B) = 1$ iff $\sigma \in \mathcal{M}(A) \cup \mathcal{M}(B)$.

The proof of (b) and (c) is left as exercise. □

Example 2.2.12 Let $V_P = \{p, q\}$. $\mathcal{M}(p) = \{pq, p\overline{q}\}$, $\mathcal{M}(q) = \{pq, \overline{p}q\}$; $\mathcal{M}(\neg p) = All_P - \mathcal{M}(p) = \{\overline{p}q, \overline{pq}\}$, and $\mathcal{M}(\neg p \vee q) = \mathcal{M}(\neg p) \cup \mathcal{M}(q) = \{pq, \overline{p}q, \overline{pq}\}$. □

2.2.3 Equivalence

In natural language, one statement can be expressed in different forms with the same meaning. For example, "If I won the lottery, I must be rich." This meaning can be also expressed "Since I am not rich, I didn't win the lottery." Introducing p for "I won the lottery" and q for "I'm rich," the first statement becomes $p \to q$ and the second statement becomes $\neg q \to \neg p$. It happens that $\mathcal{M}(p \to q) = \mathcal{M}(\neg q \to \neg p)$, as Table 2.2 shows that both formulas have the same set of models. In logic, these formulas are considered to be *equivalent*.

The formula $\neg q \to \neg p$ is called the *contrapositive* of the implication $p \to q$. The truth table shows that an implication and its contrapositive are equivalent: They are just different ways of saying the same thing. In contrast, the *converse* of $p \to q$ is the formula $q \to p$, which has a different set of models (as shown in Table 2.2).

Table 2.2 The equivalence of $p \to q$, $\neg p \vee q$, and the contrapositive $\neg q \to \neg p$.

p	q	¬p	¬q	p → q	¬p ∨ q	¬q → ¬p	q → p
1	1	0	0	1	1	1	1
1	0	0	1	0	0	0	1
0	1	1	0	1	1	1	0
0	0	1	1	1	1	1	1

Definition 2.2.13 Given two formulas A and B, A and B are *logically equivalent* if $\mathcal{M}(A) = \mathcal{M}(B)$, denoted by $A \equiv B$.

$A \equiv B$ means that, for every interpretation σ, $\sigma(A) = \sigma(B)$, so $\sigma \in \mathcal{M}(A)$ iff $\sigma \in \mathcal{M}(B)$. The relation $\equiv$ over formulas is an equivalence relation as $\equiv$ is reflexive, symmetric, and transitive. The relation $\equiv$ is also a congruence relation as $A \equiv C$ and $B \equiv D$ imply that $\neg A \equiv \neg C$ and $A \ o \ B \equiv C \ o \ D$ for any binary operator o. This property allows us to obtain an equivalent formula by replacing a part of the formula by an equivalent one. The relation $\equiv$ plays a very important role in logic as it is used to simplify formulas, to convert formulas into equivalent normal forms, or to provide alternative definitions for logical operators.

In arithmetic one writes simply $s = t$ to express that the terms s, t represent the same function. For example, "$(x+y)^2 = x^2 + 2xy + y^2$" expresses the equal values of the terms on the two sides of "=". In propositional logic, however, we use the equality sign like $A = B$ only for the syntactic identity of the formulas A and B. Therefore, the equivalence of formulas must be denoted differently, such as $A \equiv B$.

Theorem 2.2.14 *For any formulas A, B, and C, where B is a subformula of A, and $B \equiv C$, then $A \equiv A[B \leftarrow C]$.*

Proof For any interpretation σ, $\sigma(B) = \sigma(C)$, since $B \equiv C$. Apply σ to the formula trees of A and $A[B \leftarrow C]$ and compare the truth values of all corresponding nodes of the two trees, ignoring the proper subtrees of B and C. Since $\sigma(B) = \sigma(C)$, they must have the same truth values, that is, $\sigma(A) = \sigma(A[B \leftarrow C])$. □

The equivalence relation is widely used to simplify formulas and has real practical importance in computer science. Formula simplification in software can make a program easier to read and understand. Simplified programs may also run faster, since they require fewer operators. In hardware design, simplifying formulas can decrease the number of logic gates on a chip because digital circuits can be expressed by logical formulas. Minimizing logical formulas corresponds to reducing the number of gates in the circuit. The payoff of gate minimization is potentially enormous: A chip with fewer gates is smaller, consumes less power, has a lower defect rate, and is cheaper to manufacture.

Suppose a formula A contains k propositional variables, then A can be viewed as one of Boolean functions $f : \{1, 0\}^k \mapsto \{1, 0\}$. For example, $\neg p \vee q$ contains two variables and can be regarded as a Boolean function $f(p, q)$. The truth table (Table 2.2) reveals that $f(p, q)$, i.e., $\neg p \vee q$, always has the same truth value as $p \rightarrow q$, so f and $\rightarrow$ are the same function. As a result, we may use $\neg p \vee q$ as the definition of $p \rightarrow q$. As another example, the if-then-else function, $ite : \{1, 0\}^3 \mapsto \{1, 0\}$, can be defined by $ite(1, B, C) = B$ and $ite(0, B, C) = C$, instead of using a truth table of eight rows.

The following proposition lists many equivalent pairs of the formulas.

2.2 Semantics

Proposition 2.2.15 *For formulas A, B, and C,*

$$A \vee A \equiv A \qquad\qquad A \wedge A \equiv A$$
$$A \vee B \equiv B \vee A \qquad\qquad A \wedge B \equiv B \wedge A$$
$$(A \vee B) \vee C \equiv A \vee (B \vee C) \qquad (A \wedge B) \wedge C \equiv A \wedge (B \wedge C)$$
$$A \vee \bot \equiv A \qquad\qquad A \wedge \top \equiv A$$
$$A \vee \neg A \equiv \top \qquad\qquad A \wedge \neg A \equiv \bot$$
$$A \vee \top \equiv \top \qquad\qquad A \wedge \bot \equiv \bot$$
$$A \vee (B \wedge C) \equiv (A \vee B) \wedge (A \vee C) \qquad A \wedge (B \vee C) \equiv (A \wedge B) \vee (A \wedge C)$$
$$\neg \top \equiv \bot; \quad \neg \bot \equiv \top \qquad\qquad \neg\neg A \equiv A$$
$$\neg(A \vee B) \equiv \neg A \wedge \neg B \qquad \neg(A \wedge B) \equiv \neg A \vee \neg B$$
$$A \rightarrow B \equiv \neg A \vee B \qquad\qquad A \rightarrow B \equiv \neg B \rightarrow \neg A$$
$$A \oplus B \equiv (A \vee B) \wedge (\neg A \vee \neg B) \qquad A \leftrightarrow B \equiv (A \vee \neg B) \wedge (\neg A \vee B)$$

Some of the above equivalences have special names. For instance, $\neg\neg A \equiv A$ is called *double negation*. $\neg(A \vee B) \equiv \neg A \wedge \neg B$ and $\neg(A \wedge B) \equiv \neg A \vee \neg B$ are called *de Morgan's law*. $A \vee (B \wedge C) \equiv (A \vee B) \wedge (A \vee C)$ and $A \wedge (B \vee C) \equiv (A \wedge B) \vee (A \wedge C)$ are called *distributive laws*.

One way to show that two formulas are equivalent is to prove the validity of a formula, as given by the following proposition:

Proposition 2.2.16 *For any formulas A and B, $A \equiv B$ iff $\models A \leftrightarrow B$.*

Proof $A \equiv B$ means, for any interpretation σ, $\sigma(A) = \sigma(B)$, which means $\sigma(A \leftrightarrow B) = 1$, i.e., any σ is a model of $A \leftrightarrow B$, i.e., $A \leftrightarrow B$ is valid. □

It follows that $A \equiv \top$ iff $\models A$ when $B = \top$. The above proposition shows the relationship between $\equiv$, which is a semantic notation, and $\leftrightarrow$, which is a syntactical symbol. Moreover, Theorem 2.2.9 allows us to prove the equivalences using the truth table method, treating A, B, and C as propositional variables. For example, from $(p \rightarrow q) \equiv (\neg p \vee q)$, we know $(p \rightarrow q) \leftrightarrow (\neg p \vee q)$ is valid; from the validity of $(p \rightarrow q) \leftrightarrow (\neg p \vee q)$, we know $(A \rightarrow B) \leftrightarrow (\neg A \vee B)$ is valid for any formulas A and B, thus $(A \rightarrow B) \equiv (\neg A \vee B)$.

2.2.4 Entailment

In natural languages, given a set of premises, we would like to know what conclusions can be drawn from the premises. If we represent these premises by a single formula, it will be the conjunction of the premises. Since $\wedge$ is commutative and associative, a conjunction of formulas can be conveniently written as a set S of formulas. That is, we assume that $A_1 \wedge A_2 \wedge \cdots \wedge A_n$ is equivalent to $S = \{A_1, A_2, \cdots, A_n\}$. For example, $\{A, B\} \equiv A \wedge B$.

To catch the relation between the premises and the conclusion in logic, we have the notion of *entailment*.

Definition 2.2.17 Given two formulas A and B, we say A *entails* B, or B is a *logical consequence* of A, denoted by $A \models B$, if $\mathcal{M}(A) \subseteq \mathcal{M}(B)$.

Thus, since $\mathcal{M}(\top) = All_P \subseteq \mathcal{M}(B)$ implies $\mathcal{M}(B) = All_P$, $\models B$ and $\top \models B$ have the same meaning, i.e., B is valid.

The above definition allows many irrelevant formulas to be logical consequences of A, including all tautologies and logically equivalent formulas. Despite this irrelevant relationship between A and B, the concept of entailment is indispensable in logic. For instance, an inference rule is a pattern which takes formulas as premises and returns a formula as conclusion. To check the soundness of the inference rule, we let the premises be represented by formula A and the conclusion represented by B, and check if $A \models B$, i.e., B is a logical consequence of A. If the number of variables is small, we may use the truth table method to check if $\mathcal{M}(A) \subseteq \mathcal{M}(B)$, that is, if every model of A remains to be a model of B.

Example 2.2.18 The propositional version of the modus ponens rule says that given the premises $p \to q$ and p, then we draw the conclusion q. Let A be $(p \to q) \wedge p$, then $\mathcal{M}(A) = \{pq\}$, and $\mathcal{M}(q) = \{pq, \overline{p}q\}$. Since $\mathcal{M}(A) \subseteq \mathcal{M}(q)$, so q is a logical consequence of A, or $A \models q$. □

Proposition 2.2.19 *The relation $\models$ is transitive, that is, if $A \models B$ and $B \models C$, then $A \models C$.*

The proof is left as an exercise.

Definition 2.2.20 Given a formula A, the set $\{B \mid A \models B\}$ is called the *theory* of A and is denoted by $\mathcal{T}(A)$. Every member of $\mathcal{T}(A)$ is called a theorem of A.

In the above definition, A can be a set of formulas to represent the conjunction of its members.

Proposition 2.2.21 *The following three statements are equivalent: (a) $A \models B$; (b) $\mathcal{M}(A) \subseteq \mathcal{M}(B)$; and (c) $\mathcal{T}(B) \subseteq \mathcal{T}(A)$.*

Proof $(a) \to (b)$: By definition, $A \models B$ iff $\mathcal{M}(A) \subseteq \mathcal{M}(B)$.

$(b) \to (c)$: Also, by definition, $B \models C$ for any $C \in \mathcal{T}(B)$. If $A \models B$, then $A \models C$ because of the transitivity of $\models$, so $C \in \mathcal{T}(A)$.

$(c) \to (a)$: If $\mathcal{T}(B) \subseteq \mathcal{T}(A)$, since $B \in \mathcal{T}(B)$, so $B \in \mathcal{T}(A)$, thus $A \models B$. □

Corollary 2.2.22 *For any formula A, $\mathcal{T}(\top) \subseteq \mathcal{T}(A) \subseteq \mathcal{T}(\bot)$.*

Proof This holds because $\bot \models A$ and $A \models \top$. Note that $\mathcal{T}(\top)$ contains every tautology and $\mathcal{T}(\bot)$ contains every formula. □

Corollary 2.2.23 *For any formulas A and B, the following statements are equivalent: (a) $A \equiv B$; (b) $\mathcal{M}(A) = \mathcal{M}(B)$; and (c) $\mathcal{T}(A) = \mathcal{T}(B)$.*

2.2 Semantics

Proof Since $A \equiv B$ iff $A \models B$ and $B \models A$, this corollary holds by applying Proposition 2.2.21 twice. □

Intuitively, we may regard formula A as a constraint to mark some interpretations as "counterexamples" of its validity. The stronger the constraint, the smaller the set of models. The strongest constraint is $\bot$ (no interpretations left) and the weakest constraint is $\top$ (no constraints). On the other hand, $\mathcal{T}(A)$ collects all constraints which do not mark any interpretation in $\mathcal{M}(A)$ as "counterexamples."

Proposition 2.2.24 *Let A and B be sets of formulas. If $A \subseteq B \subseteq \mathcal{T}(A)$, then $\mathcal{M}(A) = \mathcal{M}(B)$.*

Proof Let $C = B - A$. Since $B \subseteq \mathcal{T}(A)$, so $C \subseteq \mathcal{T}(A)$, or $A \models C$, which implies $\mathcal{M}(A) \subseteq \mathcal{M}(C)$. Thus, $\mathcal{M}(B) = \mathcal{M}(A \cup C) = \mathcal{M}(A \wedge C) = \mathcal{M}(A) \cap \mathcal{M}(C) = \mathcal{M}(A)$, because $\mathcal{M}(A) \subseteq \mathcal{M}(C)$. □

Corollary 2.2.25 *If $C \in \mathcal{T}(A)$, then $\mathcal{T}(A) = \mathcal{T}(A \wedge C)$.*

Proof Let $B = A \cup \{C\}$, then $A \subseteq B \subseteq \mathcal{T}(A)$, so $\mathcal{M}(A) = \mathcal{M}(B) = \mathcal{M}(A \wedge C)$. Hence, $\mathcal{T}(A) = \mathcal{T}(A \wedge C)$ by Corollary 2.2.23. □

Corollary 2.2.26 *If $\bot \in \mathcal{T}(A)$, then A is unsatisfiable.*

Proof Let $B = A \cup \{\bot\}$, then $A \subseteq B \subseteq \mathcal{T}(A)$, so $\mathcal{M}(A) = \mathcal{M}(B) = \mathcal{M}(A \wedge \bot) = \mathcal{M}(\bot) = \emptyset$. □

One task of theorem proving is to show that, given a set A of axioms, B is a theorem of A, that is, $B \in \mathcal{T}(A)$. If we use the operator $\rightarrow$, then $A \rightarrow B$ is valid iff B is a logical consequence of A.

Proposition 2.2.27 *Let A and B be formulas, $A \models B$ iff $\models A \rightarrow B$.*

Proof $A \models B$ means, for any interpretation σ, if $\sigma(A) = 1$, then $\sigma(B)$ must be 1; if $\sigma(A) = 0$, we do not care about the value of $\sigma(B)$. The two cases can be combined as $\sigma(A) \rightarrow \sigma(B) = 1$, or $\sigma(A \rightarrow B) = 1$, which means $A \rightarrow B$ is valid, because σ is any interpretation. □

The above proposition shows the relationship between $\models$ (a semantic notation) and $\rightarrow$ (a syntactical symbol). The closure of the entailment relation under substitution generalizes the fact that from $p \wedge q \models p$, all entailment of the form $A \wedge B \models A$ arise from substituting A for p and B for q, just like the closure of the equivalence relation under substitution.

2.2.5 Theorem Proving and the SAT Problem

The concepts of validity, unsatisfiability, or entailment give rise to difficult problems of computation. These three problems have the same degree of difficulty and can be represented by theorem proving, which asks to check if a formula $A \in \mathcal{T}(\top)$ (A is valid), or $\bot \in \mathcal{T}(A)$ (A is unsatisfiable), or A is in $\mathcal{T}(B)$ (B entails A). Theorem proving by truth table runs out of steam pretty quickly: A formula with n variables has a truth table of 2^n interpretations, so the effort grows exponentially with the number of variables. For a formula with just 30 variables, there are over a billion interpretations to check!

The general problem of deciding whether a formula A is satisfiable is called the SAT problem. One approach to solving SAT is to construct a truth table and check each row to find a model. As with testing validity, this approach quickly bogs down for formulas with many variables.

The good news is that SAT belongs to the class of NP problems, whose solutions can be checked in polynomial time. For example, given an interpretation σ, we can check in linear time (in terms of the size of A) if σ is a model of A, thus showing A is satisfiable. This can be done using $eval(A, \sigma)$. If a problem can be solved in polynomial time (with respect to the size of inputs), it belongs to the class P. SAT is not known to be in P, and theorem proving is not known to be in NP.

The bad news is that SAT is known to be the hardest problem in NP. That is, if there exists a polynomial-time algorithm to solve SAT, then every problem in NP can be solved in polynomial time. These problems include many other important problems involving scheduling, routing, resource allocation, and circuit verification across multiple disciplines including programming, algebra, finance, and political theory. This is the famous result of Cook and Levin who proved that SAT is the first NP-complete problem. Does this mean that we cannot find an algorithm which takes $O(n^{100})$ time, where n is the size of A, to decide A is satisfiable? No one knows. $P = NP$? This is the most famous open problem in computer science. It is also one of the seven millennium problems: The Clay Institute will award you \$1,000,000 if you solve the $P = NP$ problem. If $P = NP$, then SAT solving will be in P; theorem proving (in propositional logic) will be also in P. If $P \neq NP$, then theorem proving won't be in NP: It belongs to the complement of the NP-complete problems. An awful lot hangs on the answer, and the general consensus is $P \neq NP$.

In the last 20 years, there has been exciting progress on SAT solving, which are software tools for propositional satisfiability, for practical applications like digital circuit verification. These software tools find satisfying assignments with amazing efficiency even for formulas with thousands of variables. In the next two chapters, we will study the methods for theorem proving and SAT solvers.

2.3 Normal Forms

By "normal form," we mean every formula can be converted into an equivalent formula with certain syntactical restriction on the use and position of Boolean operators. If all equivalent formulas have a unique normal form, that normal form is also called "canonical form."

Since it is expensive to use truth table to prove the equivalence relation when the number of variables is big, an alternative approach is to use algebra to prove

2.3 Normal Forms

equivalence. A lot of different operators may appear in a propositional formula, so a useful first step is to get rid of all but three: $\wedge$, $\vee$ and $\neg$. This is easy because each of the operators is equivalent to a simple formula using only these three. For example, $A \rightarrow B$ is equivalent to $\neg A \vee B$.

Using the equivalence relations discussed in the previous section, any propositional formula can be proved equivalent to a canonical form. What has this got to do with equivalence? That's easy: To prove that two formulas are equivalent, convert them both to canonical forms over the set of variables that appear in any formula. Now if two formulas are equivalent to the same canonical form, then the two formulas are certainly equivalent. Conversely, if two formulas are equivalent, they will have the same canonical form. We can also use canonical form to show a formula is valid or unsatisfiable: It is valid if its canonical form is $\top$; it is unsatisfiable if its canonical form is $\bot$.

In this section, we present four normal forms:

- Negation normal form (NNF)
- Conjunctive normal form (CNF)
- Disjunctive normal form (DNF)
- ITE (if-then-else) normal form (INF)

Of the four normal forms, there exist canonical forms for the last three. However, the canonical forms derived from DNF or CNF are essentially copies of truth tables, thus not effective enough for showing the equivalence relation. Only the last one, INF, may have a compact size so that equivalence of two formulas can be conveniently established in this canonical form.

We will define a normal form by giving a set of transformation rules, which is a set of pairs of formulas, one is the left side and the other is the right side of the rule. Most importantly, the left side of each rule is logically equivalent to its right side. Each rule deals with a formula which is not in normal form and the application of the rule on the formula will result in a new formula closer to normal form. To convert any formula into a formula in normal form, we pick a rule, matching the left side of the rule to a subformula and replacing the subformula by the right side of the rule.

A general procedure for obtaining normal forms goes as follows:

1. If the formula is already in normal form, stop with success.
2. Otherwise, it contains a subformula violating the normal form criteria.
3. Pick such a subformula and find the left side of a rule that matches.
4. Replace the subformula by the right side of the rule, go to 1 and continue.

2.3.1 Negation Normal Form (NNF)

Definition 2.3.1 A propositional formula A is a *negation norm form* (NNF) if the argument of any negation symbol $\neg$ is a propositional variable.

Formula A is said to be a *literal* if A is a variable (positive literal) or A is $\neg p$ (negative literal), where p is a variable. In an NNF, $\neg$ appears only in negative literals (as subformulas of A); if we replace negative literals by a variable, there will be no $\neg$ in the resulting formula.

Proposition 2.3.2 *Every propositional formula can be transformed into an equivalent NNF.*

The proof is done by the following algorithm, which uses only equivalence relations in the transformation.

Equivalence Rules for Obtaining NNF

Algorithm 2.3.3 The algorithm to convert a formula into NNF takes the following three groups of rules:

1. Use the following equivalences to remove $\rightarrow$, $\oplus$, and $\leftrightarrow$ from the formula.

$$A \rightarrow B \equiv \neg A \vee B.$$
$$A \oplus B \equiv (A \vee B) \wedge (\neg A \vee \neg B).$$
$$A \leftrightarrow B \equiv (A \vee \neg B) \wedge (\neg A \vee B).$$

2. Besides $\neg\top \equiv \bot$, $\neg\bot \equiv \top$, and $\neg\neg A \equiv A$, use the de Morgan laws to push $\neg$ down to variables:

$$\neg(A \vee B) \equiv \neg A \wedge \neg B.$$
$$\neg(A \wedge B) \equiv \neg A \vee \neg B.$$

3. (Optional) Use $A \vee \bot \equiv A$, $A \vee \top \equiv \top$, $A \wedge \bot \equiv \bot$, and $A \wedge \top \equiv A$, to remove extra $\top$ and $\bot$ from the formula.

For example, $\top$ and $\neg p \vee q$ are NNF. $\neg(\neg p \vee q)$ is not NNF; we may transform it into $p \wedge \neg q$, which is NNF. If the input formula contains only $\neg$, $\vee$, $\wedge$, and $\rightarrow$, it takes only linear time to obtain NNF by working recursively top-down. If $\leftrightarrow$ or $\oplus$ are present, the computing time and the size of the output formula may be exponential in terms of the original size, because the arguments of $\leftrightarrow$ and $\oplus$ are duplicated during the transformation.

Definition 2.3.4 Given an NNF A, if we replace every occurrence of $\wedge$ by $\vee$, $\vee$ by $\wedge$, positive literal by its negation, and negative literal $\neg p$ by its counterpart p, we obtain a formula B, which is called the *dual* of A, denoted by $B = dual(A)$.

For example, if A is $p \wedge (q \vee (\neg p \wedge \neg r))$, then $dual(A) = \neg p \vee (\neg q \wedge (p \vee r))$. Note that both A and $dual(A)$ are NNF.

Proposition 2.3.5 *If A is in NNF, then $dual(A)$ is also in NNF and $dual(A) \equiv \neg A$.*

The proposition can be easily proved by de Morgan's law.

2.3.2 Conjunctive Normal Form (CNF)

Definition 2.3.6 In propositional logic, a *clause* is either $\perp$ (the empty clause), a literal (unit clause), or a disjunction of literals.

A clause is also called *maxterm* in the community of circuit designs.

Definition 2.3.7 A formula is a *conjunctive normal form* (CNF) if it is a conjunction of clauses.

A CNF is also called a *product of sums* (POS) expression. The idea of conjunctive normal form is that each type of connective appears at a distinct height in the formula. The negation symbol, $\neg$, is lowest, with only variables as their arguments. The disjunction symbol, $\vee$, is in the middle, with only literals as their arguments. The conjunction symbol, $\wedge$, is at the top, taking the disjunctions or literals as their arguments. Some operators may be missing in CNF. For example, $p \wedge q$ is a CNF, which contains two unit clauses: p and q.

Equivalence Rules for Obtaining CNF

- Stage 1: The equivalence rules for obtaining NNF.
- Stage 2: $A \vee (B \wedge C) \equiv (A \vee B) \wedge (A \vee C)$, where the arguments of $\wedge$ and $\vee$ can be switched.

Proposition 2.3.8 *Every propositional formula can be transformed into an equivalent CNF.*

To show the proposition is true, we first convert the formula into NNF and then apply the distribution law to move $\vee$ under $\wedge$. This process will terminate because we have only a finite number of $\vee$ and $\wedge$.

Example 2.3.9 To convert $p_1 \leftrightarrow (p_2 \leftrightarrow p_3)$ into CNF, we first use $A \leftrightarrow B \equiv (\neg A \vee B) \wedge (A \vee \neg B)$ to remove the second $\leftrightarrow$:

$$p_1 \leftrightarrow ((\neg p_2 \vee p_3) \wedge (p_2 \vee \neg p_3)),$$

and then remove the first $\leftrightarrow$:

$$(\neg p_1 \vee ((\neg p_2 \vee p_3) \wedge (p_2 \vee \neg p_3))) \wedge (p_1 \vee \neg((\neg p_2 \vee p_3) \wedge (p_2 \vee \neg p_3))).$$

Its NNF is

$$(\neg p_1 \vee ((\neg p_2 \vee p_3) \wedge (p_2 \vee \neg p_3))) \wedge (p_1 \vee ((p_2 \wedge \neg p_3) \vee (\neg p_2 \wedge p_3))).$$

We then obtain the following non-tautology clauses:

$$\neg p_1 \vee \neg p_2 \vee p_3, \quad \neg p_1 \vee p_2 \vee \neg p_3, \quad p_1 \vee p_2 \vee p_3, \quad p_1 \vee \neg p_2 \vee \neg p_3.$$

□

Since $\vee$ is commutative and associative and $X \vee X \equiv X$, a clause can be represented by a set of literals. In this case, we use $|$ for $\vee$ and $\overline{p}$ for $\neg p$ in a clause. For example, the above clauses will be displayed as

$$(\overline{p_1} \mid \overline{p_2} \mid p_3), \quad (\overline{p_1} \mid p_2 \mid \overline{p_3}), \quad (p_1 \mid p_2 \mid p_3), \quad (p_1 \mid \overline{p_2} \mid \overline{p_3}).$$

Definition 2.3.10 If a clause includes every variable exactly once, the clause is *full*. A CNF is called a *full CNF* if every clause of the CNF is full.

If a clause C misses a variable r, we may replace C by two clauses: $\{(C \mid r), (C \mid \overline{r})\}$. Repeating this process for every missing variable, we will obtain an equivalent full CNF. Transforming a set of clauses into an equivalent full CNF is sound because for any formula A and B, the following relation is true:

$$A \equiv (A \vee B) \wedge (A \vee \neg B),$$

which can be proved by truth table.

Equivalence Rules for Obtaining Full CNF

- Stage 1: The equivalence rules for NNF.
- Stage 2: $A \vee (B \wedge C) \equiv (A \vee B) \wedge (A \vee C)$, where the arguments of $\wedge$ and $\vee$ can be switched.
- Stage 3: $A \equiv (A \vee p) \wedge (A \vee \neg p)$, if A is a clause and p does not appear in A.

Proposition 2.3.11 *Every propositional formula can be transformed into an equivalent and unique full CNF (up to the commutativity and associativity of $\wedge$ and $\vee$). That is, full CNF is a canonical form.*

Given a CNF A, each clause can be regarded as a constraint on what interpretations cannot be a model of A. For example, if A contains the unit clause $(\neg p)$, then all the interpretations where $p \mapsto 1$ (half of all interpretations) cannot be a model of A. If A contains a binary clause $(p \mid q)$, then all interpretations where $p \mapsto 0, q \mapsto 0$ (there are a quarter of such interpretations) cannot be a model of A. If A has a clause C containing every variable exactly once, then this clause removes only one interpretation, i.e., $\neg C$, as a model of A. If A has m models over V_P, i.e., $|\mathcal{M}(A)| = m$, then a full CNF of A contains $2^{|V_P|} - m$ clauses, as each clause removes exactly one interpretation as a model of A. That is exactly the idea for the proof of the above proposition.

2.3 Normal Forms

The concept of full CNF is useful for theoretical proofs. Unfortunately, the size of a full CNF is exponential in terms of number of variables, too big for any practical application. CNFs are useful to specify many problems in practice, because practical problems are often specified by a set of constraints. Each constraint can be specified by a set of clauses, and the conjunction of all clauses from each constraint gives us a complete specification of the problem.

2.3.3 Disjunctive Normal Form (DNF)

If we regard $\wedge$ as multiplication and $\vee$ as addition, then a CNF is a "product of sums." On the other hand, a "sum of products" is a disjunctive normal form (DNF). Formula A is said to be a *product* if A is either $\top$, a literal, or a conjunction of literals. A product is also called *minterm* or *product term* in the community of circuit designs.

Like clauses, we remove duplicates in a product and represent them by a set of literals.

Definition 2.3.12 A formula A is a *disjunctive normal form* (DNF) if A is a disjunction of products.

A DNF is also called a *sum of products* (SOP) expression. A formula like $p \wedge \neg q$ or $p \vee \neg q$ can be both a CNF and a DNF.

Equivalence Rules for Obtaining DNF

- Stage 1: The equivalence rules for NNF.
- Stage 2: $A \wedge (B \vee C) \equiv (A \wedge B) \vee (A \wedge C)$, where the arguments of $\wedge$ and $\vee$ can be switched.

Proposition 2.3.13 *Every propositional formula can be transformed into an equivalent DNF.*

To show the proposition is true, we first convert the formula into NNF in the first stage and then apply the distribution law in the second stage to move $\wedge$ under $\vee$.

Once a formula A is transformed into DNF, it is very easy to check if A is satisfiable: If DNF contains a product in which each variable appears at most once, then A is satisfiable, because we can simply assign 1 to each literal of this product and the product becomes true. Since it is a hard problem to decide if a formula is satisfiable, it is also hard to obtain DNF, due to its large size.

Like CNF, DNF is not a canonical form. For example, the following equivalence is true:

$$p \wedge \neg q \vee \neg p \wedge \neg r \vee q \wedge r \equiv \neg p \wedge q \vee p \wedge r \vee \neg q \wedge \neg r$$

However, we cannot convert both sides to the same formula.

Definition 2.3.14 A product is *full* if it contains every variable exactly once. A DNF is *full* if every product in the DNF is full.

A full product in a DNF A is equivalent to a model of A. For example, if $V_P = \{p, q, r\}$, then the product $p \land q \land \neg r$ specifies the model $\{p \mapsto 1, q \mapsto 1, r \mapsto 0\}$ or $pq\bar{r}$. Obviously, a full DNF A has m full products iff A has m models. Like full CNF, every propositional formula is equivalent to a unique full DNF, that is, full DNF is a canonical form.

Equivalence Rules for Obtaining Full DNF

- Stage 1: The equivalence rules for NNF.
- Stage 2: $A \land (B \lor C) \equiv (A \land B) \lor (A \land C)$, where the arguments of $\land$ and $\lor$ can be switched.
- Stage 3: $A \equiv (A \land p) \lor (A \land \neg p)$, if A is a product and p does not appear in A.

Proposition 2.3.15 *Every propositional formula can be transformed into an equivalent and unique full DNF (up to the commutativity and associativity of $\land$ and $\lor$).*

2.3.4 Full DNF and Full CNF from Truth Table

A truth table is often used to define a Boolean function. In the truth table, the truth value of the function is given for every interpretation of the function. Each interpretation corresponds to a full product and the function can be defined as a sum (disjunction) of all the products for which the function is true in the corresponding interpretation.

Example 2.3.16 Suppose we want to define a Boolean function $f : B^3 \mapsto B$, where $B = \{0, 1\}$ and the truth value of $f(a, b, c)$ is given in the following truth table. Following the convention of EDA (electronic design automation), we will use $+$ for $\lor$, $\cdot$ for $\land$, and $\bar{A}$ for $\neg A$.

a	b	c	f	Products	Clauses
0	0	0	0	$m_0 = \bar{a}\bar{b}\bar{c}$	$M_0 = a + b + c$
0	0	1	1	$m_1 = \bar{a}\bar{b}c$	$M_1 = a + b + \bar{c}$
0	1	0	0	$m_2 = \bar{a}b\bar{c}$	$M_2 = a + \bar{b} + c$
0	1	1	1	$m_3 = \bar{a}bc$	$M_3 = a + \bar{b} + \bar{c}$
1	0	0	0	$m_4 = a\bar{b}\bar{c}$	$M_4 = \bar{a} + b + c$
1	0	1	1	$m_5 = a\bar{b}c$	$M_5 = \bar{a} + b + \bar{c}$
1	1	0	0	$m_6 = ab\bar{c}$	$M_6 = \bar{a} + \bar{b} + c$
1	1	1	0	$m_7 = abc$	$M_7 = \bar{a} + \bar{b} + \bar{c}$

2.3 Normal Forms

For each interpretation σ of (a, b, c), we define a product m_i, where i is the decimal value of σ, when σ is viewed as a binary number. For each m_i, we also define a clause M_i and the relation between m_i and M_i is that $\neg m_i \equiv M_i$. For the function f defined in the truth table, $f = m_1 + m_3 + m_5 = \overline{a}\overline{b}c + \overline{a}bc + a\overline{b}c$. □

In the above example, for each model of f, we create a product (product term) in its full DNF: If the variable is true in the model, the positive literal of this variable is in the product; if a variable is false, the negative literal of the variable is in the product. f is then defined by a full DNF consisting of these products. In other words, every Boolean function can be defined by a DNF. We can also define the same function by a full CNF. This can be processed as follows: At first, we define the negation of f by a full DNF from the truth table:

$$\overline{f} = m_0 + m_2 + m_4 + m_6 + m_7$$

Apply the negation on both sides of the above equation, we have

$$f = M_0 \cdot M_2 \cdot M_4 \cdot M_6 \cdot M_7$$

where $\cdot$ stands for $\wedge$ and $\overline{m_i} \equiv M_i$. In other words, f can be defined by a full CNF. Note that $M_0 \cdot M_2 \cdot M_4 \cdot M_6 \cdot M_7$ is the dual of $m_0 + m_2 + m_4 + m_6 + m_7$. In general, the dual of DNF is CNF and the dual of CNF is DNF.

Once we know the principle, we can construct a full CNF directly from the truth table of a formula. That is, for each non-model interpretation in the truth table, we create one full clause: If a variable is true in the interpretation, the negative literal of this variable is in the clause; if a variable is false, the positive literal of the variable is in the clause.

Example 2.3.17 The truth table of $(p \vee q) \to \neg r$ has eight rows and three interpretations of $\langle p, q, r \rangle$ are non-models: $\langle 1, 1, 1 \rangle$, $\langle 1, 0, 1 \rangle$, and $\langle 0, 1, 1 \rangle$. From these three interpretations, we obtain three full clauses: $(\neg p \vee \neg q \vee \neg r) \wedge (\neg p \vee q \vee \neg r) \wedge (p \vee \neg q \vee \neg r)$, or $\{(\overline{p} \mid \overline{q} \mid \overline{r}), (\overline{p} \mid q \mid \overline{r}), (p \mid \overline{q} \mid \overline{r})\}$. □

2.3.5 Binary Decision Diagram (BDD)

Let $ite : \{1, 0\}^3 \mapsto \{1, 0\}$ be the "if-then-else" operator such that, for any formula A and B, $ite(1, A, B) \equiv A$ and $ite(0, A, B) \equiv B$. In fact, ite, 1, and 0 can be used to represent any logical operator. For example, $\neg y \equiv ite(y, 0, 1)$. Table 2.3 shows that every binary logical operator can be represented by ite. Following the convention on BDD, we will use 1 for $\top$ and 0 for $\bot$ in every propositional formula.

Definition 2.3.18 A formula is said to be an *ITE normal form* (INF) if the formula is 0, 1, or of form $ite(p, A, B)$ where p is a propositional variable, and A and B are INFs.

Table 2.3 Binary logical operator by ite. The first column shows the output of the function on the input pairs $(x, y) = (0, 0), (0, 1), (1, 0)$, and $(1, 1)$. The second column shows its conventional form and the last column gives the equivalent of INF. In column 3, Y stands for $ite(y, 1, 0)$ and $\neg Y$ for $ite(y, 0, 1)$

Output	Formula	ite formula
0000	0	0
0001	$x \wedge y$	$ite(x, Y, 0)$
0010	$x \wedge \neg y \ (x > y)$	$ite(x, \neg Y, 0)$
0011	x	$ite(x, 1, 0)$
0100	$\neg x \wedge y \ (x < y)$	$ite(x, 0, Y)$
0101	y	$ite(y, 1, 0)$
0110	$x \oplus y$	$ite(x, \neg Y, Y)$
0111	$x \vee y$	$ite(x, 1, Y)$
1000	$x \downarrow y$	$ite(x, 0, \neg Y)$
1001	$x \leftrightarrow y$	$ite(x, Y, \neg Y)$
1010	$\neg y$	$ite(y, 0, 1)$
1011	$x \vee \neg y \ (x \geq y)$	$ite(x, 1, \neg Y)$
1100	$\neg x$	$ite(x, 0, 1)$
1101	$\neg x \vee y \ (x \leq y)$	$ite(x, Y, 1)$
1110	$x \uparrow y$	$ite(x, \neg Y, 1)$
1111	1	1

An INF uses at most three operators: 0 (for $\bot$), 1 (for $\top$), and ite. Note that 1 and 0 are INFs, but p is not; $ite(p, 1, 0)$ is.

Proposition 2.3.19 *Every propositional formula can be transformed into an equivalent INF.*

For example, let A be $\neg p \vee q \wedge \neg r$; its equivalent INF is $ite(p, ite(q, ite(r, 0, 1), 0), 1)$. To prove the above proposition, we provide the algorithm below to do the transformation.

Algorithm 2.3.20 The algorithm to convert a formula into INF is a recursive procedure *convertINF*:

proc *convertINF*(A)
    if $A = 1$ **return** 1;
    if $A = 0$ **return** 0;
    else if A contains $p \in V_P$
        return $ite(p, convertINF(A[p \leftarrow 1]), convertINF(A[p \leftarrow 0]))$;
    else return *simplify*(A).

Note that $A[p \leftarrow 1]$ (or $A[p \leftarrow 0]$) stands for the formula resulting from replacing every occurrence of p by 1 (or 0) in A. The subroutine *simplify*(A) will return the truth value of formula A which has no propositional variables. The formula returned by *convertINF* contains only ite, 0, and 1 as the logical operators. All the propositional variables appear as the first argument of ite.

Example 2.3.21 Let A be $\neg p \vee q \wedge \neg r$, then $A[p \leftarrow 1] = \neg 1 \vee q \wedge \neg r \equiv q \wedge \neg r$; $A[p \leftarrow 0] = \neg 0 \vee q \wedge \neg r \equiv 1$. So $convertINF(A) = ite(p, convertINF(q \wedge \neg r), 1)$. $convertINF(q \wedge \neg r) = ite(q, convertINF(\neg r), 0)$,

2.3 Normal Forms

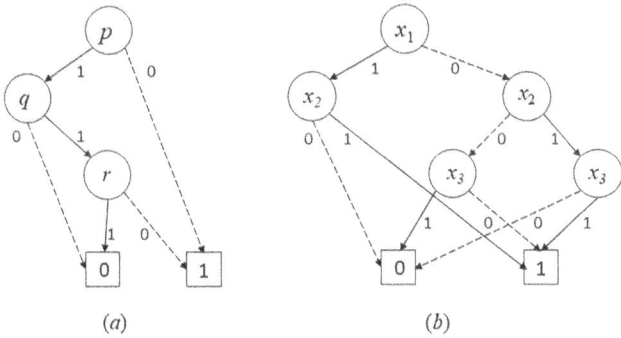

Fig. 2.2 (**a**) The BDD of $ite(p, ite(q, ite(r, 0, 1), 0), 1)$ from the last example; (**b**) the BDD of INF $ite(x_1, ite(x_2, 1, 0), ite(x_2, ite(x_3, 1, 0), ite(x_3, 0, 1)))$ derived from $\overline{x_1 x_2 x_3} \vee x_1 x_2 \vee x_2 x_3$

and $convertINF(\neg r) = ite(r, 0, 1)$. Combining all, we have $convertINF(A) = ite(p, ite(q, ite(r, 0, 1), 0), 1)$. □

A *binary decision diagram* (BDD) is a directed acyclic graph (DAG) G, which represents an INF by maximally sharing common subformulas. G has exactly two leaf nodes (no outgoing edges) with labels 1 and 0, respectively, and a set of internal nodes (with outgoing edges); each internal node represents INF $ite(p, A, B)$, that is, the node is labeled with p and has the two outgoing edges to the nodes representing A and B, respectively. Fig. 2.2 provides two examples of BDD.

Given an order on the propositional variables, if we use this order to choose each variable in the procedure *convertINF*, the resulting INF is an *ordered BDD* (OBDD). For the above example, the order $p > q > r$ is used. If we use $r > q > p$, then for $A = \neg p \vee q \wedge \neg r$, $convertINF(A) = ite(r, ite(p, 0, 1), ite(q, 1, ite(p, 1, 0)))$. The choice of orders has a big impact on the sizes of OBDD, as one order gives you the linear size (in terms of number of variables); another order may give you an exponential size. The example in Fig. 2.3 shows the two OBDDs for $(a_1 \wedge b_1) \vee (a_2 \wedge b_2) \vee (a_3 \wedge b_3)$ using two orders. If we generalize this example to $(a_1 \wedge b_1) \vee (a_2 \wedge b_2) \vee \ldots \vee (a_n \wedge b_n)$, the size of the first OBDD will be $O(n)$, while the second OBDD will have an exponential size.

Since $ite(C, A, A) \equiv A$, we may use this equivalence repeatedly to simplify an INF, until no more simplification can be done. The resulting INF is called *reduced*. The corresponding BDD is called *reduced* if, additionally, identical subformulas of an INF are represented by a unique node.

Proposition 2.3.22 *All equivalent INFs, using the same order on propositional variables, have a unique reduced OBDD (ROBDD).*

Proof Since an ROBDD is a graph representing an INF A, we may do an induction proof on the number of variables in A. If A contains no variables, then 1 and 0 are represented by the unique nodes. Let $p_n > p_{n-1} > \cdots > p_2 > p_1$ be the order on the n variables in A. Then $A \equiv ite(p_n, B, C)$ for INFs B, C, and $B \not\equiv C$. If

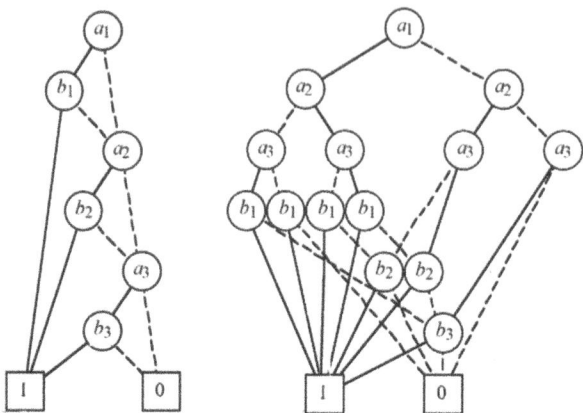

Fig. 2.3 The first BDD uses $a_1 > b_1 > a_2 > b_2 > a_3 > b_3$ and the second BDD uses $a_1 > a_2 > a_3 > b_1 > b_2 > b_3$ for the same formula $(a_1 \wedge b_1) \vee (a_2 \wedge b_2) \vee (a_3 \wedge b_3)$

$n = 1$, then $\{B, C\} = \{0, 1\}$, and A is represented by a unique node. If $n > 1$, because B and C are INFs without variable p_n, by induction hypothesis, B and C are represented by the unique ROBDDs, respectively. Thus, $A = ite(p_n, B, C)$ has a unique ROBDD representation. □

It is easy to apply $ite(p, A, A) \equiv A$ as a reduction rule on INF. To implement the sharing of common subformulas, we may use a hash table which contains the triple $\langle v, x, y \rangle$ for each node $ite(v, x, y)$ in a ROBDD, where v is a propositional variable and both x and y, $x \neq y$, are nodes in the ROBDD. The hash table provides two operations: *lookupHashTable*(v, x, y) checks if $\langle v, x, y \rangle$ exists in the hash table: If yes, the node is returned; otherwise, *null* is returned. The other operation is *saveCreateNode*(v, x, y), which creates a new node for the triple $\langle v, x, y \rangle$ and saves the node in the hash table. The pseudo-code of the algorithm for creating ROBDD is given below.

Algorithm 2.3.23 The algorithm *ROBDD* takes a formula A and returns a ROBDD node for A. *vars*(A) returns the set of propositional variables of A; *topVariable*(S) returns the maximal variable of S by the given variable order. *simplify*(A) returns the truth value of a formula without variables.

proc *ROBDD*(A)
  if $vars(A) = \emptyset$ **return** *simplify*(A)
  $v := topVariable(vars(A))$
  $x := ROBDD(A[v \leftarrow 1])$
  $y := ROBDD(A[v \leftarrow 0])$
  if $(x = y)$ **return** x // reduction
  $p := lookupHashTable(v, x, y)$
  if $(p \neq null)$ **return** p // sharing
  return *saveCreateNode*(v, x, y) // a new node is created

The input A to *ROBDD* can contain the nodes of a ROBDD, so that we can perform various logical operations on ROBDDs.

Example 2.3.24 Assume $a > b > c$, and let A be $ite(F, G, H)$, where $F = ite(a, 1, B)$, $B = ite(b, 1, 0)$, $G = ite(a, C, 0)$, $C = ite(c, 1, 0)$, $H = ite(b, 1, D)$, and $D = ite(d, 1, 0)$. $ROBDD(A)$ will call $ROBDD(A[a \leftarrow 1])$ and $ROBDD(A[a \leftarrow 0])$, which returns C and $J = ite(b, 0, H)$, respectively. Finally, $ROBDD(A)$ returns $ite(a, C, J)$. □

ROBDDs have been used for presenting Boolean functions, symbolic simulation of combinational circuits, equivalence checking and verification of Boolean functions, and finding and counting models of propositional formulas.

2.4 Optimization Problems

There exist many optimization problems in propositional logic. For example, one of them is to find a variable order for a formula, so that the resulting ROBDD is minimal. In this section, we introduce three optimization problems.

2.4.1 Minimum Set of Operators

To represent the sentences like A provided B, we do not need a formula like $A \Leftarrow B$, because we may use simply $B \rightarrow A$. This and similar reasons explain why only a few of the sixteen binary Boolean functions require notation. In CNF or DNF, we need only three Boolean operators: $\neg$, $\vee$, and $\wedge$; all other Boolean operators can be defined in terms of them.

Definition 2.4.1 A set S of Boolean operators is *sufficient* if every other Boolean operator can be defined using only the operators in S. S is *minimally sufficient* if S is sufficient and no proper subset of S is a sufficient set of operators.

Obviously, $\{\neg, \vee, \wedge\}$ is a sufficient set of operators. For example, $A \oplus B \equiv (A \vee \neg B) \wedge (\neg A \vee B)$ and $ite(A, B, C) \equiv (A \wedge B) \vee (\neg A \wedge C)$. Is the set $\{\neg, \vee, \wedge\}$ minimally sufficient? The answer is no, because $A \vee B \equiv \neg(\neg A \wedge \neg B)$. Thus, $\{\neg, \wedge\}$ is a sufficient set of operators. $\{\neg, \vee\}$ is another sufficient set, because $A \wedge B \equiv \neg(\neg A \wedge \neg B)$. To show that $\{\neg, \wedge\}$ is minimally sufficient, we need to show that neither $\{\neg\}$ nor $\{\wedge\}$ is sufficient. Since $\neg$ takes only one argument and cannot be used alone to define $\wedge$, $\{\neg\}$ cannot be sufficient. We will see later that $\{\wedge\}$ is not sufficient.

From INF (ITE normal form), we might think $\{ite\}$ is minimally sufficient. That is not true, because we also use 1 and 0 (stand for $\top$ and $\bot$) which are nullary operators. For the set $\{\neg, \wedge\}$, $\bot \equiv p \wedge \neg p$ for any propositional variable p.

Do we have a minimally sufficient set of Boolean operators which contains a single operator? The answer is yes and there are two binary operators as candidates. They are ↑ (nand, the negation of and) and ↓ (nor, the negation of or):

$$A \uparrow B \equiv \neg(A \wedge B) \qquad A \downarrow B \equiv \neg(A \vee B)$$

Proposition 2.4.2 *Both {↑} and {↓} are minimally sufficient.*

Proof To show {↑} is sufficient, we just need to define ¬ and ∧ using ↑: $\neg(A) \equiv A \uparrow A$ and $A \wedge B \equiv \neg(A \uparrow B)$. The proof that {↓} is sufficient is left as an exercise. □

In fact, among all the 16 binary operators, only ↑ and ↓ have such property.

Theorem 2.4.3 *For any binary operator which can be used alone to define both ¬ and ∧, that operator must be either ↑ or ↓.*

Proof Let o be any binary operator. If ¬ can be defined by o, then $\neg A \equiv t(o, A)$, where $t(o, A)$ is the formula represented by a binary tree t whose leaf nodes are labeled with A and whose internal nodes are labeled with o. Furthermore, we assume that t is such a binary tree with the least number of nodes. If we replace A by 1, then $t(o, 1)$ should have the truth value 0, because $\neg 1 = t(o, 1) = 0$. If any internal node of $t(o, 1)$ has the truth value 1, we may replace that node by a leaf node 1; if any internal node other than the root of t has truth value 0, we may use the subtree at this internal node as t. In both cases, this is a violation because we assumed that t is such a tree with the least number of nodes. As a result, t should have only one internal node, i.e., $1 \, o \, 1 = \neg 1 = 0$. Similarly, we should have $0 \, o \, 0 = \neg 0 = 1$.

There are four cases regarding the truth values of $1 \, o \, 0$ and $0 \, o \, 1$.

- Case 1: $1 \, o \, 0 = 0$ and $0 \, o \, 1 = 1$. Then $A \, o \, B = \neg A$ and we cannot use this o to define $A \wedge B$, as the output from o on A and B will be one of A, $\neg A$, B, or $\neg B$.
- Case 2: $1 \, o \, 0 = 1$ and $0 \, o \, 1 = 0$. Then $A \, o \, B = \neg B$, and we still cannot use this o to define $A \wedge B$ for the same reason as in case 1.
- Case 3: $1 \, o \, 0 = 1$ and $0 \, o \, 1 = 1$. o is ↑.
- Case 4: $1 \, o \, 0 = 0$ and $0 \, o \, 1 = 0$. o is ↓.

The last two cases give us the definitions of ↑ and ↓, as shown in Table 2.4. □

Table 2.4 $p \, o_1 \, q \equiv \neg p$ (the negation of projection on the first argument); $p \, o_2 \, q \equiv \neg q$ (the negation of projection on the second argument); $p \, o_3 \, q \equiv p \uparrow q$ (NAND); and $p \, o_4 \, q \equiv p \downarrow q$ (NOR)

p	q	$p \, o_1 \, q$	$p \, o_2 \, q$	$p \, o_3 \, q$	$p \, o_4 \, q$
1	1	0	0	0	0
0	1	1	0	1	0
1	0	0	1	1	0
0	0	1	1	1	1

2.4 Optimization Problems

Corollary 2.4.4 *Both $\{\neg, \wedge\}$ and $\{\neg, \vee\}$ are minimally sufficient.*

In Boolean algebra, the property of any sufficient set of operators can be specified using equations with variables and the operator. To specify the property of $\uparrow$, we may use three equations: $0 \uparrow Y = 1$, $X \uparrow 0 = 1$, and $1 \uparrow 1 = 0$. For $\uparrow$, what is the minimal number of equations needed? And what is the minimal number of symbols in the equations? These questions have been answered firmly: Only one equation, three variables (for a total of eight occurrences), and six copies of $\uparrow$ are needed. We state below the result without proof.

Proposition 2.4.5 *The Boolean algebra can be generated by the following equation:*

$$((x \uparrow y) \uparrow z) \uparrow (x \uparrow ((x \uparrow z) \uparrow x)) = z$$

The above equation contains 3 variables and 14 symbols (excluding parentheses). For the set $\{\neg, \vee\}$, a single axiom also exists to specify the Boolean algebra:

Proposition 2.4.6 *The Boolean algebra can be generated by the following equation:*

$$\neg(\neg(\neg(x \vee y) \vee z) \vee \neg(x \vee \neg(\neg z \vee \neg(z \vee u)))) = z$$

The above equation contains 4 variables and 21 symbols (excluding parentheses).

2.4.2 Logic Minimization

Logic minimization, also called *logic optimization*, is a part of circuit design process and its goal is to obtain the smallest combinational circuit that is represented by a Boolean formula. Logic minimization seeks to find an equivalent representation of the specified logic circuit under one or more specified constraints. Generally, the circuit is constrained to minimum chip area meeting a pre-specified delay. Decreasing the number of gates will reduce the power consumption of the circuit. Choosing gates with fewer transistors will reduce the circuit area. Decreasing the number of nested levels of gates will reduce the delay of the circuit. Logic minimization will reduce substantially the cost of circuits and improve its quality.

Karnaugh Maps for Minimizing DNF and CNF

In the early days of electronic design automation (EDA), a Boolean function is often defined by a truth table and we can derive full DNF (sum of products) or full CNF (product of sums) directly from the truth table, as shown in Example 2.3.16. The obtained DNF or CNF needs to be minimized, in order to use the least number of gates to implement this function.

Following the convention of EDA, we will use $+$ for $\vee$, $\cdot$ for $\wedge$, and $\overline{A}$ for $\neg A$.

Example 2.4.7 Let $f : \{0, 1\}^2 \mapsto \{0, 1\}$ be defined by $f(A, B) = \overline{AB} + A\overline{B} + \overline{A}B$. We need three AND gates and one OR gate of 3-input. However, $f(A, B) \equiv \overline{A} + \overline{B}$. Using $\overline{A} + \overline{B}$, we need only one OR gate. □

The equivalence relations used in the above simplification process are $AB + A\overline{B} \equiv A$ and $A + \overline{A}B \equiv A + B$ (or $\overline{A} + AB \equiv \overline{A} + B$). For the above example, $\overline{AB} + A\overline{B} \equiv \overline{B}$ and $\overline{B} + \overline{A}B \equiv \overline{B} + \overline{A}$.

We want to find an equivalent circuit of the minimum size possible, and this is a difficult computation problem. There are many tools such as *Karnaugh maps* available to achieve this goal.

The Karnaugh map (K-map) is a popular method of simplifying DNF or CNF formulas, originally proposed by Maurice Karnaugh in 1953. The Karnaugh map reduces the need for extensive calculations by displaying the output of the function graphically and taking advantage of humans' pattern-recognition capability. It is suitable for Boolean functions with two to four variables. When the number of variables is greater than 4, it is better to use an automated tool.

The Karnaugh map uses a two-dimensional grid where each cell represents an interpretation (or equivalently, a full product) and the cell contains the truth value of the function for that interpretation. The position of each cell contains all the information of an interpretation, and the cells are arranged such that adjacent cells differ by exactly one truth value in the interpretation. Adjacent ones in the Karnaugh map represent opportunities to simplify the formula. The products for the final formula are found by encircling groups of ones in the map. Product groups must be rectangular and must have an area that is a power of two (i.e., 1, 2, 4, 8, ...). Product rectangles should be as large as possible without containing any zeros (it may contain don't-care cells). Groups may have overlapping ones. The least number of groups that cover all ones will represent the products of a DNF of the function. These products can be used to write a minimal DNF representing the required function and thus implemented by the least number of AND gates feeding into an OR gate. The map can also be used to obtain a minimal CNF that is implemented by OR gates feeding into an AND gate.

Let $f(x, y) = x\overline{y} + \overline{x}y + \overline{xy}$. There are four products of x and y, and the K-map has four cells, that is, four products corresponding to the four interpretations on x and y. As the output of f, three cells have value 1 and one cell (i.e., xy) has value 0. Initially, each cell belongs to a distinct group of one member (Fig. 2.4).

Fig. 2.4 K-map for $f(x, y) = x\overline{y} + \overline{x}y + \overline{xy}$

x	y	f
0	0	1
0	1	1
1	0	1
1	1	0

x \ y	0	1
0	1	1
1	1	0

2.4 Optimization Problems

The merge operation on K-map: Merge two adjacent and disjoint groups of the same size and of the same truth value into one larger group.

This operation creates groups of size 2, 4, 8, etc., and cells are allowed to appear in different groups at the same time. For example, cells $x\overline{y}$ and $\overline{xy}$ in Fig. 2.4 are adjacent; they are merged into one group $\{x\overline{y}, \overline{xy}\}$, which can be represented by $\overline{y}$ as a shorthand of the group. Similarly, cells $\overline{x}y$ and $\overline{xy}$ can be merged into $\{\overline{x}y, \overline{xy}\}$, which can be represented by $\overline{x}$. Note that $\overline{xy}$ is used twice in the two merge operations. Now no more merge operations can be performed, and the final result is $f = \overline{x} + \overline{y}$ (f is the NAND function). This result can be proved logically as follows:

$$\begin{aligned} f &= x\overline{y} + \overline{x}y + \overline{xy} \\ &= (x\overline{y} + \overline{xy}) + (\overline{x}y + \overline{xy}) \\ &= (x + \overline{x})\overline{y} + \overline{x}(y + \overline{y}) \\ &= \overline{y} + \overline{x} \end{aligned}$$

Example 2.4.8 Consider the function f defined by the truth table in Fig. 2.5. The corresponding K-map has four cells with truth value 1 as f has four models shown in the truth table by the rows numbered 1, 3, 6, and 7. There are three possible merge operations: (i) $\overline{xy}z$ and $\overline{x}yz$ merge into $\overline{x}z$; (ii) $xy\overline{z}$ and xyz merge into xy; and (iii) $\overline{x}yz$ and xyz merge into yz. Since all the cells with value 1 are in the groups represented by $\overline{x}z$ and xy, yz is redundant. So, the simplified function is $f = xy + \overline{x}z$, instead of $f = xy + \overline{x}z + yz$. On the other hand, if $g = xy + xz + yz$, then none of the three products in g is redundant, because if you delete one, you will miss some ones, even though the full product xyz is covered by all the three products. □

From the above examples, it is clear that after each merge operation, two products are merged into one product containing one less variable. That is, the larger the group, the less the number of variables in the product to represent the group. This is possible because adjacent cells represent interpretations which differ by the truth value of one variable.

Because K-maps are two-dimensional, some adjacent relations of cells are not shown on K-maps. For example, in Fig. 2.5, $m_0 = \overline{xyz}$ and $m_4 = x\overline{yz}$ are adjacent,

Fig. 2.5 K-map for obtaining $f(x, y, z) = xy + \overline{x}z$

	x	y	z	f
0	0	0	0	0
1	0	0	1	1
2	0	1	0	0
3	0	1	1	1
4	1	0	0	0
5	1	0	1	0
6	1	1	0	1
7	1	1	1	1

z \ xy	00	01	11	10
0	0 (0)	0 (2)	1 (6)	0 (4)
1	1 (1)	1 (3)	1 (7)	0 (5)

Fig. 2.6 K-map for
$f(x, y, z, w) =$
$m_0 + m_2 + m_5 + m_7 + m_8 + m_{10} + m_{13} + m_{15}$. The simplified formula is
$f = \overline{yw} + yw$

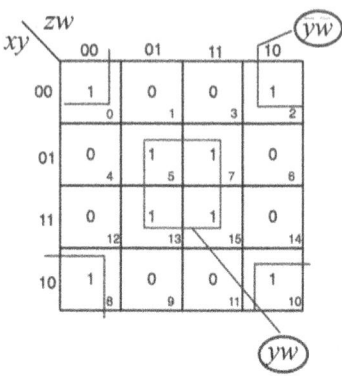

and $m_1 = \overline{x}\overline{y}z$ and $m_5 = x\overline{y}z$ are adjacent. Strictly speaking, K-maps are *toroidally connected*, which means that rectangular groups can wrap across the edges. Cells on the extreme right are actually "adjacent" to those on the far left, in the sense that the corresponding interpretations only differ by one truth value. We need a three-dimensional graph to completely show the adjacent relation of three variables, a four-dimensional graph to completely show the adjacent relation of four variables, etc.

Based on the idea of K-maps, an automated tool will create a graph of 2^n nodes for n variables, each node represents one interpretation, and two nodes have an edge iff their interpretations differ by one truth value. The merge operation is then implemented on this graph to generate products for a least size DNF. The same graph is also used for generating Gray code (by finding a Hamiltonian path in the graph).

Example 2.4.9 The K-map for $f(x, y, z, w) = m_0 + m_2 + m_5 + m_7 + m_8 + m_{10} + m_{13} + m_{15}$, is given in Fig. 2.6, where $m_0 = \overline{x}\overline{y}\overline{z}\overline{w}$, $m_2 = \overline{x}\overline{y}z\overline{w}$, $m_5 = \overline{x}y\overline{z}w$, $m_7 = \overline{x}yzw$, $m_8 = x\overline{y}\overline{z}\overline{w}$, $m_{10} = x\overline{y}z\overline{w}$, $m_{13} = xy\overline{z}w$, and $m_{15} = xyzw$ (i in m_i is the decimal value of $xyzw$ as a binary number). First, m_0 and m_2 are adjacent and can be merged into $\overline{x}\overline{y}\overline{w}$; m_8 and m_{10} are also adjacent and can be merged into $x\overline{y}\overline{w}$. Then $\overline{x}\overline{y}\overline{w}$ and $x\overline{y}\overline{w}$ can be further merged into $\overline{y}\overline{w}$. Similarly, m_5 and m_7 merge into $\overline{x}yw$; m_{13} and m_{15} merge into xyw. Then $\overline{x}yw$ and xyw merge into yw. The final result is $f = \overline{yw} + yw$. It would also have been possible to derive this simplification by carefully applying the equivalence relations, but the time it takes to do that grows exponentially with the number of products. □

Note that the formula obtained from a K-map is not unique in general. For example, $g(x, y, z, w) = x\overline{y} + z\overline{w} + \overline{x}y\overline{w}$ or $g(x, y, z, w) = x\overline{y} + z\overline{w} + \overline{yzw}$ are possible outputs for g from K-maps. This means the outputs of K-maps are not canonical.

K-maps can be used to generate simplified CNF for a Boolean function. In Example 2.3.16, we have shown how to obtain a full CNF from a full DNF. Using the same idea, if we simplify the full DNF first, the CNF obtained by negating the simplified DNF will be a simplified one.

2.4 Optimization Problems

Fig. 2.7 K-map for $f(x, y, z) = \overline{x}\,\overline{y}z + \overline{x}y\overline{z} + x\overline{y}\overline{z} + x\overline{y}z$. The simplified CNF is $f = (x+y+z)(\overline{x}+\overline{y})(\overline{y}+\overline{z})$

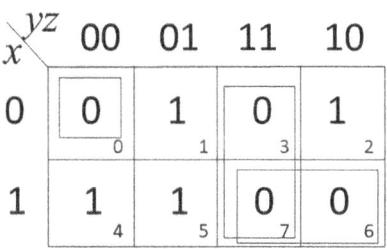

Example 2.4.10 Let $f(x, y, z) = \overline{x}\,\overline{y}z + \overline{x}y\overline{z} + x\overline{y}\overline{z} + x\overline{y}z$ and its K-map is given in Fig. 2.7. The negation of f is

$$\overline{f} = \overline{x}\,\overline{y}\,\overline{z} + \overline{x}yz + xy\overline{z} + xyz$$

Since $\overline{x}yz$ and xyz can merge into yz (shown in the figure), $xy\overline{z}$ and xyz can merge into xy, the simplified $\overline{f}$ is $\overline{f} = \overline{x}\,\overline{y}\,\overline{z} + xy + yz$. The simplified CNF for f is thus $f = (x + y + z)(\overline{x} + \overline{y})(\overline{y} + \overline{z})$. □

Karnaugh maps can also be used to handle "don't care" values and can simplify logic expressions in software design. Boolean conditions, as used, for example, in conditional statements, can get very complicated, which makes the code difficult to read and to maintain. Once minimized, sum-of-products and product-of-sums expressions can be implemented directly using AND and OR logic operators.

The above discussion illustrates the basic idea of circuit optimization. A presentation of practical optimization methods used for VLSI (very large-scale integration) chips using CMOS (complementary metal–oxide–semiconductor) technology is beyond the scope of this book.

2.4.3 Maximum Satisfiability

When a CNF is unsatisfiable, we are still interested in finding an interpretation which makes most clauses true. This is so-called the maximum satisfiability problem (Max-SAT) in which we determine the maximum number of true clauses of a given CNF under any interpretation. It is a generalization of the Boolean satisfiability problem, which asks whether there exists a truth assignment that makes all clauses true.

Example 2.4.11 Let A be a set of unsatisfiable clauses $\{(\overline{p}), (p \vee \overline{q}), (p \vee q), (p \vee \overline{r}), (p \vee r)\}$. If we assign 0 to p, two clauses will be false no matter what values are assigned to q and r. On the other hand, if p is assigned 1, only clause $(\overline{p})$ will be false. Therefore, if A is given as an instance of the MAX-SAT problem, the solution to the problem is 4 (four clauses are true in one interpretation). □

More generally, one can define a weighted version of MAX-SAT as follows: given a conjunctive normal form formula with non-negative weights assigned to each clause, find truth values for its variables that maximize the combined weight of the satisfied clauses. The MAX-SAT problem is an instance of weighted MAX-SAT where all weights are 1.

Finding the final exam schedule for a big university is a challenging job. There are hard constraints such as no students can take two exams at the same time, and no classrooms can host two exams at the same time. There are also soft constraints such that one student is preferred to take at most two exams in 1 day, or two exams are separated by at least 2 hours. To specify such a problem as an instance of weighted MAX-SAT is easy: We just give a very large weight to hard constraints and a small weight to soft constraints.

Various methods have been developed to solve MAX-SAT or weighted MAX-SAT problems. These methods are related to solving SAT problems and we will present them later in the book.

2.5 Using Propositional Logic

Propositions and logical operators arise all the time in computer programs. All programming languages such as C, C++, and Java use symbols like "&&", "||", and "!" in place of $\wedge$, $\vee$, and $\neg$. Consider the Boolean expression appearing in an if-statement:

```
if (x > 0 || (x <= 0 && y > 100)) ...
```

If we let p denote x > 0 and q be y > 100, then the Boolean expression is abstracted as $p \vee \neg p \wedge q$. Since $(p \vee \neg p \wedge q) \equiv (p \vee q)$, the original code can be rewritten as follows:

```
if (x > 0 || y > 100) ...
```

In other words, the knowledge about logic may help us to write neater code.

Consider another piece of code:

```
if (x >= 0 && A[x] == 0) ...
```

Let p be x >= 0 and q be A[x] == 0, then the Boolean expression is $p \wedge q$. Since $p \wedge q \equiv q \wedge p$, it is natural to think that (x >= 0 && A[x] == 0) can be replaced by (A[x] == 0 && x >= 0) and the program will be the same. Unfortunately, this is not the case: The execution of (A[x] == 0 && x >= 0) will be aborted when x < 0. This example shows that "&&" in programming languages are not commutative in general. Neither is "||" as (x < 0 || A[x] == 0) is different from (A[x] == 0 || x < 0). These abnormalities of the logic not only appear in array indices but also when assignments are allowed in Boolean expressions. For example, x > 0 || (x = c) > 10 is different from (x = c) > 10 || x > 0 in C, where x = c is an assignment command. As a programmer we need to be careful about these abnormalities.

2.5.1 Bitwise Operators

In programming languages like C, C++, and Java, an integer is represented by a list of 16, 32, or 64 bits (for C, a char is regarded as an 8-bit integer or a byte). A bitwise operation operates on one or two integers at the level of their individual bits and several of them are logical operations on bits. It is a fast and simple action, directly supported by the processor, and should be used whenever possible.

For illustration purpose, we will use 8-bit integers as examples. In general, for any 8-bit integer x, the binary representation of x is an 8-bit vector $(x)_2 = (b_7, b_6, b_5, b_4, b_3, b_2, b_1, b_0)$, where $b_i \in \{0, 1\}$ and $x = \sum_{i=0}^{7} b_i * 2^i$. For example, for $a = 67$, its bit vector is $(a)_2 = (0, 1, 0, 0, 0, 0, 1, 1)$, because $67 = 64 + 2 + 1 = 2^6 + 2^1 + 2^0$, and for $b = 18$, $(b)_2 = (0, 0, 0, 1, 0, 0, 1, 0)$, since $18 = 16 + 2 = 2^4 + 2^1$. In fact, a and b are stored as $(a)_2$ and $(b)_2$ in computer.

Table 2.5 provides some popular bitwise operators with examples, treating each bit as a Boolean value and performing on them simultaneously. We will use the bit vectors of $a = 67$ and $b = 18$ as examples.

If necessary, other bitwise operations can be implemented using the four operations in Table 2.5. For example, a bitwise implication on a and b can be implemented as $\sim a \mid b$.

The bit shifts are also considered bitwise operations, because they treat an integer as a bit vector rather than as a numerical value. In these operations, the bits are moved, or shifted, to the left or right. Registers in a computer processor have a fixed width, so some bits will be "shifted out" of the register at one end, while the same number of bits are "shifted in" from the other end. Typically, the "shifted out" bits are discarded and the "shifted in" bits are all 0's. When the leading bit is the sign of an integer, the left shift operation becomes more complicated, and its discussion is out of the scope of this book. In the table, we see that the shift operators take two integers: $x << y$ will shift $(x)_2$ left by y bits; $x >> y$ will shift $(x)_2$ right by y bits.

The value of the bit vector $x << y$ is equal to $x * 2^y$, provided that this multiplication does not cause an overflow. E.g., for $b = 18$, $b << 2 = 18 * 4 = 72$. Similarly, the value of $x >> y$ is equal to the integer division $x/2^y$. E.g., for $a = 67$, $a >> 2 = \lfloor 67/4 \rfloor = 16$. Shifting provides an efficient way of multiplication and

Table 2.5 Examples of bitwise operators

$(a)_2$	=	$(0, 1, 0, 0, 0, 0, 1, 1)$
$(b)_2$	=	$(0, 0, 0, 1, 0, 0, 1, 0)$
Name	Symbol	Example
NOT	$\sim$	$(\sim a)_2 = (1, 0, 1, 1, 1, 1, 0, 0)$
OR	$\mid$	$(a \mid b)_2 = (0, 1, 0, 1, 0, 0, 1, 1)$
AND	&	$(a \& b)_2 = (0, 0, 0, 0, 0, 0, 1, 0)$
XOR	^	$(a \wedge b)_2 = (0, 1, 0, 1, 0, 0, 0, 1)$
shift right	>>	$(a >> 2)_2 = (0, 0, 0, 1, 0, 0, 0, 0)$
shift left	<<	$(b << 2)_2 = (0, 1, 0, 0, 1, 0, 0, 0)$

division when the second operand is a power of two. Typically, bitwise operations are substantially faster than division, several times faster than multiplication.

Suppose you are the manager for a social club of 64 members. For each gathering, you need to record the attendance of all members. If each member has a unique member id $\in \{0, 1, \ldots, 63\}$, you may use a set of member ids to record the attendance. Instead of a list of integers, you may use a single 64-bit integer, say s, to store the set. That is, if a member attended and his id is i, then the ith bit of s is 1; otherwise, it is 0 to show his absence. For this purpose, we need a way to set a bit in s to be 1. Suppose $s = 0$ initially; we may use in Java or C the following command:

$$s = s \mid (1 << i)$$

to set the ith bit of s to be 1. To set the value of the ith bit of s to be 0, use

$$s = s \ \& \ \sim(1 << i).$$

To get the value of the ith bit of s, use $s \ \& \ (1 << i)$. For another gathering, we use another integer t. Now it is easy to compute the union or intersection of these sets: $s \mid t$ is the union and $s \ \& \ t$ is the intersection. For example, if you want to check that nobody attended both gatherings, you can check if $s \ \& \ t = 0$, which is more efficient than using lists for sets.

2.5.2 Decision Problems in Propositional Logic

Propositional logic is a very simple language but rich enough to specify most decision problems in computer science, because these problems belong to the class of NP problems and SAT is NP-complete. In fact, propositional logic is a formal language much more rigorous than natural languages. Since natural languages are ambiguous, expressing a sentence in logic helps us to understand the exact meaning of the sentence. For example, the following two sentences appear to have the same meaning.

- Students and seniors pay half price.
- Students or seniors pay half price.

One sentence uses "and" and the other uses "or," as if "and" and "or" are synonym. How to explain this? Let "being student" and "being senior" be abbreviated by p and q, respectively, and "pay half price" by r. Let A be $(p \to r) \land (q \to r)$ and B be $(p \lor q) \to r$, then A and B express somewhat more precisely the factual content of the above two sentences, respectively. It is easy to show that the formulas A and B are logically equivalent. The everyday-language statements of A and B obscure the structural difference of A and B through an apparently synonymous use of the words "and" and "or."

2.5 Using Propositional Logic

We will show how to specify various problems in propositional logic through examples.

Example 2.5.1 The so-called n-queen problem is stated as follows: Given a chessboard of size $n \times n$, where $n > 0$, how to place n queens on the board such that no two queens attack each other? For $n = 2, 3$, there is no solution; for other n, there exists a solution. To specify this problem in logic, we need n^2 propositional variables. Let the variables be $p_{i,j}$, where $1 \leq i, j \leq n$, and the meaning of $p_{i,j}$ is that $p_{i,j}$ is true iff there is a queen at row i and column j. The following properties ensure that there are n queens on the board such that no two queens attack each other:

1. There is at least one queen in each row.

$$\wedge_{1 \leq i \leq n} (p_{i,1} \vee p_{i,2} \vee \cdots \vee p_{i,n})$$

2. No more than one queen in each row.

$$\wedge_{1 \leq i \leq n} (\wedge_{1 \leq j < k \leq n} (\neg p_{i,j} \vee \neg p_{i,k}))$$

3. No more than one queen in each column.

$$\wedge_{1 \leq i \leq n} (\wedge_{1 \leq j < k \leq n} (\neg p_{j,i} \vee \neg p_{k,i}))$$

4. No more than one queen on each diagonal.

$$\wedge_{1 \leq i < k \leq n} (\wedge_{1 \leq j, l \leq n, k-i=|l-j|} (\neg p_{i,j} \vee \neg p_{k,l}))$$

The formula following each property expresses the property formally and can be easily converted into clauses. When $n = 2, 3$, the formulas are unsatisfiable. For $n \geq 4$, the formulas are satisfiable, and each model gives us a solution. For example, when $n = 4$, we have two solutions: $\{p_{1,2}, p_{2,4}, p_{3,1}, p_{4,3}\}$ and $\{p_{1,3}, p_{2,1}, p_{3,4}, p_{4,2}\}$. If we assign the four variables in a solution to be true and the other variables false, we obtain a model. Note that we ignore the property that "there is at least one queen in each column," because it is redundant. □

The next example is a logical puzzle.

Example 2.5.2 Three people are going to the beach, each using a different mode of transportation (car, motorcycle, and boat) in a different color (blue, orange, green). Who's using what? The following clues are given:

1. Abel loves orange, but he hates travel on water.
2. Bob did not use the green vehicle.
3. Carol drove the car.

The first step to solve a puzzle using propositional logic is to make sure what properties are assumed and what conditions are provided by clues. For this puzzle, the first assumed property is that "everyone uses a unique transportation tool."

To specify this property in logic, there are several ways of doing it. One way is to use a set of propositional variables to specify the relation of "who uses what transportation": Let $p_{x,y}$ denote a set of nine variables, where $x \in \{a, b, c\}$ (a for Abel, b for Bob, and c for Carol), and $y \in \{c, m, b\}$ (c for car, m for motorcycle, and b for boat). Now the property that "everyone uses a unique transportation tool" can be specified formally as the following clauses:

(1) $(p_{a,c} \mid p_{a,m} \mid p_{a,b})$ Abel uses c, m, or b.
(2) $(p_{b,c} \mid p_{b,m} \mid p_{b,b})$ Bob uses c, m, or b.
(3) $(p_{c,c} \mid p_{c,m} \mid p_{c,b})$ Carol uses c, m, or b.
(4) $(\overline{p_{a,c}} \mid \overline{p_{b,c}})$ Abel and Bob cannot both use c.
(5) $(\overline{p_{a,m}} \mid \overline{p_{b,m}})$ Abel and Bob cannot both use m.
(6) $(\overline{p_{a,b}} \mid \overline{p_{b,b}})$ Abel and Bob cannot both use b.
(7) $(\overline{p_{a,c}} \mid \overline{p_{c,c}})$ Abel and Carol cannot both use c.
(8) $(\overline{p_{a,m}} \mid \overline{p_{c,m}})$ Abel and Carol cannot both use m.
(9) $(\overline{p_{a,b}} \mid \overline{p_{c,b}})$ Abel and Carol cannot both use b.
(10) $(\overline{p_{b,c}} \mid \overline{p_{c,c}})$ Bob and Carol cannot both use c.
(11) $(\overline{p_{b,m}} \mid \overline{p_{c,m}})$ Bob and Carol cannot both use m.
(12) $(\overline{p_{b,b}} \mid \overline{p_{c,b}})$ Bob and Carol cannot both use b.

It appears quite cumbersome that 12 clauses are needed to specify one property. Note that the goal of using logic is to find a solution automatically. For computers, 12 million clauses are not too many.

From the clues of the puzzle, we have the following two unit clauses:

(13) $(p_{c,c})$ Carol drove the car.
(14) $(\overline{p_{a,b}})$ Abel hates travel on water.

These two unit clauses force $p_{c,c}$ to be assigned 1 and $p_{a,b}$ to be 0. Then we can decide the truth values of the rest seven variables:

$p_{a,c} \mapsto 0$ from $p_{c,c} \mapsto 1, (7)$.
$p_{b,c} \mapsto 0$ from $p_{c,c} \mapsto 1, (11)$.
$p_{a,m} \mapsto 1$ from $p_{a,c} \mapsto 0, p_{a,b} \mapsto 0, (1)$.
$p_{b,m} \mapsto 0$ from $p_{a,m} \mapsto 1, (5)$.
$p_{c,m} \mapsto 0$ from $p_{a,m} \mapsto 1, (8)$.
$p_{b,b} \mapsto 1$ from $p_{b,c} \mapsto 0, p_{b,m} \mapsto 0, (2)$.
$p_{c,b} \mapsto 0$ from $p_{b,b} \mapsto 1, (12)$.

2.5 Using Propositional Logic   71

The second assumed property is that "every transportation tool has a unique color." If we read the puzzle carefully, the clues involve the relation that "who will use what color of the transportation." So, it is better to use $q_{x,z}$ to represent this relation that person x uses color z, where $x \in \{a, b, c\}$ and $z \in \{b, o, g\}$ (b for blue, o for orange, and g for green). We may specify the property that "everyone uses a unique color" in a similar way as we specify "everyone uses a unique transportation tool." The clues give us two unit clauses: $(q_{a,o})$ and $(\overline{q_{b,g}})$. Starting from these two unit clauses, we can derive the truth values of all nine $q_{x,z}$ variables.   □

The next puzzle looks quite different from the previous one, but its specification is similar.

Example 2.5.3 Four people sit on a bench of four seats for a group photo: two Americans (A), a Briton (B), and a Canadian (C). They asked that (i) two Americans do not want to sit next to each other; (ii) the Briton likes to sit next to the Canadian. Suppose we use propositional variables X_y, where $X \in \{A, B, C\}$ and $1 \le y \le 4$, with the meaning that "X_y is true iff the people of nationality X sits at seat y of the bench." How do you specify the problem in CNF over X_y such that the models of the CNF match exactly all the sitting solutions of the four gentlemen?

The CNF will specify the following conditions:

1. Every seat takes at least one person: $(A_1 \mid B_1 \mid C_1)$, $(A_2 \mid B_2 \mid C_2)$, $(A_3 \mid B_3 \mid C_3)$, $(A_4 \mid B_4 \mid C_4)$. The truth of these clauses implies that there are at least four people using these seats.
2. Each seat can take at most one person: $(\neg A_1 \mid \neg B_1)$, $(\neg A_1 \mid \neg C_1)$, $(\neg B_1 \mid \neg C_1)$, $(\neg A_2 \mid \neg B_2)$, $(\neg A_2 \mid \neg C_2)$, $(\neg B_2 \mid \neg C_2)$, $(\neg A_3 \mid \neg B_3)$, $(\neg A_3 \mid \neg C_3)$, $(\neg B_3 \mid \neg C_3)$, $(\neg A_4 \mid \neg B_4)$, $(\neg A_4 \mid \neg C_4)$, $(\neg B_4 \mid \neg C_4)$. Combining with the first condition, this condition implies that there are exactly four people sitting on the bench.
3. At most two Americans on the bench: $(\neg A_1 \mid \neg A_2 \mid \neg A_3)$, $(\neg A_1 \mid \neg A_2 \mid \neg A_4)$, $(\neg A_1 \mid \neg A_3 \mid \neg A_4)$, $(\neg A_2 \mid \neg A_3 \mid \neg A_4)$. These clauses say that for every three seats, one of them must not be taken by an American.
4. At most one Briton on the bench: $(\neg B_1 \mid \neg B_2)$, $(\neg B_1 \mid \neg B_3)$, $(\neg B_1 \mid \neg B_4)$, $(\neg B_2 \mid \neg B_3)$, $(\neg B_2 \mid \neg B_4)$, $(\neg B_3 \mid \neg B_4)$. These clauses say that for every two seats, one of them must not be taken by the Briton.
5. At most one Canadian on the bench: $(\neg C_1 \mid \neg C_2)$, $(\neg C_1 \mid \neg C_3)$, $(\neg C_1 \mid \neg C_4)$, $(\neg C_2 \mid \neg C_3)$, $(\neg C_2 \mid \neg C_4)$, $(\neg C_3 \mid \neg C_4)$.
6. The Americans do not want to sit next to each other: $(\neg A_1 \mid \neg A_2)$, $(\neg A_2 \mid \neg A_3)$, $(\neg A_3 \mid \neg A_4)$.
7. The Briton likes to sit next to a Canadian: $(\neg C_1 \mid B_2)$, $(\neg C_2 \mid B_1 \mid B_3)$, $(\neg C_3 \mid B_2 \mid B_4)$, $(\neg C_4 \mid B_3)$, $(\neg B_1 \mid C_2)$, $(\neg B_2 \mid C_1 \mid C_3)$, $(\neg B_3 \mid C_2 \mid C_4)$, $(\neg B_4 \mid C_3)$.
If we read $(\neg C_2 \mid B_1 \mid B_3)$ as $C_2 \to B_1 \lor B_3$, it means that if the Canadian takes the second seat, then the Briton must sit at either seat 1 or seat 3.

Some conditions such as there are at least two Americans, and at least one Briton and one Canadian, can also be easily specified by clauses. However, these clauses are logical consequence of the other clauses, as conditions 1 and 2 ensure that there are four people sitting on the bench. If there are at most one Briton and one Canadian, then condition 2 is also redundant, and we may remove them safely. □

The input clauses of a problem serve as a set of axioms. This set is minimum if no clauses are logical consequence of other clauses. Of course, it is very expensive to ensure that a set of clauses is minimum. That is, without a thorough search, we cannot say each clause is independent, i.e., not a logical consequences of the other clauses. When specifying a problem in propositional logic, we often ignore the minimum property of the input clauses.

The next problem is called *Knights and Knaves*, which was coined by Raymond Smullyan in his 1978 work *What Is the Name of This Book?* The problem is actually a collection of similar logic puzzles where some characters can only answer questions truthfully, and others only falsely. The puzzles are set on a fictional island where all inhabitants are either *knights*, who always tell the truth, or *knaves*, who always lie. The puzzles involve a visitor to the island who meets small groups of inhabitants. Usually, the aim is for the visitor to deduce the inhabitants' type from their statements, but some puzzles of this type ask for other facts to be deduced. The puzzles may also be to determine a yes-no question which the visitor can ask in order to discover a particular piece of information.

Example 2.5.4 You meet two inhabitants: Zoey and Mel. Zoey tells you that "Mel is a knave." Mel says, "Neither Zoey nor I am a knave." Can you determine who is a knight and who is a knave? □

Let p stand for "Zoey is knight," with the understanding that if p is true, then Zoey is a knight; if p is false, then Zoey is a knave. Similarly, define q as "Mel is knight." Using a function *says*, what they said can be expressed as

$$(1)\ says(Zoey, \neg q); \qquad (2)\ says(Mel, p \wedge q),$$

together with the two axioms:

$$says(knight, X) \equiv X; \qquad says(knave, X) \equiv \neg X;$$

Now using a truth table to check all the truth values of p and q, only when $p \mapsto 1$ and $q \mapsto 0$, both (1) and (2) are true. That is, we replace Zoey by knight and Mel by knave in (1) and (2), to obtain $\neg q$ and $\neg(p \wedge q)$, respectively. Then $\{p \mapsto 1, q \mapsto 0\}$ is a model of $\neg q \wedge \neg(p \wedge q)$. Thus, the solution is "Zoey is a knight and Mel is a knave."

Example 2.5.5 You meet two inhabitants: Peggy and Zippy. Peggy tells you that "of Zippy and I, exactly one is a knight." Zippy tells you that only a knave would say that Peggy is a knave. Can you determine who is a knight and who is a knave?

Let p be "Peggy is knight" and q be "Zippy is knight." Then what they said can be expressed as

(1) $says(Peggy, p \oplus q)$; (2) $says(Zippy, says(knave, \neg p))$.

The only solution is $p \mapsto 0$ and $q \mapsto 0$, i.e., both Peggy and Zippy are knaves. □

Exercises

1. How many different logical operators have five arguments? That is, how many Boolean functions of type $f : B^5 \mapsto B$, where $B = \{0, 1\}$?
2. Suppose $V_P = \{p, q, r\}$. Then $\mathcal{M}((p \to q) \wedge r) =?$ and $\mathcal{M}((p \to q) \wedge \neg q) =?$
3. (Theorem 2.2.11) Prove that, given any two propositional formulas A and B, $\mathcal{M}(A \wedge B) = \mathcal{M}(A) \cap \mathcal{M}(B)$ and $\mathcal{M}(\neg A) = All_P - \mathcal{M}(A)$.
4. Answer the following questions with explanation:

    (a) Suppose that you have shown that whenever X is true, then Y is true, and whenever X is false, then Y is false. Have you now demonstrated that X and Y are logically equivalent?
    (b) Suppose that you have shown that whenever X is true, then Y is true, and whenever Y is false, then X is false. Have you now demonstrated that X and Y are logically equivalent?
    (c) Suppose you know that X is true iff Y is true, and you know that Y is true iff Z is true. Is this enough to show that X, Y, and Z are all logically equivalent?
    (d) Suppose you know that whenever X is true, then Y is true; that whenever Y is true, then Z is true; and whenever Z is true, then X is true. Is this enough to show that X, Y, and Z are all logically equivalent?

5. Provide the truth table for defining $ite : B^3 \mapsto B$, the if-then-else operator.
6. Provide a BNF grammar which defines the well-formed formulas without unnecessary parentheses, assuming the precedence relation from the highest to the lowest: $\neg$, $\wedge$, $\vee$, $\to$, $\{\oplus, \leftrightarrow\}$.
7. Prove by the truth table method that the following formulas are valid.

    (a) $(A \wedge (A \to B)) \to B$
    (b) $(\neg B \to \neg A) \to (A \to B)$
    (c) $(A \oplus B) \leftrightarrow ((A \vee B) \wedge (\neg A \vee \neg B))$
    (d) $(A \leftrightarrow B) \leftrightarrow ((A \vee \neg B) \wedge (\neg A \vee B))$

8. Prove by the truth table method that the following entailments hold:

   (a) $(A \land (A \to B)) \models B$
   (b) $(A \to (B \to C)) \models (A \to B) \to (A \to C)$
   (c) $(\neg B \to \neg A) \models (A \to B)$
   (d) $((A \lor B) \land (\neg A \lor C)) \models (B \lor C)$

9. Prove by the truth table method that the following logical equivalences hold:

   (a) $\neg(A \lor B) \equiv \neg A \land \neg B$
   (b) $(A \lor B) \land (A \lor \neg B) \equiv A$
   (c) $(A \leftrightarrow B) \equiv ((A \land B) \lor (\neg A \land \neg B))$
   (d) $(A \oplus B) \equiv ((A \land \neg B) \lor (\neg A \land B))$
   (e) $(A \downarrow B) \uparrow C \equiv A \uparrow (B \downarrow C)$
   (f) $(A \uparrow B) \downarrow C \equiv A \downarrow (B \uparrow C)$

10. Prove the equivalence of the following formulas by using other equivalence relations:

    (a) $A \oplus B \equiv \neg(A \leftrightarrow B) \equiv \neg A \leftrightarrow B \equiv A \leftrightarrow \neg B$
    (b) $(A \to B) \land (A \to C) \equiv \neg A \lor (A \land B \land C)$
    (c) $A \to (B \to (C \lor D)) \equiv (A \to C) \lor (B \to D)$.

11. Prove Theorem 2.2.4. (*Hint*: Induction on the structure of A.)
12. Prove that for any formulas A, B, and C, if $A \models B$, then $A \land C \models B \land C$.
13. Define all the 16 binary Boolean functions by formulas using only $\top$, $\bot$, $\neg$, and $\land$.
14. Prove that the nor operator $\downarrow$, where $x \downarrow y = \neg(x \lor y)$, can serve as a minimally sufficient set of Boolean operators.
15. Convert the following formulas into equivalent CNFs:

    (a) $p \to (q \land r)$.
    (b) $(p \to q) \to r$.
    (c) $\neg(\neg p \lor q) \lor (r \to \neg s)$.
    (d) $p \lor (\neg q \land (r \to \neg p))$.
    (e) $\neg(((p \to q) \to p) \to q)$.
    (f) $(p \to q) \leftrightarrow (p \to r)$.

16. Convert the following formulas into equivalent DNFs:

    (a) $p \to (q \land r)$.
    (b) $(p \to q) \to r$.
    (c) $\neg(\neg p \lor q) \lor (r \to \neg s)$.
    (d) $p \lor (\neg q \land (r \to \neg p))$.
    (e) $\neg(((p \to q) \to p) \to q)$.
    (f) $(p \to q) \leftrightarrow (p \to r)$.

Exercises

17. Construct full CNF for the following formulas, where $\bar{a}$ is $\neg a$, $+$ is $\vee$ and $\wedge$ is omitted:
    (a) $F = \bar{a}\bar{b}c + \bar{a}b\bar{c} + a\bar{b}\bar{c} + \bar{a}\bar{b}\bar{c}$.
    (b) $G = abc + \bar{a}bc + a\bar{b}c + ab\bar{c}$.
    (c) $H = a \oplus b \oplus c$.
    (d) $I = (a \wedge b) \rightarrow (b \rightarrow c)$.

18. Construct full DNF for the following formulas, where $\bar{a}$ is $\neg a$, $+$ is $\vee$ and $\wedge$ is omitted:
    (a) $F = (\bar{a} + \bar{b} + c)(\bar{a} + b + \bar{c})(a + \bar{b} + \bar{c})(\bar{a} + \bar{b} + \bar{c})$.
    (b) $G = (a + b + c)(\bar{a} + b + c)(a + \bar{b} + c)(a + b + \bar{c})$.
    (c) $H = a \oplus b \oplus c$.
    (d) $I = (a \wedge b) \rightarrow (b \rightarrow c)$.

19. Assume $a > b > c$; construct ROBDDs for the following formulas, where $\bar{a}$ is $\neg a$, $+$ is $\vee$ and $\wedge$ is omitted as production:
    (a) $F = abc + \bar{a}bc + a\bar{b}\bar{c}$.
    (b) $G = \bar{a}bc + abc + a\bar{b}c$.
    (c) $H = a \oplus b \oplus c$.
    (d) $I = (a \wedge b) \rightarrow (b \rightarrow c)$.

20. Let $\mathcal{F}$ be the set of all formulas built on the set of n propositional variables. Prove that there are exactly 2^{2^n} classes of equivalent formulas among $\mathcal{F}$.

21. (a) Prove that $\{\rightarrow, \neg\}$ is a minimally sufficient set of Boolean operators.
    (b) Prove that $\{\rightarrow, \bot\}$ is a minimally sufficient set of Boolean operators.

22. Prove that $\{\wedge, \vee, \rightarrow, \top\}$ is not a sufficient set of Boolean operators.

23. Prove that (a) $\{\wedge, \vee, \top, \bot\}$ is not a sufficient set of Boolean operators and (b) if we add any function o which cannot be expressed in terms of $\wedge, \vee, \top$, and $\bot$, then $\{\wedge, \vee, \top, \bot, o\}$ is sufficient.

24. Identify the pairs of equivalent formulas from the following candidates:

    (a) $p \uparrow (q \uparrow r)$, (b) $(p \uparrow q) \uparrow r$, (c) $\uparrow (p, q, r)$,
    (d) $p \downarrow (q \downarrow r)$, (e) $(p \downarrow q) \downarrow r$,
    (f) $\downarrow (p, q, r)$, (g) $p \uparrow (q \downarrow r)$, (h) $(p \uparrow q) \downarrow r$,
    (i) $p \downarrow (q \uparrow r)$, (j) $(p \downarrow q) \uparrow r$

25. Try to find an equivalent DNF of the minimal size for the following DNF formulas using K-maps:
    (a) $f(x, y, z) = \bar{x}\bar{y}\bar{z} + \bar{x}y + x y \bar{z} + x z$.
    (b) $f(x, y, z) = \bar{x} y + y \bar{z} + y z + x \bar{y} \bar{z}$.
    (c) $f(x, y, z, w) = \bar{x}\bar{y}zw + \bar{x} yz + y \bar{z}\bar{w} + y \bar{z} w$.
    (d) $f(x, y, z, w) = m_0 + m_1 + m_5 + m_7 + m_8 + m_{10} + m_{14} + m_{15}$, where m_i is a full product of x, y, z and w, and i is the decimal value of the binary string $xyzw$.

26. Try to find an equivalent CNF of the minimal size for the following Boolean functions using K-maps:

    (a) $f(x, y, z) = \bar{x}\bar{y} + \bar{x}y + xy\bar{z} + xz$.
    (b) $f(x, y, z) = \bar{x}yz + xy\bar{z} + x\bar{y}z + \bar{x}\bar{y}\bar{z}$.
    (c) $f(x, y, z, w) = m_0 + m_1 + m_3 + m_4 + m_5 + m_7 + m_{12} + m_{13} + m_{15}$.
    (d) $f(x, y, z, w) = m_0 + m_1 + m_2 + m_4 + m_5 + m_6 + m_8 + m_9 + m_{12} + m_{13} + m_{14}$.

27. A function $f : B^n \mapsto B$, where $B = \{0, 1\}$, is called *linear* if $f(x_1, x_2, \ldots, x_n) = a_0 + a_1x_1 + \cdots + a_nx_n$ for suitable coefficients $a_0, \ldots, a_n \in B$. Here $+$ denotes the addition modulo 2, and the default multiplication (not shown) is the multiplication over integers.

    (a) Show that the above representation of a linear function f is unique.
    (b) Determine the number of n-ary linear functions.

28. Let $a = 31$ and $b = 19$. Please provide the 8-bit vectors $(a)_2$ and $(b)_2$ for a and b and show the results (both in bit vectors and decimal values) of $(\sim a)_2$, $(\sim b)_2$, $(a \mid b)$, $(a \& b)$, $(a \wedge b)_2$, $a >> 2$, $b >> 2$, $a << 2$, and $b << 2$.

29. Suppose x is a 32-bit positive integer. Provide a Java program for each of the following programming tasks, using as much bitwise operations as possible:

    (a) Check whether x is even or odd.
    (b) Get the position of the highest bit of 1 in x.
    (c) Get the position of the lowest bit of 1 in x.
    (d) Set the ith of x to 1, where i is an integer.
    (e) Set the ith of x to 0, where i is an integer.
    (f) Count trailing zeros in x as a binary number.
    (g) Shift x right by 3 bits and put the last 3 bits of x at the beginning of the result from the shifting. The whole process is called "rotate x right" by 3 bits.
    (h) Convert a string of decimal digits into a binary number without using multiplication by 10.

30. Suppose a formula A contains 15 variables, each variable is represented by an integer from 0 to 14. We use the last 15 bits of an integer variable x to represent an interpretation of A, such that ith bit is 1 iff we assign 1 to variable i. Modify the algorithm $eval(A, \sigma)$, where σ is replaced by x using bitwise operations. Design an algorithm which lists all the interpretations of A and apply each interpretation to A using $eval(A, x)$. The pseudo-code of the algorithm should be like a Java program.

31. The meaning of the following propositional variables is given below:

    t: "taxes are increased"
    e: "expenditures rise"
    d: "the debt ceiling is raised"
    c: "the cost of collecting taxes rises"
    g: "the government borrows more money"
    i: "interest rates increase"

(a) Express the following statements as propositional formulas; (b) convert the formulas into CNF.

> Either taxes are increased or if expenditures rise then the debt ceiling is raised. If taxes are increased, then the cost of collecting taxes rises. If a rise in expenditures implies that the government borrows more money, then if the debt ceiling is raised, then interest rates increase. If taxes are not increased and the cost of collecting taxes does not increase then if the debt ceiling is raised, then the government borrows more money. The cost of collecting taxes does not increase. Either interest rates do not increase, or the government does not borrow more money.

32. Specify the following puzzle in propositional logic and convert the specification into CNF. Find the model of your CNF and use this model to construct a solution of the puzzle.

> Four kids, Abel, Bob, Carol, and David, are eating lunch on a hot summer day. Each has a big glass of water, a sandwich, and a different type of fruit (apple, banana, orange, and grapes). Which fruit did each child have? The given clues are:
>
> (a) Abel and Bob have to peel their fruit before eating.
> (b) Carol doesn't like grapes.
> (c) Abel has a napkin to wipe the juice from his fingers.

33. Specify the following puzzle in propositional logic and convert the specification into CNF. Find the model of your CNF and use this model to construct a solution of the puzzle.

> Four electric cars are parked at a charge station and their mileages are 10K, 20K, 30K, and 40K, respectively. Their charge costs are $15, $30, $45, and $60, respectively. Figure out how the charge cost for each car from the following hints:
>
> (a) The German car has 30K miles.
> (b) The Japanese car has 20K miles and it charged $15.
> (c) The French car has less miles than the Italian car.
> (d) The vehicle that charged $30 has 10K more mileages than the Italian car.
> (e) The car that charged $45 has 20K less mileage than the German car.

34. Specify the following puzzle in propositional logic and convert the specification into CNF. Find the model of your CNF and use this model to construct a solution of the puzzle.

> Four boys are waiting for Thanksgiving dinner and each of them has a unique favorite food. Their names are Larry, Nick, Philip, and Tom. Their ages are all different and fall in the set of $\{8, 9, 10, 11\}$. Find out which food they are expecting to eat and how old they are.
>
> (a) Larry is looking forward to eating turkey.
> (b) The boy who likes pumpkin pie is 1 year younger than Philip.
> (c) Tom is younger than the boy that loves mashed potato.
> (d) The boy who likes ham is 2 years older than Philip.

35. Six people sit on a long bench for a group photo. Two of them are Americans (A), two are Canadians (C), one Italian (I), and one Spaniard (S). Suppose we use propositional variables X_y, where $x \in \{A, C, I, S\}$ and $1 \le y \le 6$, and X_y is true iff the people of nationality X sits at seat y of the bench. Please

specify the problem formally in a propositional formula using X_y, and convert the formula into CNF such that each model of the CNF gives us one solution for all the six people to sit on the bench, enforcing the rules that (*a*) the people from the same country will not sit beside one another, (*b*) the Italian must be surrounded by Americans, and (*c*) each American sits next to a Canadian. Please list all the models of your CNF.
36. A very special island is inhabited only by knights and knaves. Knights always tell the truth, and knaves always lie. You meet two inhabitants: Sally and Zippy. Sally claims, "I and Zippy are not the same." Zippy says, "Of I and Sally, exactly one is a knight." Can you determine who is a knight and who is a knave?
37. A very special island is inhabited only by knights and knaves. Knights always tell the truth, and knaves always lie. You meet two inhabitants: Homer and Bozo. Homer tells you, "At least one of the following is true: that I am a knight or that Bozo is a knight." Bozo claims, "Homer could say that I am a knave." Can you determine who is a knight and who is a knave?
38. A very special island is inhabited only by knights and knaves. Knights always tell the truth, and knaves always lie. You meet two inhabitants: Bart and Ted. Bart claims, "I and Ted are both knights or both knaves." Ted tells you, "Bart would tell you that I am a knave." Can you determine who is a knight and who is a knave?

References

1. Kevin C. Klement, *Propositional Logic*, in James Fieser and Bradley Dowden (eds.) Internet Encyclopedia of Philosophy, iep.utm.edu/propositional-logic-sentential-logic, retrieved Nov. 11, 2023
2. Franks, Curtis, "Propositional Logic", *The Stanford Encyclopedia of Philosophy*, Edward N. Zalta & Uri Nodelman (eds.), Fall 2023, retrieved Nov. 11, 2023

Chapter 3
Reasoning in Propositional Logic

In Chap. 1, we introduced the concept of "proof procedure." This chapter will present many proof procedures for propositional logic [1]. The goal of this presentation is not to introduce efficient proof procedures but to present a formalism so that other logics can use.

3.1 Proof Procedures

Given a set A of axioms (i.e., the formulas assumed to be true) and a formula B, a proof procedure $P(A, B)$ will answer the question whether B is a theorem of A. If $P(A, B)$ returns true, we say B is proved; if $P(A, B)$ returns false, we say B is disproved. In general, the axiom set A is regarded to be equivalent to the conjunction of the formulas in A. For propositional logic, a theorem is any formula B such that $A \models B$, i.e., $B \in \mathcal{T}(A)$. That is, $P(A, B)$ is also called a *theorem prover*. The same procedure P can be used to show that B is valid (i.e., B is a tautology) by calling $P(\top, B)$, or that C is unsatisfiable by calling $P(C, \bot)$. That is, proving the entailment relation is no harder than proving validity or proving unsatisfiability. In propositional logic, the three problems are equivalent and belong to the class of *co-NP-complete problems*, i.e., the complement of the class of NP-complete problems. A co-NP-complete problem is regarded "harder" than an NP-complete problem, because the latter has a polynomial-time algorithm to verify its solution, but the former is not known to have such an algorithm.

3.1.1 Types and Styles of Proof Procedures

In practice, each proof procedure is designed to solve one of the three problems and belongs to one of the three types of proof procedures.

- **Theorem prover**
  $P(X, Y)$ returns true iff formula Y is a theorem (an entailment) of X (assuming a set of formulas is equivalent to the conjunction of all its members).
- **Tautology prover**
  $T(A)$ returns true iff formula A is valid (a tautology).
- **Refutation prover**
  $R(B)$ returns true iff formula B is unsatisfiable.

If we have one of the above three procedures, i.e., $P(X, Y)$, $T(A)$, or $R(B)$, we may implement the other two as follows:

- If $P(X, Y)$ is available, implement $T(A)$ as $P(\top, A)$ and $R(B)$ as $P(B, \bot)$.
- If $T(A)$ is available, implement $P(X, Y)$ as $T(X \to Y)$ and $R(B)$ as $T(\neg B)$.
- If $R(B)$ is available, implement $P(X, Y)$ as $R(X \wedge \neg Y)$ and $T(A)$ as $R(\neg A)$.

The correctness of the above implementations is based on the following theorem, whose proof is left as an exercise.

Theorem 3.1.1 *Given formulas A and B, the following three statements are equivalent:*

1. $A \models B$.
2. $A \to B$ *is valid.*
3. $A \wedge \neg B$ *is unsatisfiable.*

Thus, one prover can do all the three, either directly or indirectly. In Chap. 2, we have discussed several concepts and some of them can be used to construct proof procedures as illustrated below.

- Truth table: It can be used to construct a theorem prover, a tautology prover, and a refutation prover.
- CNF: It can be used to construct a tautology prover, because a CNF is valid iff every clause is valid, and it is very easy to check if a clause is valid.
- DNF: It can be used to construct a refutation prover, because a DNF is unsatisfiable iff every product is unsatisfiable, and it is very easy to check if a product is unsatisfiable.
- ROBDD or INF: It can be used to construct a tautology prover or a refutation prover, because ROBDD is a canonical form. Formula A is valid (unsatisfiable) iff A's canonical form is $\top$ ($\bot$).
- Algebraic substitution: It can be used to construct a tautology prover or a refutation prover, as substituting "equal by equal" preserves the equivalence relation. If we arrive at $\top$ or $\bot$, we can claim the input formula is valid or unsatisfiable.

3.1 Proof Procedures

In terms of its working principle, a proof procedure may have one of three styles: enumeration, reduction, and deduction.

- *enumeration-style*: enumerate all involved interpretations.
- *reduction-style*: use the equivalence relations to transform a set of formulas into a simpler set.
- *deduction-style*: use inference rules to deduce new formulas.

By definition, $A \models B$ iff $\mathcal{M}(A) \subseteq \mathcal{M}(B)$, that is, every model of A is a model of B. We may use a truth table to enumerate all the interpretations involving A and B, to see if every model of A is a model of B. A proof procedure based on truth tables is an enumeration-style proof procedure which uses exhaustive search. This style of proof procedures is easy to describe but highly inefficient. For example, if A contains 30 variables, we need to construct a truth table of 2^{30} rows.

Reduction-style proof procedures use the equivalence relations to transform a set of formulas into desired simplified formulas, such as reduced ordered binary decision diagrams (ROBDD). Since ROBDD is canonical, we decide that a formula is valid if its ROBDD is $\top$ (or 1), or unsatisfiable if its ROBDD is $\bot$ (or 0). We may also use disjunctive normal forms (DNF) to show a formula is unsatisfiable: Its simplified DNF is $\bot$. In the following, we will introduce the semantic tableau method, which is a reduction-style proof procedure and works in the same way as obtaining DNF from the input formula.

Deduction-style proof procedures use inference rules to generate new formulas, until the desired formula is generated. An inference rule typically takes one or two formulas as input and generates one or two new formulas as output. Some of these rules come from the equivalence relations between the input and the output formulas. In general, we require that the output of an inference rule be entailed by the input.

Many proof procedures are not clearly classified as one of enumeration, reduction, or deduction styles. Yet, typically their working principles can be understood as performing enumeration, or reduction, or deduction, or a hybrid of all implicitly. For example, converting a formula into DNF can be a reduction; however, converting a formula into full DNF can be an enumeration, as every term in a full DNF corresponds to one model of the formula. In practice, reduction is often used in most deduction-style proof procedures for efficiency.

In Chap. 1, we say that a procedure is a decision procedure if the procedure terminates for every input and answers the question correctly. Fortunately for propositional logic, every proof procedure introduced in this chapter is a decision procedure. Other logics may not have a decision procedure for certain questions.

3.1.2 Semantic Tableau

The semantic tableau method, also called the "truth tree" method, is a reduction-style proof procedure not just for propositional logic, but also for other logic, including first-order logic, temporal, and modal logic. A tableau for a formula is a tree structure; each node of the tree is associated with a set of formulas derived from the input formula. The tableau method constructs this tree and uses it to prove/refute the satisfiability of the input formula. The tableau method can also determine the satisfiability of finite sets of formulas of various logics. It is the most popular proof procedure for modal logic.

For propositional logic, the semantic tableau checks whether a formula is satisfiable or not, by "breaking" complex formulas into smaller ones as we do for transforming the formula into DNF (disjunctive normal form). The tableau used by the method is a tree structure which records the transforming process. The method starts with the root node which contains the input formula and grows the tree by expanding a leaf node and assigning a set of formulas to new nodes. In each expansion, one or two successors are added to the chosen leaf node.

- Each node of the tree is assigned a set of formulas and the comma "," in the set is a synonym to $\wedge$.
- A node is called *closed* if it contains a complementary pair of literals (or formulas), or $\bot$, or $\neg\top$.
- A node is called *open* if it contains only a set of consistent literals, which is equivalent to a product in DNF.
- If a node is neither closed nor open and has no children (a leaf in the current tree), then it is *expandable*. We may apply the transformation rules for DNF on one of the formulas in an expandable node to create children. Note that the rules apply only to the topmost logical operator of a formula. The expansion stops when the tree has no expandable nodes.
- The tree links indicate the equivalence relation among formula sets. If a node x has two successors, say y and z, the formula set of x is equivalent to the disjunction of the two formula sets of y and z. The transformation rules used to obtain two successors are called β-rules. If x has only one successor, the used rules are called α-rules, and the formula set of x is equivalent to that of the successor node.

In the following, we will use tree and tableau as synonyms. To prove the unsatisfiability of a formula, the method starts by generating the tree as described above and checks that every leaf node is closed, that is, no open nodes and no expandable nodes as leaves. In this case, we say the tree is *closed*.

Example 3.1.2 For formula A to be $p \wedge (\neg q \vee \neg p)$ and B be $(p \vee q) \wedge (\neg p \wedge \neg q)$, their tableaux are shown in Fig. 3.1. For A, the tableau has one open node (marked by $\odot$) and one closed node (marked by $\times$). For B, it has only two closed nodes. Thus, A is satisfiable as it has an open node; and B is unsatisfiable as its tableau is closed. □

The tableaux can be displayed linearly, using strings over $\{1, 2\}$ to label the parent–child relations: The root node has the empty label. If a node has label x, its first child has label $x1$ and its second child has label $x2$, and so on. For example, we may display the tableaux of A and B in Fig. 3.1 as follows:

3.1 Proof Procedures

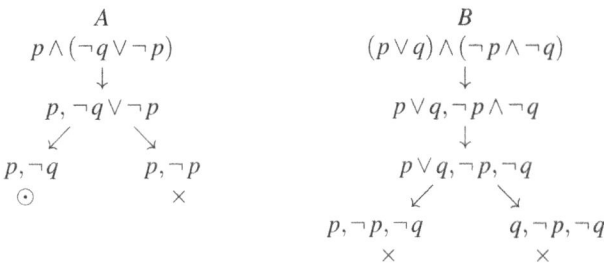

Fig. 3.1 The tableaux for $A = p \wedge (\neg q \vee \neg p)$ and $B = (p \vee q) \wedge (\neg p \wedge \neg q)$

$$\begin{array}{ll} & p \wedge (\neg q \vee \neg p) \\ 1: & p, (\neg q \vee \neg p) \\ 11: & p, \neg q \;\; \odot \\ 12: & p, \neg p \;\; \times \end{array} \qquad \begin{array}{ll} & (p \vee q) \wedge (\neg p \wedge \neg q) \\ 1: & (p \vee q), (\neg p \wedge \neg q) \\ 11: & (p \vee q), \neg p, \neg q \\ 111: & p, \neg p, \neg q \;\; \times \\ 112: & q, \neg p, \neg q \;\; \times \end{array}$$

(a) Tableau of A (b) Tableau of B

The labels of nodes can uniquely identify the positions (or addresses) of these nodes in a tableau.

3.1.3 α-Rules and β-Rules

Here is the listing of α-rules and β-rules, which apply to the top logical operator in a formula.

α	α_1, α_2
$A \wedge B$	A, B
$\neg(A \vee B)$	$\neg A, \neg B$
$\neg(A \rightarrow B)$	$A, \neg B$
$\neg(A \oplus B)$	$(A \vee \neg B), (\neg A \vee B)$
$A \leftrightarrow B$	$(A \vee \neg B), (\neg A \vee B)$
$\neg(A \uparrow B)$	A, B
$A \downarrow B$	$\neg A, \neg B$
$\neg\neg A$	A

β	β_1	β_2
$\neg(A \wedge B)$	$\neg A$	$\neg B$
$A \vee B$	A	B
$A \rightarrow B$	$\neg A$	B
$A \oplus B$	$A, \neg B$	$\neg A, B$
$\neg(A \leftrightarrow B)$	$A, \neg B$	$\neg A, B$
$A \uparrow B$	$\neg A$	$\neg B$
$\neg(A \downarrow B)$	A	B

Those rules are what we used for transforming a formula into DNF. If we start with a formula in NNF (negation normal form), we need only one α-rule for $A \wedge B$ and one β-rule for $A \vee B$. All other α-rules and β-rules are not needed for NNF, because from NNF to DNF, only the distributive law is needed.

For every α-rule, we transform formula α into α_1, α_2; for every β-rule, we transform formula β into β_1 and β_2. It is easy to check that $\alpha \equiv \alpha_1 \wedge \alpha_2$ and $\beta \equiv \beta_1 \vee \beta_2$. Since a tableau maintains the property that the formula set of the current node is equivalent to the disjunction of the formula sets of its children, a tableau is a tree-like representation of a formula that is equivalent to the disjunction of all the formula sets of its leaf nodes, i.e., a DNF, assuming a set of formulas is equivalent to the conjunction of its members.

Example 3.1.3 To show that Frege's formula, $(p \to (q \to r)) \to (p \to q) \to (p \to r)$, is valid, we show that the semantic tableau of its negation is closed.

	$\neg((p \to (q \to r)) \to (p \to q) \to (p \to r))$	$\alpha \neg \to$
1 :	$(p \to (q \to r)), \neg(p \to q) \to (p \to r)$	$\alpha \neg \to$
11 :	$(p \to (q \to r)), (p \to q), \neg(p \to r)$	$\alpha \neg \to$
111 :	$(p \to (q \to r)), (p \to q), p, \neg r$	$\beta \to$
1111 :	$\neg p, (p \to q), p, \neg r$	$\times$
1112 :	$(q \to r), (p \to q), p, \neg r$	$\beta \to$
11121 :	$(q \to r), \neg p, p, \neg r$	$\times$
11122 :	$(q \to r), q, p, \neg r$	$\beta \to$
111221 :	$\neg q, q, p, \neg r$	$\times$
111222 :	$r, q, p, \neg r$	$\times$

$\square$

In Chap. 1, we pointed out that there are two important properties concerning a proof procedure: soundness and completeness. If semantic tableau is used as a refutation prover, the soundness means that if the procedure says the formula is unsatisfiable, then the formula must be unsatisfiable. The completeness means that if the formula is unsatisfiable, then the procedure should be able to give us this result.

Theorem 3.1.4 *The semantic tableau method is a decision procedure for propositional logic.*

Proof Suppose without loss of generality the formula set at one node is $\{X, Y, Z\}$, where X represents an α formula, Y a β formula, and Z any other formula, and this set is equivalent to $A = X \wedge Y \wedge Z$. If X is transformed by α-rule to X_1 and X_2, then from $X \equiv X_1 \wedge X_2$, $A \equiv X_1 \wedge X_2 \wedge Y \wedge Z$. If Y is transformed by β-rule to Y_1 and Y_2, then from $Y \equiv Y_1 \vee Y_2$, $A \equiv (X \wedge Y_1 \wedge Z) \vee (X \wedge Y_2 \wedge Z)$. In other words, the logical equivalence is maintained for every parent–child link in the tree. By a simple induction on the structure of the tree, the formula at the root of the tree is equivalent to the disjunction of the formula sets of all leaf nodes. That is, the procedure modifies the tree in such a way that the disjunction of the formula sets of all leaf nodes of the resulting tableau is equivalent to the input formula. One of these formula sets may contain a pair of complementary literals, i.e., the corresponding node is closed, in which case that set is equivalent to false. If the sets of all leaf nodes are equivalent to false, i.e., every lead node is closed, the input formula is unsatisfiable. This shows the soundness of the semantic tableau method.

If a formula is unsatisfiable, then every term in its DNF must be equivalent to false and the corresponding node in the tableau is closed. Thus, we will have a closed tableau. This shows the completeness of the semantic tableau method.

Since neither α-rules nor β-rules can be applied forever (each rule reduces either one occurrence of a binary operator or two negations), the semantic tableau method must be terminating. □

The semantic tableau method can be used to show the satisfiability of a formula. If there is an open node in the tableau, we may assign 1 to every literal in this node to create an interpretation, and the formula set of this node is true under this interpretation. The input formula will also be true under this interpretation because the input formula is equivalent to the disjunction of the formula sets of all leaf nodes. Thus, this interpretation is a model of the input formula. This shows that the procedure can be used to find a model when the formula is satisfiable.

Semantic tableau specifies what rules can be used but it does not specify which rule should be used first or which leaf node should be expanded first. In general, α-rules are better applied before β-rules, so that the number of nodes in a tree grows slowly. The user has the freedom to apply β-rules to any formula in the set, resulting trees of different shapes. The procedure may terminate once an open node is found, if the goal is to show that the input formula is satisfiable.

Semantic tableaux are much more expressive and easier to use than truth-tables, though that is not the reason for their introduction. Sometimes, a tableau uses more space than a truth table. For example, an unsatisfiable full CNF of n variables will generate a tableau of more than $n!$ nodes, which is larger than a truth table of 2^n rows. The beauty of semantic tableaux lies in the simplicity of presenting a proof procedure using a set of rules. We will see in later chapters how new rules are added into this procedure so that a proof procedure for other logics can be obtained.

3.2 Deductive Systems

In logic, a deductive system S consists of a set of inference rules, where each rule takes premises as input and returns a conclusion (or conclusions) as output. Popular inference rules in propositional logic include modus ponens (MP), modus tollens (MT), and contraposition (CP), which can be displayed, respectively, as follows:

$$\frac{A \to B \quad A}{B} \text{ (MP)} \qquad \frac{A \to B \quad \neg B}{\neg A} \text{ (MT)} \qquad \frac{A \to B}{\neg B \to \neg A} \text{ (CP)}$$

3.2.1 Inference Rules and Proofs

In general, an inference rule can be specified as

$$\frac{P_1 \quad P_2 \quad \cdots \quad P_k}{C}$$

where $P_1, P_2, \ldots, P_k$ are the premises and C is the conclusion. We may also write it as

$$P_1, P_2, \ldots, P_k \vdash C$$

If the premises are empty ($k = 0$), we say this inference rule is an *axiom rule*.

An inference rule is *sound* if its premises entail the conclusion, that is,

$$\{P_1, P_2, \ldots, P_k\} \models C$$

for every inference rule "$P_1, P_2, \ldots, P_k \vdash C$" (or equivalently, $(P_1 \wedge P_2 \wedge \cdots \wedge P_k) \to C$ is a tautology). For example, MP (modus ponens) is sound because $\{A \to B, A\} \models B$; MT (modus tollens) is sound because $\{A \to B, \neg B\} \models \neg A$.

Given a set A of axioms, which are the formulas assumed to be true, a *proof* of formula B in S is a sequence of formulas $F_1, F_2, \ldots, F_n$, such that $F_n = B$ and each F_i is either a formula in A or can be generated by an inference rule of S, using the formulas before F_i in the sequence as the premises. If such a proof exists, we denote it by $A \vdash_S B$; the subscript S can be dropped if S is understood from the context. The proof procedure $P(A, B)$ based on S is simply to show $A \vdash_S B$. Obviously, this is a deduction-style proof procedure.

Example 3.2.1 Let A be the axiom set $\{p \to (q \to r), p, \neg r\}$, and $S = \{MP, CT\}$. A proof of $A \vdash_S \neg q$ is given below:

1. $p \to (q \to r)$ axiom
2. p axiom
3. $q \to r$ MP, 1, 2
4. $\neg r$ axiom
5. $\neg q$ MT, 3, 4

where "MP, 1, 2" means formula 3. is obtained by the MP rule with formulas 1. and 2. as premises. □

The formulas in a proof can be rearranged so that all axioms appear in the beginning of the sequence. For example, the above proof can be rewritten as the following.

3.2 Deductive Systems

$$
\begin{aligned}
&1.\ p \to (q \to r) &&\text{axiom} \\
&2.\ p &&\text{axiom} \\
&3.\ \neg r &&\text{axiom} \\
&4.\ q \to r &&\text{MP, 1, 2} \\
&5.\ \neg q &&\text{MT, 3, 4}
\end{aligned}
$$

Every proof can be represented by a directed graph where each node is a formula in the proof, and if a formula B is derived from $A_1, A_2, \ldots, A_k$ by an inference rule, then there are edges (A_i, B), $1 \le i \le k$, in the graph. This proof graph must be acyclic as we assume the premises must be present in the sequence before the new formula is derived by an inference rule. Every topological sort of the nodes in this graph should give us a linear proof. Since every acyclic direct graph (DAG) can be converted into a tree (by duplicating some nodes) and still preserve the parent–child relation, we may present a proof by a tree as well. The proof tree for the above proof example is given in Fig. 3.2. Note that all axioms do not have the incoming edges. The final formula in the proof does not have outgoing edges. All intermediate formulas have both incoming and outgoing edges.

The soundness of an inference system comes from the soundness of all inference rules. In general, we require all inference rules be sound to preserve the truth of all the derived formulas. For a sound inference system S, every derived formula B is a theorem of A, i.e., $B \in \mathcal{T}(A)$. If A is a set of tautologies, B is tautology, too.

Theorem 3.2.2 *If an inference system S is sound and $A \vdash_S B$, then $A \models B$.*

Proof $A \vdash_S B$ means there is a proof of B in S. Suppose the proof is $F_1, F_2, \ldots, F_n$ and $F_n = B$. By induction on n, we show that $A \models F_i$ for all $i = 1, \ldots, n$. If $F_i \in A$, then $A \models F_i$. If F_i is derived from an inference rule "$P_1, P_2, \ldots, P_m \vdash C$", using $F_{j_1}, F_{j_2}, \ldots, F_{j_m}$ as premises, by induction hypotheses, for $1 \le k \le m$, $A \models F_{j_k}$, because $j_k < i$. That is, $A \models \bigwedge_{k=1}^m F_{j_k}$.

Since the inference rule is sound, $\bigwedge_{k=1}^m P_k \models C$. Applying the substitution theorem, $\bigwedge_{k=1}^m F_{j_k} \models F_i$. Because $\models$ is transitive, $A \models F_i$. □

Different inference systems are obtained by changing the axioms or the inference rules. In propositional logic, all these systems are equivalent in the sense that they are sound and complete.

Fig. 3.2 The proof tree of $\neg q$

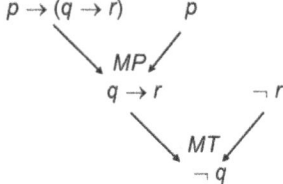

3.2.2 Hilbert System

Hilbert system, sometimes called Hilbert calculus, or Hilbert-style inference system, is a type of deduction system attributed to Gottlob Frege and David Hilbert. Variants of Hilbert system for propositional logic exist and these deductive systems are extended to first-order logic and other logic.

Definition 3.2.3 Hilbert system $\mathcal{H}$ contains three axiom rules plus MP (modus ponens):

$$H_1 : \vdash (A \to (B \to A)),$$
$$H_2 : \vdash (A \to (B \to C)) \to ((A \to B) \to (A \to C)),$$
$$H_3 : \vdash (\neg B \to \neg A) \to (A \to B),$$
$$\text{MP} : A, A \to B \vdash B.$$

The axiom rule H_3 is not needed if $\to$ is the only operator in all formulas. Other logical operators can be defined using $\to$ and $\neg$. For example, $A \wedge B \equiv \neg(A \to \neg B)$ and $A \vee B \equiv (\neg A \to B)$. $\neg A$ can be replaced by $A \to \bot$ in H_3.

Example 3.2.4 To show that $\vdash_{\mathcal{H}} (p \to p)$, we have the following proof in $\mathcal{H}$:

1. $p \to ((p \to p) \to p)$ $H_1, A = p, B = (p \to p)$
2. $(p \to ((p \to p) \to p)) \to$
   $((p \to (p \to p)) \to (p \to p))$ $H_2, A = C = p, B = p \to p$
3. $(p \to (p \to p)) \to (p \to p)$ MP, 1, 2
4. $(p \to (p \to p))$ $H_1, A = p, B = p$
5. $p \to p$ MP, 3, 4

All tautologies are derivable in Hilbert system. □

Proposition 3.2.5 *The Hilbert system $\mathcal{H}$ is sound: If $A \vdash_{\mathcal{H}} B$, then $A \models B$.*

Proof It is easy to check that every inference rule of $\mathcal{H}$ is sound. That is, every axiom rule H_i, $i = 1, 2, 3$, is tautology, so any instance of the rule is tautology. For MP, $(p \wedge (p \to q)) \to q$ is tautology. By Theorem 3.2.2, the proposition holds. □

Hilbert system originally assumed that $A = \emptyset$ (or $A \equiv \top$); thus, every derived formula is tautology in a proof. This restriction is relaxed today so that we may take a set of formulas as axioms. The truth of the formulas becomes conditional on the truth of the axioms.

Example 3.2.6 Assume $X = \{p \to q, q \to r\}$; to show that $X \vdash_{\mathcal{H}} (p \to r)$, we have the following proof in $\mathcal{H}$:

3.2 Deductive Systems

1. $p \to q$ axiom
2. $q \to r$ axiom
3. $((q \to r) \to (p \to (q \to r)))$ H_1
4. $p \to (q \to r)$ MP, 2, 3
5. $(p \to (q \to r)) \to ((p \to q) \to (p \to r))$ H_2
6. $(p \to q) \to (p \to r)$ MP, 4, 5
7. $p \to r$ MP, 1, 6

□

Since $\mathcal{H}$ is sound, from $A \vdash_\mathcal{H} (p \to r)$, we have $(p \to q) \wedge (q \to r) \models_\mathcal{H} (p \to r)$, or $((p \to q) \wedge (q \to r)) \to (p \to r)$ is valid.

Hilbert system is one of the earliest inference systems and has great impact to the creation of many inference systems. However, theorem proving in this system has always been considered a challenge because of its high complexity. Since Hilbert system can hardly be a practical tool for problem solving, we state without proof the completeness result of Hilbert system.

Theorem 3.2.7 *Hilbert system is sound and complete, that is, $A \models B$ iff $A \vdash_\mathcal{H} B$.*

3.2.3 Natural Deduction

Both semantic tableaux and Hilbert system for constructing proofs have their disadvantages. Hilbert system is difficult to construct proofs and its main uses are metalogical. The small number of rules makes it easier to prove soundness about logic. The tableau method on the other hand is easy to use mechanically, but, because of the form of the rules and the fact that a tableau starts from the negation of the formula to be proved, the resulting proofs are not a natural sequence of easily justifiable steps. Likewise, very few proofs in mathematics are from axioms directly. Mathematicians in practice usually reason in a more flexible way.

Natural deduction is a deductive system in which logical reasoning is expressed by inference rules closely related to the "natural" way of reasoning. This contrasts with Hilbert system, which instead use axiom rules as much as possible to express the logical laws of deductive reasoning. Natural deduction in its modern form was independently proposed by the German mathematician Gerhard Gentzen in 1934. In Gentzen's natural deductive system, a formula is represented by a set A of formulas, $A = \{A_1, A_2, \ldots, A_n\}$, where the comma in A is understood as $\vee$. To avoid confusion, we will write A as

$$(A_1 \mid A_2 \mid \cdots \mid A_n)$$

A is called a *sequent* by Gentzen. If $n = 0$, then $A = (\ )$, which is equivalent to $\bot$. Since A is a set, duplicated formulas are removed and the order of formulas in A is irrelevant.

Definition 3.2.8 A natural deductive system $\mathcal{G}$ consists of the following three inference rules, followed by the logical rules which guide the introduction and elimination of logical operators in the sequents.

name	inference rule
axiom	$\vdash (A \mid \neg A \mid \alpha)$
cut	$(A \mid \alpha), (\neg A \mid \beta) \vdash (\alpha \mid \beta)$
thinning	$(\alpha) \vdash (A \mid \alpha)$

op	introduction	elimination
$\neg$	$(A \mid \alpha) \vdash (\neg\neg A \mid \alpha)$	$(\neg\neg A \mid \alpha) \vdash (A \mid \alpha)$
$\vee$	$(A \mid B \mid \alpha) \vdash (A \vee B \mid \alpha)$	$(A \vee B \mid \alpha) \vdash (A \mid B \mid \alpha)$
$\wedge$	$(A \mid \alpha), (B \mid \alpha) \vdash (A \wedge B \mid \alpha)$	(a) $(A \wedge B \mid \alpha) \vdash (A \mid \alpha)$
		(b) $(A \wedge B \mid \alpha) \vdash (B \mid \alpha)$
$\rightarrow$	$(\neg A \mid B \mid \alpha) \vdash (A \rightarrow B \mid \alpha)$	$(A \rightarrow B \mid \alpha) \vdash (\neg A \mid B \mid \alpha)$
$\neg\vee$	$(\neg A \mid \alpha), (\neg B \mid \alpha) \vdash (\neg(A \vee B) \mid \alpha)$	(a) $(\neg(A \vee B) \mid \alpha) \vdash (\neg A \mid \alpha)$
		(b) $(\neg(A \vee B) \mid \alpha) \vdash (\neg B \mid \alpha)$
$\neg\wedge$	$(\neg A \mid \neg B \mid \alpha) \vdash (\neg(A \wedge B) \mid \alpha)$	$(\neg(A \wedge B) \mid \alpha) \vdash (\neg A \mid \neg B \mid \alpha)$
$\neg\rightarrow$	$(A \mid \alpha), (\neg B \mid \alpha) \vdash (\neg(A \rightarrow B) \mid \alpha)$	(a) $(\neg(A \rightarrow B) \mid \alpha) \vdash (A \mid \alpha)$
		(b) $(\neg(A \rightarrow B) \mid \alpha) \vdash (\neg B \mid \alpha)$

where α and β, possibly empty, denote the rest formulas in a sequent.

In natural deduction, a sequent is deduced from a set of axioms by applying inference rules repeatedly. Each sequent is inferred from other sequents on earlier lines in a proof according to inference rules, giving a better approximation to the style of proofs used by mathematicians than Hilbert system.

Example 3.2.9 The following is a proof of $\neg(p \wedge q) \rightarrow (\neg p \vee \neg q)$ in $\mathcal{G}$:

1. $(p \mid \neg p \mid \neg q)$ axiom, $A = p, \alpha = \neg q$
2. $(q \mid \neg q \mid \neg p)$ axiom, $A = q, \alpha = \neg p$
3. $(p \wedge q \mid \neg p \mid \neg q)$ $\wedge_I, 1, 2, A = p, B = q, \alpha = \neg p \mid \neg q$
4. $(\neg\neg(p \wedge q) \mid \neg p \mid \neg q)$ $\neg_I, 4, A = (p \wedge q), \alpha = (\neg p \vee \neg q)$
5. $(\neg\neg(p \wedge q) \mid (\neg p \vee \neg q))$ $\vee_I, 3, A = \neg p, B = \neg q, \alpha = \neg\neg(p \wedge q)$
6. $(\neg(p \wedge q) \rightarrow (\neg p \vee \neg q))$ $\rightarrow_I, 5, A = \neg(p \wedge q), B = (\neg p \vee \neg q), \alpha = \bot$

Note that I in $\wedge_I$ means the *introduction* rule for $\wedge$ is used. For better understanding, we added the substitution of variables in each step. □

Proposition 3.2.10 *The deductive system $\mathcal{G}$ is sound, i.e., if $A \vdash_\mathcal{G} B$, then $A \models B$.*

Proof It is easy to check that every inference rule of $\mathcal{G}$ is sound. In particular, the rules of operator introduction preserve the logical equivalence. In the rules of operator elimination, some sequents generate two results. For example, $(A \wedge B \mid \alpha)$

3.2 Deductive Systems 91

infers both (a) $(A \mid \alpha)$ and (b) $(B \mid \alpha)$. If we bind the two results together by $\wedge$, these rules preserve the logical equivalence, too. On the other hand, the cut rule and the thinning rule do not preserve the logical equivalence; only the entailment relation holds. That is, for the cut rule, $(A \mid \alpha), (\neg A \mid \beta) \vdash (\alpha \mid \beta)$, we have

$$(A \vee \alpha) \wedge (\neg A \vee \beta) \models \alpha \vee \beta,$$

and for the thinning rule, $(\alpha) \vdash (A \mid \alpha)$, we have $\alpha \models A \vee \alpha$. □

Note that the proof in the example above uses only the axiom rule (twice) and the introduction rules for logical operators. This is the case when we are looking for a proof of a tautology. Recall how we would prove that $\neg(p \wedge q) \to (\neg p \vee \neg q)$ is tautology by semantic tableau. Since semantic tableau is a refutation prover, we need to show that the negation of a tautology has a closed tableau.

id	label	formulas of node	rule
a.		$: \neg(\neg(p \wedge q) \to (\neg p \vee \neg q))$	$\alpha \neg \to$
b.	1 :	$\neg(p \wedge q), \neg(\neg p \vee \neg q)$	$\alpha \neg \vee$
c.	11 :	$\neg(p \wedge q), \neg\neg p, \neg\neg q$	$\alpha \neg$
d.	111 :	$\neg(p \wedge q), p, \neg\neg q$	$\alpha \neg$
e.	1111 :	$\neg(p \wedge q), p, q$	$\beta \neg \wedge$
f.	11111 :	$\neg p, p, q$	×
g.	11112 :	$\neg q, p, q$	×

Comparing the proof of $\mathcal{G}$ and the tableau above, we can see a clear correspondence: Closed nodes (f) and (g) correspond to (1) and (2), which are axioms; (e) to (3), (c) and (d) to (4), (b) to (5), and finally (a) to (6), respectively. That is, the formula represented by each sequent in a proof of $\mathcal{G}$ can always find its negation in the corresponding node of the semantic tableau.

We show below that $\mathcal{G}$ can be used as a refutation prover by the same example.

Example 3.2.11 Let $A = \{\neg(\neg(p \wedge q) \to (\neg p \vee \neg q))\}$. A proof of $A \models \bot$ in $\mathcal{G}$ is given below.

1. $(\neg(\neg(p \wedge q) \to (\neg p \vee \neg q)))$ assumed
2. $(\neg(p \wedge q))$ $\neg \to_E (a), 1, A = \neg(p \wedge q)$
3. $(\neg(\neg p \vee \neg q))$ $\neg \to_E (b), 1, B = (\neg p \vee \neg q)$
4. $(\neg p \mid \neg q)$ $\neg \wedge_E, 2, A = p, B = q$
5. $(\neg\neg p)$ $\neg \vee_E (a), 3, A = \neg p$
6. $(\neg\neg q)$ $\neg \vee_E (b), 3, B = \neg q$
7. (p) $\neg_E, 5, A = p$
8. (q) $\neg_E, 6, A = q$
9. $(\neg q)$ cut, 4, 7, $A = p, \alpha = \bot, \beta = \neg q$
10. $\bot$ cut, 8, 9, $A = q, \alpha = \bot, \beta = \bot$

Note that $\perp$ in the last step stands for the empty sequent. □

In this proof the used inference rules are the cut rule and the rules for eliminating logical operators. If we apply repeatedly the rules of operator elimination to a formula, we will obtain a list of sequents, each of which represents a clause. In other words, these rules transform the formula into CNF, then the cut rule works on them. The cut rule for clauses has another name, i.e., *resolution*, which is the topic of the next section.

Indeed, $\mathcal{G}$ contains more rules than necessary: Using only the axiom rule and the rules for operator introduction, it can imitate the dual of a semantic tableau upside down. Using only the cut rule and the rules for operator elimination, it can do what a resolution prover can do. Both semantic tableau and resolution prover are decision procedures for propositional logic. The implication is that natural deduction $\mathcal{G}$ is a decision procedure, too. We give the main result without a proof.

Theorem 3.2.12 *The natural deduction $\mathcal{G}$ is sound and complete, that is, $A \models B$ iff $A \vdash_\mathcal{G} B$.*

3.2.4 Inference Graphs

We have seen two examples of deductive systems: Hilbert system $\mathcal{H}$ and Gentzen's natural deduction $\mathcal{G}$. To show the completeness of these deductive systems, we need strategies on how to use the inference rules in these systems and the completeness depends on these strategies. To facilitate the discussion, let us introduce a few concepts.

Definition 3.2.13 Given an inference system $\mathcal{S}$, the *inference graph* of $\mathcal{S}$ over the axiom set A is defined as a directed graph $G = (V, E)$, where each vertex of V is a set $\mathbf{x}$ of formulas, where $A \in V$, $\mathbf{x} \subseteq \mathcal{T}(A)$, and $(\mathbf{x}, \mathbf{y}) \in E$ iff $\mathbf{y} = \mathbf{x} \cup \{C\}$, where $C \notin \mathbf{x}$ is the result of an inference rule $\mathbf{r} \in \mathcal{S}$ using some formulas in $\mathbf{x}$ as the premises of $\mathbf{r}$.

Graph $G = (V, E)$ defines the search space for $A \vdash_\mathcal{S} B$, which becomes a search problem: To find a proof of $A \vdash_\mathcal{S} B$ is the same as to find a path from A to a node of V containing B. That is, if all the axioms of A in a proof appear in the beginning of the proof sequence, then this proof presents a path in the inference graph. Suppose the proof is a sequence of formulas $F_1, F_2, \ldots, F_n$ such that the first k formulas are from A, i.e., $F_i \in A$ for $1 \leq i \leq k$ and $F_i \notin A$ for $i > k$. Then this proof represents a directed path in $\mathbf{x}_0, \mathbf{x}_1, \ldots, \mathbf{x}_{n-k}$ in the inference graph such that $\mathbf{x_0} = A$, $F_i \in A$ for $1 \leq i \leq k$, and $F_j \in \mathbf{x_{j-k}}$ for $k < j \leq n$. That is, a proof is a succinct way of presenting a directed path in the inference graph G.

3.3 Resolution

Example 3.2.14 In Example 3.2.1, the second proof is copied here:

 1. $p \to (q \to r)$ axiom
 2. p axiom
 3. $\neg r$ axiom
 4. $q \to r$ MP, 1, 2
 5. $\neg q$ MT, 3, 4

This proof corresponds to the path $(\mathbf{x_0}, \mathbf{x_1}, \mathbf{x_2})$ in the inference graph, where $\mathbf{x_0} = \{1., 2., 3.\}$, $\mathbf{x_1} = \mathbf{x_0} \cup \{4.\}$, and $\mathbf{x_2} = \mathbf{x_1} \cup \{5.\}$. □

Given a set of inference rules, a proof procedure can be easily constructed if we use a *fair strategy* to search the inference graph. For example, starting from A in the inference graph, a breadth-first search is a fair strategy, but a depth-first search is not if the graph is infinite. Fair strategy requires that if an inference rule is applicable at some point, this application will happen eventually. Any fair search strategy, including heuristic search, best-first search, A*, etc., can be used for a proof procedure.

If every inference rule is sound, the corresponding proof procedure will be sound. Extra requirement like fair strategy is needed in general to show that the proof procedure is complete or halting. If the inference graph is finite, the termination of the proof procedure is guaranteed, but it does not imply the completeness of the procedure due to lack of certain inference rules.

3.3 Resolution

In the previous section, we introduced two deductive systems, Hilbert system and natural deduction. These two systems are exemplary for the development of many deductive systems but are not practical for theorem proving in propositional logic. One of the reasons is the use of axiom rules which provide infinite possibilities. Unlike reduction rules, inference rules may be applied without termination. For automated theorem proving, we need a deductive system which contains very few rules as computers can do simple things fast. Resolution $\mathcal{R}$ is such a system, which contains a single inference rule, called *resolution*. Note that resolution is a special case of the cut rule when all sequents are clauses.

3.3.1 Resolution Rule

To use the resolution rule, formula A is first transformed into CNF, which is a conjunction of clauses (disjunction of literals). We will use $A = \{C_1, C_2, \ldots, C_m\}$

to denote the conjunction of the clauses; each clause C_i is denoted by

$$(l_1 \mid l_2 \mid \cdots \mid l_k),$$

where each l_j is a literal (propositional variable or its negation). If $k = 0$, clause C_i is *empty*, denoted by $\perp$; if $k = 1$, C_i is called a *unit clause*; if $k = 2$, C_i is called a *binary clause*. A clause is said to be *positive* if it is not empty and contains only positive literals; it is *negative* if it is not empty and contains only negative literals. A clause is tautology if it contains a pair of complementary literals, i.e., it contains both p and $\neg p$ for some variable p.

The total number of clauses, of which duplicated literals are removed, on n propositional variables is bound by 3^n, because each variable has one of three cases in a non-tautology clause: missing, as a positive literal, or as a negative literal.

Formally, the resolution rule is defined by the following schema, where A is a propositional variable, α and β are the rest literals of the two clauses as premises, respectively, and the result is a new clause $(\alpha \mid \beta)$.

$$\frac{(A \mid \alpha) \quad (\neg A \mid \beta)}{(\alpha \mid \beta)}$$

Alternatively, we may write the resolution rule as

$$(A \mid \alpha), (\neg A \mid \beta) \vdash (\alpha \mid \beta).$$

The clause $(\alpha \mid \beta)$ produced by the resolution rule is called *resolvent* by resolving off A from C_1 and C_2. C_1 and C_2 are the *parents* of the resolvent. This resolution rule is also called *binary resolution*, as it involves two clauses as premises.

The resolution rule for propositional logic can be traced back to Davis and Putnam (1960). Resolution is extended to first-order logic in 1965 by John Alan Robinson, using the unification algorithm. If we use $\to$ instead of $\mid$ in the resolution rule, we obtain

$$\frac{(\neg \alpha \to A) \quad (A \to \beta)}{(\neg \alpha \to \beta)}$$

which is the transitivity of $\to$. If α is empty, then we obtain MP:

$$\frac{A \quad A \to \beta}{\beta}$$

Example 3.3.1 Let $C = \{(p \mid q), (p \mid \neg q), (\neg p \mid q \mid r), (\neg p \mid \neg q), (\neg r)\}$. A proof of $C \vdash_\mathcal{R} \perp$ is given below:

3.3 Resolution

1. $(p \mid q)$ axiom
2. $(p \mid \neg q)$ axiom
3. $(\neg p \mid q \mid r)$ axiom
4. $(\neg p \mid \neg q)$ axiom
5. $(\neg r)$ axiom
6. $(q \mid r)$ $\mathcal{R}, 1, 3, A = p, \alpha = q, \beta = (q \mid r)$
7. $(\neg q)$ $\mathcal{R}, 2, 4, A = p, \alpha = \neg q, \beta = \neg q$
8. (r) $\mathcal{R}, 6, 7, A = q, \alpha = r, \beta = \bot$
9. $(\,)$ $\mathcal{R}, 8, 5, A = r, \alpha = \bot, \beta = \bot$

Note that we assume that duplicated literals are removed and the order of literals is irrelevant in a clause. Thus, the resolvent generated from clauses 1 and 3 is $(q \mid q \mid r)$, but we keep it as $(q \mid r)$ by removing a copy of q. From clauses 1 and 4, we may obtain a resolvent $(q \mid \neg q)$ on p, or another resolvent $(p \mid \neg p)$ on q. These two resolvents are tautologies and are not useful in the search for $\bot$. You cannot resolve on two variables at the same time to get $(\,)$ directly from clauses 1 and 4. □

Definition 3.3.2 A *resolution proof* P from the set of input clauses C is a refutation proof of $C \vdash_{\mathcal{R}} \bot$ and the length of P is the number of resolvents in P, denoted by $|P|$. The set of the input clauses appearing in P is called the *core* of P, denoted by $core(P)$.

The length of the resolution proof in Example 3.3.1 is 4, which is the length of the list minus the number of axiom clauses in the proof. The core of this proof is the entire input.

Proposition 3.3.3 *The resolution $\mathcal{R}$ is sound, i.e., if $C \vdash_{\mathcal{R}} B$, then $C \models B$.*

Proof Using truth table, it is easy to check that $((A \vee \alpha) \wedge (\neg A \vee \beta)) \to (\alpha \vee \beta)$ is tautology. Thus, the resolution rule is sound. The proposition holds following Theorem 3.2.2. □

Corollary 3.3.4 *If $C \vdash_{\mathcal{R}} \bot$, then C is unsatisfiable.*

Proof Since $C \vdash_{\mathcal{R}} \bot$ means $C \models \bot$, so $\mathcal{M}(C) = \mathcal{M}(C \wedge \bot) = \mathcal{M}(C) \cap \mathcal{M}(\bot) = \emptyset$, because $\mathcal{M}(\bot) = \emptyset$ (see Corollary 2.2.26). □

This corollary allows us to design a refutation prover: To show A is valid, we convert $\neg A$ into an equivalent set C of clauses and show that $C \vdash_{\mathcal{R}} \bot$, where C is called the *input clauses*. Since the total number of clauses is finite, the inference graph of $\mathcal{R}$ is finite. It means that the proof procedure will terminate but it does not mean the graph is small. Before we establish the completeness of $\mathcal{R}$, let us consider how to build an effective resolution prover.

A successful implementation of a resolution prover needs the integration of search strategies that avoid many unnecessary resolvents. Some strategies remove redundant clauses as soon as they appear in a derivation. Some strategies avoid generating redundant clauses in the first place. Some strategies sacrifice the completeness for efficiency.

3.3.2 Resolution Strategies

Here are some well-known resolution strategies which add restriction to the use of the resolution rule.

- **Unit resolution**: One of the two parents is a unit clause.
- **Input resolution**: One of the two parents is an input clause (given as the axioms).
- **Ordered resolution**: Given an order on propositional variables, the resolved variable must be maximal in both parent clauses.
- **Positive resolution**: One of the two parents is positive.
- **Negative resolution**: One of the two parents is negative.
- **Set-of-support**: The input clauses are partitioned into two sets, S and T, where S is called the *set of support*, and a resolution is not allowed if both parents are in T.
- **Linear resolution**: The latest resolvent is used as a parent for the next resolution (no restrictions on the first resolution).

A resolution proof is called *X resolution proof* if every resolution in the proof is an X resolution, where X is "unit," "input," "ordered," "positive," "negative," "set-of-support," or "linear."

Example 3.3.5 Let $C = \{1. (p \mid q), 2. (\neg p \mid r), 3. (p \mid \neg q \mid r), 4. (\neg r)\}$. In the following proofs, the input clauses are omitted and the parents of each resolvent are shown as a pair of numbers following the resolvent.

(a)		(b)		(c)	
5. $(\neg p)$	$(2, 4)$	5. $(p \mid r)$ $(1, 3)$		5. $(q \mid r)$	$(1, 2)$
6. $(p \mid \neg q)$	$(3, 4)$	6. (r)	$(2, 5)$	6. $(\neg q \mid r)$	$(2, 3)$
7. $(\neg q)$	$(5, 6)$	7. $()$	$(4, 6)$	7. (r)	$(5, 6)$
8. (q)	$(1, 5)$			8. $()$	$(4, 7)$
9. $()$	$(7, 8)$				

Note that (a) is a unit and negative resolution proof; (b) is an input, positive and linear resolution proof; (c) is an ordered resolution proof, assuming $p > q > r$.

The length of proof (a) is 5; the length of (b) is 3; and the length of (c) is 4. It is easy to check that (a) is not an input resolution proof, because clause 7 is not obtained by input resolution. (b) is not a unit resolution proof. (c) is neither a unit nor an input resolution proof; (c) is neither positive nor negative resolution proof.

□

3.3 Resolution

Efficiency is an important consideration in automated reasoning, and one may sometimes be willing to trade completeness for speed. In unit resolution, one of the parent clauses is always a unit clause; in input resolution, one of the parent clauses is always selected from the input set. Albeit efficient, neither strategy is complete. For example, $C = \{(p \mid q), (p \mid \neg q), (\neg p \mid q), (\neg p \mid \neg q)\}$ is unsatisfiable, though no unit resolutions are available. We may use input resolutions to obtain p and $\neg p$ but cannot obtain ( ) because p and $\neg p$ are not input clauses.

Ordered resolution imposes a total ordering on the propositional variables and treats each clause not as a set of literals but a sorted list of literals. Ordered resolution is efficient and complete.

Set-of-support resolution is one of the powerful strategies employed by Wos, Carson, and Robinson in 1965, and its completeness depends on the choice of set of support. To prove $A \models B$ by resolution, we convert $A \wedge \neg B$ into clauses. The clauses obtained from $\neg B$ serve in general as a set of support S and the clauses derived from A are in T. Set-of-support resolution dictates that the parents of any resolved are not both from T. The motivation behind set-of-support is that since A is usually satisfiable it might be wise not to resolve two clauses originally from A.

It must be noted that some strategies improve certain aspects of the deduction process at the expense of others. For instance, a strategy may reduce the size of the proof search space at the expense of increasing the length of the shortest proofs.

3.3.3 Preserving Satisfiability

The resolution rule does not preserve the logical equivalence, that is, if D is a resolvent of C_1 and C_2, then $C_1 \wedge C_2 \models D$, or $C_1 \wedge C_2 \equiv C_1 \wedge C_2 \wedge D$, but not $C_1 \wedge C_2 \equiv D$.

Example 3.3.6 Let q be the resolvent of p and $\neg p \mid q$. $\mathcal{M}(q) = \{pq, \overline{p}q\}$ has two models, where $\overline{p}$ denotes $\neg p$. However, $\mathcal{M}(p \wedge (\neg p \mid q))$ has only one model, i.e., pq. Thus, $q \not\equiv (p \wedge (\neg p \mid q))$. However, both formulas are satisfiable. □

Definition 3.3.7 Given two formulas A and B, A and B are said to be *equally satisfiable* or *equisatisfiable*, if whenever A is satisfiable, so is B, and vice versa. We denote this relation by $A \approx B$.

Obviously, $A \approx B$ means $\mathcal{M}(A) = \emptyset$ iff $\mathcal{M}(B) = \emptyset$. $A \approx B$ is weaker than the logical equivalence $A \equiv B$, which requires $\mathcal{M}(A) = \mathcal{M}(B)$.

Proposition 3.3.8 *Let* $S = C \cup \{(\alpha \mid \beta)\}$ *and* $S' = C \cup \{(\alpha \mid x), (\neg x \mid \beta)\}$ *be two sets of clauses, where x is a variable not appearing in S. Then $S \approx S'$.*

Proof If S is satisfiable and σ is a model of S, then $\sigma(C) = 1$ and $\sigma(\alpha \mid \beta) = 1$. If $\sigma(\alpha) = 0$, then $\sigma(\beta) = 1$. We define $\sigma(x) = 1$ if $\sigma(\alpha) = 0$; otherwise $\sigma(x) = 0$. Thus, both $\sigma(\alpha \mid x) = 1$ and $\sigma(\neg x \mid \beta) = 1$. So σ is a model of S'.

On the other hand, if σ is a model of S', then σ is also a model of S without modification, because $(\alpha \mid \beta)$ is a resolvent of $(\alpha \mid x)$ and $(\neg x \mid \beta)$. □

In the above proposition, the condition that x does not appear in C is necessary. Without it, for example, if $C = \{(\neg p)\}$, $\alpha = p$, $\beta = q$ and $x = p$, then $S = \{(\neg p), (q)\}$ and $S' = \{(\neg p), (p), (\neg p \mid q)\}$. Then S is satisfiable but S' is not.

Using the above proposition, we may break a long clause into a set of shorter clauses by introducing some new variables. For example, the clause $(l_1 \mid l_2 \mid l_3 \mid l_4 \mid l_5)$ can be transformed into three clauses: $(l_1 \mid l_2 \mid x)$, $(\neg x \mid l_3 \mid y)$, and $(\neg y \mid l_4 \mid l_5)$. This transformation does not preserve the logical equivalence, but preserves the satisfiability.

Another usage of the above proposition is to use resolution to remove x from a clause set. If we exhaust all resolutions on x, keep all the resolvents and remove all the clauses containing x, then the resulting set of clauses will preserve the satisfiability, as we will see in Lemma 3.3.12.

Proposition 3.3.9 *Let B be a subformula of A and x is a new variable. Then $A \approx (A[B \leftarrow x] \wedge (x \leftrightarrow B))$.*

Proof If A is satisfiable and σ is a model of A, then we define $\sigma(x) = \sigma(B)$. When applying σ to the formula trees of A and $A[B \leftarrow x]$, all the common nodes have the same truth value, so $\sigma(A) = \sigma(A[B \leftarrow x])$. Since $\sigma(x) = \sigma(B)$, so $\sigma(x \leftrightarrow B) = 1$, thus σ is a model of $A[B \leftarrow x] \wedge (x \leftrightarrow B)$.

On the other hand, if σ is a model of $A[B \leftarrow x] \wedge (x \leftrightarrow B)$, σ is also a model of A. □

The above proposition plays an important role when we convert a formula into CNF and keep the size of CNF as a linear function of the size of the input formula. When we use the distributive law

$$(A \wedge B) \vee C \equiv (A \vee C) \wedge (B \vee C)$$

to transform a formula into CNF, C becomes duplicated in the result. Because of this, there exist examples where the resulting CNF has an exponential size of the input size. Using the above proposition and introducing new variables, we may keep the size of CNF as four times of the input size, and every clause contains at most three literals. The CNF obtained this way is not logically equivalent to the input formula, but it preserves the satisfiability of the input formula: The CNF is satisfiable iff the input formula is satisfiable. This way of transforming a formula into CNF is called *Tseitin transformation*.

From now on, we will use $\overline{p}$ for $\neg p$ if a compact form is preferred.

Example 3.3.10 Let A be $(p \wedge (q \vee \neg p)) \vee \neg(\neg q \wedge (p \vee q))$. We search the formula tree of A bottom-up to replace each subformula whose topmost symbol is a binary operator by a new variable and introduce an $\leftrightarrow$-formula for each new variable. We then convert each $\leftrightarrow$-formula into at most three clauses and each clause contains at most three literals.

3.3 Resolution

$$x_1 \leftrightarrow (q \vee \neg p) \qquad (\overline{x_1} \mid q \mid \overline{p}), (\overline{q} \mid x_1), (p \mid x_1),$$
$$x_2 \leftrightarrow (p \wedge x_1) \qquad (\overline{x_2} \mid p), (\overline{x_2} \mid x_1), (\overline{p} \mid \overline{x_1} \mid x_2),$$
$$x_3 \leftrightarrow (p \vee q) \qquad (\overline{x_3} \mid p \mid q), (\overline{p} \mid x_3), (\overline{q} \mid x_3),$$
$$x_4 \leftrightarrow (\neg q \wedge x_3) \qquad (\overline{x_4} \mid \overline{q}), (\overline{x_4} \mid q), (q \mid \overline{x_3} \mid x_4),$$
$$x_2 \vee \neg x_4 \qquad (x_2 \mid \overline{x_4}).$$

The input size of A is 14 (excluding commas and parentheses). The total number of literals of the resulting clause set is 30. □

Example 3.3.11 In Example 2.3.9, we have converted $p_1 \leftrightarrow (p_2 \leftrightarrow p_3)$ into four clauses:

$$(\overline{p_1} \mid \overline{p_2} \mid p_3), \quad (\overline{p_1} \mid p_2 \mid \overline{p_3}), \quad (p_1 \mid p_2 \mid p_3), \quad (p_1 \mid \overline{p_2} \mid \overline{p_3})$$

If we like to convert $p_1 \leftrightarrow (p_2 \leftrightarrow (p_3 \leftrightarrow p_4))$ into clauses, we may replace p_3 by $(p_3 \leftrightarrow p_4)$ in the above clauses and then convert each of them into clauses. That is, from $(\overline{p_1} \mid \overline{p_2} \mid (p_3 \leftrightarrow p_4))$, $(\overline{p_1} \mid p_2 \mid (\overline{p_3 \leftrightarrow p_4}))$, $(p_1 \mid p_2 \mid (p_3 \leftrightarrow p_4))$, and $(p_1 \mid \overline{p_2} \mid \overline{p_3 \leftrightarrow p_4})$, we obtain eight clauses, twice as before:

$$(\overline{p_1} \mid \overline{p_2} \mid \overline{p_3} \mid p_4), \quad (\overline{p_1} \mid \overline{p_2} \mid p_3 \mid p_4), \quad (\overline{p_1} \mid \overline{p_2} \mid p_3 \mid \overline{p_4}), \quad (\overline{p_1} \mid p_2 \mid \overline{p_3} \mid \overline{p_4}),$$
$$(p_1 \mid p_2 \mid \overline{p_3} \mid p_4), \quad (p_1 \mid \overline{p_2} \mid p_3 \mid p_4), \quad (p_1 \mid p_2 \mid p_3 \mid \overline{p_4}), \quad (p_1 \mid \overline{p_2} \mid \overline{p_3} \mid \overline{p_4})$$

If we replace p_4 by $(p_4 \leftrightarrow p_5)$, we will get a set of 16 clauses of length 5. In general, $p_1 \leftrightarrow (p_2 \leftrightarrow (p_3 \leftrightarrow (\ldots (p_{n-1} \leftrightarrow p_n) \ldots)))$ will produce 2^{n-1} clauses of length n. Instead, if we introduce $n - 3$ new variables for this formula of n variables, we obtain only $4(n - 1)$ clauses of length 3. For example, for $n = 5$, we introduce two new variables, q_1 and q_2:

$$(p_1 \leftrightarrow (p_2 \leftrightarrow q_1)) \wedge (q_1 \leftrightarrow (p_3 \leftrightarrow q_2)) \wedge (q_2 \leftrightarrow (p_4 \leftrightarrow q_5)),$$

which will generate $3 \times 4 = 12$ clauses of length 3. □

Lemma 3.3.12 *Let C be a set of clauses and x be a variable in C. Let $D \subset C$ such that each clause of D contains neither x nor $\overline{x}$. Let $X = \{(\alpha \mid \beta) : (x \mid \alpha), (\overline{x} \mid \beta) \in C\}$, then $C \approx D \cup X$.*

Proof If C is satisfiable and σ is a model of C, then $\sigma(C) = 1$ and $\sigma(D) = 1$. If $\sigma(X) = 0$, then there exists a clause $(\alpha \mid \beta) \in X$ such that $\sigma(\alpha) = \sigma(\beta) = 0$. It means either $\sigma(x \mid \alpha) = 0$ or $\sigma(\overline{x} \mid \beta) = 0$, that is a contradiction to the fact that $\sigma(C) = 1$, because both $(x \mid \alpha)$ and $(\overline{x} \mid \beta)$ are in C. So $\sigma(X) = 1$. It means σ is a model of $D \cup X$.

On the other hand, if $D \cup X$ is satisfiable and σ is its model, we may define $\sigma(x) = 1$ if $\sigma(\alpha) = 0$ for some $x \mid \alpha \in C$; in this case, we must have $\sigma(\beta) = 1$ for all $(\overline{x} \mid \beta) \in C$; otherwise, if $\sigma(\beta) = 0$ for some $(\overline{x} \mid \beta) \in C$, then $\sigma(\alpha \mid \beta) = 0$, that is a contradiction to the fact that $\sigma(X) = 1$. Similarly, if $\sigma(\alpha) = 1$ for all $x \mid \alpha \in C$, we define $\sigma(x) = 0$. Then σ becomes a model of C. □

3.3.4 Completeness of Resolution

Lemma 3.3.12 plays an important role in the completeness proof of resolution: The set X can be obtained from C by resolution. From C to $D \cup X$, we have eliminated one variable from C. Replacing C by $D \cup X$, we repeat the same process and remove another variable, until no more variable in C: either $C = \emptyset$ or $C = \{\bot\}$. From the final result, we can tell if the set of input clauses is satisfiable or not.

The following proof is based on the above idea for the completeness of ordered resolution.

Theorem 3.3.13 *Ordered resolution is complete with any total ordering on the variables.*

Proof Suppose there are n variables in the input clauses S_n with the order $x_n > x_{n-1} > \cdots > x_2 > x_1$. For i from n to 1, we apply Lemma 3.3.12 with $C = S_i$ and $x = x_i$, and obtain $S_{i-1} = D \cup X$. Every resolvent in X is obtained by ordered resolution on x_i, which is the maximal variable in S_i. Finally, we obtain S_0 which contains no variables. Since $S_n \approx S_{n-1} \approx \cdots \approx S_1 \approx S_0$, if S_n is unsatisfiable, then $S_0 = \{\bot\}$, where $\bot$ denotes the empty clause; if S_n is satisfiable, we arrive at $S_0 = \emptyset$, i.e., $S_0 \equiv \top$. In this case, we may use the idea in the proof of Lemma 3.3.12 to assign a truth value for each variable, from x_1 to x_n, to obtain a model of S_n. □

Example 3.3.14 Let $S = \{c_1 : (\overline{a} \mid \overline{b} \mid \overline{c}), c_2 : (a \mid \overline{e}), c_3 : (b \mid \overline{c}), c_4 : (\overline{b} \mid d), c_5 : (c \mid \overline{d} \mid \overline{e}), c_6 : (\overline{d} \mid e) \}$ and $a > b > c > d > e$. By ordered resolution, we get $c_7 : (\overline{b} \mid \overline{c} \mid \overline{e})$ from c_1 and c_2; $c_8 : (\overline{c} \mid d)$ from c_3 and c_4; $c_9 : (\overline{c} \mid \overline{e})$ from c_3 and c_7; and $c_{10} : (\overline{d} \mid \overline{e})$ from c_5 and c_9. No other clauses can be generated by ordered resolution, that is, c_1–c_{10} are saturated by ordered resolution. Since the empty clause is not generated, we may construct a model from c_1–c_{10} by assigning truth values to the variables from the least to the greatest. Let S_x be the set of clauses from c_1–c_{10} such that x appears in each clause of S_x as the maximal variable.

- $S_e = \emptyset$ and we can assign either 0 or 1 to e, say $\sigma_e = \{e\}$.
- $S_d = \{c_6, c_{10}\}$. From c_{10}, $\overline{d}$ has to be 1, so $\sigma_d = \sigma_e \cup \{\overline{d}\} = \{e, \overline{d}\}$.
- $S_c = \{c_5, c_8, c_9\}$. From c_8 or c_9, $\overline{c}$ has to be 1, so $\sigma_c = \sigma_d \cup \{\overline{c}\} = \{e, \overline{d}, \overline{c}\}$.
- $S_b = \{c_3, c_4, c_7\}$. Both c_3 and c_7 are satisfied by σ_c. From c_4, we have to assign 0 to b, so $\sigma_b = \sigma_c \cup \{\overline{b}\} = \{e, \overline{d}, \overline{c}, \overline{b}\}$.
- $S_a = \{c_1, c_2\}$. From c_2, a has to be 1, so $\sigma_a = \sigma_b \cup \{a\} = \{e, \overline{d}, \overline{c}, \overline{b}, a\}$.

It is easy to check that σ_a is a model of S. Hence, S is satisfiable. □

The above example illustrates that if a set of clauses is saturated by ordered resolution and the empty clause is not there, we may construct a model from these clauses. In other words, we may use ordered resolution to construct a decision procedure for deciding if a formula A is satisfiable or not.

Algorithm 3.3.15 The algorithm *orderedResolution* will take a set C of clauses and a list V of variables, $V = (p_1 < p_2 < \ldots < p_n)$. It returns true iff C is satisfiable.

3.3 Resolution

It uses the procedure $sort(A)$, which places the maximal literal of clause A as the first one in A.

proc $orderedResolution(C, V)$
    $C := \{sort(A) \mid A \in C\}$
    for $i := n$ **downto** 1 **do**
        $X := \emptyset$
        for $(p_i \mid \alpha) \in C$ **do** // p_i is maximal in $(p_i \mid \alpha)$
            for $(\overline{p_i} \mid \beta) \in C$ **do** // p_i is maximal in $(\overline{p_i} \mid \beta)$
                $A := (\alpha \mid \beta)$
                if $A = ()$ **return** false
                $X := X \cup \{sort(A)\}$
        $C := C \cup X$
    return true

Before returning true at the last line, this algorithm can call *createModel* to construct a model of C when C is *saturated* by ordered resolution. That is, C contains the results of every possible ordered resolution among the clauses of C.

proc $createModel(C)$
    $\sigma := \emptyset$ // the empty interpretation
    for $i := 1$ **to** n **do**
        $v := 0$ // default value for p_i
        for $(p_i \mid \alpha) \in C$ **do** // p_i is maximal
            if $eval(\alpha, \sigma) = 0$ **do**
                $v := 1$ // p_i has to take 1
        $\sigma := \sigma \cup \{p_i \mapsto v\}$
    return σ

The resolution algorithm is regarded as a deduction-style proof procedure when the empty clause is its goal. If finding a model is the goal of the resolution algorithm, it is a *saturation-style* proof procedure, because the set of clauses is *saturated* by ordered resolution, that is, all the possible results of the inference rules are generated, before we can build a model.

Corollary 3.3.16 *Resolution is a sound and complete decision procedure for propositional logic.*

Proof We have seen that resolution is sound. If C is unsatisfiable, we may obtain an ordered resolution proof from C. This proof is also a resolution proof. Thus, $C \models \bot$ implies $C \vdash_\mathcal{R} \bot$. If C is satisfiable, we may construct a model for C as we do for ordered resolution. □

Theorem 3.3.17 *Positive resolution is complete for refutation.*

Proof Suppose C is unsatisfiable and there is a resolution proof $C \vdash_\mathcal{R} \bot$. We prove by induction on the length of the proof on how to convert this proof into a positive resolution proof.

Let P be a resolution proof of $C \vdash_\mathcal{R} \bot$ of minimal length. If $|P| = 1$, then the only resolution uses a positive unit clause to obtain $\bot$ and it is a positive resolution. If $|P| > 1$, suppose the last resolution in P resolves on variable q. Then $C \vdash_\mathcal{R} q$ and $C \vdash_\mathcal{R} \overline{q}$. Let P_1 and P_2 be the proofs of $C \vdash_\mathcal{R} q$ and $C \vdash_\mathcal{R} \overline{p}$ presented in P, respectively.

If we remove all the occurrences of q from P_1, we obtain a resolution proof P_1' and $core(P_1')$ contains clauses from $core(P)$ with q being removed. Since $|P_1'| < |P|$, by induction hypothesis, we have a positive resolution proof of $core(P_1') \vdash_R \bot$. If we add q back to $core(P_1')$, we obtain a positive resolution proof P_1" of $core(P) \vdash_\mathcal{R} q$.

We then remove all the occurrences of $\overline{q}$ from P_2 and obtain a resolution proof P_2' and $core(P_2')$ contains clauses from $core(P)$ with $\overline{q}$ being removed. By induction hypothesis, there exists a positive resolution proof P_2" of $core(P_2') \vdash_\mathcal{R} \bot$. For each clause $\alpha \in core(P_2')$, if α is obtained from $(\alpha \mid \overline{q}) \in core(P)$, we add α as the resolvent of $(\alpha \mid \neq q)$ and q at the beginning of P_2". Finally, we append P_1", which generates q by positive resolution, and P_2" into one proof; the result is a positive resolution proof of $C \vdash_\mathcal{R} \bot$. □

The following result can be proved using a similar approach.

Corollary 3.3.18 *Negative resolution is complete for refutation.*

The completeness of resolution strategies is important to ensure that if the empty clause is not found, the input clauses are satisfiable. In this case, we have to compute all possible permitted resolutions before claiming the input clauses are satisfiable. We say that the set of clauses, old and new, is *saturated* by a resolution strategy, if no new resolvents can be computed by the given resolution strategy. Restricted resolutions like ordered resolution, positive resolution, and set-of-support resolution work much better than unrestricted resolution, when the saturation is needed to show the satisfiability of a set of clauses.

3.3.5 A Resolution-Based Decision Procedure

If we like to employ heuristics in a decision procedure based on resolution, the following procedure fits the purpose.

Algorithm 3.3.19 The algorithm *resolution* will take a set C of clauses and returns true iff C is satisfiable. It uses the procedure *resolvable*(A, B), which decides if clauses A and B are allowed to do restricted resolution, and if yes, *resolve*(A, B) is their resolvent. It uses *pickClause*(C) to pick out a clause according to a heuristic.

proc *resolution*(C)
　　$G := C // G$: given clauses
　　$K := \emptyset // K$: kept clauses
　　while $G \neq \emptyset$ **do**

3.3 Resolution

$A := pickClause(G)$
$G := G - \{A\}$
for $B \in K$ **if** $resolvable(A, B)$ **do**
   $res := resolve(A, B)$
   if $res = ()$ **return** false
   if $res \notin (G \cup K)$
      $G := G \cup \{res\}$
$K := K \cup \{A\}$
return true

In the above algorithm, we move each clause from G to K, and compute the resolution between this clause and all the clauses in K. When G is empty, resolution between any two clauses of K is done.

With the exception of linear resolution, the above procedure can be used to implement all the resolution strategies introduced in this section. The restriction will be implemented in the procedure *resolvable*. For set-of-resolution, a better implementation is that G is initialized with the set of support and K is initialized with the rest of the input clauses.

The termination of this procedure is guaranteed because only a finite number of clauses exist. When the procedure returns false, the answer is correct because resolution is sound. When it returns true, the correctness of the answer depends on the completeness of the resolution strategy implemented in *resolvable*.

In *pickClause*, we may implement various heuristics for selecting a clause in G to do resolutions with clauses in K. Popular heuristics include preferring shorter clauses or older clauses. Preferring shorter clauses because shorter clauses are stronger as constraints on the acceptance of interpretations as models. An empty clause prohibits all interpretations as models; a unit clause prohibits half of the interpretations as models; a binary clause prohibits a quarter of the interpretations as models; and so on. Preferring older clauses alone would give us the breadth-first strategy. If we wish to find a shorter proof, we may mix this preference with other heuristics.

Despite the result that the algorithm *resolution* with a complete strategy can serve as a decision procedure for propositional logic, and some strategies can even build a model when the input clauses are satisfiable, resolution provers are in general not efficient for propositional logic, because the number of possible clauses is exponential in terms of the number of variables. The computer may quickly run out of memory before finding a solution for real application problems.

3.3.6 Simplification Strategies

There are several strategies which can safely remove redundant clauses inside the *resolution* algorithm. They are *tautology deletion, subsumption deletion*, and *pure literal deletion*.

A clause is tautology if it contains a complementary pair of literals, say, $(p \mid \overline{p})$. A tautology clause is true in every interpretation and is not needed in any resolution proof.

Proposition 3.3.20 *Given any set S of clauses, let S be partitioned into $S' \cup T$, where T is the set of tautology clauses in S. Then $S \equiv S'$.*

Proof For any interpretation σ, $\sigma(S) = \sigma(S') \wedge \sigma(T) = \sigma(S')$ as $\sigma(T) = 1$. □

Definition 3.3.21 A clause C *subsumes* another clause B if $B = (C \mid \alpha)$ for some literals α. That is, every literal of B appears in C.

For example, $(p \mid q)$ subsumes $(p \mid q \mid r)$. Intuitively, resolution tries to remove literals by resolution one by one to obtain the empty clause. If we use $(p \mid q)$ instead of $(p \mid q \mid r)$, we avoid the job of removing r.

Proposition 3.3.22 *If C subsumes B, then $S \cup \{C, B\} \equiv S \cup \{C\}$.*

Proof Since C subsumes B, let B be $(C \mid \alpha)$, $\mathcal{M}(C) \subseteq \mathcal{M}(C) \cup \mathcal{M}(\alpha) = \mathcal{M}(B)$. $\mathcal{M}(S \cup \{C, B\}) = \mathcal{M}(S) \cap (\mathcal{M}(C) \cap \mathcal{M}(B)) = \mathcal{M}(S) \cap \mathcal{M}(C) = \mathcal{M}(S \cup \{C\})$. □

The above proposition shows that if C subsumes B, then B can be dropped as a constraint in the presence of C. The following proposition shows that B is not needed in a resolution proof.

Proposition 3.3.23 *If C and B are two clauses appearing in the resolution proof of $S \vdash_\mathcal{R} \bot$ such that C subsumes B, then B is no longer needed once C is present.*

Proof If B is used in a resolution proof after the presence of C, we may replace B by C to obtain a proof of shorter or equal length, by checking all the resolutions which remove every literal from B: Suppose the first resolution is between B and γ and the resolvent is B'. There are two cases to consider: (1) The resolved variable is not in C. This resolution is unnecessary, because C subsumes B', and we replace B' by C. (2) the resolved variable is in C, then let the resolvent of C and γ be C'. It is easy to check that C' subsumes B', and we replace B' by C'. We replace the descendants of B by the corresponding descendants of C, until the last resolution. The modified resolution proof will have shorter or equal length. □

Proposition 3.3.24 *If there exists a resolution proof using a tautology clause, then there exists a resolution proof without the tautology clause.*

Proof Suppose the tautology clause appearing in the proof is $(p \mid \overline{p} \mid \alpha)$. We need another clause $(\overline{p} \mid \beta)$ (or $(p \mid \beta)$) to resolve off p. The resolvent of $(p \mid \overline{p} \mid \alpha)$ and $(\overline{p} \mid \beta)$ is $(\overline{p} \mid \alpha \mid \beta)$, which is subsumed by $(\overline{p} \mid \beta)$. By Proposition 3.3.23, $(\overline{p} \mid \alpha \mid \beta)$ is not needed in the resolution proof. Since every resolvent of a tautology is not needed, we do not need any tautology in a resolution proof. □

Definition 3.3.25 A literal l is said to be *pure* in a set C of clauses if its complement does not appear in C.

3.3 Resolution

Proposition 3.3.26 *If l is pure in C, let D be the set of clauses obtained from C by removing all clauses containing l from C, then $C \approx D$.*

Proof If σ is a model of C, then σ is also a model of D. If σ is a model of D, define $\sigma(l) = 1$, then σ is a model of C. □

For example, if $C = \{c_1 : (p \mid q),\ c_2 : (\overline{p} \mid r),\ c_3 : (p \mid \overline{q} \mid r),\ c_4 : (\overline{p} \mid q)\}$, then r is pure. Removing c_2 and c_3 from C, q becomes pure. Removing c_1 and c_4, we obtain an empty set of clauses, which is equivalent to $\top$. By the proposition, it means C is satisfiable. A model can be obtained by setting $q = r = 1$ in the model.

The above propositions allow to use the following clause deletion strategies without worrying about missing a proof:

- **Tautology deletion**: Tautology clauses are discarded.
- **Subsumption deletion**: Subsumed clauses are discarded.
- **Pure deletion**: Clauses containing pure literals are discarded.

These deletion strategies can be integrated into the algorithm *resolution* as follows.

Algorithm 3.3.27 The algorithm *resolution* will take a set S of clauses and return true iff S is satisfiable. In *preprocessing(S)*, we may implement tautology deletion, pure-literal check, or subsumption to simplify the input clauses as a preprocessing step. *subsumedBy(C, S)* checks if clause C is subsumed by a clause in S.

proc *resolution(S)*
1 $G := $ *preprocessing(S)* // G: given clauses
2 $K := \emptyset$ // K: kept clauses
3 **while** $G \neq \emptyset$ **do**
4      $C := $ *pickClause(G)*
5      $G := G - \{C\}$
6      $N := \{\}$ // new clauses from C and K
7   **for** $B \in K$ **if** *resolvable(C, B)* **do**
8          $res := $ *resolve(C, B)*
9   **if** $res = (\,)$ **return** false
10   **if** res is tautology **continue**
11   **if** *subsumedBy(res, $G \cup K$)* **continue**
12         $N := N \cup \{res\}$
13     $K := K \cup \{C\}$
14   **for** $C \in G$ **if** *subsumedBy(C, N)* **do**
15         $G := G - \{C\}$
16   **for** $B \in K$ **if** *subsumedBy(B, N)* **do**
17         $K := K - \{B\}$
18     $G := G \cup N$
19 **return** true // K is saturated by resolution.

A life cycle of clause C goes as follows in the algorithm *resolution*:

1. If C is an input clause, it passes the preprocessing check, such as pure-literal check and subsumption check, and goes into G (line 1).
2. If C is picked out of G (lines 4, 5), it will pair with every clause B in K to try resolution on C and B. Once this is done, C will go to K as a kept clause (line 13).
3. C may be subsumed by a new clause when it is in G (line 14) or in K (line 16) and will be thrown away.
4. If C is a new clause (line 8), it has to pass the subsumption check (line 11) and parks in N (line 12).
5. All new clauses in N will be used to check if they can subsume some clauses in G (lines 14, 16), then merge into G (line 18), and start a life cycle the same way as an input clause.
6. If the algorithm stops without finding the empty clause (line 9), the set G will be empty (line 3), and all the resolutions among the clauses in K are done. That is, K is saturated by resolution.

In this algorithm, unit clauses are given high priority to do unit resolution and subsumptions. Dealing with unit clauses can be separated from the main loop of the algorithm as we will see in the next section.

3.4 Boolean Constraint Propagation (BCP)

A unit clause (A) subsumes every clause $(A \mid \alpha)$, where A appears as a literal. Unit resolution between (A) and $(\overline{A} \mid \beta)$, where $\overline{A} = p$ if A is $\overline{p}$, generates the resolvent (β), which will subsume $(\overline{A} \mid \beta)$. The rule for replacing $(\overline{A} \mid \beta)$ by (β) in the presence of (A) is called *unit deletion*. That is, when subsumption deletion is used in resolution, a unit clause (A) allows us to remove all the occurrences of A (with the clauses containing A) and $\overline{A}$ (only $\overline{A}$ in the clauses containing $\overline{A}$), with the unit clause itself as the only exception. New unit clauses may be generated by unit resolution, and we can continue to simplify the clauses by the new unit clause and so on. This process is traditionally called *Boolean constraint propagation* (BCP), or *unit propagation*.

3.4.1 BCP: A Simplification Procedure

Clause deletion strategies allow us to remove unnecessary clauses during resolution. BCP also allows us to simplify clauses by removing unnecessary clauses and shortening clauses. In the following, we will describe BCP as an algorithm which takes a set C of clauses as input and returns a pair (U, S), where U is a set of unit clauses and S is a set of non-unit clauses such that U and S share no variables and

3.4 Boolean Constraint Propagation (BCP)

$C \equiv U \cup S$. Typically, S contains a subset of C or clauses from C by removing some literals.

To facilitate the discussion, we assume that no clauses contain duplicated literals (so a unit clause cannot be hidden as $(p \mid p)$) and no clauses are tautology. We also assume that a set of unit clauses does not contain complementary literals (so the empty clause cannot be hidden as (p) and $(\neg p)$).

Algorithm 3.4.1 The algorithm *BCP* will take a set C of clauses and apply unit resolution and subsumption deletion repeatedly, until no more new clauses can be generated by unit resolution. It will return $\perp$ if the empty clause is found; otherwise, it returns a simplified set of clauses equivalent to C.

proc $BCP(C)$
    $S := C$ // S: simplified clauses
    $U := \emptyset$ // U: unit clauses
    while S has a unit clause (A) **do**
        $S := S - \{(A)\}$
        $U := U \cup \{A\}$
        for $(A \mid \alpha) \in S$ **do**
            $S := S - \{(A \mid \alpha)\}$
        for $(\overline{A} \mid \alpha) \in S$ **do**
            if $\alpha = ()$ **return** $\perp$
            $S := S - \{(\overline{A} \mid \alpha)\} \cup \{(\alpha)\}$
    return (U, S)

Example 3.4.2 Let $C = \{c_1 : (x_2 \mid x_5), c_2 : (\overline{x_1} \mid \overline{x_4}), c_3 : (\overline{x_2} \mid x_4), c_4 : (x_1 \mid x_2 \mid \overline{x_3}), c_5 : (\overline{x_5})\}$. Then $BCP(C)$ will return

$$(\{\overline{x_5}, x_2, x_4, \overline{x_1}\}, \emptyset)$$

because we first add $\overline{x_5}$ into U; c_1 becomes (x_2), and x_2 is then added into U; c_4 is deleted because $x_2 \in U$; c_3 becomes (x_4) and x_4 is added into U; c_2 becomes $(\overline{x_1})$ and $\overline{x_1}$ is added into U. The set U allows us to construct two models of C:

$$\{x_1 \mapsto 0, x_2 \mapsto 1, x_3 \mapsto v, x_4 \mapsto 1, x_5 \mapsto 0\}$$

where $v \in \{0, 1\}$ as x_3 can take either value.   □

Proposition 3.4.3

(a) If $BCP(C)$ returns $\perp$, then C is unsatisfiable.
(b) If $BCP(C)$ returns (U, S), then $C \equiv U \cup S$, where U is a set of unit clauses and S is a set of non-unit clauses, and U and S share no variables.

Proof In $BCP(C)$, S is initialized with C. S is updated through three ways: (1) unit clause (A) is removed from S into U; (2) clause $(A \mid \alpha)$ is removed from S; and (3) clause $(\overline{A} \mid \alpha)$ is replaced by (α). (α) is the resolvent of (A) and $(\overline{A} \mid \alpha)$, so $C \models \alpha$.

(a) If $\alpha = (\ )$ in (3), $\perp$ will be returned and $C \equiv \perp$ because $C \models \perp$.
(b) Initially, $C \equiv U \cup S$. Since A subsumes $(A \mid \alpha)$, and α subsumes $(\overline{A} \mid \alpha)$, by Proposition 3.3.22, it justifies that both $(A \mid \alpha)$ and $(\overline{A} \mid \alpha)$ can be safely removed, as it maintains $C \equiv U \cup S$ for each update of U and S. When (A) is moved from S to U, both A or $\overline{A}$ are removed from S, so the variable of A does not appear in S anymore. The procedure stops only when S does not have any unit clause. □

The procedure *BCP* can serve as a powerful simplification procedure when unit clauses are present in the input. *BCP* plays an important role in deciding if a formula is satisfiable, which is the topic of the next chapter. We will show later that *BCP(C)* can be implemented in time $O(n)$, where n is the number of literals in C.

3.4.2 A Decision Procedure for Horn Clauses

A *Horn clause* is a clause with at most one positive literal. Horn clauses are named after Alfred Horn, who first pointed out their significance in 1951. A Horn clause is called *definite clause* if it has exactly one positive literal. Thus, Horn clauses can be divided into definite clauses and negative clauses. Definite clauses can be further divided into *fact* (a positive unit clause) and *rule* (non-unit definite clauses). Horn clauses have important applications in logic programming, formal specification, and model theory, as they have very efficient decision procedures, using unit resolution or input resolution. That is, unit and input resolutions are incomplete in general, but they are complete for Horn clauses.

Theorem 3.4.4 BCP *is a decision procedure for the satisfiability of Horn clauses.*

Proof Suppose H is a set of Horn clauses and $\perp \notin H$. It suffices to show that procedure $BCP(H)$ returns $\perp$ iff $H \equiv \perp$. If $BCP(H)$ returns $\perp$, by Proposition 3.4.3, $H \equiv \perp$. If $BCP(H)$ returns (U, S), by Proposition 3.4.3, $H \equiv U \cup S$, U is a set of unit clauses, S is a set of non-unit clauses, and U and S share no variables. In this case, we create an assignment σ in which every literal in U is true, and every variable in S is false. Note that each variable in U appears only once; U and S do not share any variable. Thus, this assignment is consistent and is a model of $U \cup S$, because every unit clause in U is true under σ; for every clause in S, since it is a non-unit Horn clause and must have a negative literal, which is true under σ. Thus, H is satisfiable when $BCP(H)$ returns (U, S). □

Proposition 3.4.5 *Unit resolution is sound and complete for Horn clauses.*

Proof Unit resolution is sound because resolution is sound. The previous theorem shows that *BCP* is a decision procedure for the satisfiability of Horn clauses. The only used inference rule in *BCP* is unit resolution. Every clause generated by *BCP* can be generated by unit resolution, and subsumption only reduces some unnecessary clauses and does not affect the completeness. □

3.4.3 Unit Resolution Versus Input Resolution

In the following, we show that input resolution and unit resolution are equivalent, in the sense that if there exists a unit resolution proof, then there exists an input resolution proof and vice versa.

Theorem 3.4.6 *For any set C of clauses, if there exists an input resolution proof from C, then there exists a unit resolution proof from C.*

Proof We will prove this theorem by induction on the length of proofs. Let P be an input resolution proof of minimal length from C. The last resolvent is $\bot$, both its parents must be unit clauses, and one of them must be an input clause, say (A). For simplicity, assume all the input clauses used in P appear in the beginning of P and (A) is the first in P. If $|P| = 1$, then P must be a unit resolution proof and we are done. If $|P| > 1$, then remove (A), the last clause in P (i.e., $\bot$), and all occurrences of $\overline{A}$ from P, where $\overline{A} = \overline{p}$ if $A = p$ and $\overline{A} = p$ if $A = \overline{p}$. The result is an input resolution proof P' from C', where C' is obtained from C by removing all occurrences of $\overline{A}$ in C. That is, $(\overline{A} \mid \alpha) \in C$ iff $(\alpha) \in C'$. Because $|P'| < |P|$, by induction hypothesis, there exists a unit resolution proof Q' from C'. If $(\alpha) \in C'$ comes from $(\overline{A} \mid \alpha) \in C$, then change the reason of $(\alpha) \in Q'$ from "axiom" to "resolution from (A) and $(\overline{A} \mid \alpha)$". Let the modified Q' be Q, then Q is a unit resolution from C. □

Example 3.4.7 We illustrate the proof of this theorem by Example 3.3.5, where

$$C = \{c_1 : (p \mid q),\ c_2 : (\overline{p} \mid r),\ c_3 : (p \mid \overline{q} \mid r),\ c_4 : (\overline{r})\}$$

Let P be the input resolution proof (b) of Example 3.3.5. Then $A = \overline{r}$, and C' is the result of removing r from C:

$$C' = \{c_1 : (p \mid q),\ c'_2 : (\overline{p}),\ c'_3 : (p \mid \overline{q})\}$$

and two input resolutions in C' are $c_5 : (p)$ from (c_1, c'_3) and $c_6 : (\ )$ from (c'_2, c_5). A unit resolution proof Q' from C' will contain three unit resolutions:

$$c_5 : (q) \text{ from } (c_1, c'_2),\quad c_6 : (\overline{q}) \text{ from } (c'_2, c'_3),\quad c_7 : (\ ) \text{ from } (c_5, c_6)$$

Adding two unit resolutions before Q', we obtain Q:

$$c'_2 : (\overline{p}) \text{ from } (c_2, c_4),\ c'_3 : (p \mid \overline{q}) \text{ from } (c_3, c_4),\ c_5 : (q) \text{ from } (c_1, c'_2),$$
$$c_6 : (\overline{q}) \text{ from } (c'_2, c'_3),\ c_7 : (\ ) \text{ from } (c_5, c_6),$$

which is a unit resolution proof from C. □

Theorem 3.4.8 *For any set C of clauses, if there exists a unit resolution proof from C, then there exists an input resolution proof from C.*

The proof of this theorem is left as an exercise.

The above two theorems establish the equivalence of input resolution and unit resolution: A unit resolution proof exists iff an input resolution proof exists. Since unit resolution is complete for Horn clauses, we have the following result.

Corollary 3.4.9 *Input resolution is sound and complete for Horn clauses.*

Theorem 3.4.10 *For any set H of Horn clauses, let $H = X \cup Y$, where X is the set of all negative clauses and $Y = H - X$. Then input and set-of-support resolution is sound and complete for the satisfiability of H, when X is used as the set of support.*

Proof Resolution is not possible among the clauses in X as resolution needs a pair of complementary literals from the two parent clauses. Set-of-support resolution does not allow resolution among the clauses in Y. Thus, resolution is possible only when one clause is from X and one from Y, and the resolvent is a negative clause, because the only positive literal from Y is resolved off. For each resolution between X and Y, we add the resolvent into X, until either the empty clause $\bot$ is found or no new resolvents can be generated. If $\bot$ is generated, then $H \equiv \bot$ because resolution is sound.

If the empty clause is not generated and X is saturated (i.e., no new resolvents can be generated), we create an assignment σ as follows: Call *BCP* on Y. Since Y is satisfiable, let $(U, S) = BCP(Y)$. For any unit clause p in U, let $\sigma(p) = 1$; for any other unassigned variable q in S, let $\sigma(q) = 0$.

We claim that σ is a model of H. At first, $\sigma(U \cup S) = 1$, because U and S do not share any variables, and the assignment will make every clause in $U \cup S$ true.

Let X be the set of negative clauses from H plus all the resolvents between X and Y. If $\sigma(X) = 0$, there must exist a clause $C \in X$ of minimal length, such that $\sigma(C) = 0$. Since C is a negative clause, let $C = (\overline{p_1} \mid \overline{p_2} \mid \cdots \mid \overline{p_k})$. Since $\sigma(C) = 0$, for $1 \leq j \leq k$, $\sigma(p_i) = 1$ and it must be the case that $p_i \in U$ because we only assign 1 to the variables of U.

Now consider p_1 of C. If $p_1 \in Y$, the resolvent of p_1 and C is $C' = (\overline{p_2} \mid \cdots \mid \overline{p_k})$. If $p_1 \notin Y$, then $Y \cup \{\overline{p_1}\}$ is unsatisfiable and there exists a negative resolution proof P from $Y \cup \{(\overline{p_1})\}$. Replacing $(\overline{p_1})$ by C in P, we obtain $C' = (\overline{p_2} \mid \cdots \mid \overline{p_k})$ by negative resolution from $Y \cup \{C\}$. It is easy to see that $\sigma(C') = 0$ and $C' \in X$. Since $|C'| < |C|$, this a contradiction to the assumption that C is of minimal length. Thus, $\sigma(X) = 1$, σ is a model of $H = X \cup Y$ and H is satisfiable. □

Note that input resolution with negative clauses as the set of support on Horn clauses is also a negative resolution. If the set of support initially contains a single clause, this input resolution is also a linear resolution. If the positive literal is maximal in each clause, then this resolution is also an ordered resolution.

Corollary 3.4.11 *If H is a set of Horn clauses which contains a single negative clause, then the negative, input and linear resolution is complete for the satisfiability of H.*

3.4 Boolean Constraint Propagation (BCP)

Proof If C is the only negative clause, let $X = \{C\}$ and $Y = H - X$. Using X as the set of support, we will get an input resolution proof. Since all negative clauses between X and Y are from C, the proof can be arranged as a linear resolution proof. □

The above result is the basis for the completeness of pure Prolog programs, which will be discussed in Chap. 8.

3.4.4 Head/Tail Literals for BCP

For a unit clause (A) to be true, the truth value of literal A needs to be 1. We can use a partial interpretation σ to remember which literals appear in unit clauses.

Definition 3.4.12 A *partial interpretation* σ is a set σ of literals, where no variables appear in σ more than once, with the understanding that for any literal A, if $A \in \sigma$, then $\sigma(A) = 1$ and $\sigma(\overline{A}) = 0$, where $\overline{A} = p$ if A is $\overline{p}$. When σ contains every variable exactly once, it is a *full interpretation*.

From the above definition, it is convenient to consider a set of unit clauses as a partial interpretation. For example, if $U = \{p, \neg q, r\}$, it represents the interpretation $\sigma = \{p \mapsto 1, q \mapsto 0, r \mapsto 1\}$.

We can evaluate the truth value of a clause or formula in a partial interpretation as we do in a full interpretation.

Definition 3.4.13 Given a partial interpretation σ and a clause c,

- c is *satisfied* if one or more of its literals is true in σ.
- c is *conflicting* if all the literals of c are false in σ.
- c is *unit* if all but one of its literals are false and c is not satisfied.
- c is *unsolved* if c is not one of the above three cases.

When c becomes unit, we have to assign 1 to the unassigned literal, say A, in c, so that c becomes satisfied. That is, c is the *reason* for literal A being true, and we record this as $reason(A) = c$. We also say literal A is *implied* by c.

Example 3.4.14 Given the clauses $\{c_1 : (\overline{x_1}),\ c_2 : (x_3),\ c_3 : (x_1 \mid x_2 \mid \overline{x_3})\}$, $reason(\overline{x_1}) = c_1, reason(x_3) = c_2$, and $reason(x_2) = c_3$. x_2 is implied to be true by c_3 because c_3 is equivalent to $(\overline{x_1} \wedge x_3) \rightarrow x_2$. □

In *BCP*, we start with an empty partial interpretation σ. If we have found a unit clause (A), we extend σ by adding A into σ so that the unit clause becomes satisfied. To save time, we do not remove clauses containing A from the clause set. We do not remove $\overline{A}$ from any clause to create a new clause. Instead, we just remember $\overline{A}$ is now false in σ.

We need to find out which clauses become unit clauses when some of literals become false. A clause c of k literals becomes unit when $k - 1$ literals of c are false. If we know that two literals of c are not false, we are certain that c is unsolved. Let us designate these two literals as *head* and *tail* literals of c, which are two distinct literals of c; we have the following property [2].

Proposition 3.4.15 *A clause c of length longer than one is neither unit nor conflicting in a partial interpretation σ iff either one of the head/tail literals of c is true or neither of them is false in σ.*

Proof Suppose c contains k literals and $k > 1$. If c is satisfied, then it contains a true literal and we may use it as one of the head/tail literals. If c is unit or conflicting, there are $k - 1$ or all literals of c are false, and no literal in c is true. In this case, there do not exist two literals which are not false for the head/tail literals. If c is unsolved, then there must exist two unassigned literals for the head/tail literals. □

Initially, let the first and last literals be the head/tail literals. When a head literal becomes false, we scan right to find in c the next literal which is not false and make it the new head literal; similarly, when a tail literal becomes false, we scan left to find the next tail literal. If we cannot find new head or tail literal, we must have found a unit clause or a conflicting clause.

To implement the above idea, we use the following data structures for clauses: For each clause c containing more than one literal:

- $lits(c)$: the array of literals storing the literals of c
- $head(c)$: the index of the first non-false literal of c in $lits(c)$
- $tail(c)$: the index of the last non-false literal of c in $lits(c)$

Let $|c|$ denote the number of literals in c, then the valid indices for $lits(c)$ are $\{0, 1, \ldots, |c| - 1\}$, and $head(c)$ is initialized with 0 and $tail(c)$ with $|c| - 1$.

The following are the data structures associated with each literal A:

- $val(A)$: the truth value of A under partial interpretation, $val(A) \in \{1, 0, \times\}$, $\times$ ("unassigned") is the initial value.
- $cls(A)$: list of clauses c such that A is in c and pointed by either $head(c)$ or $tail(c)$ in $lits(c)$. We use insert$(c, cls(A))$ and remove$(c, cls(A))$ to insert and remove clause c into/from $cls(A)$, respectively.

Assuming global variables are used for clauses and literals, the procedure *BCP* can be implemented as follows:

proc *BCP(C)*
    // initialization
1    $U := \emptyset$ // U: stack of unit clauses
2   **for** $c \in C$ **do**
3   **if** $c = ()$ **return** $\bot$
4   **if** $(|c| > 1)$
5            $lits(c) := $ makeArray(c) // create an array of literals
6            $head(c) := 0$; insert$(c, cls(lits(c)[0]))$
7            $tail(c) := |c| - 1$; insert$(c, cls(lits(c)[|c| - 1]))$
8   **else** push(c, U)
9   **for** each literal A, $val(A) := \times$ // $\times$ = "unassigned"

3.4 Boolean Constraint Propagation (BCP)

```
            // major work
    10    res := BCPht(U)
            // finishing
    11    if (res ≠ "SAT") return ⊥
    12    U := {A | val(A) = 1}
    13    S := {clean(c) | val(lits(c)[head(c)]) = ×, val(lits(c)[tail(c)]) = ×}
    14    return (U, S)
```

Note that the procedure $clean(c)$ at line 13 will remove from c those literals which are false (under val) and return ⊤ if one of the literals in c is true. The procedure *BCPht* will do the major work of *BCP*.

Algorithm 3.4.16 *BCPht(U)* assumes the input clauses $C = U \cup S$, where U is a stack of unit clauses and S is a set of non-unit clauses stored using the head/tail data structure, and can be accessed through $cls(A)$. *BCPht* returns a conflict clause if an empty clause is found during unit propagation; it returns "SAT" if no empty clause is found.

```
proc BCPht(U)
 1    while U ≠ ∅ do
 2          A := pop(U)
 3          if val(A) = 0 return reason(Ā) // an empty clause is found
 4          else if (val(A) = ×) // × is "unassigned"
 5              val(A) := 1; val(Ā) := 0
 6              for c ∈ cls(Ā) do // Ā is either head or tail literal of c
 7                  if (Ā = lits(c)[head(c)])
 8                      e₁ := head(c); e₂ := tail(c); step := 1 // scan from left to right
 9                  else
10                      e₁ := tail(c); e₂ := head(c); step := −1 // scan from right to left
11                  while true do
12                      x := e₁ + step // x takes values from e₁ + step to e₂
13                      if x = e₂ break // exit the inner while loop
14                      B := lits(c)[x]
15                      if (val(B) ≠ 0) // new head/tail literal is found
16                          remove(c, cls(Ā)); insert(c, cls(B))
17                          if (step = 1) head(c) := x else tail(c) := x
18                          break // exit the inner while loop
19                  if (x = e₂) // no new head/tail available
20                      A := lits(c)[e₂] // check if c is unit or conflicting
21                      if val(A) = 0 return c // c is conflicting
22                      else if (val(A) = ×)    // c is unit
23                          push(A, U); reason(A) := c; val(A) := 1; val(Ā) := 0
24    return "SAT" // no empty clauses are found
```

Example 3.4.17 For the clauses in Example 3.4.2, we have the following data structure after the initialization (the head and tail indices are given after the literal list in each clause):

$$C = \{c_1 : \langle [x_2, x_5], 0, 1 \rangle, c_2 : \langle [\overline{x_1}, \overline{x_4}], 0, 1 \rangle, c_3 : \langle [\overline{x_2}, x_4], 0, 1 \rangle,$$
$$c_4 : \langle [x_1, x_2, \overline{x_3}], 0, 2 \rangle\}$$

$U = \{\overline{x_5}\}$; $cls(A)$ contains these clauses:

$cls(x_1) = \{c_4\}, \quad cls(\overline{x_1}) = \{c_2\}$
$cls(x_2) = \{c_1\}, \quad cls(\overline{x_2}) = \{c_3\}$
$cls(x_3) = \emptyset, \quad cls(\overline{x_3}) = \{c_4\}$
$cls(x_4) = \{c_3\}, \quad cls(\overline{x_4}) = \{c_2\}$
$cls(x_5) = \{c_1\}, \quad cls(\overline{x_5}) = \emptyset$

and $val(A) = \times$ (unassigned) for all literal A. Note that each clause appears twice in the collection of $cls(A)$.

Inside of $BCPht(U)$, when $\overline{x_5}$ is popped off U, we assign $val(\overline{x_5}) = 1$, $val(x_5) = 0$ (line 5) and check $c_1 \in cls(x_5)$ (line 6). c_1 generates unit clause x_2 and is added into U (line 23) with the assignments $val(x_2) = 1$, $val(\overline{x_2}) = 0$. When x_2 is popped off U, we check $c_3 \in cls(\overline{x_2})$. c_3 generates unit clause x_4, we push x_4 into U, and assign $val(x_4) = 1$, $val(\overline{x_4}) = 0$. When x_4 is popped from U, we check $c_2 \in cls(\overline{x_4})$. c_2 generates unit clause $\overline{x_1}$; we push $\overline{x_1}$ into U and assign $val(\overline{x_1}) = 1$, $val(x_1) = 0$. When $\overline{x_1}$ is popped off U, we check $c_4 \in cls(x_1)$. c_4 is true because $val(x_2) = 1$. Now the head index of c_4 is updated to 1, and we add $c_4 = \langle [x_1, x_2, \overline{x_3}], 1, 2 \rangle$ into $cls(x_2)$. Since the empty clause is not found, $BCPht$ returns "SAT." Finally, BCP returns $(U, S) = (\{\overline{x_5}, x_2, x_4, \overline{x_1}\}, \emptyset)$. □

Note that the two assignments "$val(a) := 1$; $val(\overline{a}) := 0$" at line 23 of $BCPht$ can be removed as the same work will be done at line 5. However, having them here often improves the performance of $BCPht$. At line 23, we also record the reason for the assignment of a literal; this information is returned at line 3 and is useful if we like to know why a clause becomes conflicting. $BCPht$ can be improved slightly by checking $(val(lits(c)[e_2]) = 1)$ after line 10: If it is true, then skip clause c as c is satisfied.

In this implementation of BCP, once all the clauses C are read in, each occurrence of literals of non-unit clauses will be visited at most once in $BCPht$.

Proposition 3.4.18 $BCP(C)$ *runs in* $O(n)$ *time, where* n *is the total number of the literal occurrences in all clauses of* C.

Proof Obviously, the initialization of BCP takes $O(n)$. For $BCPht$, we access a non-unit clause c when its head (or tail) literal becomes false (line 6). We skip the head (tail) literal (line 11) and check the next literal b (line 13); if it is false, we continue (line 14); otherwise, we remove c from $cls(\overline{A})$ into $cls(B)$ (line 16) and update the head (tail) index (line 17). If we cannot find a new head (tail) literal (line 19), we have found a new unit clause (line 20). None of the literals of c is visited more than once by *BCPht*. In particular, we never access the clauses from $cls(A)$ when literal A is true. □

Combining the above proposition with Theorem 3.4.4, we have the following result:

Theorem 3.4.19 BCP *is a linear time decision procedure for the satisfiability of Horn clauses.*

Exercises

1. Prove formally that the following statements are equivalent: For any two formulas A and B, (a) $A \models B$; (b) $A \to B$ is valid; (c) $A \wedge \neg B$ is unsatisfiable.
2. Use the semantic tableaux method to decide if the following formulas are valid or not:

    (a) $(A \leftrightarrow (A \wedge B)) \to (A \to B)$
    (b) $(B \leftrightarrow (A \vee B)) \to (A \to B)$
    (c) $(A \oplus B) \to ((A \vee B) \wedge (\neg A \vee \neg B))$
    (d) $(A \leftrightarrow B) \to ((A \vee \neg B) \wedge (\neg A \vee B))$

3. Use the semantic tableaux method to decide if the following entailment relations are true or not:

    (a) $(A \wedge B) \to C \models (A \to C) \vee (B \to C)$
    (b) $(A \wedge B) \to C \models A \to C$
    (c) $A \to (B \to C) \models (A \to B) \to (A \to C)$
    (d) $(A \vee B) \wedge (\neg A \vee C) \models B \vee C$

4. Prove by the semantic tableau method that the following statement, "either the debt ceiling isn't raised or expenditures don't rise" ($\neg d \vee \neg e$), follows logically from the statements in Problem 31 of Chap. 2.
5. Prove that if all the axioms are tautologies and all the inference rules are sound, then all the derived formulas from the axioms and the inference rules are tautologies.
6. Specify the introduction and elimination rules for the operators $\oplus$ and $\leftrightarrow$ in a natural deductive system.
7. Estimate a tight upper bound of the size of the inference graph for resolution, if the input clauses contain n variables.
8. Given a non-unit clause c, we may split c into two clause by introducing a new variable: Let c be $(\alpha \mid \beta)$, where α and β are lists of literals, and $S = \{(\alpha \mid y), (\overline{y} \mid \beta)\}$, where y is a new variable. Show that $S \models c$ and $S \approx c$. Using this

technique, show that a clause of $n \geq 4$ literals can be split into an equisatisfiable set of $n - 2$ clauses of 3 literals each (3CNF).

9. Use the Tseitin encoding to convert the following formulas into 3CNF (CNF in which every clause has at most three literals):

   (a) $(A \leftrightarrow (A \wedge B)) \rightarrow (A \rightarrow B)$
   (b) $(A \oplus B) \rightarrow (A \vee B)$
   (c) $((A \wedge B) \rightarrow C) \rightarrow (A \rightarrow C)$
   (d) $((A \vee B) \wedge (\neg A \vee C)) \rightarrow (B \vee C)$

10. Use the resolution method to decide if the following formulas are valid or not. If the formula is valid, provide a resolution proof; if it is not valid, provide an interpretation which falsifies the formula.

    (a) $(A \leftrightarrow (A \wedge B)) \rightarrow (A \rightarrow B)$
    (b) $(B \leftrightarrow (A \vee B)) \rightarrow (A \rightarrow B)$
    (c) $(A \oplus B) \rightarrow ((A \vee B) \wedge (\neg A \vee \neg B))$
    (d) $((\overline{A} \downarrow B) \uparrow \overline{C}) \rightarrow (A \uparrow (B \downarrow C))$

11. Use the resolution method to decide if the following entailment relations are true or not. If the relation holds, provide a resolution proof; if it doesn't, provide an interpretation such that the premises are true, but the consequence of the entailment is false.

    (a) $(A \rightarrow C) \wedge (B \rightarrow D) \models (A \vee B) \rightarrow (C \vee D)$
    (b) $(A \wedge B) \rightarrow C \models (A \rightarrow C)$
    (c) $A \rightarrow (B \rightarrow C) \models (A \rightarrow B) \rightarrow (A \rightarrow C)$
    (d) $(A \vee B) \wedge (\neg A \vee C) \models B \vee C$

12. Apply clause deletion strategies to each of the following sets of clauses and obtain a simpler equisatisfiable set.

    $S_1 = \{(p \mid q \mid r), (\overline{p} \mid q \mid \overline{r} \mid s), (p \mid \overline{q} \mid s), (\overline{q} \mid r), (q \mid \overline{r})\}$
    $S_2 = \{(p \mid q \mid r \mid s), (\overline{p} \mid q \mid \overline{r} \mid \overline{s}), (p \mid q \mid s), (q \mid \overline{r}), (q \mid s \mid \overline{r})\}$
    $S_3 = \{(p \mid q \mid s), (\overline{p} \mid q \mid \overline{r} \mid \overline{s}), (p \mid r \mid s), (\overline{p} \mid \overline{r}), (\overline{p} \mid q \mid \overline{r})\}$

13. Given the clause set S, where $S = \{$1. $(t \mid \overline{e} \mid d)$, 2. $(\overline{t} \mid c)$, 3. $(e \mid \overline{d} \mid i)$, 4. $(\overline{g} \mid \overline{d} \mid i)$, 5. $(t \mid c \mid \overline{d} \mid g)$, 6. $(\overline{c})$, 7. $(\overline{i} \mid \overline{g})$, 8. $(c \mid d)$, 9. $(c \mid e)$ $\}$, find the following resolution proofs for S: (a) unit resolution, (b) input resolution, (c) positive resolution, and (d) ordered resolution using the order $t > i > g > e > d > c$.

14. Apply various resolution strategies to prove that the following clause set S entails $\neg s$: unit resolution, input resolution, negative resolution, positive resolution, and ordered resolution using the order of $p > q > r > s$.

    $S = \{1.\ (p \mid q \mid \overline{r}), 2.\ (p \mid \overline{q} \mid s), 3.\ (\overline{p} \mid q \mid \overline{r}), 4.\ (\overline{p} \mid \overline{q} \mid r), 5.\ (\overline{p} \mid q \mid r),$
    $6.\ (\overline{q} \mid \overline{r}), 7.\ (r \mid \overline{s})\}$

    Please show the proofs of these resolution strategies.

15. Decide by ordered resolution if the clause set S is satisfiable or not, where $S = \{1.\ (a \mid c \mid \overline{d} \mid f), 2.\ (a \mid d \mid \overline{e}), 3.\ (\overline{a} \mid c), 4.\ (b \mid \overline{d} \mid e), 5.\ (b \mid d \mid \overline{f}),$
    $6.\ (\overline{b} \mid \overline{f}), 7.\ (c \mid \overline{e}), 8.\ (c \mid e), 9.\ (\overline{c} \mid \overline{d})\ \}$, and the order is $a > b > c > d > e > f$. If the clause set is satisfiable, please display all the resolvents from ordered resolution and construct a model from these clauses. If S unsatisfiable, please display an ordered resolution tree of the proof.

16. Prove by resolution that $S \models \neg f$, where S is the set of clauses given in the previous problem. Please display the resolution tree of this proof.

17. If head/tail literals are used for implementing BCP, what literals of the clause set S in the previous problem will be assigned a truth value during unit resolution before an empty clause is found? In what order?

18. A set C of clauses is said to be a *2CNF* if every clause contains at most two literals. The *2SAT* problem is to decide if a 2CNF formula is satisfiable or not. Show that *orderedResolution* takes $O(n^3)$ time and $O(n^2)$ space to solve 2SAT, where n is the number of propositional variables in 2CNF.

19. Prove Theorem 3.4.8: For any set C of clauses, if there exists a unit resolution proof from C, then there exists an input resolution proof from C.

References

1. Franks, Curtis, "Propositional Logic", *The Stanford Encyclopedia of Philosophy*, Edward N. Zalta & Uri Nodelman (eds.), Fall 2023, retrieved Nov. 11, 2023
2. Hantao Zhang, Mark Stickel, "Implementing the Davis-Putnam method", *J. of Automated Reasoning* 24: 277–296, 2000

Chapter 4
Propositional Satisfiability

Propositional satisfiability (SAT) is the problem of deciding if a propositional formula is satisfiable, i.e., if the formula has a model. Software programs for solving SAT are called SAT solvers. SAT has long enjoyed a special status in computer science. On the theoretical side, it is the first NP-complete problem ever discovered. NP problems are those problems whose solutions can be verified in polynomial times (time), and NP-complete problems are the hardest NP problems. In the worst case, the computation time of any known algorithm for an NP-complete problem increases exponentially with the input size. On the practical side, since every NP problem can transform into SAT, an efficient SAT solver can be used to solve many practical problems. SAT found several important applications in the design and verification of hardware and software systems, and in many areas of artificial intelligence and operation research. Thus, there is a strong motivation to develop practically useful SAT solvers. However, the NP-completeness is a cause for pessimism, since it is unlikely that we will be able to scale the solutions to large practical instances.

In theory, the decision procedures presented in the previous chapter, such as semantic tableau or resolution, can be used as SAT solvers. However, they served largely as academic exercises with little hope of practical use. Resolution as a refutation prover searches for an empty clause. Semantic tableau imitates the transformation of a formula into DNF. None of these decision procedures aims at searching for a model.

Fortunately, in the last two decades or so, several research developments have enabled us to tackle SAT instances with dozens of thousands of variables and millions of clauses. SAT researchers introduced techniques such as conflict-driven clause learning, novel branching heuristics, and efficient unit propagation. These techniques form the basis of many modern SAT solvers. Using these ideas, contemporary SAT solvers can often handle very large SAT instances.

Annual competitions of SAT solvers promote the continuing advances of SAT solvers. SAT solvers are the engines of modern model checking tools. Contemporary automated verification techniques such as bounded model checking, proof-based abstraction, and interpolation-based model checking, are all based on SAT solvers and their extensions. These tools are used to check the correctness of software and hardware designs.

4.1 The DPLL Algorithm

In 1960, Martin Davis and Hilary Putnam proposed the inference rule of resolution for propositional logic. For realistic problems, the number of clauses generated by resolution grows quickly. In 1962, to avoid this explosion, Davis, George Logemann, and Donald Loveland suggested replacing the resolution rule with a case split: Pick a variable p and consider the two cases $p \mapsto 1$ and $p \mapsto 0$ separately. This modified algorithm is commonly referred to as the *DPLL* algorithm, where *DPLL* stands for Davis, Putnam, Logemann, and Loveland.

DPLL can be regarded as a search procedure: The search space is all the interpretations, partial or full. A partial interpretation can be represented by a set of literals, where each literal in the set is assigned *true* and each literal outside of the set is assumed unassigned. *DPLL* starts with the empty interpretation and tries to extend the interpretation by adding either p or $\overline{p}$ through the case split. *DPLL* stops when every variable is assigned without contradiction and the partial interpretation it maintains becomes a model.

4.1.1 Recursive Version of DPLL

In the following, we give a recursive version of *DPLL*. The first call to *DPLL* is made with the set C of input clauses and the empty interpretation $\sigma = \{\}$. It is one of the oldest backtracking procedures. *DPLL* uses a procedure named *BCP* (Boolean constraint propagation) heavily for handling unit clauses. From the previous chapter, we know that *BCP* implements unit resolution and subsumption deletion. Given a set C of clauses, $BCP(C)$ returns $\bot$ if the empty clause is found by unit resolution; otherwise, $BCP(C)$ returns (U, S), such that $C \equiv (U \cup S)$, U is a set of unit clauses, S does not contain any unit clauses, and U and S share no variables. Note that $BCP(C)$ takes $O(n)$ time to run, where n is the number of literal occurrences of C.

4.1 The DPLL Algorithm

Algorithm 4.1.1 $DPLL(C, \sigma)$ takes as input a set C of clauses and a partial interpretation σ, C and σ do not share any variables. It returns a model of C if C is satisfiable; it returns $\bot$ if C is unsatisfiable. Procedure *pickLiteral(S)* will pick a literal appearing in S.

proc $DPLL(C, \sigma)$
1    $res := BCP(C)$
2   **if** $(res = \bot)$ **return** $\bot$ // C has no models.
     // It must be the case $res = (U, S)$
3    $\sigma := \sigma \cup U$
4   **if** $(S = \emptyset)$ **return** σ // σ is a model of C.
5    $A := pickLiteral(S)$
6    $res := DPLL(S \cup \{(A)\}, \sigma)$
7   **if** $(res \neq \bot)$ **return** res // a model is found.
8   **return** $DPLL(S \cup \{(\overline{A})\}, \sigma)$ // either $\bot$ or a model is returned.

In the above procedure, we assume again that $\overline{A} = p$ if A is literal $\overline{p}$. If $BCP(C)$ does not return $\bot$, it simplifies C to $U \cup S$, where U is a set of literals, S is a set of non-unit clauses, and U and S do not share any variables. By Proposition 3.4.3, $C \equiv U \cup S$. We identify two types of literals in U.

Definition 4.1.2 In *DPLL*, a literal $A \in U$ is said to be a *decision literal* if A is the one returned by *pickLiteral(S)* at line 5 of *DPLL*; otherwise, A is said to be an *implied literal*.

Since U is a set of literals without complementary literals, U is regarded as a partial interpretation σ such that $\sigma(U) = 1$. Since U is satisfiable and $C \equiv U \cup S$, we conclude that $C \approx S$. Furthermore, checking the satisfiability of S is transformed recursively to checking $S \wedge A$ and $S \wedge \neg A$ separately, because $S \equiv (S \wedge A) \vee (S \wedge \neg A)$ for a literal A appearing in S. This transformation, applied recursively, yields a complete decision procedure for satisfiability. The execution of *DPLL* can be depicted by a tree of which the root is the first call to *DPLL* and each non-root node is a recursive call to *DPLL*. In the following, this recursion tree will be called *decision tree*, because a decision literal is involved in each recursive call.

Example 4.1.3 Let $C = \{c_1 : (x_2 \mid x_3 \mid x_5),\ c_2 : (\overline{x_1} \mid \overline{x_4}),\ c_3 : (\overline{x_2} \mid x_4),\ c_4 : (\overline{x_1} \mid x_2 \mid \overline{x_3}),\ c_5 : (\overline{x_5})\}$. If *pickLiteral* picks a variable in the order of $x_1, x_2, \ldots, x_5$, then

$DPLL(C, \{\})$ calls
     $BCP(C)$, which returns $(\{\overline{x_5}\}, C_1 = \{c_1' : (x_2 \mid x_3), c_2, c_3, c_4\})$;
     x_1 is picked as the decision literal of level 1;
     $DPLL(C_1 \cup \{(x_1)\}, \{\overline{x_5}\})$, which calls
         $BCP(C_1 \cup \{(x_1)\})$, which returns $\bot$;
         and returns $\bot$
     $DPLL(C_1 \cup \{(\overline{x_1})\}, \{\overline{x_5}\})$, which calls
         $BCP(C_1 \cup \{(\overline{x_1})\})$, which returns $(\{\overline{x_1}\}, C_2 = \{c_1', c_3\})$;

x_2 is picked as the decision literal of level 2;
$DPLL(C_2 \cup \{(x_2)\}, \{\overline{x_1}, \overline{x_5}\})$, which calls
  $BCP(C_2 \cup \{(x_2)\})$, which returns $(\emptyset, \{\overline{x_1}, x_2, x_4, \overline{x_5}\})$
  and returns $\{\overline{x_1}, x_2, x_4, \overline{x_5}\}$
and returns $\{\overline{x_1}, x_2, x_4, \overline{x_5}\}$
and returns $\{\overline{x_1}, x_2, x_4, \overline{x_5}\}$

Thus, a model of C can be constructed from $\{\overline{x_1}, x_2, x_4, \overline{x_5}\}$ by adding $x_3 \mapsto v$ for any value $v \in \{0, 1\}$. □

In the above example, the height of the decision tree is two, x_1 and x_2 are decision literals, and x_4 and $\overline{x_5}$ are implied literals. We have called *BCP* four times: one call at the root (level 0); two calls at the nodes of level one; and one call at the node of level two. If we want to search for all models, we may choose the branch with $x_2 \mapsto 0$, as shown in Fig. 4.1.

Definition 4.1.4 The *level* of an assigned variable is the depth of a node of the decision tree of $DPLL(C, \{\})$ at which the variable is assigned a truth value. We write $x@l$ if $x \mapsto 1$ at level l, and $\overline{x}@l$ if $x \mapsto 0$ at level l.

In the above example, for the model returned by *DPLL*, we may write it as

$$\{\overline{x_1}@1,\ x_2@2,\ x_4@2,\ \overline{x_5}@0\}$$

Theorem 4.1.5 *$DPLL(C, \{\})$ is a decision procedure for the satisfiability of C.*

Proof Assuming $\bot \notin C$. To prove the theorem, we just need to show that *DPLL* is sound and terminating. $DPLL(C, \sigma)$ will terminate because for each recursive call, the number of variables in C is decreased by at least one. In $DPLL(C, \sigma)$, since σ and C share no variables, $(C \cup \sigma) \approx C$. So, in the soundness proof of *DPLL*, we ignore the presence of σ.

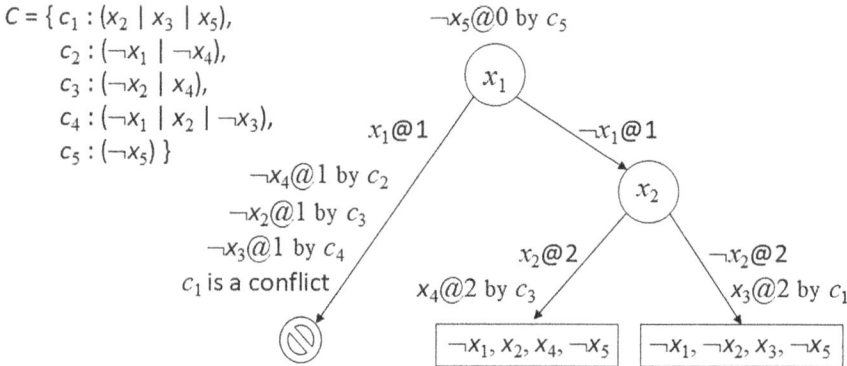

Fig. 4.1 The decision tree of DPLL for Example 4.1.3. The result of BCP is shown along the tree links

4.1 The DPLL Algorithm

We will prove by induction on the structure of the decision tree of *DPLL*. As a base case, if $DPLL(C, \sigma)$ returns $\bot$, it means that $\bot$ is generated by unit resolution inside $BCP(C)$. By Proposition 3.4.3, $C \equiv \bot$. If $BCP(C, \sigma)$ returns (U, S), by Proposition 3.4.3, $C \equiv U \cup S$. As another base case, if $S = \emptyset$, then $C \equiv U$ and U is the model for the unit clause set U if U is regarded as an interpretation, so C is satisfiable. If $S \neq \emptyset$, then since

$$S \equiv (S \wedge A) \vee (S \wedge \neg A)$$

for any literal A of S, if either $DPLL(S \cup \{p\}, \sigma)$ or $DPLL(S \cup \{\overline{p}\}, \sigma)$ returns a model, by induction hypotheses, it means $(S \wedge p) \vee (S \wedge \neg p)$ is satisfiable, so is S. If both $DPLL(S \cup \{p\}, \sigma)$ and $DPLL(S \cup \{\overline{p}\}, \sigma)$ return $\bot$, by induction hypotheses, it means $(S \wedge p) \vee (S \wedge \neg p)$ is unsatisfiable, so is S. □

4.1.2 All-SAT and Incremental SAT Solvers

Given a satisfiable set of clauses, *DPLL* returns a single satisfying assignment. Some applications, however, require us to enumerate all satisfiable assignments of a formula, and this problem is called *All-SAT*. It is easy to modify *DPLL* so that all the models can be found. That is, we replace line 4 of *DPLL* as follows:

4:    **if** $(S = \emptyset)$ save(σ); **return** $\bot$; // save model and continue

When the modified *DPLL* finishes, all the models are saved. Note that these models are partial models. One partial model represents 2^k full models, where k is the number of unassigned variables in the partial model.

During the execution of *DPLL*, the list of decision literals (labeled with 1) and their complements (labeled with 2) decides the current position in the decision tree of *DPLL*. This list of literals, called *guiding path*, can be very useful for designing an incremental SAT solver: The solver takes the guiding path as input, uses the literals in the guiding path to skip the branches already explored, and explores the new branches in the decision tree for a new model.

Example 4.1.6 In Example 4.1.3, a model is found with the decision literals $\overline{x_1}$ and x_2. The guiding path for this model is $[\overline{x_1} : 2, x_2 : 1]$, where the label 2 after $\overline{x_1}$ means this is the second branch of x_1. The label 1 after x_2 means this is the first branch of x_1. If the next model is needed, it must be in the subtree with the initial guiding path $[\overline{x_1} : 2, x_2 : 2]$. Using $[\overline{x_1} : 2, x_2 : 2]$, the SAT solver avoids the search space it has traveled before. □

Since modern SAT solvers are often designed as a black box, it is cumbersome to use them as an incremental solver for enumerating each model. We can force a SAT solver to find a new model by subsequently blocking all models previously found. A model can be viewed as a conjunction of true literals, one for every variable. The negation of this conjunction is a clause, called *blocking clause*. If this clause is added into the set of input clauses, the model will be blocked by this clause and cannot be generated by the SAT solver. Since a blocking clause contains every variable, it is unlikely to become unit and it is expensive to keep all the blocking clauses. Therefore, it is desirable to reduce the size of blocking clauses, i.e., to construct a shorter clause which still blocks the model. If we know the guiding path of this model, the negation of the conjunction of all the decision literals in the guiding path will block the model.

Many applications of SAT solvers require solving a sequence of instances by adding more clauses. Incremental SAT solvers with the guiding path can support this application without any modification, because the search space skipped by a guiding path does not contain any new model if more clauses are added.

In many application scenarios, it is beneficial to be able to make several SAT checks on the same input clauses under different forced partial interpretations, called "*assumptions*." For instance, people may be interested in questions like "Is the formula F satisfiable under the assumption $x \mapsto 1$ and $y \mapsto 0$?" In such applications, the input formula is read in only once; the user implements an iterative loop that calls the same solver instantiation under different sets of assumptions. The calls can be adaptive, i.e., assumptions of future SAT solver calls can depend on the results of the previous solver calls. The SAT solver can keep its internal state from the previous call to the next call. *DPLL* with a guiding path can be easily modified to support such applications, treating the assumption as a special set of unit clauses which can be added into or removed from the solver. The guiding path allows us to backtrack to the new search space but never revisit the search space visited before.

To obtain a high-performance SAT solver, we may use the deletion techniques presented in Sect. 3.3.6 as preprocessing techniques. They enable us to reduce the size of the formula before passing it to *DPLL*.

Inside the body of *DPLL*, it calls itself twice, once with $S \cup \{(A)\}$ and once with $S \cup \{(\overline{A})\}$. A persistent data structure for S would quickly run out of memory as S is very large and the decision tree is huge for practical SAT instances. A better approach is to use a destructive data structure for S: We remember all the operations performed onto S inside the first call, and then undo them, if necessary, when we enter the second call.

4.1.3 BCPw: *Use of Watch Literals in DPLL*

In Algorithm 3.4.16, we presented an efficient implementation of *BCP* as a decision procedure for Horn clauses. This version of *BCP* calls *BCPht*, which uses the head/tail data structure for each non-unit clause. Naturally, this *BCP* can be used to support *DPLL* with minor modification. That is, *BCP* can implement the idea of *watch literals*, which is based on head/tail literals and used in SAT solver chaff by Matthew Moskewicz et al. at Princeton University.

The function of watch literals is the same as head/tail literals: Use two literals to decide if a clause becomes unit or conflicting, as stated in Proposition 3.4.15.

4.1 The DPLL Algorithm

Head/tail literals are watch literals; when head/tail literals are allowed to appear anywhere in a clause, head/tail literals are identical to watch literals. As watch literals, we do not know the literals in a clause before or after the head/tail literals are false or not when *BCP* is used to support *DPLL*. Thus, we need to search the whole clause for locating the next head/tail literal.

Since the idea of watch literals is based on that of head/tail literals, the implementation of watch literals is easy if we have an implementation of head/tail literals. In *BCPht*, when a head or tail literal of c becomes false, we look for its replacement at index x, where $head(c) < x < tail(c)$. There are two alternative ways to modify *BCPht*. One solution is to allow $x \in \{0, 1, \ldots, |c| - 1\} - \{head(c), tail(c)\}$; we will introduce an implementation of this solution shortly. The other solution is to fix $head(c) = 0$ and $tail(c) = 1$ (or $tail(c) = |c| - 1$) and swap the elements of $lits(c)$ if necessary.

The second solution saves the space for $head(c)$ and $tail(c)$ but may cost more for visiting literals of $lits(c)$. For example, if a clause has k literals which become false in the order from left to right during *BCP*. The second solution may take $O(k^2)$ time to visit the literals of the clause, while the first solution takes $O(k)$ time. As a result, *BCP* cannot be a linear time decision procedure for Horn clauses if the second solution is used. Of course, we may introduce a pointer in the clause to remember the position of the last visited literal, so that the second solution still gives us a linear time algorithm.

Algorithm 4.1.7 *BCPw(U)* supports the implementation of *BCP(C)* based on the head/tail data structure. *BCPw(U)* works the same way as *BCPht(U)*, where U is a stack of unit clauses and non-unit clauses are stored using the head/tail data structure. Like *BCPht*, *BCPw* returns a conflicting clause if the empty clause is generated; it returns "SAT" otherwise. Unlike *BCPht*, values of $head(c)$ and $tail(c)$ can be any valid indices of $lits(c)$.

proc *BCPw(U)*
1   **while** $U \neq \{\}$ **do**
2       $A := \text{pop}(U)$
3   **if** $val(A) = 0$ **return** $reason(\overline{A})$ // a conflicting clause is found.
4   **else if** $(val(A) \neq 1)$
5          $val(A) := 1; val(\overline{A}) := 0$
6   **for** $c \in cls(\overline{A})$ **do** // $\overline{A}$ is either head or tail of c.
7   **if** $(\overline{A} = lits(c)[head(c)])$
8                $e_1 := head(c); e_2 := tail(c); step := 1$ // scan from left to right.
9   **else**
10               $e_1 := tail(c); e_2 := head(c); step := -1$ // scan from right to left.
11   **while** true **do**
12               $x := (x + step) \bmod |c|$ // x takes any valid index of $lits(c)$.
13   **if** $x = e_1$ **break** // exits the inner while loop.
13'   **if** $x = e_2$ **continue** // go to line 12.
14               $B := lits(c)[x]$

```
15          if (val(B) ≠ 0)
16              remove(c, cls(A̅)); insert(c, cls(B))
17              if (step = 1) head(c) := x else tail(c) := x
18              break // exit the inner while loop.
19      if (x = e₁) // no new head or tail found.
20          A := lits(c)[e₂] // c is unit or conflicting.
21          if val(A) = 0 return c // c is conflicting.
22          else if (val(A) ≠ 1) // c is unit.
23              push(A, U); reason(A) := c; val(A) := 1; val(A̅) := 0
24  return "SAT" // no empty clauses are found.
```

The above algorithm comes from *BCPht* (Algorithm 3.4.16) by the following modifications, which allow the values of $head(c)$ and $tail(c)$ to be any valid indices of $lits(c)$:

- At line 12, "$x := x + step$" is replaced by "$x := (x + step) \bmod |c|$."
- At line 13, the condition "$x = e_2$" is replaced by "$x = e_1$."
- Line 13' is added to skip the literal watched by e_2; this is the literal to be in the unit clause if the clause becomes unit.
- At line 19, the condition "$x = e_2$" is replaced by "$x = e_1$." When $x = e_1$ is true, every literal of the clause is false, with the exception of the literal watched by e_2.

In *DPLL*, the book-keeping required to detect when a clause becomes unit can involve a high computational overhead if implemented naively. Since it is sufficient to watch in each clause two literals that have not been assigned yet, assignments to the non-watched literals can be safely ignored. When a variable p is assigned 1, the SAT solver only needs to visit clauses watched by $\bar{p}$. Each time one of the watched literals becomes false, the solver chooses one of the remaining unassigned literals to watch. If this is not possible, the clause is necessarily unit, or conflicting, or already satisfied under the current partial assignment. Any sequence of assignments that makes a clause unit will include an assignment of one of the watched literals. The computational overhead of this strategy is relatively low: In a formula with m clauses and n variables, $2m$ literals need to be watched, and m/n clauses are visited per assignment on average. This advantage is inherited from the idea of head/tail literals.

The key advantage of the watch literals over the head/tail literals is that the watched literals do not need to be updated upon backtracking in *DPLL*. That is, the key advantage of *BCPw* over *BCPht* is its use inside *DPLL*: When you backtrack from a recursive call of *DPLL*, you do not need to undo the operations on $head(c)$ and $tail(c)$, because any two positions of c will be sufficient for checking if c becomes unit.

4.1.4 Iterative Implementation of DPLL

Since the compiler will convert a recursion procedure into an iterative one using a stack, to avoid the expense of this conversion, in practice, the procedure *DPLL* is not implemented by means of recursion but in an iterative manner using a stack for the partial interpretation σ. If the head/tail data structure is used, σ can be obtained from $val(a)$, but we still need a stack of literals to record the time when a literal is assigned. This stack is called *trail*, which can also serve as a partial interpretation. Using the trail, we keep track of the literals, either decision literals or implied literals, from the root of the decision tree to the current node. Every time a literal is assigned a value, either by a decision or an implication inside *BCP*, the literal is pushed into the trail and its level in the decision tree is remembered.

A trail may lead to a dead-end, i.e., resulting in a conflicting clause, in which case we have to explore the alternative branch of a case split. This corresponds to backtracking which reverts the truth value of a decision literal. When we backtrack from a recursive call, we pop those literals off the trail and make them as "unassigned" (denoted by $\times$). When backtracking enough times, the search algorithm eventually exhausts all branches of the decision tree, either finding all models or claiming no models available.

To implement the above idea, we use the trail σ which is a stack of literals and serves as a partial interpretation. σ will keep all the assigned literals. For each literal A, we use $lvl(A)$ to record the level at which A is assigned a truth value. This is actually done at line 5 of *BCPw*. Line 5 of *BCPw* now is read as follows:
5:     $val(A) := 1; val(\overline{A}) := 0; lvl(A) := lvl(\overline{A}) := level;$ push(A, σ)
where $val, lvl, \sigma,$ and $level$ are global variables.

In practice, it is more convenient to set the level of a right child the same as the level of its parent in the decision tree, to indicate that the right child has no alternatives to consider. In the following implementation of *DPLL*, we have used this idea.

In short, $BCP(C)$ used in *DPLL* (Algorithm 4.1.1) simplifies C to (U, S), where U is a set of unit clauses and S is a set of non-unit clauses. Now, the clauses in S are stored in the head/tail data structure and the clauses in U are stored in the trail σ. We replace $BCP(C)$ by $BCPw(X)$ to obtain the following version of *DPLL*, where X is initially with the unit clauses from the input and contains the unit clause created by each decision literal during the execution of *DPLL*.

Algorithm 4.1.8 Non-recursive *DPLL(C)* takes as input a set C of clauses and returns a model of C if C is satisfiable; it returns $\bot$ if C is unsatisfiable. Procedure *initialize(C)* will return the unit clauses of C and store the rest clauses in the head/tail data structure. Global variables $val, lvl, \sigma,$ and $level$ are used in all procedures.

proc *DPLL(C)*
1     $U :=$ initialize(C) // initialize the head/tail data structure
2     $\sigma := \{\}; level := 0$ // initialize σ (partial model of C) and $level$

```
3    while (true) do
4      if (BCPw(U) = ⊥)
5        if (level = 0) return ⊥  // C is unsatisfiable
6        level := level − 1       // level of the last decision literal
7        A := undo()              // A is the last decision literal picked at line 10
8        U := {Ā}                 // the second branch of A
9      else
10       A := pickLiteral()       // pick an unassigned A as new decision literal
11       if (A = nil) return σ    // all literals assigned, a model is found
12       level := level + 1       // new level in the first branch of A
13       U := {A}                 // create a new unit clause from A
```

proc *undo*()
  // Backtrack to *level*, undo all assignments to literals of higher levels.
  // Return the last literal of *level* + 1 in the trail σ, or nil.
```
1    A := nil
2    while (σ ≠ {}) ∧ (lvl(top(σ)) > level) do
3      A := pop(σ)
4      val(A) := val(Ā) := ×  // × means "unassigned"
5    return A
```

Procedure *undo*() will undo all the assignments (made at line 5 of *BCP*) to the literals of levels higher than *level* (including their complements), and return the last literal of *level* + 1 in σ, which must be the last decision literal A picked at line 10 of *DPLL*.

Procedure *pickLiteral*() (line 10) is the place for implementing various heuristics of best search strategies. The procedure will choose an unassigned literal and assign a truth value. As we know from Theorem 4.1.5, this choice has no impact on the completeness of the search algorithm. It has, however, a significant impact on the performance of the solver, since this choice is instrumental in pruning the search space. We will address this issue later.

4.2 Conflict-Driven Clause Learning (CDCL)

This section introduces conflicting clauses as a source information to prevent the repeated exploration of futile search space. We start with a motivating example.

Example 4.2.1 Suppose C is a set of clauses over the variable set $\{x_1, x_2, \ldots, x_n\}$ and C has more than a million models. Let $C' = C \cup S$, where $S = \{(y_1 \mid y_2), (y_1 \mid \overline{y_2}), (\overline{y_1} \mid y_2), (\overline{y_1} \mid \overline{y_2})\}$. Then C' has no models, because the added four clauses are unsatisfiable. Now we feed C' to *DPLL* and assume that pickLiteral will prefer x_i over y_1 and y_2. Every time right before *DPLL* picks y_1 for case split, a model of C was found at that point. If y_1 is true, then one of $(\overline{y_1} \mid y_2)$ and $(\overline{y_1} \mid \overline{y_2})$ will

4.2 Conflict-Driven Clause Learning (CDCL)

be conflicting. If y_1 is false, then one of $(y_1 \mid y_2)$ and $(y_1 \mid \overline{y_2})$ will be conflicting. Either value of y_1 will lead to a conflicting clause in S, so *DPLL* will backtrack. Since C has more than a million models, it means *DPLL* will try to show that S is unsatisfiable more than a million times. □

If we can learn from the conflicting clauses of S, we may avoid repeatedly doing the same work again and again. This is the motivation for the technique of conflict-driven clause learning (CDCL): Learn new information from conflicting clauses and use the new information to avoid futile search effort. Using CDCL, new information is obtained in the form of clauses which are then added back into the search procedure. For the above example, when y_1 is true by case split, $(\overline{y_1} \mid y_2)$ becomes unit, so y_2 must be set to 1. Now $(\overline{y_1} \mid \overline{y_2})$ becomes conflicting. A resolution between these two clauses on y_2 will generate a new clause, i.e., $(\overline{y_1})$. The new clause has nothing to do with the assignments of x_i, so we can jump back to the level 0 of *DPLL* and the subsequent *BCP* will find the empty clause from $S \wedge \overline{y_1}$ and *DPLL* returns $\bot$, thus avoiding finding models in C a million times.

4.2.1 Generating Clauses from Conflicting Clauses

Recall that in Algorithm 4.1.7, when a non-unit clause c becomes unit, where c contains an unassigned literal a, we do $reason(a) := c$ at line 23, i.e., c is the reason for the unassigned literal a to be true. When a conflicting clause is found in *DPLL*, all of its literals are false, and we pick the literal a, which is assigned last in c, and do resolution on a between this clause and $reason(\overline{a})$, hoping the resolvent will be useful to cut the search space. Since resolution is sound, new clauses generated this way are logical consequences of the input and it is safe to keep them. For practical reasons, we must selectively do resolution and keep at most one clause per conflicting clause.

Algorithm 4.2.2 *conflictAnalysis(c)* takes as input a conflicting clause c in the current partial interpretation at the current *level*, generates a new clause by resolution, and returns it as output. Procedure *latestAssignedLiteral(c)* will return a literal a of c such that a is the last literal assigning a truth value in c. Procedure *countLevel(c, lvl)* will return the number of literals in c which are assigned at level *lvl*. Procedure *resolve*(c_1, c_2) returns the resolvent of c_1 and c_2.

proc *conflictAnalysis(c)*
    // c is conflicting in the partial interpretation $val()$
1    $\alpha := c$
2   **while** $(|\alpha| > 1)$ **do** // as α has more than one literal
3       $a := latestAssignedLiteral(\alpha)$ // a is assigned last in α
4       $\alpha := resolve(\alpha, reason(\overline{a}))$ // resolution on a
5   **if** $(countLevel(\alpha, level) \leq 1)$ **return** α
6   **return** α

Example 4.2.3 Let the input clauses $C = \{c_1 : (x_2 \mid x_3),\ c_2 : (\overline{x_1} \mid \overline{x_4}),\ c_3 : (\overline{x_2} \mid x_4),\ c_4 : (\overline{x_1} \mid x_2 \mid \overline{x_3})\}$. If we pick x_1 as the first decision literal in *DPLL*, it will result in the following assignments:

assignment	reason
$x_1@1$	(x_1)
$\overline{x_4}@1$	$c_2 : (\overline{x_1} \mid \overline{x_4})$
$\overline{x_2}@1$	$c_3 : (\overline{x_2} \mid x_4)$
$x_3@1$	$c_1 : (x_2 \mid x_3)$

Clause c_4 becomes conflicting and contains three literals of level 1. We call *conflictAnalysis*(c_4): Let α_0 be c_4; the following resolutions are performed:

1. $\alpha_1 : (\overline{x_1} \mid x_2) = resolve(\alpha_0, c_1)$.
2. $\alpha_2 : (\overline{x_1} \mid x_4) = resolve(\alpha_1, c_3)$.
3. $\alpha_3 : (\overline{x_1}) = resolve(\alpha_2, c_2)$.

α_3 will be returned by *conflictAnalysis*(c_4). □

Proposition 4.2.4 *Assuming c is conflicting in the current interpretation:*

- (*a*) *The procedure* conflictAnalysis(*c*) *will terminate and return a clause α which is an entailment of the input clauses.*
- (*b*) *α is a conflicting clause in the current interpretation.*
- (*c*) *If we backtrack off the current level, α will be a unit clause in the interpretation after the backtrack.*

Proof

(*a*): We assigned only a finite number of literals a truth value at the current level. The literals we chose to resolve off are in the backward order when they are assigned. Once resolved off, these literals can never come back into the new clause; thus, there is no chance for a loop. Since α is generated by resolution from the input clauses, α must be an entailment of the input clauses.

(*b*): In fact, any resolvent in *conflictAnalysis*(*c*) is a conflicting clause in the current interpretation. Since *c* is a conflicting clause, every literal is false in the current interpretation. *c* is resolved with a reason clause which has only one true literal in the interpretation, and this true literal was resolved off during the resolution. The resulting resolvent will contain only false literals in the current interpretation. α is the last resolvent produced in the procedure.

(*c*): The procedure *conflictAnalysis*(*c*) will terminate when one literal, say B, of the current level remains in α. If we backtrack off the current level, B will be unassigned and the rest literals are false by (*b*); thus, α is a unit clause in the interpretation after the backtrack. □

4.2.2 DPLL with CDCL

The new *DPLL* procedure with CDCL (conflict-driven clause learning) can be described as follows: We will replace $BCPw(U)$ by $BCP(U)$ which is a minor modification of $BCPw(U)$: Instead of returning $\bot$ when a conflicting clause is found, just return that conflicting clause.

The major difference of the new *DPLL* algorithm lies on lines 7–8: In Algorithm 4.1.8, a new unit clause is created from the last decision literal; in the new version, a new clause is created from a conflicting clause and this new clause will become unit after the *undo* operation.

Algorithm 4.2.5 $DPLL(C)$ takes as input a set C of clauses and returns a model of C if C is satisfiable; it returns $\bot$ if C is unsatisfiable. Procedure *initialize(C)* will return the unit clauses of C and store the rest clauses in the head/tail data structure. Global variables val, lvl, σ, and $level$ are used as in Algorithm 4.1.8.

proc $DPLL(C)$
1   $U :=$ initialize(C) // initialize the head/tail data structure
2   $\sigma := \{\}$; $level := 0$ // initialize σ (partial model of C) and $level$
3   **while** (true) **do**
4     $res := BCP(U)$ // U is a set of unit clauses.
5   **if** $(res \neq$ "SAT") // res contains a conflicting clause.
6   **if** $(level = 0)$ **return** $\bot$ // C is unsatisfiable.
7       $U :=$ *insertNewClause*(*conflictAnalysis*(res)) // new level is set.
8   *undo*() // undo to the new level
9   **else**
10      $A :=$ pickLiteral() // pick an unassigned A as new decision literal
11   **if** $(A = nil)$ **return** σ // all literals assigned, a model is found
12      $level := level + 1$ // new level in the first branch of A
13      $U := \{A\}$ // create a new unit clause from A

proc *insertNewClause*(α)
1   **if** $(|\alpha| = 1)$ // α is a unit clause
2     $level := 0$
3     $A :=$ literal(α); $reason(A) := \alpha$ // assume $\alpha = (A)$
4   **return** $\{A\}$
5   **else** // insert α into the head/tail data structure for clauses
6     $lits(\alpha) :=$ makeArray(α) // create an array of literals
7   secondHigh $:= 0$ // look for the second highest level in α
8   **for** $x := 0$ **to** $|\alpha| - 1$ **do**
9   **if** $(lvl(lits(\alpha)[x]) = level)$ $head(\alpha) := x$; $b := lits(\alpha)[x]$ // $lvl(b) = level$
10   **else if** $(lvl(lits(\alpha)[x]) >$ secondHigh) secondHigh $:= lvl(lits(\alpha)[x])$; $t := x$
11    $tail(\alpha) := t$
12    $level :=$ secondHigh
13    $reason(b) := \alpha$
14   **return** $\{b\}$ // the head literal of α

Procedure *insertNewClause* will return a singleton set of literals served as a new unit clause and reset the *level* variable of *DPLL*. If the new clause α generated by *conflictAnalysis* is a unit clause, this clause will be returned, and the level is set to 0. If the new clause is not unit, by Proposition 4.2.4, there is only one literal of the current level in α, and this literal will be the head literal of α when the level is decreased. We then pick one literal of the highest level other than the head literal as the tail literal. The level will be reset to that of the tail literal. When we backtrack to this level, by Proposition 4.2.4, α becomes unit and is the reason for the head literal, unassigned at this point, to be assigned true.

Example 4.2.6 Consider the clause set C of the following clauses:

$$c_1 : (x_1 \mid x_2) \qquad c_2 : (x_1 \mid x_3 \mid \overline{x_7})$$
$$c_3 : (\overline{x_2} \mid \overline{x_3} \mid x_4) \qquad c_4 : (\overline{x_4} \mid x_5 \mid x_6)$$
$$c_5 : (\overline{x_3} \mid \overline{x_4}) \qquad c_6 : (x_5 \mid \overline{x_6})$$
$$c_7 : (\overline{x_2} \mid x_3) \qquad c_8 : (\overline{x_1} \mid x_3)$$
$$c_9 : (\overline{x_1} \mid x_2 \mid \overline{x_3})$$

If the first three literals chosen by *pickLiteral* in *DPLL* are $\overline{x_7}, \overline{x_6}$, and $\overline{x_5}$, then the only implied literal is $\overline{x_4}@3$ by c_4. The next decision literal is $\overline{x_1}$, which implies the following assignments: $x_2@4$ by c_1 and $\overline{x_3}@4$ by c_3. c_7 becomes conflicting at level 4.

Calling *conflictAnalysis*(c_7), we obtain $(\overline{x_2} \mid x_4)$ by resolution between c_7 and c_3, which contains $\overline{x_2}$ as the only literal at level 4. We add $(\overline{x_2} \mid x_4)$ into *DPLL* as c_{10}.

The *level* of *DPLL* is set to 3 as $lvl(x_4) = 3$. So, we backtrack to level 3, and $c_{10} = (\overline{x_2} \mid x_4)$ becomes unit. The next call to *BCP* will make the following assignments at level 3: $\overline{x_2}@3$ by c_{10}, $x_1@3$ by c_1, $x_3@3$ by c_8, and c_9 becomes conflicting.

Calling *conflictAnalysis*(c_9), we obtain $c_{11} = (x_2)$ by two resolutions between c_8, c_9, and c_1. The new clause is unit and we set $level = 0$. Backjumping to level 0 by *undo*, the next call to *BCP* will imply the following assignments: $x_2@0$ by c_{11}, $x_3@0$ by c_7, $x_4@0$ by c_{10}, and c_5 becomes conflicting.

Calling *conflictAnalysis*(c_5), we obtain $c_{12} = (\overline{x_2})$ by two resolutions between c_5, c_{10}, and c_7. Clauses c_{11} and c_{12} will generate the empty clause. Thus, the input clauses are unsatisfiable. □

Clause learning with conflict analysis does not impair the completeness of the search algorithm: Even if the learned clauses are removed at a later point during the search, the trail guarantees that the solver never repeatedly enters a decision level with the same partial assignment. We have shown the correctness of clause learning by demonstrating that each conflicting clause is implied by the original formula.

The idea of CDCL was used first in solver `grasp` of Marques-Silva and Sakallah, and solver `relsat` of Bayardo and Schrag. In their solvers, they introduced independently a novel mechanism to analyze the conflicts encountered during the

4.2 Conflict-Driven Clause Learning (CDCL)

search for a satisfying assignment. There are many ways to generate new clauses from conflicting clauses and most of them are based on *implication graphs*. The method of learning new clauses through resolution is the easiest one to present.

CDCL brings a revolutionary change to *DPLL* so that *DPLL* is no longer a simple backtrack search procedure, because the level reset by *insertNewClause* may not be $level - 1$ and *DPLL* will backjumping to a level less than $level - 1$ (to a node closer to the root of the decision tree of *DPLL*), thus avoiding unnecessary search space. If the new clause learned contains a single literal, the new level is set to 0. *DPLL* with CDCL can learn from failures and backjump to the level where the source of failures is originated. Several new techniques are proposed based on CDCL, including generating a resolution proof when the input clauses are unsatisfiable, or randomly restarting the search.

4.2.3 Unsatisfiable Cores

The existence of a resolution proof for a set of unsatisfiable clauses is guaranteed by the completeness of resolution. Finding a resolution proof is a difficult job. It has been shown that CDCL as practiced in today's SAT solvers corresponds to a proof system substantially more powerful than resolution. Specifically, each learning step is in fact a sequence of resolution steps, of which the learned clause is the final resolvent; conversely, a resolution proof can be obtained by regarding the learning procedure as a guided resolution procedure. In Example 4.2.6, we have seen how the empty clause is generated. The resolution proof leading to the empty clause can be displayed as follows:

$$
\begin{aligned}
&c_1 : (x_1 \mid x_2) &&\text{assumed} \\
&c_3 : (\overline{x_2} \mid \overline{x_3} \mid x_4) &&\text{assumed} \\
&c_5 : (\overline{x_3} \mid \overline{x_4}) &&\text{assumed} \\
&c_7 : (\overline{x_2} \mid x_3) &&\text{assumed} \\
&c_8 : (\overline{x_1} \mid x_3) &&\text{assumed} \\
&c_9 : (\overline{x_1} \mid x_2 \mid \overline{x_3}) &&\text{assumed} \\
&c_{10} : (\overline{x_2} \mid x_4) &&\text{resolvent of } c_7, c_3 \\
&c_{11} : (\overline{x_1} \mid x_2) &&\text{resolvent of } c_8, c_9 \\
&c_{12} : (x_2) &&\text{resolvent of } c_{11}, c_1 \\
&c_{13} : (\overline{x_2} \mid \overline{x_3}) &&\text{resolvent of } c_5, c_{10} \\
&c_{14} : (\overline{x_2}) &&\text{resolvent of } c_{13}, c_7 \\
&c_{15} : (\,) &&\text{resolvent of } c_{12}, c_{14}
\end{aligned}
$$

This example shows the general idea of obtaining a resolution proof when *DPLL* finds a set of clauses unsatisfiable. During the search inside *DPLL*, conflicting clauses are generated due to decision literals. The procedure *conflictAnalysis* produces a new clause to show why the previous decisions were wrong and we go up in the decision tree according to this new clause. *DPLL* returns $\bot$ when it finally finds a conflicting clause at level 0. In this case, a resolution proof is generated by *conflictAnalysis*, and this proof uses either the input clauses or the clauses generated previously by *conflictAnalysis*.

Given an unsatisfiable set C of clauses, we can use all the resolution steps inside the procedure *conflictAnalysis* to construct a resolution proof. Obviously, such a proof provides evidence for the unsatisfiability of C. The clauses used as the premises of the proof are a subset of the clauses of C and are called the unsatisfiable *core* of the proof. Note that a formula typically does not have a unique unsatisfiable core. Any unsatisfiable subset of C is an unsatisfiable core. Resolution proofs and unsatisfiable cores have applications in hardware verification.

An unsatisfiable core is *minimal* if removing any clause from the core makes the remaining clauses in the core satisfiable. We may use this definition to check if every clause in the core is necessary, and this is obviously a difficult job when the core is huge. The core of the above example contains every input clause except c_2, c_4, and c_6; it is a minimal core.

4.2.4 Random Restart

Suppose we are looking for a model of C by $DPLL(C)$ and the first decision literal is A. Unfortunately, $C \wedge A$ is a hard unsatisfiable instance and *DPLL* stucks there without success. Randomly restarting *DPLL* may be your only option. By "restart," we mean that *DPLL* throws away all the previous decision literals (this can be easily done by $level := 0$; $undo()$ in *DPLL*). By "random," we mean that when you restart the search, the first decision literal will most likely be a different literal from A. Intuitively, randomly restarting means there is a chance of avoiding bad luck and getting luckier with guessing the right literal assignments that would lead to a quick solution.

Without CDCL, random restart makes *DPLL* incomplete. With CDCL, the input and generated clauses keep *DPLL* from choosing the same sequence of decision literals and cannot generate the same clause again. Since the number of possible clauses is finite, *DPLL* cannot run forever. This is a theoretical view. In practice, we cannot keep all generated clauses as the procedure will run out of memory. Managing generated clauses in a *DPLL* with CDCL is an important issue not covered in this book.

The same intuition suggests random restart should be much more effective when the problem instance is in fact satisfiable. Experimental results showed that random restart also helps when the input clauses are unsatisfiable. If restart is frequent, this assumes a deviation from standard practice, CDCL can be as powerful as general resolution, while *DPLL* without restart has been known to correspond to the weaker tree-like resolution. The theoretical justification for the speedup is that a restart allows the search to benefit from the knowledge gained about persistently troublesome conflicting variables sooner than backjumping would otherwise allow the partial assignment to be similarly reset. In effect, restarts may allow the discovery of shorter proofs of unsatisfiability.

To implement the restart strategy, when *DPLL* has found a certain number of failures (conflicting clauses) without success (a model), it is time for a restart. Today's SAT solvers often use the following restart policy: Let $r_i = r_0 \gamma^{i-1}$ be the number of failures for the ith restart, where r_0 is a given integer, and $1 \leq \gamma < 2$. If $r_0 = 300$ and $\gamma = 1.2$, it means the first restart happens when *DPLL* has 300 failures, the second restart happens after another 360 failures, and so on. If $\gamma = 1$, it means *DPLL* restarts after a fixed number of failures.

Given the common use of restarts in today's clause learning SAT solvers, the task of choosing a good restart policy appears appealing. While experimental results showed that no restart policy is better than others for a wide range of problems, a clause learning SAT solver could benefit substantially from a carefully designed restart policy. Specifically, experiments show that nontrivial restart policies did significantly better than if restarts were disabled and exhibited considerably different performance among themselves. This provides motivation for the design of better restart policies, particularly dynamic ones based on problem types and search statistics.

4.2.5 Branching Heuristics for DPLL

Using CDCL or not, every implementation of *DPLL* needs a function for selecting literals for case split. We assume that when a literal is chosen for case split, the literal will be assigned true in the first branch of DPLL. Prior to the development of the CDCL techniques, branching heuristics were the primary method used to reduce the size of the search space. It seems likely, therefore, that the role of branching heuristics is likely to change significantly for algorithms that prune the search space using clause learning.

It is conventional wisdom that it is advantageous to assign first the most tightly constrained variables, i.e., variables that occur in a large number of clauses. One representative of such a selection strategy is known as the MOMS rule, which branches on a literal which has the *maximum occurrences in clauses of minimum size*. If the clauses contain binary clauses, then the MOMS rule will choose a literal in a binary clause. By assigning true to this literal, it is likely that a maximal number of binary clauses will become satisfied. On the other hand, if the primary criterion for the selection of a branch literal is to pick one that would enable a cascade of unit propagation (the result of such a cascade is a smaller subproblem), we would assign false to the literal chosen by the MOMS rule in a binary clause; assigning this literal false will likely create a maximal number of unit clauses from all the binary clauses. MOMS provides a rough but easily computed approximation to the number of unit clauses (i.e., implied literals) that a particular variable assignment might cause.

Alternatively, one can call BCP multiple times on a set of promising variables in turn and compute the exact number of unit clauses that would be caused by a branching choice. Each chosen variable is assigned to be true and to be false in turn and BCP is executed for each choice. The precise number of unit propagation caused is then used to evaluate possible branching choices. Unlike the MOMS heuristic, this rule is obviously exact in its attempt to judge the number of unit clauses caused by a potential variable assignment. Unfortunately, it is also considerably more expensive to compute because of the multiple executions of BCP and undoing them. It has been shown that, using MOMS to choose a small number of promising candidates, each of which is then evaluated exactly using BCP, it outperforms other heuristics on randomly generated problems.

Another strategy is to branch on variables that are likely to be *backbone literals*. A *backbone literal* is one that must be true in all models of the input clauses. The likelihood that any particular literal is a backbone literal is approximated by counting the appearances of that literal in the satisfied clauses during the execution of DPLL. This heuristic outperforms those discussed in the previous paragraphs on many examples.

The development of CDCL techniques enabled solvers to attack more structured, realistic problems. There are no formal studies comparing the previously discussed heuristics on structured problems when CDCL techniques are used. CDCL techniques create many new clauses and make the occurrence counts of each literal difficult or less significant. Branching techniques and learning are deeply related, and the addition of learning to a *DPLL* implementation will have a significant effect on the effectiveness of any of these branching strategies. As new clauses are learned, the number of unit clauses caused by an assignment can be expected to vary; the reverse is also true in that the choice of decision literal can affect the generated clauses.

Branching heuristics that are designed to function well in the context of clause learning generally try to branch on literals among the new clauses which have been learned recently. This tends to allow the execution of *DPLL* to keep "making progress" on a single section of the search space as opposed to moving from one area to another; an additional benefit is that existing learned clauses tend to remain relevant, avoiding the inefficiencies associated with losing the information present in learned clauses that become irrelevant and are discarded.

One of the popular branch heuristics for *DPLL* with clause learning is called "dynamic largest individual sum (DLIS) heuristic" and it behaves similarly as the MOMS rule. At each decision point, it chooses the assignment that satisfies the most unsatisfied clauses. Formally, let p_x be the number of unresolved clauses containing x and n_x be the number of unresolved clauses containing $\bar{x}$. Moreover, let x be the variable for which p_x is maximal, and let y be variable for which n_y is maximal. If $p_x > n_y$, choose x as the next decision literal; otherwise, choose $\bar{y}$. The disadvantage of this strategy is that the computational overhead is high: The algorithm needs to visit all clauses that contain a literal that has been set to true in order to update the values p_x and n_x for all variables contained in these clauses. Moreover, the process needs to be reversed upon backtracking.

A heuristic commonly used in contemporary SAT solvers favors literals in recently added conflict clauses. Each literal is associated with a count, which is initialized with the number of times the literal occurs in the clause set. When a learned clause is added, the count associated with each literal in the clause is incremented. Periodically, all counters are divided by a constant greater than 1, resulting in a decay causing a bias towards branching on variables that appear in recently learned clauses. At each decision point, the solver then chooses the unassigned literal with the highest count (where ties are broken randomly by default). This approach, known as the variable state independent decaying sum (VSIDS) heuristics, was first implemented in zchaff. Zchaff maintains a list of unassigned literals sorted by count. This list is only updated when learned clauses are added, resulting in a very low overhead. Decisions can be made in constant time. The heuristic used in berkmin builds on this idea but responds more dynamically to recently learned clauses. The berkmin heuristic prefers to branch on literals that are unassigned in the most recently learned clause that is not yet satisfied.

The emphasis on variables that are involved in recent conflicting clauses leads to a locality-based search, effectively focusing on a subspace. The subspaces induced by this decision strategy tend to coalesce, resulting in more opportunities for resolution of conflicting clauses, since most of the variables are common. Representing count using integer variables leads to a large number of tied counts. Solver minisat avoids this problem by using a floating-point number to represent the weight. Another possible (but significantly more complex) strategy is to concentrate only on unresolved conflicting clauses by maintaining a stack of conflicting clauses.

When restart is used, the branching heuristic should be a combination of random selection and other heuristics, so that DPLL will select a different branch literal after each restart. All told, there are many competing branching heuristics for satisfiability solvers, and there is still much to be done in evaluating their relative effectiveness with clause learning on realistic, structured problems.

4.3 Use of SAT Solvers

There exist excellent implementations of *DPLL* and you can find these SAT solvers at the international SAT competition web page (satcompetition.org). Most of them are available, free of charge, such as minisat, glucose, lingeling, maple_LCM, to name a few. These SAT solvers are also called general-purpose model generators, because they accept CNF as input and many problems can be specified in CNF as SAT is NP-complete.

If you have a hard search problem to solve, either solving a puzzle or finding a solution under certain constraints, you may either write a special-purpose program or describe your problem in CNF and use one of these general-purpose model generators. While both special-purpose tools and general-purpose model generators rely on exhaustive search, there are fundamental differences. For the latter, every problem has a uniform internal representation (i.e., clauses for SAT solvers). This uniform representation may introduce redundancies and inefficiencies. However, since a single search engine is used for all the problems, any improvement to the search engine is significant. Moreover, we have accumulated knowledge of three decades on how to make such a uniform search engine efficient. Using general-purpose model generators to solve hard combinatorial problems deserves attention for at least two reasons.

- It is much easier to specify a problem in CNF than to write a special program to solve it. Similarly, fine-tuning a specification is much easier than fine-tuning a special program.
- General-purpose model generators can provide competitive and complementary results for certain combinatorial problems. Several examples will be provided in the following.

4.3.1 Specify SAT Instances in DIMACS Format

Today SAT solvers use the format suggested by DIMACS (the Center for Discrete Mathematics and Theoretical Computer Science) many years ago for CNF formulas. The DIMACS format uses a text file: If the first character of a line is "c," it means a comment. The CNF starts with a line "p cnf n m," where n is the number of propositional variables and m is the number of clauses. Most SAT solvers only require m to be an upper bound for the number of clauses. Each literal is represented by an integer: Positive integers are positive literals and negative integers are negative literals. A clause is a list of integers ending with 0.

Example 4.3.1 To represent $C = \{(x_1 \mid x_2 \mid \overline{x_3}), (\overline{x_1} \mid x_2), (\overline{x_2} \mid \overline{x_3}), (x_2 \mid x_3)\}$ in the DIMACS format, the text file will look as follows: □

```
c a tiny example
c
p cnf 3 4
1 2 -3 0
-1 2 0
-2 -3 0
2 3 0
```

To use today's SAT solvers, you need to prepare your SAT instances in the DIMACS format. This is done typically by writing a small program called *encoder*. Once a model is found by a SAT solver, you may need another program, called *decoder*, which translates the model into desired format. For example, the encoder for a Sudoku puzzle will generate a formula A in CNF from the puzzle and the decoder will take the model of A and generate the Sudoku solution.

Fig. 4.2 A typical example of Sudoku puzzle and its solution

	2		5		1		9	
8			2		3			6
	3			6			7	
		1				6		
5	4						1	9
		2				7		
	9			3			8	
2			8		4			7
	1		9		7		6	

Unsolved Sudoku

4	2	6	5	7	1	3	9	8
8	5	7	2	9	3	1	4	6
1	3	9	4	6	8	2	7	5
9	7	1	3	8	5	6	2	4
5	4	3	7	2	6	8	1	9
6	8	2	1	4	9	7	5	3
7	9	4	6	3	2	5	8	1
2	6	5	8	1	4	9	3	7
3	1	8	9	5	7	4	6	2

Solved Sudoku

4.3.2 Sudoku Puzzle

The standard Sudoku puzzle is to fill a 9×9 board with the digits 1 through 9, such that each row, each column, and each of nine 3×3 blocks contain all the digits from 1 to 9. A typical example of Sudoku puzzle and its solution is shown in Fig. 4.2.

To solve this problem by using a SAT solver, we need to specify the puzzle in CNF. First, we need to define the propositional variables. Let $p_{i,j,k}$ be the propositional variable such that $p_{i,j,k} = 1$ iff digit k is at row i and column j of the Sudoku board. Since $1 \leq i, j, k \leq 9$, there are $9^3 = 729$ variables.

We may encode $p_{i,j,k}$ as $81(i-1)+9(j-1)+k$, then $p_{1,1,1} = 1$ and $p_{9,9,9} = 729$. Another encoding is $p_{i,j,k} = 100i + 10j + k$, then $p_{1,1,1} = 111$ and $p_{9,9,9} = 999$. The latter wasted some variables, but it allows us to see clearly the values of i, j and k from the integer. We will use the latter; since this is a very easy SAT problem, a tiny waste does not hurt. The first three lines of the CNF file may read as

```
c Sudoku puzzle
p cnf 999 1000000
```

where 1000000 is an estimate for the number of clauses from the constraints on the puzzle.

The general constraints of the puzzle can be stated as follows:

- Each cell contains exactly one digit of any value.
- Each row contains every digit exactly once, i.e., there are no duplicate copies of a digit in a row.
- Each column contains every digit exactly once.
- Each 3×3 grid contains every digit exactly once.

A solution to a given board will say which digit should be placed in which cell.

The first constraint above can be divided into two parts: (1) Each cell contains at least one digit, and (2) each cell contains at most one digit. If every cell in a row contributes one digit and each digit appears at most once, then it implies that every digit appears exactly once in this row. The same reasoning applies to columns and grids, so the next three constraints can be simplified by dropping the constraints like "every row contains every digit at least once."

1. Each cell contains at least one digit.
2. Each cell contains at most one digit.
3. Each row contains every digit at most once.
4. Each column contains every digit at most once.
5. Each 3×3 grid contains every digit at most once.

Now it is easy to convert the above constraints into clauses:

1. For $1 \leq i, j \leq 9$, $(p_{i,j,1} \mid p_{i,j,2} \mid \cdots \mid p_{i,j,9})$.
2. For $1 \leq i, j \leq 9, 1 \leq k < c \leq 9$, $(\overline{p_{i,j,k}} \mid \overline{p_{i,j,c}})$.
3. For $1 \leq i, k \leq 9, 1 \leq j < b \leq 9$, $(\overline{p_{i,j,k}} \mid \overline{p_{i,b,k}})$.
4. For $1 \leq j, k \leq 9, 1 \leq i < a \leq 9$, $(\overline{p_{i,j,k}} \mid \overline{p_{a,j,k}})$.
5. For $0 \leq a, b \leq 2, 1 \leq i < i' \leq 3, 1 \leq j < j' \leq 3, 1 \leq k \leq 9$, $(\overline{p_{3a+i,3b+j,k}} \mid \overline{p_{3a+i',3b+j',k}})$.

Finally, we need to add the initial configuration of the board as clauses. If digit c is placed at row a and column b, we create a unit clause $(p_{a,b,c})$ for it. Each Sudoku puzzle should have at least 17 digits placed initially on the board.

You may write a small program to convert the above clauses in the DIMACS format and use one of the SAT solvers available on the internet to solve the Sudoku puzzle. A SAT solver takes no time to find the solution. If you would like to design a Sudoku puzzle for your friend, you could use a SAT solver to check if your puzzle has no solutions or has more than one solution.

4.3.3 Latin Square Problems

Definition 4.3.2 Given a set S of n elements, a *Latin square* L over S is an $n \times n$ matrix such that each row and each column of L is a permutation of S. The size of S, $|S|$, is called the order of L.

A Sudoku puzzle is a Latin square of order 9 over $S = \{1, 2, \ldots, 9\}$. People usually use $S = Z_n = \{0, 1, \ldots, n-1\}$ for a Latin square of order n.

Every Latin square can be regarded as a definition of the function $* : S \times S \mapsto S$ such that $(a * b) = c$ iff the cell at row a and column b contains c. The pair $(S, *)$ is a special structure in abstract algebra, called *quasigroup*.

Latin squares of special properties have been extensively studied in design theory and SAT solvers have been used to solve many open problems there. For example, SAT solvers have been used successfully to show the non-existence of Latin squares of orders $n = \{9, 10, 12, 13, 14, 15\}$ satisfying the constraint $((y * x) * y) * y = x$ for $x, y \in Z_n$. It remains open if there exists such a Latin square of order 18.

4.3 Use of SAT Solvers

To represent the constraint $((y*x)*y)*y = x$ in CNF, we need to "flatten" the constraint by introducing new variables: The constraint $((y*x)*y)*y = x$ is equivalent to

$$(y*x = z) \land (z*y = w) \rightarrow w*y = x$$

If we replace $y*x = z$ by $p_{y,x,z}$, etc., we have the following clause:

$$(\overline{p_{y,x,z}} \mid \overline{p_{z,y,w}} \mid p_{w,y,x})$$

for $x, y, z, w \in Z_n$. Since a Latin square can be specified in CNF as we did for a Sudoku puzzle, now searching for a Latin square of order n satisfying $((y*x)*y)*y = x$ is the same as searching a model of the following clauses:

1. For $0 \le i, j < n$, $(p_{i,j,0} \mid p_{i,j,2} \mid \cdots \mid p_{i,j,n-1})$.
2. For $0 \le i, j < n, 0 \le k < c < n$, $(\overline{p_{i,j,k}} \mid \overline{p_{i,j,c}})$.
3. For $0 \le i, k < n, 0 \le j < b < n$, $(\overline{p_{i,j,k}} \mid \overline{p_{i,b,k}})$.
4. For $0 \le j, k < n, 0 \le i < a < n$, $(\overline{p_{i,j,k}} \mid \overline{p_{a,j,k}})$.
5. For $0 \le x, y, z, w < n$, $(\overline{p_{y,x,z}} \mid \overline{p_{z,y,w}} \mid p_{w,y,x})$.

The first four sets of clauses are copied from those for the Sudoku puzzle, and the last set is the specification of $((y*x)*y)*y = x$ in propositional logic.

4.3.4 Graph Problems

Many application problems can be modeled as a graph problem, such as representing the exam scheduling problem as graph coloring problem. In the study of NP-completeness theory, we learned to reduce SAT to some graph problems. That is, we use the algorithms for the graph problems to solve SAT. Now, since we have a SAT solver, the reduction is reversed: Graph problems are reduced to SAT, i.e., we use SAT solvers to solve graph problems. The challenge is to ensure that the reduction produces a CNF of minimal size, or at least, the size of CNF is a polynomial of the input size.

Definition 4.3.3 (Graph Coloring) Given an undirected simple graph $G = (V, E)$ and a positive integer K, each number k, $1 \le k \le K$, represents a color. The *graph coloring problem* is to assign a color k, to each $v \in V$ such that the two ends of every edge of E receive different colors.

Suppose $V = \{v_1, v_2, \ldots, v_n\}$. We introduce kn propositional variables $p_{i,k}$, $1 \le i \le n, 1 \le k \le K$, with the meaning that $p_{i,k} = 1$ iff v_i receives color k. The constraints for the graph coloring are given below:

1. Every vertex receives exactly one color.
2. The two ends of every edge receive different color.

These constraints translate into the following clauses:

1. For $1 \le i \le |V|$, $(p_{i,1} \mid p_{i,2} \mid \cdots \mid p_{i,K})$.
2. For $1 \le i \le |V|$, $1 \le j < k \le K$, $(\overline{p_{i,j}} \mid \overline{p_{i,k}})$.
3. For $(v_i, v_j) \in E$, $1 \le k \le K$, $(\overline{p_{i,k}} \mid \overline{p_{j,k}})$.

The first two sets of clauses specify "at least one" and "at most one" for the constraint "exactly one," as we did in the Sudoku puzzle. The size of the first set of clauses is $K|V|$ (counting the number of literals); the size of the second set is $(K-1)K/2|V|$; the size of the third set is $2K|E|$. Strictly speaking, this is not a polynomial reduction if K is not a constant. If K is large, we may assume $K = s * t$ for some integers $s, t > 1$ and introduce s new variables $q_{i,j}$, $0 \le j < s$, such that $q_{i,j}$ is true if node i receives color c for $j * t < c \le (j+1) * t$, and use these variables to reduce the size of the second set of clauses.

Definition 4.3.4 (Clique) Given an undirected simple graph $G = (V, E)$ and a positive integer k, the *clique problem* is to find a subset $S \subseteq V$ such that $|S| = k$ and for any two vertices x, y of S, $(x, y) \in E$.

Suppose $V = \{v_1, v_2, \ldots, v_n\}$. As a first try, let us define n propositional variables, x_i, $1 \le i \le n$, with the understanding that $x_i = 1$ iff $v_i \in S$. It is easy to specify if $v_i, v_j \in S$, then $(v_i, v_j) \in E$. That is, if $(v_i, v_j) \notin E$, then we create clause $(\overline{x_i} \mid \overline{x_j})$. However, it will be inefficient to specify that $|S| = k$. To ensure that $|S| = k$, we may consider any subset $T \subset V$ with $n - k + 1$ vertices. One of them must be in S. To specify this constraint in logic, we have the following clauses:

For $T \subset V$ and $|T| = n - k + 1$, $(\vee_{v_i \in T} x_i)$.

This is a very large set of very long clauses. Evidently the reduction from the clique to CNF is not polynomial. For example, if $n = 100$ and $k = 5$, then there are $3,921,225 \ (= 100!/(4!96!))$ clauses of length 96. Obviously, any SAT solver will have difficulty to handle them.

An alternative solution is to consider that S is stored as a list L of k vertices. We introduce additionally $k * n$ propositional variables, $p_{i,j}$, $1 \le i \le n, 1 \le j \le k$, with the understanding that $p_{i,j} = 1$ iff v_i is at position j of L. We then have the following constraints:

1. There is exactly one vertex at each position of L.
2. If $v_i \in L$, then $v_i \in S$.
3. If $v_i \in S$, then $v_i \in L$ (optional).

The above constraints, plus the old constraint, can be specified in CNF as follows:

(a_1) For $1 \le j \le k$, $(p_{1,j} \mid p_{2,j} \mid \cdots \mid p_{n,j})$.
(a_2) For $1 \le j \le k$, $1 \le i < i' \le n$, $(\overline{p_{i,j}} \mid \overline{p_{i',j}})$.
(a_3) For $1 \le j \le k$, $1 \le i \le n$, $(\overline{p_{i,j}} \mid x_i)$.
(a_4) For $1 \le i \le n$, $(\overline{x_i} \mid p_{i,1} \mid p_{i,2} \mid \cdots \mid p_{i,k})$.
(a_5) For $1 \le i, j \le n$, if $(v_i, v_j) \notin E$, $(\overline{x_i} \mid \overline{x_j})$.

Without loss of generality, we may assume that L is a sorted list by adding more clauses, to reduce some symmetric solutions.

If $n = 100, k = 5$, (a_1) produces 5 clauses of length 100; (a_2) produces 24,750 ($=5*100*99/2$) clauses of length 2; (a_3) produces 500 clauses of length 2; and (a_4) produces 100 clauses of length 6, for a total of 25,355 clauses. This is a substantial reduction from 3,921,225 at the cost of introducing 500 new variables.

4.4 Maximum Satisfiability

Given a set C of clauses and a full interpretation σ, we may count how many clauses are false under σ. Let

$$f(C, \sigma) = |\{c : c \in C, \sigma(c) = 0\}|,$$

then $0 \leq f(C, \sigma) \leq |C|$, where $|C|$ is the number of clauses in C. When $f(C, \sigma) = 0$, σ is a model of C. The maximum satisfiability problem *MAX-SAT* is thus the optimization problem where f serves as the objective function: Find σ such that $f(C, \sigma)$ is minimal (or $|C| - f(C, \sigma)$ is maximal).

4.4.1 2SAT Versus Max2SAT

Let Max-kSAT denote the decision version of MaxSAT which takes a positive integer m and a set C of clauses as input, where each clause has no more than k literals, and returns true iff there exists an interpretation σ such that at least m clauses are satisfied under σ. When $m = |C|$, then Max-kSAT becomes the satisfiability problem kSAT. It is well-known that 2SAT can be solved in polynomial time, while Max2SAT is NP-complete. Thus, MaxSAT is substantially more difficult than SAT.

Theorem 4.4.1 *There is a linear time decision procedure for 2SAT.*

Proof This sketch of the proof comes from (Aspvall, Plass, Tarjan, 1979). Let C be in 2CNF without pure literals. We construct a directed graph $G = (V, E)$ for C as follows: V contains the two literals for each variable appearing in C. For each clause $(A \mid B) \in C$, we add two edges, $(\overline{A}, B)$ and $(\overline{B}, A)$, into E. Obviously, the edges represent the implication relation among the literals, because $(A \mid B) \equiv (\overline{A} \to B) \wedge (\overline{B} \to A)$. Since the implication relation is transitive and $(A \to B) \equiv (\overline{B} \to \overline{A})$, the following three statements are equivalent for any pair of literals x and y in G:

1. $C \models (x \to y)$.
2. There is a path from x to y in G.
3. There is a path from $\overline{y}$ to $\overline{x}$ in G.

Fig. 4.3 The implication graph for 2CNF C and the SCC

$C = \{ c_1 : (p \mid q), c_2 : (-p \mid -r), c_3 : (p \mid -r),$
$c_4 : (-q \mid r), c_5 : (q \mid s), c_6 : (-r \mid -s) \}$

Implication graph of C:

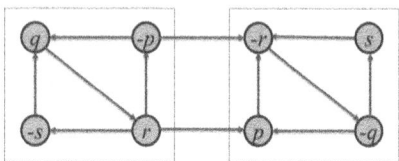

We then run the algorithm for strongly connected components (SCC), which can be done in linear time, on G. It is easy to see that $C \models (x \leftrightarrow y)$ for any two literals x and y in the same component, because x and y share a cycle in G. If both a variable, say p, and its complement, $\overline{p}$, appear in the same component, then C is unsatisfiable, because $C \models (p \leftrightarrow \overline{p})$. If no such cases exist, then C must be satisfiable. A model can be obtained by assigning the same truth value to each node in the same component, following the topological order of the components induced by G: Always assign 0 to a component, unless one of the literals in the component had already received 1. □

Example 4.4.2 Given $C = \{c_1 : (p \mid q), c_2 : (\overline{p} \mid \overline{r}), c_3 : (p \mid \overline{r}), c_4 : (\overline{q} \mid r), c_5 : (q \mid s), c6 : (\overline{r} \mid \overline{s})\}$, its implication graph is shown in Fig. 4.3. There are two SCCs: $S_1 = \{\overline{p}, q, r, \overline{s}\}$ and $S_2 = \{p, \overline{q}, \overline{r}, s\}$. We have to assign 0 to every node in S_1 (and 1 to every node in S_2) to obtain a model of C. □

Theorem 4.4.3 *Max2SAT is NP-complete.*

Proof This proof comes from Garey, Johnson, and Stockmeyer [2]. NP is a class of decision problems whose solutions can be checked in polynomial time. Max2SAT is in NP because a solution candidate is an interpretation σ and we can check how many clauses are true under σ in polynomial time.

To show Max2SAT is NP-hard, we reduce 3SAT to Max2SAT. That is, if there exists an algorithm A to solve Max2SAT, we can construct algorithm B, which uses A, to solve 3SAT, and the computing times of A and B differ by a polynomial of the input size.

Given any instance C of 3SAT, we assume that every clause of C has exactly three literals. For every clause $c_i = (l_1 \mid l_2 \mid l_3) \in C$, where l_1, l_2, and l_3 are arbitrary literals, we create 10 clauses in Max2SAT:

$$(l_1), (l_2), (l_3), (x_i), (\overline{l_1} \mid \overline{l_2}), (\overline{l_1} \mid \overline{l_3}), (\overline{l_2} \mid \overline{l_3}), (l_1 \mid \overline{x_i}), (l_2 \mid \overline{x_i}), (l_3 \mid \overline{x_i}),$$

where x_i is a new variable. Let C' be the collection of these clauses created from the clauses in C, then $|C'| = 10|C|$ and the following statements are true:

- If an interpretation satisfies c_i, then we can make seven of the ten clauses satisfied.
- If an interpretation falsifies c_i, then at most six of the ten clauses can be satisfied.

4.4 Maximum Satisfiability

The above statements can be verified by a truth table on l_1, l_2, l_3, and x_i:

l_1	l_2	l_3	x_i	$(\overline{l_1}\|\overline{l_2})$	$(\overline{l_1}\|\overline{l_3})$	$(\overline{l_2}\|\overline{l_3})$	$(l_1\|\overline{x_i})$	$(l_2\|\overline{x_i})$	$(l_3\|\overline{x_i})$	sum
0	0	0	0	1	1	1	1	1	1	6
0	0	0	1	1	1	1	0	0	0	4
0	0	1	0	1	1	1	1	1	1	7
0	0	1	1	1	1	1	0	0	1	6
0	1	1	0	1	1	0	1	1	1	7
0	1	1	1	1	1	0	0	1	1	7
1	1	1	0	0	0	0	1	1	1	6
1	1	1	1	0	0	0	1	1	1	7

The first four columns serve both the truth values of $\{l_1, l_2, l_3, x_i\}$ and the first four clauses of the ten clauses. The first two lines of the table shows that when $(l_1 \mid l_2 \mid l_3)$ is false, the sum of the truth values of the ten clauses (the last column) is at most 6. The rest of the table shows the sums of the truth values of the ten clauses when one, or two, or three literals of $\{l_1, l_2, l_3\}$ are true (some cases are ignored due to the symmetry of $l_1, l_2,$ and l_3). It is clear from the table that when $(l_1 \mid l_2 \mid l_3)$ is true, we may choose the value of x_i to obtain *seven* true clauses out of the ten clauses. Then C is satisfiable iff C' has an interpretation which satisfies $K = 7|C|$ clauses of C'. If we have an algorithm to solve C' with $K = 7|C|$, then the output of the algorithm will tell if C is satisfiable or not. □

4.4.2 Weighted and Hybrid MaxSAT

Definition 4.4.4 Given a set C of clauses and a function $w : C \mapsto \mathcal{N}$, $w(c)$ is called the *weight (or cost)* of $c \in C$. (C, w) is called *weighted CNF (WCNF)*. Given (C, w), the *weighted MaxSAT* problem is to find a full interpretation σ such that $\sum_{c \in C, \sigma(c)=0} w(c)$ is minimal (or equivalently, $\sum_{c \in C, \sigma(c)=1} w(c)$ is maximal). Any clause in C is called a *weighted clause* and denoted by $(c; w(c))$.

Example 4.4.5 Let C be a weighted MaxSAT instance, where $C = \{c_1 : (x_1; 5),$ $c_2 : (\overline{x_1} \mid x_2; 4), c_3 : (x_1 \mid \overline{x_2} \mid x_3; 3), c_4 : (\overline{x_1} \mid \overline{x_2}; 2), c_5 : (x_1 \mid x_2 \mid \overline{x_3}; 4), c_6 : (\overline{x_1} \mid x_3; 1), c_7 : (\overline{x_1} \mid \overline{x_2} \mid \overline{x_3}; 2)\}$. Then $\sigma = \{x_1, x_2, \overline{x_3}\}$ is the optimal interpretation for C, where c_4 and c_6 are false, with the total weights of falsified clauses being 3; the sum of weights of true clauses is 18. □

Definition 4.4.6 Given a WCNF (C, w), let (H, S) be a partition of the set C, where each clause $c \in H$ has weight $w(c) = \infty$, a *solution* of (H, S) is a full interpretation σ such that every clause of H is true in σ. The *hybrid MaxSAT* problem is to find an optimal solution such that $\sum_{c \in S, \sigma(c)=0} w(c)$ is minimal. The clauses in H are called *hard clauses* and the clauses in S are called *soft clauses*.

Hybrid MaxSAT is also called *weighted partial MaxSAT*. The definition of hybrid MaxSAT assumes that hybrid MaxSAT is a special case of weighted MaxSAT. Weighted MaxSAT can be also viewed as a special case of hybrid MaxSAT when $H = \emptyset$. In other words, the two definitions are equivalent in expressive power. In practice, ∞ can be replaced by the sum of all weights in S. Separating H from S will allow us to handle H efficiently, using advanced techniques for SAT.

Since MaxSAT is a special case of weighted MaxSAT or hybrid MaxSAT, weighted MaxSAT or hybrid MaxSAT must be NP-complete, too. In fact, it is very easy to reduce NP-complete problems to weighted MaxSAT or hybrid MaxSAT.

The clique problem is a well-known NP-complete problem in graph theory. Given a graph $G = (V, E)$, let $V = \{v_1, v_2, \ldots, v_n\}$. We define n propositional variables x_i, $1 \le i \le n$, with the meaning that $x_i = 1$ iff v_i is in the solution of a maximum clique. Then we may reduce the clique problem to the weighted MaxSAT problem by converting $G = (V, E)$ into a formula of 2WCNF, which consists of the following two sets of weighted clauses.

- For $1 \le i \le n$, create $(x_i; 1)$.
- For $v_i, v_j \in V, i \neq j$ and $(v_i, v_j) \notin E$, create $(\overline{x_i} \mid \overline{x_j}; n)$.

The first set of clauses says that if a vertex is chosen in the solution, there is a reward of 1 (equivalently, if the vertex is not in the solution, there is a penalty of 1). The second set of clauses specifies the property of a clique: If two vertices are chosen in the solution and there is no edge between them, there is a penalty of $|V|$.

Example 4.4.7 For the graph in Fig. 4.4, the clauses are

$$C = \{(\overline{x_a} \mid \overline{x_e}; 5), (\overline{x_b} \mid \overline{x_c}; 5), (\overline{x_c} \mid \overline{x_e}; 5), (x_a; 1), (x_b; 1), (x_c; 1), (x_d; 1), (x_e; 1)\}.$$

One of the optimal solutions is $\sigma = \{x_a \mapsto 1, x_b \mapsto 1, x_c \mapsto 0, x_d \mapsto 1, x_e \mapsto 0\}$, under which, two clauses, i.e., (x_c) and (x_e), are false and the total penalty is 2. □

Consider an interpretation σ in which one variable x_i is true; all the other variables are false. Given σ, all the clauses in the second set will be true and all the clauses in the first set will be false, except one. That is, the sum of weights of the falsified clauses is $n - 1$ under σ. Thus, the total weights of all falsified clauses in any optimal solution cannot be greater than $n - 1$. In other words, all the clauses in the second set must be true in an optimal solution. This implies that there must be an edge between two vertices in the solution. The optimal solution picks a maximal

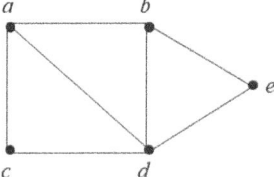

Fig. 4.4 Finding a maximal clique in a graph

number of x_i's and thus gives us a maximal clique. It is natural to specify the second set of clauses as hard clauses in the hybrid MaxSAT format.

4.4.3 Local Search Methods

Local search methods are widely used for solving optimization problems. In these methods, we first define a *search space*, and for each point in the search space, we define the *neighbors* of the point. Starting with any point (often randomly chosen), we replace repeatedly the current point by one of its neighbors, usually a better one under the objective function, until the point fits the search criteria, or we run out of time.

For SAT or MAX-SAT, the search space of common local search methods is the set of full interpretations. If C contains n variables, then the search space size is $O(2^n)$. In comparison, for exhaustive search procedures like *DPLL*, the search space size is $O(3^n)$, because all partial interpretations are considered. For convenience, we assume that a full interpretation σ is represented by a set of n literals (or equivalently, a full product), where each variable appears exactly once. A *neighbor* of σ is obtained by flipping the truth value of one variable in σ. Thus, every σ has n neighbors:

$$neighbors(\sigma) = \{\sigma - \{A\} \cup \{\neg A\} : A \in \sigma\}$$

Selman et al. proposed a greedy local search procedure called *GSAT* for SAT. The running time of *GSAT* is controlled by two parameters: *MaxTries* is the maximal number of random starting points, and *MaxFlips* is the maximal number of flips allowed for any starting point.

Algorithm 4.4.8 $GSAT(C)$ takes as input a set C of clauses and returns a model of C if it finds one; it returns "unknown" if it terminates without finding a model.

proc $GSAT(C, \sigma)$
1   **for** $i := 1$ **to** $MaxTries$ **do**
2         $\sigma :=$ a random full interpretation of C // a starting point
3   **for** $j := 1$ **to** $MaxFlips$ **do**
4   **if** $\sigma(C) = 1$ **return** σ // a model is found.
5   pick $A \in \sigma$ // pick a literal of σ for flipping.
6             $\sigma' := \sigma - \{A\} \cup \{\neg A\}$
7             $\sigma := \sigma'$
8   **return** "unknown"

Example 4.4.9 Let $C = \{c_1 : (x_1 \mid x_3 \mid x_4),\ c_2 : (\overline{x_1} \mid \overline{x_4}),\ c_3 : (\overline{x_2} \mid \overline{x_4}),\ c_4 : (\overline{x_1} \mid x_2 \mid \overline{x_3}),\ c_5 : (x_2 \mid \overline{x_4})\}$. If the starting interpretation is $\sigma_0 = \{x_1, x_2, x_3, x_4\}$, $f(C, \sigma_0) = 2$ as $\sigma_0(c_2) = \sigma_0(c_3) = 0$. If x_1 is chosen for flipping, we obtain $\sigma_1 = \{\overline{x_1}, x_2, x_3, x_4\}$ and $f(C, \sigma_1) = 1$ since $\sigma_1(c_3) = 0$. If x_2 is chosen next for flipping, then $\sigma_2 = \{\overline{x_1}, \overline{x_2}, x_3, x_4\}$ and $f(C, \sigma_2) = 1$ since $\sigma_2(c_5) = 0$. If x_4 is chosen next for flipping, then $\sigma_3 = \{\overline{x_1}, \overline{x_2}, x_3, \overline{x_4}\}$ and $f(C, \sigma_3) = 0$, i.e., σ_3 is a model of C. □

There exist many variations of *GSAT*, as there are many ways to pick a literal at line 5. In general, we need to pick a literal $A \in neighbors(\sigma)$ such that the resulting interpretation σ', here $\sigma' = \sigma - \{A\} \cup \{\neg A\}$, satisfies $(f(C, \sigma') \leq f(C, \sigma))$. We may randomly pick one of such literals or pick the best one such that $f(C, \sigma')$ is minimal.

When $f(C, \sigma') = f(C, \sigma)$, *GSAT* still chooses to move from σ to σ', and such moves are called "sideways" moves; they are moves that do not increase or decrease the total number of unsatisfied clauses. To avoid *GSAT* going back and forth between two neighboring interpretations of the same f value, the idea of *tabu search* is needed, so that recent visited interpretations are not allowed to be the next interpretation. The search of *GSAT* typically begins with a rapid greedy descent towards a better truth assignment, followed by long sequences of "sideways" moves. Experiments indicate that on many formulas, *GSAT* spends most of its time on sideways moves.

We can also allow moves that increase the value of f (called *upward moves*) occasionally, especially in the beginning of the execution. For example, using the idea of "simulated annealing," upward moves can be allowed with a probability, and this probability decreases as the number of flips increases. To implement this idea, we replace line 7 by the following two lines:

7    $x :=$ random(0, 1) // x is a random number in [0, 1)
7'   **if** $(x < p \vee (f(C, \sigma') \leq f(C, \sigma))$ $\sigma := \sigma'$

Selman et al. proposed a variation of *GSAT*, called walksat, which select the best variable to flip in a randomly chosen conflicting clause. Walksat's strategy for picking a literal to flip is as follows:

1. Randomly pick a conflicting clause, say α.
2. If there is a literal $B \in \alpha$ such that the flipping of B does not turn any currently satisfied clauses to unsatisfied, return B as the picked literal.
3. With probability p, randomly pick $B \in \alpha$ and return B.
4. With probability $1 - p$, pick $B \in \alpha$ such that the flipping of B turns the least number of currently satisfied clauses to unsatisfied, return B.

Since the early 1990s, there has been active research on designing, understanding, and improving local search methods for SAT. In the framework of *GSAT*, various ideas are proposed, and some of them are given below.

- When picking a literal, take the literal that was flipped longest ago for breaking ties.
- A weight is given to each clause, incrementing the weight of unsatisfied clauses by one for each interpretation. One of the literals occurring in more clauses of maximum weight is picked.

- A weight is given to each variable, which is increased in variables that are often flipped. The variable with minimum weight is chosen.

Selman et al. showed through experiments that *GSAT* as well as `walksat` substantially outperformed the best DPLL-based SAT solvers on some classes of formulas, including randomly generated formulas and graph coloring problems.

GSAT returns either a model or "unknown," as it can never declare that "the input clauses are unsatisfiable." This is a typical feature of local search methods as they are incomplete in the sense that they do not provide the guarantee that it will eventually either report a satisfying assignment or declare that the given formula is unsatisfiable.

There have also been attempts at hybrid approaches that explore the combination of ideas from *DPLL* methods and local search techniques. There has also been work on formally analyzing local search methods, yielding some of the best $O(2^n)$ time algorithms for SAT. For instance, the expected running time, ignoring polynomial factors, of a simple local search method with restarts after every $3n$ "flips" has been shown to be $((k+1)/k)^n$ for kCNF instances of n variables, where each clause contains at most k literals. When $k = 3$, the result yields a complexity of $O((4/3)^n)$ for 3SAT.

The algorithm *GSAT* can be used to solve the MaxSAT problem, which asks to find an interpretation for a set of clauses such that a maximal number of clauses are true. Since *GSAT* is incomplete, the answer found by *GSAT* is not guaranteed to be optimal. This is a typical feature of local search methods which always find local optimal solutions. *GSAT* can also be used to solve weighted MaxSAT and hybrid MaxSAT.

4.4.4 The Branch-and-Bound Algorithm

To obtain an optimal solution of MaxSAT or hybrid MaxSAT, the traditional approach is to use the branch-and-bound algorithm. In these algorithms, we often cast MaxSAT as a minimization problem: Find a solution with minimum (optimal) cost. The branching part of the branch-and-bound algorithm is very much like *DPLL*, doing case splits on selected literals. The bounding part of the algorithm checks if an estimation of the current solution is better or not than the solution found so far. If not better, the algorithm cuts this futile branch and backtracks. If we look for a minimal solution, the estimation produces a lower bound; for a maximal solution, the estimation produces an upper bound. For example, if we look for a minimal solution, and already have a solution of value 30, if the estimation says the lower bound is 32, then there is no point to continue, because the value of the solution from the current branch cannot be lower than 32.

Here we present the branch-and-bound algorithm for hybrid MaxSAT in the style of recursive *DPLL* (Algorithm 4.1.1).

Algorithm 4.4.10 *HyMaxSATBB*($H, F, \sigma, soln$) takes as input a set H of hard clauses, a set F of weighted soft clauses, a partial interpretation σ, and the best solution *soln* found so far. H and σ do not share any variables. It looks for a full interpretation σ which is a model of H and the sum of the weights of the false clauses in F is minimal under σ; this weight sum will be returned by the algorithm. Procedure *pickLiteral*(S) will pick a literal appearing in S.

proc *HyMaxSATBB*($H, F, \sigma, soln$)
1   $res := BCP(H)$
2   **if** ($res = \bot$) **return** ∞ // hard causes violated
3   // Assume $res = (U, S)$
4   $\sigma := \sigma \cup U$
5   $F := simplifySoft(F, \sigma)$
6   **if** $soln \leq lowerBound(F)$ **return** $soln$
7   $A := pickLiteral(S \cup F)$
8   **if** ($A = nil$) **return** $countFalseClauseWeight(F, \sigma)$
9   $soln := min(soln, HyMaxSATBB(S \cup \{(A)\}, F, \sigma, soln))$
10 **return** $min(soln, HyMaxSATBB(S \cup \{(\overline{A})\}, F, \sigma, soln))$

We have seen *BCP* in Algorithm 4.2.5; the following four procedures are called inside *HyMaxSATBB*:

- *simplifySoft*(F, σ) at line 5: Simplify soft clauses F using the partial interpretation σ and keep the total weights of false clauses unchanged under any interpretation. Details will follow.
- *lowerBound*(F) at line 6: Estimate the minimal total weights of false clauses when σ is extended to a full interpretation in all possible ways. Details will be discussed later.
- *pickLiteral*($S \cup F$) at line 7: Pick an unassigned literal for case splits; it returns *nil* if every literal has a truth value. Here is the place heuristics for branching are implemented.
- *countFalseClauseWeight*(F, σ) at line 8: Compute the sum of all the weights of false clauses in $F\sigma$. Here, σ is a full interpretation and can be saved for late use.

For an efficient implementation of *HyMaxSATBB*, the first solution is always obtained by a local search algorithm such as *GSAT*. We can use all techniques used for *DPLL*, including iterative procedure, destructive data structure, and conflict-driven clause learning for hard clauses.

For soft clauses, the treatment is different. For example, unit resolution is a powerful simplification rule for hard clauses, but it is unsound for soft clauses. If $C = \{(\overline{x_1}; 1), (x_1 \mid x_2; 1), (x_1 \mid \overline{x_2}; 1), (x_1 \mid x_3; 1), (x_1 \mid \overline{x_3}; 1)\}$, the optimal solution has one false clause when $x_1 \mapsto 1$. If we apply unit resolution to C to get $C' = \{(\overline{x_1}; 1), (x_2; 1), (\overline{x_2}; 1), (x_3; 1), (\overline{x_3}; 1)\}$, an optimal solution of C' will contain two false clauses. Of course, if a unit clause is hard, we may apply unit resolution to both hard and soft clauses.

4.4 Maximum Satisfiability

Simplification Rules and Lower Bounds

Definition 4.4.11 An inference rule or simplification rule applicable to a set C of weighted clauses is said to be *sound*, if an optimal solution of the resulting set C' of weighted clauses has the same value as an optimal solution of C.

In other words, sound rules should preserve the value of optimal solutions. Most inference rules of propositional logic are unsound for weighted clauses because they generate extra clauses. One exception is when a hard clause can be generated from soft clauses. For example, given a soft clause $(c; w)$, if c becomes false, w will be added as a portion of the lower bound. If the addition of w causes the lower bound to exceed the value of the current solution, then we can generate the hard clause c to prevent c from being false. In the following, we focus on simplification rules, which replace some clauses with other clauses when we transform C to C'. Each simplification rule can be denoted by

$$C_1, C_2, ..., C_m \Longrightarrow C'_1, C'_2, ..., C'_n$$

when we want to replace $\{C_1, C_2, ..., C_m\} \subseteq C$ by $\{C'_1, C'_2, ..., C'_n\}$, where C_i, C'_j are weighted clauses.

In the following rules, α and β denote arbitrary, possibly empty, disjunction of literals.

1. **Zero elimination**: Weighted clauses with zero weights can be discarded.

   $$(\alpha; 0) \Longrightarrow \top$$

2. **Tautology elimination**: Valid clauses are always true and can be discarded.

   $$(p \mid \overline{p} \mid \alpha; w) \Longrightarrow \top$$

3. **Identical merge**: Identical clauses are merged into one.

   $$(\alpha; w_1), (\alpha; w_2) \Longrightarrow (\alpha; w_1 + w_2)$$

4. **Unit clash**: Unit clauses with complement literals can be reduced to one.

   $$(A; w_1), (\overline{A}; w_2) \Longrightarrow (A; w_1 - w_2), (\bot; w_2)$$

   where A is a literal and $w_1 \geq w_2$. For example, given $(p; 2)$ and $(\overline{p}; 3)$, if $p \mapsto 1$, we get a cost of 3 from $(\overline{p}; 3)$; if $p \mapsto 0$, we get a cost of 2 from $(p; 2)$. The same result can be obtained from $(\overline{p}, 1)$ and $(\bot; 2)$; the latter is added into the cost function.

5. **Weighted resolution**:

$$(p \mid \alpha; w_1), (\overline{p} \mid \beta; w_2) \implies \begin{array}{l}(\alpha \mid \beta; u), (p \mid \alpha, w_1 - u), (\overline{p} \mid \beta; w_2 - u), \\ (p \mid \alpha \mid \overline{\beta}; u), (\overline{p} \mid \overline{\alpha} \mid \beta; u)\end{array}$$

where $w_1 > 0$, $w_2 > 0$, and $u = min(w_1, w_2)$. This rule is proposed by Larrosa et al. (2007). Since either $w_1 - u = 0$ or $w_2 - u = 0$, the above rule cannot be used forever. For example, from $(p \mid q; 2)$ and $(\overline{p} \mid q; 3)$, $\alpha = \beta = q$, $u = min(2, 3)$, we obtain five clauses: $(q; 2)$, $(p \mid q; 0)$, $(\overline{p} \mid q; 1)$, $(p \mid q \mid \overline{q}; 2)$, and $(\overline{p} \mid \overline{q} \mid q; 2)$. The second clause is removed by zero elimination, and the last two clauses by tautology elimination, we obtain two clauses: $(q; 2)$, $(\overline{p} \mid q; 1)$. Note that $\overline{\beta}$ and $\overline{\alpha}$ in the last two clauses are not disjunction of literals in general. If this is the case, we need to convert $(p \mid \alpha \mid \overline{\beta}; u)$ (or $(\overline{p} \mid \overline{\alpha} \mid \beta; u)$) into a set of clauses, each having the weight u. In practice, this rule often applies to short clauses or when $\alpha = \beta$.

It is easy to check all of the above simplification rules are sound for weighted clauses, as they preserve optimal solutions.

To efficiently implement *simplifySoft*(F, σ), F will be simplified first by σ, as σ represents a set of hard unit clauses. Like BCP, every true clause under σ will be removed from F and every false literal under σ will be removed from the clauses in F. After this step, F and σ do not share any variable. The second step is to apply the above simplification rules. Some rules like zero elimination and tautology elimination, are applied eagerly. Some rules like weighted resolution are applied selectively.

To implement *lowerBound*$(F, soln)$, at first, we assume that F is simplified and we count the total weights of all false clauses. If the sum exceeds the current solution, we exit.

For non-false clauses of F, there exist various techniques to estimate the lower bound. One technique is to formulate F as an instance of integer linear programming (ILP). ILP solvers are common optimization tools for operation research. ILP solvers solve problems with linear constraints and linear objective function where some variables are integers. State-of-the-art ILP solvers are powerful and effective and can attack MaxSAT directly. In practice, ILP solvers are effective on many standard optimization problems, e.g., vertex cover, but for problems where there are many Boolean constraints, ILP is not as effective. More information on ILP can be found in Chap. 12.

To obtain a lower bound of ILP, the constraints on integer solutions are relaxed so that ILP becomes an instance of the linear programming (LP), which is known to have polynomial-time solutions.

If we view $\top$ as 1 and $\bot$ as 0, it is well-known that a clause can be converted into a linear inequation, such as $(\overline{x_1} \mid x_2 \mid \overline{x_3})$ becomes $(1 - x_1) + x_2 + (1 - x_3) \geq 1$, so that the clauses are satisfiable iff the set of inequations has a 0/1 solution for all variables.

4.4 Maximum Satisfiability

For any clause c, let $P(c)$ and $N(c)$ be the sets of positive literals and negative literals in c, respectively. Let $F = \{(c_i; w_i) : 1 \leq i \leq m\}$ over n variables, say $\{x_1, x_2, \ldots, x_n\}$. F can be formulated as the following instance of ILP:

minimize $\sum_{1 \leq i \leq m} w_i(1 - y_i)$ // minimize the weights of falsified clauses
subject to $\sum_{x \in P(c_i)} x + \sum_{\bar{x} \in N(c_i)} (1 - x) \geq y_i, 1 \leq i \leq m$,   // c_i is true iff $y_i = 1$
$y_i \in \{0, 1\}, 1 \leq i \leq m$,   // every clause is falsified or satisfied
$x_j \in \{0, 1\}, 1 \leq j \leq n$.   // every variable is false or true
where y_i is a new variable for each clause c_i in F.

Intuitively, $y_i = 1$ implies that $\sum_{x \in P(c_i)} x + \sum_{\bar{x} \in N(c_i)} (1 - x) \geq 1$, i.e., c_i is true. Thus, the more y_i's with $y_i = 1$, the less the value of $\sum_{1 \leq i \leq m} w_i(1 - y_i)$.

The above ILP instance can be relaxed to an instance L of LP:

minimize $\sum_{1 \leq i \leq m} w_i(1 - y_i)$
subject to $\sum_{x \in P(c_i)} x + \sum_{\bar{x} \in N(c_i)} (1 - x) \geq y_i, 1 \leq i \leq m$,
$0 \leq y_i \leq 1, 1 \leq i \leq m$, and
$0 \leq x_j \leq 1, 1 \leq j \leq n$.

If L has a solution, then $\sum_{1 \leq i \leq m} w_i(1 - y_i)$ is a lower bound for the solutions of F. The LP can be used to obtain an approximate solution for F when we round up/down the values of x_i from reals to integers.

Other techniques for lower bounds look for inconsistencies that force some soft clause to be falsified. Some people suggested learning falsified soft clauses; some people use the concepts of *clone*, *minibcukets*, or width restricted BDD in relaxation. For instance, we may treat the soft clauses as if they were hard and then run BCP to locate falsified clauses. This can help us to find unsatisfiable subsets of clauses quickly and use them to estimate the total weights of falsified clauses. These techniques can be effective on small combinatorial problems. Once the number of variables in F gets to 1,000 or more, these lower bound techniques become weak or too expensive. Strategies for turning on/off these techniques in *HyMaxSATBB* remain to be research topics.

Besides the algorithm based on branch-and-bound, some MaxSAT solvers convert a MaxSAT instance into a sequence of SAT instances where each instance encodes a decision problem of the form, for different values of k, "is there an interpretation that falsifies soft clauses of total weights at most k?"

For example, if we start with a small value of k, the SAT instance will be unsatisfiable. By increasing k gradually, the first k when the SAT instance becomes satisfiable will be the value of an optimal solution. Of course, this linear strategy on k does not provide necessarily an effective MaxSAT solver.

There are much ongoing research works in this direction. The focus of this approach is mainly doing two things:

1. Develop economic ways to encode the decision problem at each stage of the sequence.
2. Exploit information obtained from the SAT solver at each stage in the next stage.

Some MaxSAT solvers use innovative techniques to obtain more efficient ways to encode and solve the individual SAT decision problems. Some solvers use unsatisfiable cores to improve SAT solving efficiency. These solvers appear promising as they are effective on some large MaxSAT problems, especially those with many hard clauses.

4.4.5 Use of Hybrid MaxSAT Solvers

Most real-world problems involve an optimization component and there is high demand for automated approaches to finding good solutions to computationally hard optimization problems. We have seen how to encode the clique problem easily and compactly into a MaxSAT instance. MaxSAT allows for compactly encoding various types of high-level finite-domain soft constraints.

There exists an extension of DIMACS format for WCNF. To specify an instance of hybrid MaxSAT, the line starting with p will contain the keyword wcnf, followed by the number of variables, the number of clauses, and the label for hard clauses. That is,

```
p wcnf <# of variables> <# of clauses> <hard clause label>
```

The hard clause label is an integer larger than the maximal weight of soft clauses. Each clause starts with the weight value, followed by literals and ending with 0. For the small clique example of Example 4.4.7, the input file to a MaxSAT solver will look like:

```
c small clique example
c xa = 1, xb = 2, xc = 3, xd = 4, xe = 5
p wcnf 5 8 9
9 -1 -5 0
9 -2 -3 0
9 -3 -5 0
1 1 0
1 2 0
1 3 0
1 4 0
1 5 0
```

A clause is hard if the weight of the clause is the label for hard clauses.

The standard format for MaxSAT instances makes the annual evaluation of MaxSAT solvers possible. The latest website for MaxSAT evaluation can be found on the internet and the site for the year 2020 is given below:

https://maxsat-evaluations.github.io/2020/

Researchers on MaxSAT can assess the state-of-the-art MaxSAT solvers and create a collection of publicly available MaxSAT benchmark instances.

These MaxSAT solvers are built on the successful techniques for SAT, evolve constantly in practical solver technology, and offer an alternative to traditional approaches, e.g., integer programming. MaxSAT solvers use propositional logic as the underlying declarative language and are especially suited for inherently "very Boolean" optimization problems. The performance of MaxSAT solvers has surpassed that of specialized algorithms on several problem domains:

- Correlation clustering
- Probabilistic inference
- Maximum quartet consistency
- Software package management
- Fault localization
- Reasoning over bionetwork
- Optimal covering arrays
- Treewidth computation
- Bayesian network structure learning
- Causal discovery
- Cutting planes for IP (integer programming)
- Argumentation dynamics

The advances of contemporary SAT and MaxSAT solvers have transformed the way we think about NP-complete problems. They have shown that, while these problems are still unmanageable in the worst case, many instances can be successfully tackled.

Exercises

1. Draw the four complete decision trees (trees of recursive calls) of *DPLL* on the following set of clauses:

$$\{(p \mid q), (p \mid \overline{r}), (\overline{p} \mid r), (\overline{p} \mid q \mid \overline{r}), (\overline{q} \mid \overline{r})\}$$

   using two orders of variables, p, q, r and r, q, p, respectively, for splitting (either value can be chosen first), and with or without BCP, respectively.

2. Draw the complete decision tree of *DPLL* on the following set of clauses:

$$\{(p \mid q \mid s), (p \mid \overline{r} \mid \overline{s}), (q \mid \overline{r} \mid s), (\overline{p} \mid q \mid \overline{r}), (\overline{q} \mid \overline{r} \mid \overline{s})\}$$

   using the orders of p, q, r, s for splitting (the value 1 is used first for each splitting variable) and report the truth values of variables at various levels. How many models are found by *DPLL*? What is the guiding path when each model is found?

3. If we feed the clause set in the previous problem to a SAT solver which works as a black box and can produce only one model at a time, what clause should be added into the clause set so that the SAT solver will produce a new model?
4. Estimate a tight upper bound of the size of the inference graph for resolution, if the input clauses contain n variables.
5. Provide the pseudo-code for implementing *BCPw* based on *BCPht* such that the space for $head(c)$ and $tail(c)$ are freed by assuming $head(c) = 0$ and $tail(c) = |c| - 1$ for any non-unit clause c, and no additional space can be used for c. In your code, you may call $swap(lits(c), i, j)$, which swaps the two elements at indices i and j in the array $lits(c)$.
6. Given a set H of $m + 1$ Horn clauses over m variables:

$$H = \{c_1 = (\overline{x_1} \mid \overline{x_2} \mid \cdots \mid \overline{x_{m-1}} \mid \overline{x_m}), c_2 = (\overline{x_1} \mid x_2), c_3 = (\overline{x_2} \mid x_3), \ldots,$$
$$c_m = (\overline{x_{m-1}} \mid x_m), c_{m+1} = (x_1)\}$$

Show that the size of H is $O(m)$, and the algorithm *BCPw* in the previous problem will take $\Omega(m^2)$ time on H, if all the clauses in $cls(A)$ must be processed before considering $cls(B)$, where $\neg A$ is assigned true before $\neg B$.
7. We apply *DPLL* (without CDCL) to the following set S of clauses: $S = \{1.\ (a \mid c \mid \overline{d} \mid f),\ 2.\ (a \mid d \mid \overline{e}),\ 3.\ (\overline{a} \mid c),\ 4.\ (b \mid \overline{d} \mid e),\ 5.\ (b \mid d \mid \overline{f}),\ 6.\ (\overline{b} \mid \overline{f}),\ 7.\ (c \mid \overline{e}),\ 8.\ (c \mid e),\ 9.\ (\overline{c} \mid \overline{d})\ \}$, assuming that *pickLiteral* picks a literal in the following order: $\overline{a}, \overline{b}, \overline{c}, \overline{d}, \overline{e}, \overline{f}$ (false is tried first for each variable). Please (a) draw the decision tree of DPLL, (b) provide the unit clauses derived by BCP at each node of the decision tree, and (c) identify the head/tail literals of each clause at the end of *DPLL* (assuming initially the first and last literals are the head/tail literals).
8. We apply *GSAT* to the set S of clauses in the previous problem. Assume that the initial node (i.e., interpretation) is $\sigma_0 = \{\overline{a}, b, \overline{c}, d, \overline{e}, f\}$, and the strategy for picking a neighbor to replace the current node is (a) the best among all the neighbors and (b) the first neighbor which is better than the current node (flipping variables in the order of $a, b, \ldots, f$). Sideways moves are allowed if no neighbors are better than the current node. If no neighbors are better than the current node, a local optimum is found and you need to restart *GSAT* with the node $\sigma_0 = \{a, \overline{b}, c, \overline{d}, e, \overline{f}\}$.
9. We apply *DPLL* with CDCL to the clause set C : $\{c_1 : (x_1 \mid x_2 \mid x_5),\ c_2 : (x_1 \mid x_3 \mid \overline{x_4} \mid x_5),\ c_3 : (\overline{x_1} \mid x_2 \mid x_3 \mid x_4),\ c_4 : (\overline{x_3} \mid x_4 \mid x_5),\ c_5 : (\overline{x_2} \mid x_3 \mid x_5),\ c_6 : (\overline{x_3} \mid \overline{x_4} \mid \overline{x_5}),\ c_7 : (\overline{x_3} \mid x_4 \mid \overline{x_5})\}$. If *pickLiteral* picks a literal for case split in the following order $\overline{x_5}, \overline{x_4}, \ldots, \overline{x_1}$ ($x_i = 0$ is tried first), please answer the following questions:

(a) For each conflicting clause found in $DPLL(C)$, what new clause will be generated by *conflictAnalysis*?
(b) What will be the head and tail literals if the new clause generated in (a) is not unit? And what will be the value of *level* at the end of the call to *insertNewClause*?

(c) Draw the *DPLL* decision tree until either a model is found or *DPLL(C)* returns *false*.

10. The pigeonhole problem is to place $n + 1$ pigeons into n holes such that every pigeon must be in a hole and no holes can hold more than one pigeon. Please write an encoder in any programming language that reads in an integer n and generates the CNF formula for the pigeonhole problem with $n + 1$ pigeons and n holes in the DIMACS format. Store the output of your encoder in a file named pigeonX.cnf, where X is n, and feed pigeonX.cnf to a SAT solver which accepts CNF in the DIMACS format. Turn in both the encoder, the result of the SAT solver for $n = 4, 6, 8, 10$, and the CNF file for $n = 4$ (i.e., pigeon4.cnf), plus a summary of the outputs of SAT solvers on the sizes of the input, the computing times, and the numbers of conflicts. What is the relation between the size of the input and the computing time?

11. The n-queen problem is to place n queens on an $n \times n$ board such that no two queens are in the same row, the same column, or the same diagonal. Please write a program called *encoder* in any programming language that reads in an integer n and generates the CNF formula for the n-queen problem in the DIMACS format. Store the output of your encoder in a file named queenX.cnf and feed queenX.cnf to a SAT solver which accepts CNF in the DIMACS format. Write another program called *decoder* which reads the output of the SAT solver and display a solution of n-queen on your computer screen. Turn in both the encoder and decoder, the result of the SAT solver for $n = 5, 10, 15$, the CNF file for $n = 5$ (i.e., queen5.cnf), and the output of your decoder for $n = 10$.

12. You are asked to write two small programs, one called *encoder* and the other called *decoder*. The encoder will read a Sudoku puzzle from a file (some examples are provided online) and generate the CNF formula for this puzzle in the DIMACS format. Store the output of your encoder in a file named sudokuN.cnf and feed sudokuN.cnf to a SAT solver which accepts CNF in the DIMACS format. The decoder will read the output of the SAT solver and displays a solution of the Sudoku puzzle on your computer screen. Turn in both the encoder and decoder, the result of the SAT solver, and the result of your decoder for the provided Sudoku puzzles.

13. A Latin square over $N = \{0, 1, \ldots, n - 1\}$ is an $n \times n$ square where each row and each column of the square is a permutation of the numbers from N. Every Latin square defines a binary function $* : N, N \mapsto N$ such that $(x * y) = z$ iff the entry at the xth row and the yth column is z. You are asked to write a small program called *encoder*. The encoder will read an integer n and generates the CNF formula for the Latin square of size n satisfying the constraints $x * x = x$ and $(x * y) * (y * x) = y$ (called *Stein's third law*) in the DIMACS format. Store the output of your encoder in a file named latinN.cnf and feed latinN.cnf to a SAT solver which accepts CNF in the DIMACS format. Turn in the encoder, and a summary of the results of the SAT solver for $n = 4, 5, \ldots, 9$.

14. Given an undirected simple graph $G = (V, E)$, an *independent set* of G is a subset $X \subseteq V$ such that for any two vertices of X, $(x, y) \notin E$. The

independent set problem is to find a maximum independent set of G. (a) Specify the independent set problem in hybrid MaxSAT; and (b) specify the decision version of the independent set problem in SAT.

15. Given an undirected simple graph $G = (V, E)$, a *vertex cover* of G is a subset $X \subseteq V$ such that for every edge $(x, y) \in E$, $\{x, y\} \cap X \neq \emptyset$. The *vertex cover problem* is to find a minimal vertex cover of G. (a) Specify the vertex cover problem in hybrid MaxSAT; and (b) specify the decision version of the vertex cover problem in SAT.

16. Given an undirected simple graph $G = (V, E)$, the *longest path problem* is to find the longest simple path (no repeated vertices) in G. (a) Specify the longest path problem in hybrid MaxSAT; and (b) specify the decision version of the longest path problem in SAT.

References

1. A. Biere, M. Heule, H. Van Maaren, T. Walsh (eds.) *Handbook of Satisfiability*, Frontiers in Artificial Intelligence and Applications, 336, April 2021
2. M. R. Garey, D. S. Johnson, and L. Stockmeyer, "Some Simplified NP-Complete Graph Problems," Theoretical Computer Science, 1976, pp. 237–267

Part II
First-Order Logic

Chapter 5
First-Order Logic

Propositional logic provides a good start at describing the general principles of logical reasoning, but it is not possible to express general properties of many important sets, especially when a set is infinite. For example, how to specify the following statements: Every natural number is either even or odd. A natural number is even iff its successor is odd. The relation $<$ is transitive over the natural numbers.

The weak expressive power of propositional logic accounts for its relative mathematical simplicity, but it is a very severe limitation, and it is desirable to have a more expressive logic. Mathematics and some other disciplines such as computer science often consider sets of elements in which certain relations and operations are singled out. When using the language of propositional logic, our ability to talk about the properties of such relations and operations is very limited. Thus, it is necessary to refine the logic language, in order to increase the expressive power of logic. This is exactly what is provided by a logical framework known as first-order logic, which will be the topic of the next three chapters. First-order logic is a considerably richer logic than propositional logic, yet enjoys many nice mathematical properties, of which many are inherited from propositional logic.

In first-order logic, many interesting and important properties about various sets can be expressed. Technically, this is achieved by allowing the propositional symbols to have arguments ranging over elements of sets, to express the relations and operations in question. These propositional symbols are called *predicate symbols*. This is the reason that first-order logic is called *predicate calculus* [1, 2].

5.1 Syntax of First-Order Languages

Logic is a collection of closely related formal languages. There are certain languages called first-order languages, and together they form first-order logic. Our study of first-order logic will parallel the study of propositional logic: We will at first define the syntax and semantics of first-order languages as an extension of propositional languages, and then present various reasoning tools for these languages.

5.1.1 Terms and Formulas

Consider some general statements about the natural numbers:
- Every natural number is even or odd, but not both.
- A natural number x is even iff $x + 1$ is odd.
- For any natural number x, $x < x + 1$ and $\neg(x < x)$.
- For any three natural numbers x, y, and z, if $x < y$ and $y < z$, then $x < z$.
- For every natural number x, there exists a natural number y, such that $x < y$.

Before any attempt to prove these statements, let us consider how to write them in a formal language. First, we need a way to describe the set of natural numbers. Like propositional logic, first-order logic does not assume the existence of any special sets, except some symbols, and everything needs to be constructed from scratch. Fortunately, from Example 1.3.8, we know that the set of natural numbers can be constructed recursively from two symbols, i.e., the constant 0 and the unary function symbol s: Natural number i is uniquely represented by the term $s^i(0)$, which means there are i copies of s in $s(s(\ldots s(0) \ldots))$. Hence, our first-order language will have 0 and s.

Next, to express "even" or "odd" properties, we may introduce the predicate symbols *even* and *odd*, which are examples of decision functions that take a natural number and return a Boolean value. The relation $<$ is used in the above statements and we take it in the first-order language. Since some variables like x, y, and z are already used in the statements, we borrow them, too.

Finally, to express "for every" and "there exists" in natural languages, we add two important symbols, $\forall$ and $\exists$, respectively, in a first-order language. The symbol $\forall$ is called *universal quantifier* and $\exists$ is called *existential quantifier*. These two symbols will be discussed extensively in the next subsection.

Definition 5.1.1 A first-order language L is built up by an *alphabet* that consists of four sets of symbols:

- P is a set of *predicate* symbols, including propositional variables, usually denoted by p, q, r, with or without subscript, or some popular binary relation symbols, such as $<, \leq, \geq, >$, etc.
- F is a set of *function* symbols, including constants, usually denoted by a, b, c (for constants), f, g and h, with or without subscript.
- X is a set of *variables*, usually denoted by x, y, and z, with or without subscript.
- Op is a set of logical operators such as $\top, \bot, \neg, \wedge, \vee, \rightarrow$, and $\leftrightarrow$, and quantifiers $\forall$ and $\exists$.

5.1 Syntax of First-Order Languages

For each symbol in $P \cup F$, the number of arguments for that symbol is fixed and is dictated by the *arity* function $arity : P \cup F \mapsto \mathcal{N}$. If $arity(p) = 0$ for $p \in P$, p is the same as a propositional variable; if $arity(f) = 0$ for $f \in F$, f is a constant.

The pair (P, F) with arity is called the *signature* of the language. We denote the language L by $L = (P, F, X, Op)$.

In general, P or F could be empty; X and Op can never be empty. By convention, we use $\{p/arity(p) \mid p \in P\}$ and $\{f/arity(f) \mid f \in F\}$ to show P and F with arity in a compact way. For instance, $even/1$ says $arity(even) = 1$. In theory, P or F or X could be infinite, but in practice, they are always finite. We also assume that no symbols appear more than once in these four sets. Rather than fixing a single language once and for all, different signatures generate different first-order languages and allow us to specify the symbols we wish to use for any given domain of interest.

Example 5.1.2 To specify the properties of the natural numbers listed at the beginning of this section, we use the signature (P, F), where $P = \{even/1, odd/1, </2\}$, $F = \{0/0, s/1\}$. The arity is given after the symbol / following each predicate or function symbol. □

Once a signature is given, we add some variables X and logical operators Op, to obtain the alphabet for a first-order language. First-order logic allows us to build complex expressions out of the alphabet. Starting with the variables and constants, we can use the function symbols to build up compound expressions which are called *terms*.

Definition 5.1.3 Given first-order language $L = (P, F, X, Op)$, the set of *terms* of L are the set of strings built up by F and X, which can be formally defined by the following BNF grammar:

$\langle Constants \rangle ::= a$ **if** $a \in F, arity(a) = 0$
$\langle Variables \rangle ::= x$ **if** $x \in X$
  $\langle Terms \rangle ::= \langle Constants \rangle \mid \langle Variables \rangle \mid$
    $f(\langle Terms \rangle_1, \langle Terms \rangle_2, \ldots, \langle Terms \rangle_k)$ **if** $f \in F, arity(f) = k$

The set of terms in $\langle Terms \rangle$ is often denoted by $T(F, X)$. A term is *ground* if it does not contain any variable. The set of all ground terms is denoted by $T(F)$. Note that $T(F) = \emptyset$ if F contains no constants.

For $F = \{0, s\}$ in the previous example, $T(F) = \{0, s(0), s(s(0)), \ldots\}$ and for $X = \{x, y\}$, $T(F, X) = \{0, s(0), s(x), s(y), s(s(0)), s(s(x)), s(s(y)), \ldots\}$.

Intuitively, the terms denote objects in the intended domain of interest. In a term, besides the symbols from F and X, the parentheses "(" and ")" and the comma "," are also used to identify its structure. Like every propositional formula has a formula tree (Definition 2.1.1), for each term we can have a *term tree* in which parentheses and commas are omitted. We may assign a *position*, which is a sequence of positive integers, to each node of the term tree and use this position to identify the subterm denoted by that node.

Definition 5.1.4 Given term t, a *position* p of t, with the *subterm* at p, denoted by t/p, is recursively defined as follows:

- If t is a constant or a variable, then $p = \epsilon$, the empty sequence, and $t/p = t$.
- If $t = f(t_1, \ldots, t_k)$, then either $p = \epsilon$ and $t/p = t$, or $p = i.q$, where $1 \leq i \leq k$ and $t/p = t_i/q$, assuming $p.\epsilon = \epsilon.p = p$ for any p.

For example, if $t = f(x, g(g(y)))$, then the legal positions of t are $\epsilon, 1, 2, 2.1$, and $2.1.1$, and the corresponding subterms are $t, x, g(g(y)), g(y)$, and y, respectively.

Now adding the usual logical operators from propositional logic plus the two quantifiers, we have pretty much all the symbols needed for a first-order language.

Definition 5.1.5 Given first-order language $L = (P, F, X, Op)$, the *formulas* of L are the set of strings built up by terms, predicate symbols, and logical operators. Let op be the binary operators used in the current application, then the *formulas* for this application can be defined by the following BNF grammar:

$$\langle op \rangle ::= \land \mid \lor \mid \rightarrow \mid \oplus \mid \leftrightarrow$$
$$\langle Atoms \rangle ::= p(\langle Terms \rangle_1, \langle Terms \rangle_2, \ldots, \langle Terms \rangle_k) \text{ if } p \in P, arity(p) = k$$
$$\langle Literals \rangle ::= \langle Atoms \rangle \mid \neg \langle Atoms \rangle$$
$$\langle Formulas \rangle ::= \top \mid \bot \mid \langle Atoms \rangle \mid \neg \langle Formulas \rangle \mid$$
$$(\langle Formulas \rangle \langle op \rangle \langle Formulas \rangle) \mid$$
$$(\forall \langle Variables \rangle \langle Formulas \rangle) \mid (\exists \langle Variables \rangle \langle Formulas \rangle)$$

A formula (clause, literal, atom) is *ground* if it does not contain any variable.

Atoms are a shorthand for *atomic formulas*. We can have a formula tree for each first-order formula.

Some parentheses are unnecessary if we use a precedence among the logical operators. When displaying a formula, it is common to drop the outmost parentheses or those following a predicate symbol of zero arity, e.g., we write p instead of $p()$. We may also define subformulas of a formula as we did for propositional formulas and terms.

Example 5.1.6 Given signature (P, F), where $P = \{child/2, love/2\}$ and $F = \{a_i/0 \mid 1 \leq i \leq 100\}$, then $T(F) = F$ and $T(F, X) = F \cup X$, because F does not have any non-constant function symbols. If $X = \{x, y, z\}$, the following are some examples of formulas:

- **Ground atoms**: $child(a_2, a_1), child(a_3, a_1), love(a_1, a_2), \ldots$
- **Non-ground atoms**: $child(x, y), love(y, x), love(z, x), \ldots$

- **Ground non-atom formulas**: $\neg child(a_3, a_4)$, $child(a_2, a_1) \lor child(a_3, a_1)$, $child(a_4, a_1) \to love(a_1, a_4), \ldots$
- **General formulas**: $\neg child(a_1, y), \neg child(x, a_2), \neg child(x, a_2), \neg child(x, y) \lor love(y, x), love(x, y) \lor \neg love(y, x), \ldots$

In these examples, if we replace each atom with a propositional variable, they become propositional formulas. This is true for all formulas without quantifiers. Examples of formulas with quantifiers are given in the next subsection. □

5.1.2 Quantifiers

What makes first-order logic powerful is that it allows us to make general assertions using quantifiers: Universal quantifier ∀ and existential quantifier ∃. In the definition of ⟨*Formulas*⟩, every occurrence of quantifiers is followed by a variable, say x, and then followed by a formula, say A. That is, the quantifiers appear in a formula like either ∀x A or ∃x A; the former says "A is true for every value of x," and the latter says "A is true for some value of x."

In middle school algebra, when we say + is commutative, we use formula $x+y = y+x$, here x and y are meant for all values. When solving an equation like $x^2 - 3x + 2 = 0$, we try to find some value of x so that the equation holds. In first-order logic, the two cases are expressed as ∀x∀y $(x + y = y + x)$ and ∃x $(x^2 - 3x + 2 = 0)$, respectively.

In natural language, the sentences using words like "every," "always," or "never," are usually translated into formulas of first-order logic with ∀:

- "Everything Ada did is perfect.": ∀x $(did(Ada, x) \to perfect(x))$, where $did(x, y)$ is true iff x did y. If we can find something that Ada did not do perfectly, then the sentence is false.
- "Bob is always late for appointment!": ∀x $(appointment(x) \to late(Bob, x))$, where $late(x, y)$ means x is late for event y. This sentence is false if Bob is on time once.
- "Cathy never likes the food cooked by Tom.": ∀x $(cook(Tom, x) \to \neg like(Cathy, x))$, where $cook(x, y)$ is true iff x cooked y.

The above three sentences are likely examples of the *hasty generalization fallacy*, sometimes called the *over-generalization fallacy*. It is basically making a claim based on evidence that is just too small. In general, people assume that "everything rule has an exception." However, in first-order logic, if an exception is not stated explicitly in a rule, then the rule cannot be true and thus cannot be used in an argument.

Example 5.1.7 From Example 5.1.6, if $child(x, y)$ means "x is a child of y," $parent(x, y)$ means "x is a parent of y," $descen(x, y)$ means "x is a descendant of y," and $love(x, y)$ means "x loves y," using these predicates, we can write many assertions in first-order logic.

- x is a child of y iff y is a parent of x: $\forall x \forall y \, (child(x, y) \leftrightarrow parent(y, x))$.
- If x is a child of y, then x is a descendant of y: $\forall x \forall y \, (child(x, y) \rightarrow descen(x, y))$.
- Every parent loves his child: $\forall x \forall y \, (parent(x, y) \rightarrow love(x, y))$.
- Everybody is loved by somebody: $\forall x \exists y \, love(y, x)$.
- The descendant relation is transitive: $\forall x \forall y \forall z \, (descen(x, y) \wedge descen(y, z) \rightarrow descen(x, z))$.

We may define many more complex relations using these predicates. □

When using $\forall$ to describe a statement under various conditions, these conditions are usually the first argument of $\rightarrow$. When using $\exists$ to describe a statement under various conditions, we cannot use $\rightarrow$. For example, let $own(x, y)$ means "x owns y," $red(x)$ means "x is red," $toy(x)$ means "x is a toy," and $car(x)$ means "x is a car." Then the statement "Tom loves every red toy he owns" can be expressed as

$$\forall x \, (own(Tom, x) \wedge red(x) \wedge toy(x) \rightarrow love(Tom, x)).$$

It is natural to read it as a restricted quantifier: "For everything x, if Tom owns x, x is red and x is a toy, then Tom loves x." Of course, if we get rid of $\rightarrow$ in the above formula, we have

$$\forall x \, (\neg own(Tom, x) \vee \neg red(x) \vee \neg toy(x) \vee love(Tom, x)),$$

which can be read as "for everything x, either Tom doesn't own x, x is not red, x is not a toy, or Tom loves x." Note that if $\rightarrow$ is replaced by $\wedge$ in the above formula, we get

$$\forall x \, (own(Tom, x) \wedge red(x) \wedge toy(x) \wedge love(Tom, x))$$

which claims that "Tom owns everything," "everything is red," "everything is a toy," and "Tom loves everything."

On the other hand, to express "Tom loves one of his red toy cars," the formula is

$$\exists x \, (own(Tom, x) \wedge toy(x) \wedge red(x) \wedge car(x) \wedge love(Tom, x)).$$

If we replace the last $\wedge$ by $\rightarrow$ in the above formula, then the resulting formula can be converted to the following equivalent formula:

$$\exists x \, (\neg own(Tom, x) \vee \neg toy(x) \vee \neg red(x) \vee \neg car(x) \vee love(Tom, x)),$$

which claims that there exists something, denoted by "it," and one (or more) of the following is true: "Tom doesn't own it," "it is not a toy," "it is not red," "it is not a car," or "Tom loves it." Since there are many things that "Tom doesn't own it," "it is not a toy," "it is not red," or "it is not a car," the above formula does not catch the meaning that "Tom loves one of his red toy cars."

5.1 Syntax of First-Order Languages

Definition 5.1.8 In $\forall x\, A$ or $\exists x\, A$, the *scope* of the variable x is A, and every occurrence of x in A is said to be *bounded*. A variable in A is *free* if it is not bounded. A formula A is *closed* if it has no free variables. A *sentence* is a closed formula.

In $love(x, y)$, both x and y are free; in $\exists y\, p(x, y)$, and x is free, y is bounded (we cannot say "y in $p(x, y)$ is bounded" as "y is free in $p(x, y)$"); in $\forall x \exists y\, p(x, y)$, neither x nor y is free; thus, it is closed; in $(\forall x\, q(x)) \vee r(x)$, the first occurrence of x is bounded and the second occurrence of x is free.

When A represents any formula, we write $A(x_1, x_2, \ldots, x_n)$ to indicate that the free variables of A are among $x_1, x_2, \ldots, x_n$.

Like propositional formulas, we may use a precedence on the logical operators to avoid writing out all the parentheses in a formula. The precedence relation used in Chap. 2 is

$$\neg,\ \wedge,\ \vee,\ \rightarrow,\ \{\oplus, \leftrightarrow\}$$

Where to place $\forall$ and $\exists$? As a common practice, we will place them right after $\neg$.

$$\neg,\ \{\forall, \exists\},\ \wedge,\ \vee,\ \rightarrow,\ \{\oplus, \leftrightarrow\}.$$

That is, in formulas, $\forall$ or $\exists$ is treated as $\neg$ when binary Boolean operators are present. For example, $\forall x\, A \vee B$ is interpreted as $(\forall x\, A) \vee B$ (the scope of x is A), just like $\neg A \vee B$ is not $\neg(A \vee B)$. We will write $\forall x\, (A \vee B)$ to show that the scope of x is $A \vee B$. As always, omitted parentheses can be added back to avoid possible confusions.

Now, using the predicates and relation symbols, we can create formulas about the statements mentioned in the beginning of this section:

- $\forall x\, (odd(x) \leftrightarrow \neg even(x))$: Every natural number is even or odd, but not both.
- $\forall x\, (even(s(x)) \leftrightarrow odd(x))$: A natural number x is odd iff $s(x)$, which denotes $x + 1$, is even.
- $\forall x\, (x < s(x) \wedge \neg(x < x))$: For any natural number x, $x < x + 1$ and $(x < x)$ is false.
- $\forall x \exists y\, x < y$: For every natural number x, there exists a natural number y, such that $x < y$.
- $\forall x \forall y \forall z\, ((x < y) \wedge (y < z) \rightarrow (x < z))$: For any three natural numbers x, y, and z, if $x < y$ and $y < z$, then $x < z$, that is, $<$ is transitive.

5.1.3 Unsorted and Many-Sorted Logic

First-order languages are generally regarded as unsorted languages. If we look at its syntax carefully, two sets of objects are defined: $\langle Terms \rangle$ and $\langle Formulas \rangle$. Obviously, $\langle Formulas \rangle$ are of sort Boolean, denoted by $Bool = \{0, 1\}$. $\langle Terms \rangle$

represent objects of interest, and the set of objects is commonly called the *universe* (also called the *universe of discourse* or *domain of discourse*) of the language. For example, we can use $T(F)$, the set of ground terms, to represent the universe.

If we use U to denote the sort of $\langle Terms \rangle$, for any signature (P, F), predicate symbol $p \in P$ represents a function $p : U^{arity(p)} \mapsto Bool$ and function symbol $f \in F$ represents a function $f : U^{arity(f)} \mapsto U$. The universal and existential quantifiers range over the universe U. For example, the first-order language in Example 5.1.6 talks about 100 people living in a certain town, with a relation $love(x, y)$ to express that x loves y. In such a language, we might express the statement that "everyone loves someone" by writing $\forall x \exists y \, love(x, y)$, assuming U is the set of 100 people.

For many applications, $Bool$ and U are disjoint, like in Examples 5.1.2 and 5.1.6. For some applications, $Bool$ can be regarded as a subset of U. For example, to model the if-then-else command in a programming language, we may use the symbol $ite : Bool \times U \times U \mapsto U$. Another example is the function symbol $says : U \times Bool \mapsto Bool$ in the Knight and Knave puzzles (Example 2.5.4).

It is not difficult to introduce sorts (or types) in first-order logic by using unary (or monadic) predicate symbols ($arity = 1$). In Example 5.1.6, we use a_i to stand for person i. If we need function $age(x)$ to tell the age of person x, we need the function symbols for natural numbers. If we use those in Example 5.1.2, then $F = \{0, s, a_i \mid 1 \leq i \leq 100\}$. In this application, we do need to tell which are persons and which are numbers by using the following predicate symbols:

- $person : U \mapsto Bool$: $person(x)$ is true iff x is a person.
- $nat : U \mapsto Bool$: $nat(x)$ is true iff x is a natural number.

$person(x)$ can be easily defined by asserting $person(a_i)$ for $1 \leq i \leq 100$. $nat(x)$ can also be defined by asserting $nat(0)$ and $nat(s(x)) \equiv nat(x)$. Using nat, we can avoid the meaningless terms like $s(a_i)$ since $nat(s(a_i))$ is false.

Now, to state general properties as in Examples 5.1.2 and 5.1.6, we have to put restrictions on the variables. For example, to express "everybody is loved by somebody," the formula now is

$$\forall x \exists y \, (person(x) \wedge person(y) \rightarrow love(y, x)).$$

To say "<" is transitive on natural numbers, we use

$$\forall x, y, z \, (nat(x) \wedge nat(y) \wedge nat(z) \wedge (x < y) \wedge (y < z) \rightarrow (x < z))$$

Dealing with such unary predicates is tedious. To get rid of such clumsiness, people introduce "*sorts.*" Let

$$Person = \{x \mid person(x) \wedge x \in U\}$$
$$Nat = \{x \mid nat(x) \wedge x \in U\}$$

Then define $love : Person^2 \mapsto Bool$ and $<:\ Nat^2 \mapsto Bool$ and the formulas become neat again: We do not need $person(x)$, $nat(x)$, etc., in the formula. For the *age* function, we have $age : Person \mapsto Nat$. That is the birth of *many-sorted logic*, which is a light extension of first-order logic.

The idea of many-sorts is very useful in programming languages where each sort is a set of similar objects. Sets are usually defined by unary predicates. From a logician's viewpoint, a set $S = \{x \mid p(x)\}$ and a predicate $p(x)$ have the same meaning. In many-sorted logic, one can have different sorts of objects—such as *persons* and *natural numbers*—and a separate set of variables ranging over each. Moreover, the specification of function symbols and predicate symbols indicates what sorts of arguments they expect (the *domain*), and, in the case of function symbols, what sort of value they return (the *range*). The concepts of *arity* and *signature* are extended to contain the domain and range information of each function and predicate symbol.

Many-sorted logic can reflect formally our intention not to handle the universe as a collection of objects, but to partition the universe in a way that is similar to types in programming languages. This partition can be carried out on the syntax level: Substitution of variables can be done only accordingly, respecting the " *sorts*." We will return to this topic in Sect. 5.2.4 after introducing the semantics of first-order logic.

5.2 Semantics

We should be aware that, at this stage, all the function and predicate symbols are just symbols. In propositional logic, a symbol p can represent "Tom runs fast," or "Mary is tall." The same holds true in first-order logic as we may give different meanings to a symbol.

In Example 5.1.7, we have seen the following two formulas:

- $\forall x \forall y \, (child(x, y) \rightarrow descen(x, y))$.
- $\forall x \forall y \, (parent(x, y) \rightarrow love(x, y))$.

The first formula says "if x is a child of y, then x is a descendant of y"; the second says "every parent loves his child." What is the difference between them? They differ only on predicate symbols. If we give different meanings to the predicate symbols, one can replace the other. That is, we have designed the language with a certain interpretation in mind, but one could also interpret the same formula differently.

In this section, we will spell out the concepts of interpretations, models, and satisfiable or valid formulas, as an extension of these concepts of propositional logic.

5.2.1 Interpretation

For propositional logic, an interpretation is an assignment of propositional variables to truth values. For a first-order language $L = (P, F, X, Op)$, we need an interpretation for every symbol of L.

Definition 5.2.1 Given a closed formula A of $L = (P, F, X, Op)$, an interpretation of A is a triple $I = (D, R, G)$, where:

- D is a non-empty set called the *universe* of I, often denoted by D^I, if we like to emphasize that D is used by I.
- For each predicate symbol $p \in P$ of arity k, there is a k-ary relation $r_p \in R$ over D, $r_p \subseteq D^k$, denoted by $r_p = p^I$. If $k = 0$, then p is a propositional variable and p^I is either 1 or 0, a truth assignment to p.
- For each function symbol $f \in F$ of arity k, there is a function $g : D^k \mapsto D \in G$, denoted by $g = f^I$. If $k = 0$, then $f^I \in D$.

Example 5.2.2 Consider $L = (P, F, X, Op)$, where $P = \{p/2\}$, $F = \{a/0\}$, and $X = \{x\}$. Then for the formula $A = \forall x\, p(a, x)$, we may have the following interpretations:

1. $I_1 = (\mathcal{N}, \{\leq\}, \{0\})$ and the meaning of A is "for every natural number x, $0 \leq x$."
2. $I_2 = (\mathcal{N}, \{|\}, \{1\})$, $x \mid y$ returns true iff x divides y, and the meaning of A is "for every natural number x, $1 \mid x$."
3. $I_3 = (\{0, 1\}^3 \cup \{\epsilon\}, \{prefix\}, \{\epsilon\})$, $prefix(x, y)$ returns true iff x is a prefix of y, the meaning of A is "for every binary string x of length 3 or x is empty, the empty string ϵ is a prefix of x."
4. $I_4 = (\mathcal{P}(\mathcal{N}), \{\subseteq\}, \{\emptyset\})$, $\mathcal{P}(\mathcal{N})$ is the power set of $\mathcal{N}$, and the meaning of A is "for every subset x of natural numbers, $\emptyset \subseteq x$."
5. $I_5 = (V, \{E\}, \{a\})$, where $V = \{a, b, c\}$, $E = \{(a, a), (a, b), (a, c), (c, c)\}$, $G = (V, E)$ is a graph, and the meaning of A is "for every vertex $x \in V$, there is an edge $(a, x) \in E$."

□

In propositional logic, once we have an interpretation, we may use a procedure like *eval* (Algorithm 2.2.4) to obtain the truth value of a formula. We can do the same in first-order logic, provided that we have a way of handling free variables and quantifiers in the formula.

Let $free(A)$ be the free variables of formula A. Given a universe D, an *assignment* is a mapping $\theta : free(A) \mapsto D$, which maps each free variable of A to an element of D. The purpose of θ is that, during the evaluation of a formula under an interpretation, if we see a free variable x, we will use the value $\theta(x)$ for x. Initially, every free variable of A is affected in θ. If A is closed, then θ is empty. As quantifiers are removed and variables become free during the evaluation, we update θ dynamically.

5.2 Semantics

Procedure 5.2.3 The procedure *eval* takes as input a formula or a term A, with or without free variables, in $L = (P, F, X, Op)$, an interpretation I for L, and an assignment $\theta : free(A) \mapsto D^I$, and returns a Boolean value for a formula and an element of D^I for a term. If $A = f(t_1, t_2, \ldots, t_k)$ and $f \in P$, then A is an atom.

proc $eval(A, I, \theta)$
    if $A = \top$ **return** 1 // 1 means true.
    if $A = \bot$ **return** 0 // 0 means false.
    if $A \in X$ **return** $\theta(A)$ // A is a free variable.
    if $A = f(t_1, t_2, \ldots, t_k)$ **return** $f^I(t'_1, t'_2, \ldots, t'_k)$, where $t'_i = eval(t_i, I, \theta)$
    if $A = \neg B$ **return** $\neg eval(B, I, \theta)$
    if $A = (B \ op \ C)$ **return** $eval(B, I, \theta) \ op \ eval(C, I, \theta)$, where $op \in Op$
    if $A = (\forall x \ B)$ **return** $allInD(B, I, \theta, x, D^I)$
    if $A = (\exists x \ B)$ **return** $someInD(B, I, \theta, x, D^I)$
    else return "unknown"

Procedure $allInD(B, I, \theta, x, S)$ takes formula B, interpretation I, assignment θ, free variable x in B, and $S \subseteq D$ and evaluates the value of B under I and $\theta \cup \{x \leftarrow d\}$ for every value $d \in S$. If $eval(B, I, \theta \cup \{x \leftarrow d\})$ returns 0 for one $d \in D^I$, return 0; otherwise, return 1.

proc $allInD(B, I, \theta, x, S)$
    if $S = \emptyset$ **return** 1
    pick $d \in S$
    if $(eval(B, I, \theta \cup \{x \leftarrow d\}) = 0)$ **return** 0
    return $allInD(B, I, \theta, x, S - \{d\})$

Procedure $someInD(B, I, \theta, x, S)$ works in the same way as $allInD$, except that if one of the evaluations, $eval(B, I, \theta \cup \{x \leftarrow d\})$, returns 1, return 1; otherwise, return 0.

proc $someInD(B, I, \theta, x, S)$
    if $S = \emptyset$ **return** 0
    pick $d \in S$
    if $(eval(B, I, \theta \cup \{x \leftarrow d\}) = 1)$ **return** 1
    return $someInD(B, I, \theta, x, S - \{d\})$

Evidently, when the domain of I, D^I, is infinite, neither $allInD(B, I, \theta, x, S)$ nor $someInD(B, I, \theta, x, S)$ will stop, and *eval* cannot be an algorithm. In this case, we view them as recursive mathematical definitions. While the values of these procedures cannot be decided by an algorithm, we can reason about them using mathematical induction. In the last section of this chapter, we will see that D^I can be represented by a Herbrand base, which is the set of all ground atomic formulas. Since every Herbrand base is countable, a well-found order exists for D^I and we may apply induction on D^I. We present *eval* as a procedure because we assume that the reader is familiar with algorithms. When D^I is finite, the termination of *eval* is easy to establish.

Proposition 5.2.4 *Given formula A of $L = (P, F, X, Op)$, an interpretation I, and an assignment $\theta : free(A) \mapsto D^I$, $eval(A, I, \theta)$ will return a Boolean value; if t is a term appearing in A, then $eval(t, I, \theta)$ will return a value of D^I.*

Proof of Sketch Let D be D_I. We do induction on the tree structure of A. The base cases when A or t is a symbol of arity 0 are trivial. If $t = f(t_1, t_2, \ldots, t_k)$ is a term in A, the returned value is $f^I(t'_1, t'_2, \ldots, t'_k) \in D^I$ because $f^I : D^k \mapsto D$ is a function. If A is an atom $p(t_1, t_2, \ldots, t_k)$, where $p \in P$, then p^I is a relation and $p^I(t'_1, t'_2, \ldots, t'_k) = 1$ iff $(t'_1, t'_2, \ldots, t'_k) \in p^I$ by our assumption. That is, $p^I(t'_1, t'_2, \ldots, t'_k)$ is a Boolean value. For non-atom formulas, the returned values are always Boolean values as shown by the next proposition. □

From now on, we will denote $eval(A, I, \theta)$ by $I(A, \theta)$ and $eval(A, I, \emptyset)$ by $I(A)$ for brevity.

Proposition 5.2.5 *For formulas A and B, interpretation I, and assignment θ,*

- $I(\neg A, \theta) = \neg I(A, \theta)$
- $I(A \vee B, \theta) = I(A, \theta) \vee I(B, \theta)$
- $I(A \wedge B, \theta) = I(A, \theta) \wedge I(B, \theta)$
- $I(A \oplus B, \theta) = I(A, \theta) \oplus I(B, \theta)$
- $I(A \leftrightarrow B, \theta) = I(A, \theta) \leftrightarrow I(B, \theta)$
- $I(\forall x\, B(x), \theta) = allInD(B(x), I, \theta, x, D^I)$
- $I(\exists x\, B(x), \theta) = someInD(B(x), I, \theta, x, D^I)$

From the definition of *eval*, it is easy to check that the above proposition holds.

Example 5.2.6 For I_5 in Example 5.2.2, $I_5(\forall x\, p(a, x))$ will call $allInD$ with parameters $(p(a, x), I_5, \emptyset, x, \{a, b, c\})$, which will call *eval* three times, with $\theta = \{x \leftarrow a\}, \{x \leftarrow b\}$, and $\{x \leftarrow c\}$, respectively. Since E contains (a, x) for $x \in \{a, b, c\}$, all three calls of *eval* return 1. The final result can be expressed as

$$I_5(\forall x\, p(a, x)) = \bigwedge_{d \in \{a,b,c\}} I_5(p(a, x), \{x \leftarrow d\})$$

We can do the same for all the interpretations in Example 5.2.2. □

Generalizing from the above example,

$$I(\forall x\, B(x), \theta) = \bigwedge_{d \in D^I} I(B(x), \theta \cup \{x \leftarrow d\})$$

which catches the meaning that "$I(\forall x\, B(x), \theta)$ returns 1 iff $B(x)$ is evaluated to 1 under I and $\theta \cup \{x \leftarrow d\}$ for every value $d \in D$."

Similarly, we have a notation for $I(\exists x\, B(x), \theta)$, or equivalently, $someInD(B(x), I, \theta, x, D)$:

$$I(\exists x\, B(x), \theta) = \bigvee_{d \in D^I} I(B(x), \theta \cup \{x \leftarrow d\})$$

5.2 Semantics

which catches the meaning that "$I(\exists x\, B(x), \theta)$ returns 1 iff $B(x)$ is evaluated to true under I and $\theta \cup \{x \leftarrow d\}$ for some value $d \in D^I$." These notations will help us to understand quantified formulas and establish their properties. Note that if the domain D^I is empty, then $I(\forall x\, B(x)) = 1$ and $I(\exists x\, B(x)) = 0$.

5.2.2 Models, Satisfiability, and Validity

Here are just straightforward extensions from propositional logic to first-order logic. Recall that a sentence is a closed formula.

Definition 5.2.7 Given sentence A and an interpretation I, I is said a *model* of A if $I(A) = 1$. If A has a model, A is said to be *satisfiable*.

The satisfiability problem of first-order logic is to decide if a set of first-order formulas is satisfiable or not. This problem is also called *constraint satisfaction* where the constraint is a set of first-order formulas. The SAT problem and the linear programming problem are special cases of constraint satisfaction.

For a propositional formula, the number of models is finite; for a first-order formula, the number of models is infinite in general, because we have an infinite number of choices for domains, relations, and functions in an interpretation. However, we can still borrow the notation $\mathcal{M}(A)$ from propositional logic.

Definition 5.2.8 Given a closed formula A, let $\mathcal{M}(A)$ be the set of all models of A. If $\mathcal{M}(A) = \emptyset$, A is *unsatisfiable*; if $\mathcal{M}(A)$ contains every interpretation, i.e., every interpretation is a model of A, then A is *valid*.

With the identical definitions from propositional logic, the following result is expected in first-order logic.

Proposition 5.2.9 *Every valid propositional formula is a valid formula in first-order logic, and every unsatisfiable propositional formula is unsatisfiable in first-order logic.*

Of course, there are more valid formulas in first-order logic and our attention is on formulas with quantifiers.

Example 5.2.10 Let us check some examples.

1. A_1 is $\forall x\, (p(a, x) \rightarrow p(a, a))$.
   Consider $I = (\{a, b\}, \{p\}, \{a\})$, where $p^I = \{\langle a, a\rangle\}$, $a^I = a$, $I(A_1) = 1$ because $p^I(a, a) = 1$ and

$$\begin{aligned}
I(A_1) &= \bigwedge_{d \in \{a,b\}} I(p(a, x) \rightarrow p(a, a)), \{x \leftarrow d\}) \\
&= I(p(a, a) \rightarrow p(a, a)) \land I(p(a, b) \rightarrow p(a, a)) \\
&= 1 \land (p^I(a, b) \rightarrow p^I(a, a)) \\
&= 1
\end{aligned}$$

Let $I' = (\{a, b\}, \{\{\langle a, b\rangle\}\}, \{a\})$, then $I'(A_1) = 0$ because $p^{I'}(a, a) = 0$ and $p^{I'}(a, b) = 1$ imply that $I'(p(a, b) \to p(a, a)) = 0$. $I'(A_1) = I'(p(a, a) \to p(a, a)) \land I'(p(a, b) \to p(a, a)) = 1 \land 0 = 0$. Thus, A_1 is satisfiable but not valid.

2. A_2 is $(\forall x \, p(a, x)) \to p(a, a)$.

    A_2 is different from A_1 because the scopes of x are different in A_1 and A_2. For any interpretation I, we consider the truth value of $p^I(a^I, a^I)$. If $p^I(a^I, a^I) = 1$, then $I(A_2, \emptyset) = 1$. If $p^I(a^I, a^I) = 0$, then $I(\forall x \, p(a, x)) = 0$ because x will take on value a^I. So, $I(A_2) = 1$ in any case. Since I is arbitrary, A_2 must be valid.

3. A_3 is $\exists x \exists y \, (p(x) \land \neg p(y))$.

    Consider $I = (\{a, b\}, \{p\}, \emptyset)$, where $p(a) = 1$ and $p(b) = 0$, then $p(a) \land \neg p(b) = 1$. From $I(p(x) \land \neg p(y), \{x \leftarrow a, y \leftarrow b\}) = 1$, we conclude that I is a model of A_3. A_3 is false in any interpretation whose domain is empty or contains a single element. Thus, A_3 is not valid.

□

Definition 5.2.11 Let $A(x)$ be a formula where x is a free variable of A. Then $A(t)$ denotes the formula $A(x)[x \leftarrow t]$, where every occurrence of x is replaced by term t. The formula $A(t)$ is called an *instance* of $A(x)$ by substituting x for t (ref. Definition 2.1.3).

We will use substitutions extensively in the next chapter. The substitution $\theta : X \mapsto D$ used in the procedure *eval* coincides with the above definition if $D = T(F)$ (the set of ground terms built on F). That is, for any interpretation $I = (D, R, G)$ of $L = (P, F, X, Op)$, if $D = T(F)$, then $I(A(x), \{x \leftarrow d\}) = I(A(d))$, i.e., $eval(A(x), I, \{x \leftarrow d\}) = eval(A(d), I, \emptyset)$, where $x \in X, d \in T(F)$, and $A(d) = A(x)[x \leftarrow d]$.

We will keep the convention that a set S of formulas denotes the conjunction of the formulas appearing in the set. If every formula is true in the same interpretation, then the set S is said to be satisfiable, and the interpretation is its model.

Example 5.2.12 Let $S = \{\forall x \exists y \, p(x, y), \ \forall x \, \neg p(x, x), \ \forall x \forall y \forall z \, (p(x, y) \land p(y, z) \to p(x, z))\}$. The second formula states that "p is irreflexive"; the third formula says "p is transitive."

Consider the interpretation $I = (\{\mathcal{N}\}, \{<\}, \emptyset)$; it is easy to check that I is a model of S, because $<$ is transitive and irreflexive, i.e., $\neg(x < x)$, and for every $x \in \mathcal{N}$, there exists $x + 1 \in \mathcal{N}$ such that $x < x + 1$.

Does S have a finite model, i.e., a model in which the universe is finite? If S has a finite model, say $I = (D, \{r_p\}, \emptyset)$, where $|D| = n$, consider the directed graph $G = (D, r_p)$. If there exists a path from a to b in G, there must exist an edge from a to b, because r_p is transitive. We cannot have an edge from a vertex to itself because r_p is irreflexive. Thus, if G has a cycle, then there exists a path from a vertex v in the cycle to itself, which leads to an edge from v to itself by transitivity, a contradiction. Without cycles, every path in $G = (D, r_p)$ has an end since D is finite. For a vertex

5.2 Semantics

x at the end of a path, we cannot find another vertex y satisfying $\forall x \exists y\, p(x, y)$. Hence, S cannot have any finite model. □

In Chap. 2, given a set P of propositional variables, we use All_P to denote all the propositional interpretations involving P. Similarly, in first-order logic, given a set P of predicate symbols and a set F of function symbols, let $All_{P,F}$ denote all the interpretations involving P and F. Then we have a copy of Theorem 2.2.11 for first-order logic, which shows a close relationship between logic and set theory.

Theorem 5.2.13 *For any closed formulas A and B of $L = (P, F, X, Op)$,*

$$(a)\ \mathcal{M}(A \vee B) = \mathcal{M}(A) \cup \mathcal{M}(B)$$
$$(b)\ \mathcal{M}(A \wedge B) = \mathcal{M}(A) \cap \mathcal{M}(B)$$
$$(c)\ \mathcal{M}(\neg A) = All_{P,F} - \mathcal{M}(A)$$

The proof of the above theorem follows the same approach as that of Theorem 2.2.11.

5.2.3 Equivalence and Entailment

The following two definitions are copied from Definitions 2.2.13 and 2.2.17.

Definition 5.2.14 Given two formulas A and B, A and B are *logically equivalent* if $\mathcal{M}(A) = \mathcal{M}(B)$, denoted by $A \equiv B$.

Definition 5.2.15 Given two formulas A and B, we say A *entails* B, or B is a *logical consequence* of A, denoted by $A \models B$, if $\mathcal{M}(A) \subseteq \mathcal{M}(B)$.

Assuming A is $\top$, We can simply write $\models B$ to denote that "B is valid."
With the same definitions, we have the same results for first-order logic.

Theorem 5.2.16 *For any two formulas A and B, (a) $A \models B$ iff $A \rightarrow B$ is valid; and (b) $A \equiv B$ iff $A \leftrightarrow B$ is valid.*

The substitution theorem from propositional logic still holds in first-order logic.

Theorem 5.2.17 *For any formulas A, B, and C, where B is a subformula of A, and $B \equiv C$, then $A \equiv A[B \leftarrow C]$.*

Proposition 5.2.18 *For any formula $A(x)$, where x is a free variable of A, variable y does not appear in $A(x)$, and $A(y)$ denotes $A(x)[x \leftarrow y]$, then $\forall x\, A(x) \equiv \forall y\, A(y)$ and $\exists x\, A(x) \equiv \exists y\, A(y)$.*

Proof For any interpretation I,

$$I(\forall x\, A(x)) = \bigwedge_{d \in D} I(A(x), \{x \leftarrow d\}) = \bigwedge_{d \in D} I(A(d))$$
$$I(\forall y\, A(y)) = \bigwedge_{d \in D} I(A(y), \{y \leftarrow d\}) = \bigwedge_{d \in D} I(A(d))$$

The variable x serves as a placeholder in $A(x)$, just like y serves as the same placeholder in $A(y)$, to tell *eval* when to replace x or y by d. Any distinct name can be used for this placeholder, thus $I(\forall x\, A(x), \theta) = I(\forall y\, A(y), \theta)$. Since I is arbitrary, it must be the case that $\forall x\, A(x) \equiv \forall y\, A(y)$. The proof for $\exists x\, A(x) \equiv \exists y\, A(y)$ is identical. □

The above proposition allows us to change the names of bounded variables safely, so that different variables are used for each occurrence of quantifiers. For example, it is better to replace $\forall x\, (p(x) \vee \exists x\, q(x, y))$ by $\forall x\, (p(x) \vee \exists z\, q(z, y))$, to improve the readability.

Proposition 5.2.19 *Let $A(x)$, $B(x)$, and $C(x, y)$ be the formulas with free variables x and y.*

1. $\neg \forall x\, A(x) \equiv \exists x\, \neg A(x)$
2. $\neg \exists x\, A(x) \equiv \forall x\, \neg A(x)$
3. $\forall x\, (A(x) \wedge B(x)) \equiv (\forall x\, A(x)) \wedge (\forall x\, B(x))$
4. $\exists x\, (A(x) \vee B(x)) \equiv (\exists x\, A(x)) \vee (\exists x\, B(x))$
5. $(\forall x\, A(x)) \vee B \equiv \forall x\, (A(x) \vee B)$ if x is not free in B
6. $(\exists x\, A(x)) \wedge B \equiv \exists x\, (A(x) \wedge B)$ if x is not free in B
7. $\forall x \forall y\, C(x, y) \equiv \forall y \forall x\, C(x, y)$
8. $\exists x \exists y\, C(x, y) \equiv \exists y \exists x\, C(x, y)$

Proof We give the proofs of 1, 3, 5, and 7, and leave the rest as exercises.

1. Consider any interpretation I, using de Morgan's law, $\neg(A \wedge B) \equiv \neg A \vee \neg B$,

$$\begin{aligned}
& I(\neg \forall x\, A(x)) \\
&= \neg I(\forall x\, A(x)) \\
&= \neg \bigwedge\nolimits_{d \in D^I} I(A(x), \{x \leftarrow d\}) \\
&= \bigvee\nolimits_{d \in D^I} \neg I(A(x), \{x \leftarrow d\}) \\
&= \bigvee\nolimits_{d \in D^I} I(\neg A(x), \{x \leftarrow d\}) \\
&= I(\exists x\, \neg A(x)).
\end{aligned}$$

Because I is arbitrary, the equivalence holds.

3. Consider any interpretation $I = (D, R, G)$, using the commutativity and associativity of $\wedge$,

$$\begin{aligned}
& I(\forall x\, (A(x) \wedge B(x))) \\
&= \bigwedge\nolimits_{d \in D} I(A(x) \wedge B(x), \{x \leftarrow d\}) \\
&= \bigwedge\nolimits_{d \in D} (I(A(x), \{x \leftarrow d\}) \wedge I(B(x), \{x \leftarrow d\})) \\
&= (\bigwedge\nolimits_{d \in D} I(A(x), \{x \leftarrow d\})) \wedge (\bigwedge\nolimits_{d \in D} I(B(x), \{x \leftarrow d\})) \\
&= I(\forall x\, A(x)) \wedge I(\forall x\, B(x)) \\
&= I((\forall x\, A(x)) \wedge (\forall x\, B(x)))
\end{aligned}$$

Because I is arbitrary, the equivalence holds.

5.2 Semantics

5. In the following proof, we assume that x does not appear in B. Consider any interpretation $I = (D, R, G)$, using the distribution law of $\vee$ over $\wedge$,

$$I((\forall x\, A(x)) \vee B)$$
$$= I(\forall x\, A(x)) \vee I(B)$$
$$= (\bigwedge_{d \in D} I(A(x), \{x \leftarrow d\})) \vee I(B)$$
$$= \bigwedge_{d \in D} (I(A(x), \{x \leftarrow d\}) \vee I(B))$$
$$= \bigwedge_{d \in D} (I(A(x), \{x \leftarrow d\}) \vee I(B, \{x \leftarrow d\}))$$
$$= \bigwedge_{d \in D} I(A(x) \vee B, \{x \leftarrow d\})$$
$$= I(\forall x\, (A(x) \vee B))$$

Because I is arbitrary, the equivalence holds.

7. Consider any interpretation $I = (D, R, G)$, using the commutativity and associativity of $\wedge$,

$$I(\forall x \forall y\, C(x, y))$$
$$= \bigwedge_{d \in D} I(\forall y\, C(x, y), \{x \leftarrow d\})$$
$$= \bigwedge_{d \in D} (\bigwedge_{e \in D} I(C(x, y), \{x \leftarrow d, y \leftarrow e\}))$$
$$= \bigwedge_{e \in D} (\bigwedge_{d \in D} I(C(x, y), \{x \leftarrow d, y \leftarrow e\}))$$
$$= \bigwedge_{e \in D} I(\forall x\, C(x, y), \{y \leftarrow e\})$$
$$= I(\forall y \forall x\, C(x, y))$$

Because I is arbitrary, the equivalence holds. $\square$

Example 5.2.20 Let us prove that $(\exists x\, p(x)) \to \forall y\, q(y)$ and $\forall x\, (p(x) \to q(x))$ are not equivalent. Let $I = (\{a, b\}, \{r_p, r_q\}, \emptyset)$, where $r_p(a) = r_q(a) = 1$ and $r_p(b) = r_q(b) = 0$. Then $I(\exists x\, p(x)) = 1$ and $I(\forall y\, q(y)) = 0$. So, $I((\exists x\, p(x)) \to \forall y\, q(y)) = 0$. On the other hand, $I(\forall x\, (p(x) \to q(x))) = 1$ because $r_p = r_q$, thus $I(p(x) \to q(x), \{x \leftarrow d\}) = 1$. Hence, $(\exists x\, p(x)) \to (\forall y\, q(y))$ and $\forall x\, (p(x) \to q(x))$ are not equivalent. $\square$

The above example is a typical proof that the two formulas are not equivalent: We provide an interpretation in which one is true and the other is false.

Example 5.2.21 Consider $\exists x \forall y\, p(x, y)$ and $\forall y \exists x\, p(x, y)$. If $p(x, y)$ stands for "x loves y," then the former means "there exists somebody who loves everyone," but the latter means "everyone is loved by somebody." Obviously, the two are not equivalent. To show

$$\exists x \forall y\, p(x, y) \models \forall y \exists x\, p(x, y),$$

let $I = (D, R, G)$ be any model of $\exists x \forall y\, p(x, y)$. There exists $d \in D$ such that $I(\forall y\, p(x, y), \{x \leftarrow d\}) = 1$. Then $\forall y \exists x\, p(x, y)$ must be true in I:

$$\begin{aligned}
&I(\forall y \exists x\, p(x, y), \emptyset) \\
&= \bigwedge_{e \in D} I(\exists x\, p(x, y), \{y \leftarrow e\}) \\
&= \bigwedge_{e \in D} I(p(x, y), \{y \leftarrow e, x \leftarrow d\}) \\
&= I(\forall y\, p(x, y), \{x \leftarrow d\}) \\
&= 1
\end{aligned}$$

To show that $\forall y \exists x\, p(x, y)$ does not entail $\exists x \forall y\, p(x, y)$, we need to find an interpretation in which $\forall y \exists x\, p(x, y)$ is true but $\exists x \forall y\, p(x, y)$ is false. The detail is left as an exercise. □

Example 5.2.22 We may use known equivalence relations to show $\exists x\, (A(x) \to B(x)) \equiv (\forall x\, A(x)) \to (\exists x\, B(x))$.

$$\begin{aligned}
&\exists x\, (A(x) \to B(x)) \\
&\equiv \exists x\, (\neg A(x) \vee B(x)) && //A \to B \equiv \neg A \vee B \\
&\equiv (\exists x\, \neg A(x)) \vee (\exists x\, B(x)) && //\text{Proposition } 5.2.19(4) \\
&\equiv \neg(\forall x\, A(x)) \vee (\exists x\, B(x)) && //\text{Proposition } 5.2.19(1) \\
&\equiv (\forall x\, A(x)) \to (\exists x\, B(x)) && //A \to B \equiv \neg A \vee B
\end{aligned}$$

The more equivalences are established, the more opportunity we can apply this method. □

Example 5.2.23 The *drinker paradox* can be stated as "There is someone in the pub such that, if he or she is drinking, then everyone in the pub is drinking." It was popularized by the logician Raymond Smullyan, who called it the "drinking principle" in his 1978 book *What Is the Name of This Book*.

Let $p(x)$ denote "x is drinking in the pub"; the drinker paradox can be represented by the formula

$$\exists x\, (p(x) \to \forall y\, p(y))$$

which can be converted into the following equivalent formula:

$$A: \quad \exists x\, \neg p(x) \vee \forall y\, p(y)$$

This formula is valid because for any interpretation I, $\forall y\, p(y)$ is either true or false in I. If it is true, then A is true in I; if it is false, then there exists an element a of I such that $p(a)$ is false, so $\exists x\, \neg p(x)$ is true in I. Note that if the domain of I is empty, $\forall y\, p(y)$ is trivially true in I.

Let $p(x)$ denote "x is rich"; from A, we obtain the following valid statement in first-order logic: "There is someone in the world such that, if he or she is rich, then everyone in the world is rich." It appears that if we have found this person, the world would be rid of poverty, although in logic, any poor guy can be this person. The apparently paradoxical nature of this statement comes from the fact that the "if then" statements often represent causation in natural language. However, in first-order logic, $A \to B$ is true if A is false and there is no causation between A and B.

The statement of the drinker paradox may be wrongly specified by the formula

$$(\exists x\, p(x)) \to (\forall y\, p(y))$$

which is equivalent to

$$B : \quad (\forall x\, \neg p(x)) \vee (\forall y\, p(y))$$

B is true only when "nobody is drinking" or "everybody is drinking." A mix-up of A and B is perhaps another reason why this paradox is puzzling. □

5.2.4 Set Constructions and Many-Sorted Logic

It is common in mathematics and computer science that we define a set by a unary (or monadic) predicate. For example, the set S_1 of natural numbers between 5 and 10 can be defined as

$$S_1 = \{x \mid x \in \mathcal{N} \wedge x \geq 5 \wedge x \leq 10\}$$

where $\mathcal{N}$ denotes the set of all natural numbers. Let $A(x)$ be "$x \in \mathcal{N} \wedge x \geq 5 \wedge x \leq 10$". Obviously, $A(x)$ is a monadic predicate (with a free variable x). Let $B(x)$ be "$x \in \mathcal{N} \wedge \exists y (y \in \mathcal{N} \wedge x = y * y)$". Then $S_2 = \{x \mid B(x)\}$ is the collection of all square numbers.

The use of monadic predicates in the construction of a set is very interesting because (1) the formula must have a single free variable and (2) the construction uses the intended interpretation of the formula. For the above two examples, the interpretation for $A(x)$ and $B(x)$ is the same: The domain is $\mathcal{N}$ with known relations (e.g., $\geq$) and functions (e.g., $*$). S_1 and S_2 are the collections of $x \in \mathcal{N}$ such that $A(x)$ and $B(x)$ are interpreted to be true, respectively. In general, if $\mathcal{N}$ is understood from the context, "$x \in \mathcal{N}$" can be dropped from $A(x)$ and $B(x)$.

Extensional and Intentional Definitions

The construction of a set using predicates can help us to understand *extensional* and *intentional* definitions in philosophical logic. An *intentional definition* gives meaning to a term by specifying necessary and sufficient conditions for when the term should be used. In the case of nouns, this is equivalent to specifying the conditions that an object needs to have in order to be counted as a referent of the term. An *extensional definition* gives meaning to a term by specifying its extension (i.e., its domain). That is, every object that falls under the definition of the term in question belongs to this domain. For instance, let $p(x)$ denote that "x is a natural number greater than 1 and its only divisors are 1 and itself." Then $p(x)$ is the intentional definition of *prime numbers* and $P = \{2, 3, 5, 7, 11, \ldots\}$ is its extensional definition.

Intentional and extensional definitions are the two important ways in which the objects or concepts are formally defined. In the construction of a set by predicates, we see both definitions clearly: Given $S = \{x \mid A(x)\}$, $A(x)$ is the intentional definition and specifies the necessary and sufficient conditions of the set; S is the extensional definition which is the collection of all objects that satisfies the conditions in question.

Example 5.2.24 Let $r(x)$ be "x is a reptile," $s(x)$ be "x has a shell-wrapped body," and $t(x)$ be "x is a turtle." Then the set T of all turtles can be defined as $T = \{x \mid r(x) \wedge s(x)\}$. Let $t(x)$ be $r(x) \wedge s(x)$, then $t(x)$ is the intentional definition of turtles and T is the extensional definition of turtles. Similarly, we may define the set R of all reptiles without using $r(x)$ to avoid self-definition. Obviously, $T \subset R$, and $\forall x \, (t(x) \rightarrow r(x))$ is valid. □

In general, we have the following result, whose proof is left as an exercise.

Theorem 5.2.25 *Given two formulas $A(x)$ and $B(x)$, where x is the only free variable, let $S_A = \{x \mid A(x)\}$ and $S_B = \{x \mid B(x)\}$.*

(a) $(S_A \cup S_B) = \{x \mid A(x) \vee B(x)\}$
(b) $(S_A \cap S_B) = \{x \mid A(x) \wedge B(x)\}$
(c) $(S_A \subseteq S_B) \equiv \forall x \, (A(x) \rightarrow B(x))$

We say $A(x)$ is *stronger* than $B(x)$ if $\forall x \, (A(x) \rightarrow B(x))$ is true. Obviously, the strongest formula is $\bot$ and the weakest is $\top$. From (c) of the above theorem, it is clear that the stronger the conditions in an intentional definition, the smaller the extension set and vice versa. This implication relation among unary predicates induces an order, i.e., the subset relation, over the sets defined by these predicates. So-called order-sorted logic pays attention to this order over these sets.

We commented that Theorem 5.2.13 shows a close relation between logic and set theory. The above theorem strengthens this relation. Understanding this relation helps us to avoid logical fallacies. For Example 5.2.24, if we know c is a turtle, we conclude that c is a reptile. This can be proved in two ways: (1) $c \in T$ and $T \subseteq R$ imply $c \in R$; and (2) $t(c)$ and $\forall x \, (t(x) \rightarrow r(x))$ imply $r(c)$.

5.2 Semantics

Many-Sorted Logic and Strong Typing

A many-sorted logic is an extension of first-order logic by introducing types (or sorts) into the logic, where each type is viewed as a subset of the universe. Today's programming languages use many-sorted logic to model data types. The meaning of a *type* is simply a set of objects satisfying a unary or monodic predicate. For instance, the meaning of type *int* of integers is the set

$$int = \{x \mid P_{int}(x)\}$$

where $P_{int}(x)$ is true iff x is an integer. Many-sorted logic assumes a universal set U and every type is a subset of U. There may exist relations between types. For example, the type *nat* of natural numbers is a subtype of *int*, whose meaning is

$$nat = \{x \mid x \in int, x \geq 0\}$$

In other words, $nat \subseteq int$. Many-sorted logic deals with a hierarchy of types and was one of the approaches invented to avoid a paradox in naive set theory (see Sect. 1.3.1).

Some modern programming languages use a *strong type system*, where types are used by compilers as "values" for type checking, an idea coming from higher-order logic. We illustrate the idea of strong typing by an example.

In Chap. 1, we introduced the *Backus–Naur form* with the following definition of natural numbers:

$$\langle Num \rangle ::= 0 \mid suc(\langle Num \rangle)$$

where suc denotes the successor function. A programming language with a strong type system may accept the above definition to define a new type Num. Moreover, it can create dynamically many new types. For instance, if Num denotes the set of natural numbers, then $suc(Num)$ is the *type* of positive integers, $suc(suc(Num))$ is the *type* of natural numbers greater than 1, and so on. If a function needs positive integer x (as a divisor), we may declare the type of x as $suc(Num)$, without explicitly adding $x > 0$ in the code. A compiler will add the condition $x > 0$ automatically before the call of the function. This way of using types may create a neat code and avoid human errors.

Strong type system is a common feature of functional programming languages based on typed lambda calculus (to be discussed next). When we write our code in a functional programming language like Haskell, we can make use of its strong type system that helps us to filter out a couple of logical errors in advance.

5.2.5 First-Order Logic Versus Higher-Order Logic

It seems that first-order logic is a natural and meaningful object of study. From the current chapter to the end of this book, seven chapters out of eight are entirely devoted to the study of first-order logic. Even for the only chapter not on it, first-order logic is indispensable. That is, it occupies the two-third space of this book, because first-order logic has long been regarded as the "right" logic for studying mathematics and computer science.

Higher-Order Logic

People may wonder the meaning of "first-order" and ask what are "second-order," "third-order," etc. The variables of first-order logic take individuals from a set called the *universe*. If we allow variables to take sets of individuals, it is called *second-order logic*. Third-order logic will allow variables to take sets of sets and so on. Higher-order logic is often referred to any of nth-order logic for $n \geq 1$ and allows quantification of variables over sets that are nested arbitrarily deep.

Higher-order logic is more expressive than first-order logic, but their model-theoretic properties are less well-behaved than those of first-order logic. In Chap. 11, we will show that some problems of first-order logic are not computable. More problems become uncomputable in higher-order logic.

Another definition of "higher-order logic" is that variables can take arbitrary functions over the universe, not just individuals of the universe. A well-known example of this kind of formal logic systems is *lambda calculus*, introduced by Alonzo Church in the 1930s. Lambda calculus uses function abstraction, variable binding, and substitution for computation. It is the foundation of type theory and functional programming. More information related to lambda calculus can be found in Chaps. 9 and 11. A detailed introduction of lambda calculus is out of the scope of this book, even though it has great impact on many fields of computer science.

Sometimes people use higher-order logic to denote the union of second-order, third-order, ..., nth-order, ..., logic as a contrast to first-order logic. Since third-order or fourth-order logic is rarely needed, higher-order may simply be second-order logic because the advantage and the disadvantage that came along with moving beyond first-order logic have already appeared in second-order logic.

"Privileged" Status of First-Order Logic

According to William Ewald [3], the history of first-order logic is anything but straightforward, and is certainly not a matter of a sudden discovery by a single researcher. First-order logic was studied by logicians like Gottlob Frege and explicitly identified by Charles Peirce in 1885, but was then forgotten. It was independently re-discovered by David Hilbert in 1917. That is, Peirce was the first to identify it, but it is Hilbert who put it on the map, very much like "the Vikings discovered America, but they forgot about it, because they did not yet need it."

Hilbert isolated first-order logic by posing questions of completeness, consistency, and decidability for systems of logic. Although these questions represented an enormous conceptual leap in logic, Hilbert did not treat first-order logic as more significant than higher-order logic. It is Gödel and others who recognized first-order logic as being importantly different from higher-order logic in the early 1930s. Even after the Gödel results were widely understood, logicians continued to work in higher-order logic, and it took years before first-order logic attained a "privileged" status.

The emerging of first-order logic is bound up with technical discoveries, with differing conceptions of what constitutes logic, with different programs of mathematical research, and with philosophical and conceptual reflection. There are reasons why first-order logic is regarded as a "privileged" logical system—that is, as the "right" logic for investigations in foundations of mathematics. In general, first-order logic is more expressive than propositional logic and less complicated than higher-order logic. It appears that our grasp on quantification over objects is firmer than our grasp on quantification over properties, even when the universe is finite. For instance, solving a Sudoku puzzle is a first-order quantification problem (see Sect. 4.3.2). Questions about Sudoku's initial configurations, such as "can it admit one and only one solution?", "what is its difficulty?", "what is the limit on the number of cells filled initially?", etc., are problems of second-order quantification. These are hard problems even for the experts of Sudoku puzzles.

Every universe of first-order logic is assumed to be countable. However, quantification over all subsets of a countably infinite set entails quantification over an uncountable set (Proposition 1.3.7). In Chap. 11, we will show that uncountable sets are uncomputable (Proposition 11.4.8). In other words, quantification in higher-order logic is uncomputable in general when the universe is infinite.

It seems that the "privileged" status of first-order logic also came from philosophical considerations: the need to avoid the set-theoretical paradoxes, a search for secure foundations for mathematics, and a sense that higher-order logic was both methodologically suspect and avoidable. All these things show the continuing influence of the "Crisis of Foundations," *Grundlagenkrise* in German, of the 1920s, which did so much to set the terms of the subsequent philosophical understanding of the foundations of mathematics [3].

Type Theory Versus Higher-Order Logic

Type theory [4] was invented by Alonzo Church in 1940, called *simple type theory*, and there are many generalizations of simple type theory. Church's type theory is based on his lambda calculus and *lambda*-notation is the only binding mechanism employed in simple type theory. Since properties and relations can be regarded as functions from the universe to truth values, the concept of a function is taken as primitive in Church's type theory, and the *lambda*-notation is incorporated into the formal language of type theory. Moreover, quantifiers and description operators are introduced in a way so that additional binding mechanisms can be avoided, and *lambda*-notation is reused instead. Nowadays type theory usually means study of systems based on typed lambda calculi.

In mathematical logic, type theory is a synonym of higher-order logic, since they allow quantification not only over the elements of the universe (as in first-order logic) but also over functions, predicates, and even higher-order variables. Type theory assigns types to entities, distinguishing, for example, between numbers, sets of numbers, functions from numbers to sets of numbers, and sets of such functions. These distinctions allow one to discuss the conceptually rich world of sets and functions without encountering the paradoxes of naive set theory. Hence, type theory and higher-order logic are equivalent and both can be viewed as the union of a family of finite-order logic.

Type theory is a formal logical language including first-order, but more expressive in a practical sense. Type theory constitutes an excellent formal language for representing the knowledge in automated information systems, sophisticated automated reasoning systems, systems for verifying the correctness of mathematical proofs, and a range of projects involving logic and artificial intelligence. It also plays an important role in the study of the formal semantics of natural language. Despite of high complexity of type theory related algorithms, there are successful and influential implementations based on type theory, including ML, Haskell, Coq, the HOL family, and Twelf.

5.3 Proof Methods

Proof methods of first-order logic have the same three tasks as in propositional logic: to prove a formula A is (a) valid, (b) unsatisfiable, or (c) satisfiable. (a) and (b) are equivalent because A is valid iff $\neg A$ is unsatisfiable. (c) is different because both A and $\neg A$ can be satisfiable at the same time. As shown in Example 5.2.22, we may use equivalence relations to transform A to $\top$ or $\bot$, thus showing A is valid or unsatisfiable. However, this approach has only very limited chance of success.

Some proof methods like those based on truth table or ROBDD do not work for first-order logic. However, many proof methods can be extended easily from propositional logic to first-order logic, even though the latter is much more expressive than the former. That means first-order logic has proof methods simple to reason about theoretically or to implement correctly on a computer. For example, the well-known modus ponens rule can be specified as follows:

Definition 5.3.1 The modus ponens rule of the first-order logic is the following inference rule:

$$(modus\ ponens) \quad \frac{\forall x\, (A(x) \to B(x)) \quad A(t)}{B(t)} \quad t \text{ is any term}$$

5.3 Proof Methods

For instance, let $A(x)$ be "x is human," $B(x)$ be "x is mortal," t be constant "Adam," then "Adam is mortal" can be deduced from "Every human is mortal" and "Adam is human."

In Chap. 3, we introduced various proof systems for propositional logic: semantic tableau, natural deduction, resolution, etc. All these systems can be extended to first-order logic. In this section, we briefly introduce the extensions of semantics tableau and natural deduction to first-order logic and leave resolution to the next chapter.

5.3.1 Semantic Tableau

Semantic tableau as given in Chap. 3 transforms a propositional formula into DNF by a set of α-rules (which handle conjunctions and generate one child in the tableau) and β-rules (which handle disjunctions and generate two children in the tableau). These rules will be inherited by first-order logic. Keep in mind that the rules apply to the top logical operators in a formula in a leaf node of the tableaux.

α	α_1, α_2
$A \wedge B$	A, B
$\neg(A \vee B)$	$\neg A, \neg B$
$\neg(A \rightarrow B)$	$A, \neg B$
$\neg(A \oplus B)$	$(A \vee \neg B), (\neg A \vee B)$
$A \leftrightarrow B$	$(A \vee \neg B), (\neg A \vee B)$
$\neg(A \uparrow B)$	A, B
$A \downarrow B$	$\neg A, \neg B$
$\neg\neg A$	A

β	β_1	β_2
$\neg(A \wedge B)$	$\neg A$	$\neg B$
$A \vee B$	A	B
$A \rightarrow B$	$\neg A$	B
$A \oplus B$	$A, \neg B$	$\neg A, B$
$\neg(A \leftrightarrow B)$	$A, \neg B$	$\neg A, B$
$A \uparrow B$	$\neg A$	$\neg B$
$\neg(A \downarrow B)$	A	B

To handle the quantifiers, we introduce two rules: one for the universal quantifier, called the $\forall$-rule, and one for the existential quantifier, called the $\exists$-rule.

Definition 5.3.2 The $\forall$-rule is an inference rule:

$$(\forall) \quad \frac{\forall x \, A(x)}{A(t)} \quad t \text{ is a ground term}$$

The $\forall$-rule works as follows: If $\forall x \, A(x)$ appears in a leaf node n, we create a child node n' of n, copy every formula of n into n', and add $A(t)$ into n', where t is a ground term (without variables) appearing in n. If n' does not have any ground term, we may introduce a new constant as t.

We emphasize that the $\forall$-rule is an inference rule, not a simplification rule, because it creates and adds something, not replacing $\forall x \, A(x)$ by something. Because of the $\forall$-rule, we may not be able to show the termination of the semantic tableau method.

Definition 5.3.3 The ∃-rule is a simplification rule:

$$(\exists) \quad \frac{\exists x\, A(x)}{A(c)} \quad c \text{ is a new constant}$$

The ∃-rule works as follows: If $\exists x\, A(x)$ appears in a leaf node n, we create a child node n' of n, copy every formula of n into n', and replace $\exists x\, A(x)$ in n' by $A(c)$. Later after finishing reading this chapter, you will know that the new constant is called *Skolem constant*, to remember Thoralf Skolem's contribution to logic.

If the formulas are not in NNF (negation normal form), we also need the ¬∃-rule, which is also an inference rule:

$$(\neg \exists) \quad \frac{\neg \exists x\, A(x)}{\neg A(t)} \quad t \text{ is a ground term}$$

and the ¬∀-rule, which is a simplification rule:

$$(\neg \forall) \quad \frac{\neg \forall x\, A(x)}{\neg A(c)} \quad c \text{ is a new constant}$$

Example 5.3.4 To show $(\exists x \forall y\, p(x, y)) \rightarrow (\forall y \exists x\, p(x, y))$ is valid by semantic tableaux, we work on the negation of the formula.

$$
\begin{array}{lll}
\epsilon: & \neg((\exists x \forall y\, p(x, y)) \rightarrow (\forall y \exists x\, p(x, y))) & \alpha \neg \rightarrow \\
1: & \exists x \forall y\, p(x, y),\ \neg \forall y \exists x\, p(x, y) & \exists \\
11: & \forall y\, p(c_1, y),\ \neg \forall y \exists x\, p(x, y) & \neg \forall \\
111: & \forall y\, p(c_1, y),\ \neg \exists x\, p(x, c_2) & \forall \\
1111: & \forall y\, p(c_1, y),\ \neg \exists x\, p(x, c_2),\ p(c_1, c_2) & \neg \exists \\
11111: & \forall y\, p(c_1, y),\ \neg \exists x\, p(x, c_2),\ p(c_1, c_2),\ \neg p(c_1, c_2) & closed
\end{array}
$$

Since the negation of the formula has a closed tableau, the formula is valid. □

Example 5.3.5 Let A be $B \wedge C \wedge D$, where $B = (\forall x\, p(x, s(x)))$, $C = (\forall x \forall y\, (q(x) \vee q(y)))$, and $D = (\neg \exists x\, q(x))$. If we choose to work on q first, we will find a closed tableau for A; thus, A is unsatisfiable. In the following, we will start with $\{B, C, D\}$ by omitting the use of the $\alpha \wedge$ rule.

$$
\begin{array}{lll}
\epsilon: & B, C, (\neg \exists x\, q(x)) & \forall \\
1: & B, C, (\neg \exists x\, q(x)), \forall y\, (q(c) \vee q(y)), & \forall \\
11: & B, C, (\neg \exists x\, q(x)), \forall y\, (q(c) \vee q(y)), q(c) \vee q(c), & \neg \exists \\
111: & B, C, (\neg \exists x\, q(x)), \forall y\, (q(c) \vee q(y)), q(c) \vee q(c), \neg q(c), & \beta \vee \\
1111: & B, C, (\neg \exists x\, q(x)), \forall y\, (q(c) \vee q(y)), q(c), \neg q(c), & closed \\
1112: & B, C, (\neg \exists x\, q(x)), \forall y\, (q(c) \vee q(y))), q(c), \neg q(c), & closed
\end{array}
$$

If we choose to work on p first, then the tableau will never terminate:

$\epsilon : (\forall x\, p(x, s(x))), (\forall x \forall y\, (q(x) \vee q(y))), (\neg \exists x\, q(x))$ $\forall$
$1 : (\forall x\, p(x, s(x))), \dots, p(c, s(c)),$ $\forall$
$11 : (\forall x\, p(x, s(x))), \dots, p(c, s(c)), p(s(c), s(s(c)))$ $\forall$
$111 : (\forall x\, p(x, s(x))), \dots, p(c, s(c)), p(s(c), s(s(c))), p(s(s(c)), s(s(s(c))))$ $\forall$
$1 \dots 1 : \dots$

One choice leads to a proof and the other leads to a loop.   □

This example shows clearly that the nondeterminism in the application of the rules (i.e., choosing a leaf node, choosing a rule, or choosing a term in the ∀ rule) does matter for obtaining a closed tableau. If we use a fair strategy for the application of the rules, then a closed tableau is guaranteed to be found when the input formula is unsatisfiable, as indicated by the following result (without proof).

Theorem 5.3.6 *A formula A is unsatisfiable iff A has a closed tableau, which can be found by a fair strategy.*

5.3.2 Natural Deduction

In natural deduction for propositional logic, we have an introduction rule and an elimination rule for each logical operator. These rules will be inherited, plus the new rules for the quantifiers over the sequents:

op	Introduction	Elimination
∀	$(A(x) \mid \alpha) \vdash (\forall x\, A(x) \mid \alpha)$	$(\forall x\, A(x) \mid \alpha) \vdash (A(t) \mid \alpha)$
∃	$(A(t) \mid \alpha) \vdash (\exists x\, A(x) \mid \alpha)$	$(\exists x\, A(x) \mid \alpha) \vdash (A(c) \mid \alpha)$

where x is a variable and does not appear in α, t is a term, and c is a new constant. From these rules, we can see that free variables have the same function as universally quantified variables. We use two examples to illustrate these rules.

Example 5.3.7 Let $A = \{\forall x\, p(x), \forall x\, q(x)\}$. A proof of $A \models \forall x\, (p(x) \wedge q(x))$ in natural deduction is given below.

  1. $(\forall x\, p(x))$ assumed
  2. $(\forall x\, q(x))$ assumed
  3. $(p(x))$ $\forall_E$ from 1
  4. $(q(x))$ $\forall_E$ from 2
  5. $(p(x) \wedge q(x))$ $\wedge_I$ from 3, 4
  6. $(\forall x\, (p(x) \wedge q(x)))$ $\forall_I$ from 5

By Theorem 5.2.16, $(\forall x\, p(x)) \wedge (\forall x\, q(x)) \to \forall x\, (p(x) \wedge q(x))$ is valid.   □

Example 5.3.8 Show that $(\exists x\, (p(x) \vee q(x))) \models (\exists x\, p(x)) \vee (\exists x\, q(x))$ in natural deduction.

1.	$(\exists x\, (p(x) \vee q(x)))$	assumed
2.	$(p(c) \vee q(c))$	$\exists_E$ from 1
3.	$(p(c) \mid q(c))$	$\vee_E$ from 2
4.	$((\exists x\, p(x)) \mid q(c))$	$\exists_I$ from 3
5.	$((\exists x\, p(x)) \mid (\exists x\, q(x)))$	$\exists_I$ from 4
6.	$((\exists x\, p(x)) \vee (\exists x\, q(x)))$	$\vee_I$ from 5

By Theorem 5.2.16, $(\exists x\, (p(x) \vee q(x))) \rightarrow ((\exists x\, p(x)) \vee (\exists x\, q(x)))$ is valid. □

5.4 Conjunctive Normal Form (CNF)

We just introduced an extension of semantic tableau and natural deduction to first-order logic and illustrated the use of these two methods for proving theorems in first-order logic. In practice, these two methods are not as effective as the resolution method, which is the focus of the next chapter.

Recall that in propositional logic, the resolution is a refutational method and the input formula must be in clausal form. This is still true for first-order logic. In this section, we discuss how to convert first-order formulas into clausal form.

Definition 5.4.1 In first-order logic, a *literal* is either an atom (positive literal) or the negation of an atom (negative literal); a *clause* is a disjunction of literals; a formula is in CNF if it is a conjunction of clauses.

From the definition, if we replace each atom in a clause by a propositional variable, we obtain a clause in propositional logic. Since clauses use only three logical operators, i.e., ¬, ∨, and ∧, we need to get rid of →, ↔, ⊕, etc., as we did in propositional logic. We also need to get rid of quantifiers and we will talk about it now.

5.4.1 Prenex Normal Form

Definition 5.4.2 A formula A is in *prenex normal form* (PNF) if

$$A = Q_1 x_1 Q_2 x_2 \cdots Q_k x_k B(x_1, x_2, \ldots, x_k),$$

where $Q_i \in \{\forall, \exists\}$ and B does not contain any quantifier.

5.4 Conjunctive Normal Form (CNF)

The idea of prenex normal form is to move all the quantifiers to the top of the formula. For example, $\forall x \forall y \, (p(x, y) \vee q(x))$ is a PNF but $\forall x ((\forall y \, p(x, y)) \vee q(x))$ is not.

To make the discussion easier, we assume that the formulas contain only $\neg$, $\wedge$, $\vee$, plus the two quantifiers, as operators. A quantifier which is not at the top must have another operator immediately above it: In this case the operator will be one of the three operators: $\neg$, $\wedge$, or $\vee$. Since the quantifier can be in either parameter position of $\wedge$ or $\vee$, and there are two quantifiers, there are 10 cases to consider. If we consider $\wedge$ and $\vee$ are commutative, we need six rules, which are provided in Proposition 5.2.19:

Equivalence Rules for PNF

1. $\neg \forall x \, A(x) \equiv \exists x \, \neg A(x)$
2. $\neg \exists x \, A(x) \equiv \forall x \, \neg A(x)$
3. $(\forall x \, A(x)) \wedge B \equiv \forall x \, (A(x) \wedge B)$ if x is not free in B
4. $(\forall x \, A(x)) \vee B \equiv \forall x \, (A(x) \vee B)$ if x is not free in B
5. $(\exists x \, A(x)) \wedge B \equiv \exists x \, (A(x) \wedge B)$ if x is not free in B
6. $(\exists x \, A(x)) \vee B \equiv \exists x \, (A(x) \vee B)$ if x is not free in B

where the arguments of $\vee$ and $\wedge$ can be switched.

To meet the condition that "x is not free in B," it is necessary first to rename all the quantified variables so that no variable appears more than once following a quantifier.

In Chap. 2, we introduced the concept of "negation normal form," where the negation appears only in literals. Using the first two of the above rules plus de Morgan's laws and the double negation law, we may push the negation down, until all negations appear in literals. The resulting formulas are said to be in *negation normal form* (NNF), a straightforward extension from propositional logic.

Proposition 5.4.3 *Every first-order formula can be transformed into an equivalent formula in NNF.*

Example 5.4.4 Given the following formula

$$(\forall x \exists y \, p(x, y)) \wedge \neg(\forall x \forall y \, (q(x, y) \wedge q(y, x))),$$

we may choose first to rename the variable names so that every quantified variable has a different name:

$$(\forall x \exists y \, p(x, y)) \wedge \neg(\forall z \forall w \, (q(z, w) \wedge q(w, z)))$$

Applying equivalence rule 1 twice,

$$(\forall x \exists y \, p(x, y)) \wedge (\exists z \exists w \, \neg(q(z, w) \wedge q(w, z))$$

Now we may apply de Morgan's law to get

$$(\forall x \exists y\, p(x, y)) \wedge (\exists z \exists w\, (\neg q(z, w) \vee \neg q(w, z)))$$

which is in NNF; this step is not required for obtaining PNF. We have two options now: Either choose equivalence rule 3 or rule 5. To use rule 3, we have

$$\forall x\, ((\exists y\, p(x, y)) \wedge (\exists z \exists w\, (\neg q(z, w) \vee \neg q(w, z))))$$

To apply rule 5 three times, we have

$$\forall x \exists y \exists z \exists w\, (p(x, y) \wedge (\neg q(z, w) \vee \neg q(w, z))),$$

which is a PNF. □

Note that the order of quantified variables in this formula is $xyzw$. There are five other possible outcomes where the order is $xzyw$, $xzwy$, $zxyw$, $zxwy$, or $zwxy$, respectively. If we are allowed to switch the quantifiers of the same type (the last two in Proposition 5.2.19), we obtain more equivalent formulas where the order is $xywz$, $xwyz$, $xwzy$, $wxyz$, $wxzy$, or $wzxy$. These formulas are all equivalent as we use the equivalence relations in obtaining PNF.

Proposition 5.4.5 *Every first-order formula can be transformed into an equivalent PNF.*

Proof By Proposition 5.2.18, renaming quantifiers variable names preserves the equivalence. The equivalence rules for PNF are terminating (an exercise problem) and preserve the equivalence. Applying these rules repeatedly, we will obtain the desired result. □

In the following, we will use *PrenexNF(A)* to denote the procedure for transforming formula A into a prenex normal form.

5.4.2 Skolemization

Skolemization is the process in which we remove all the quantifiers, both universal and existential, leaving a formula with only free variables. This process is made easier by having the original formula in prenex normal form, and we assume this in this section.

Removing the universal quantifiers while preserving meaning is easy since we assume that the meaning of a free variable is the same as a universally quantified variable: The formula will be true for any value of a free variable. Thus, we simply discard all the top universal quantifiers from a PNF.

Doing the same for an existentially quantified variable is not possible because a free variable can have only one default meaning. The key idea is to introduce a new

5.4 Conjunctive Normal Form (CNF)

function to replace that variable. These functions are called *Skolem functions*. When the functions are nullary (i.e., the arity is 0), they are called *Skolem constants*.

Example 5.4.6 Consider the formula $\forall x \exists y\, (y^2 \leq x \land \neg((y+1)^2 \leq x))$. This asserts the existence of a maximal number y whose square is no more than x for any value of x. We may drop the quantifier $\forall$ so that x becomes free and keep the meaning of the formula. The same effect could be achieved by asserting the existence of a function f, depending on x, satisfying

$$f(x)^2 \leq x \land \neg((f(x)+1)^2 \leq x)$$

The function f is indeed a Skolem function. We might then use our knowledge of algebra to prove that $\lfloor \sqrt{x} \rfloor$ is a solution for $y = f(x)$, but this is beyond the scope of our present concerns. □

In general, the parameters to a Skolem function will be free variables, which have the same meaning as universally quantified variables, thus allowing a different value for each different values of free variables.

Algorithm 5.4.7 $Sko(A)$ takes as input a formula A and returns a formula without quantifiers. Procedure $PrenexNF(A)$ returns a formula in PNF and equivalent to A. $free(B)$ returns the list of free variables in B.

proc $Sko(A)$
1    $A := PrenexNF(A)$
2   **while** A has a quantifier **do**
3   **if** $(A = \exists x\, B)\ A := B[x \leftarrow f(free(B))]$ // f is a new Skolem function
4   **else if** $(A = \forall x\, B)\ A := B$
5   **return** A

Algorithm $Sko(A)$ first converts formula A into a prenex normal form by $PrenexNF(A)$ (discussed in the previous section), then peels off each quantifier from left to right, until they are all eliminated.

Example 5.4.8 Let A be $\forall x \exists y \exists z \exists w\, (p(x,y) \land (\neg q(z,w) \lor \neg q(w,z)))$ from Example 5.4.4. A is a PNF, so $PrenexNF(A) = A$. Lines 2–4 of $Sko(A)$ will first remove $\forall x$, making x free, and then replace y by $f_1(x)$, z by $f_2(x)$, and w by $f_3(x)$, where f_1, f_2, and f_3 are Skolem functions introduced when eliminating y, z, and w. So, $Sko(A)$ returns $p(x, f_1(x)) \land (\neg q(f_2(x), f_3(x)) \lor \neg q(f_3(x), f_2(x)))$. For

$$B = \exists z \exists w \forall x \exists y\, (p(x,y) \land (\neg q(z,w) \lor \neg q(w,z))),$$

$Sko(B) = p(x, f_3(x)) \land (\neg q(c_1, c_2) \lor \neg q(c_2, c_1))$. We know that $A \equiv B$ from Example 5.4.4. However, $Sko(A) \not\equiv Sko(B)$, because they contain different function symbols. □

The above example illustrates that Skolemization does not preserve the equivalence. The Skolemization process, applied to a formula in prenex normal form, terminates with a weakly equivalent formula. To see that the process terminates is easy, since the process of converting a formula into PNF is terminating and the number of quantifiers is reduced in the body of the while loop of $Sko(A)$. The loop continues while a quantifier remains at the top of the formula, so the final formula will have no quantifiers. The input and output formulas are not necessarily equivalent, but their meanings are closely related, and we will capture this relationship with the notion "equally satisfiable" or "equisatisfiable."

Recall that in propositional logic (Definition 3.3.7), we say two formulas, A and B, are equisatisfiable, denoted by $A \approx B$, if $\mathcal{M}(A) = \emptyset$ whenever $\mathcal{M}(B) = \emptyset$ and vice versa. This relation is weaker than the logical equivalence, which requires $\mathcal{M}(A) = \mathcal{M}(B)$. We will reuse $\approx$ with the same definition for first-order formulas.

It is easy to see that elimination of universal quantifiers preserves equivalence. The difficulties arise with the introduction of Skolem functions/constants.

Lemma 5.4.9 *Let A be $\forall x \exists y\, B(x, y)$. Then there exists a function $f(x)$ such that (a) $\forall x\, B(x, f(x)) \models A$ and (b) $\forall x\, B(x, f(x)) \approx A$.*

Proof

(a): For any model I of $\forall x\, B(x, f(x))$, and for every $d \in D^I$, it must be the case that $I(B(x, f(x)), \{x \leftarrow d\}) = 1$. Let $f^I(d) = e \in D^I$, where $f^I : D^I \mapsto D^I$ is the interpretation of f in I, then $I(B(x, y), \{x \leftarrow d, y \leftarrow e\}) = 1$. So, we have

$$I(\exists y\, B(x, y), \{x \leftarrow d\}) = 1$$

Since d is arbitrary in D^I, $I(\forall x \exists y\, B(x, y)) = 1$. In other words, I is a model of A.

(b): Since (a) says "if $\forall x\, B(x, f(x))$ is satisfiable, so is A," we need only to show that if A is satisfiable, so is $\forall x\, B(x, f(x))$. Suppose $I = (D, R, G)$ is a model of A. Then for any $d \in D$, there exists $e \in D$ such that:

$$I(B(x, y), \{x \leftarrow d, y \leftarrow e\}) = 1$$

Let $g : D \mapsto D$ such that $g(d) = e$, and $I' = (D, R, G \cup \{g\})$ with $g = f^{I'}$. Then I' is a model of $\forall x\, B(x, f(x))$.

□

Corollary 5.4.10 *Let $A = \forall x_1 \forall x_2 \cdots \forall x_k \exists y\, B(x_1, x_2, \ldots, x_k, y)$. Then there exists a function $f(x_1, x_2, \ldots, x_k)$ such that*

(a) $\forall x_1 \forall x_2 \cdots \forall x_k\, B(x_1, x_2, \ldots, x_k, f(x_1, x_2, \ldots, x_k)) \models A$.
(b) $\forall x_1 \forall x_2 \cdots \forall x_k\, B(x_1, x_2, \ldots, x_k, f(x_1, x_2, \ldots, x_k)) \approx A$.

Proof Replace x by $(x_1, x_2, \ldots, x_k)$ in the proof of Lemma 5.4.9. □

Theorem 5.4.11 *(a) $Sko(A) \models A$; and (b) $Sko(A) \approx A$.*

5.4 Conjunctive Normal Form (CNF)

Proof Since a free variable and a universally quantified variable have the same interpretation, dropping a universal quantifier does not change the semantics of a formula. When we remove an existential quantifier, we introduce a Skolem function. By Corollary 5.4.10, the resulting formula is equally satisfiable to the original formula. Applying Corollary 5.4.10 to every existential quantifier, we have the desired result, because both $\models$ and $\approx$ are transitive:

- (a): $Sko(A) \models A$ by Corollary 5.4.10(a).
- (b): $A \approx Sko(A)$ by Corollary 5.4.10(b).

$\square$

Relation $\approx$ is what we call "weakly equivalent," and we will see that it is sufficient for our purposes.

Skolemizing Non-prenex Formulas

Although Skolemization is easy to explain and justify on formulas in PNF, it can actually be applied directly to any formula. The involved process is not easily explained as in the case of PNF. Instead, we will provide a recursive algorithm for Skolemization which takes two parameters: The first parameter is a formula to be Skolemized, which must be in NNF (negation normal form), and the second parameter is the list of free variables collected so far.

Algorithm 5.4.12 $Skol(A, V)$ takes as input a formula A which is in NNF (containing no Boolean operators other than $\neg, \wedge, \vee$ and each quantified variable is distinct) and a list V of variables which are free in A, and returns a formula without quantifiers.

proc $Skol(A, V)$
1   **if** A is a literal or $A = \top$ or $A = \bot$ **return** A // base case
2   **if** $(A = B \text{ op } C)$ **return** $Skol(B, V) \text{ op } Skol(C, V)$ // op is either $\wedge$ or $\vee$.
3   **if** $(A = \exists x\, B)$ **return** $Skol(B[x \leftarrow f(V)], V)$ // f is a new Skolem function.
4   **if** $(A = \forall x\, B)$ **return** $Skol(B, V \cup \{x\})$

When A is a literal, i.e., either an atom, or the negation of an atom, A is returned. When the topmost operator is $\vee$ or $\wedge$, the situation is simple: $Skol$ is merely called recursively on each of the arguments of the operator. If the topmost symbol is $\exists$, $Skol$ introduces a new Skolem function f (a Skolem constant if V is empty) and replaces every occurrence of x by $f(V)$. If the topmost symbol is $\forall$, the variable x is collected in the second parameter for the recursive call on B. Assuming computing $B[x \leftarrow f(V)]$ takes linear time, $Skol$ will take quadratic time in terms of the size of the input formula.

Example 5.4.13 Let A be the first NNF in Example 5.4.4

$$A = (\forall x \exists y\, p(x, y)) \wedge (\exists z \exists w\, (\neg q(z, w) \vee \neg q(w, z)))$$

Then $Skol(A, \emptyset)$ calls itself on the two arguments of $\wedge$:

- $Skol(\forall x \exists y\, p(x, y), \emptyset)$ calls $Skol(\exists y\, p(x, y)), \{x\})$ and returns $p(x, f(x))$.
- $Skol(\exists z \exists w\, \neg q(z, w) \vee \neg q(w, z), \emptyset)$ returns $\neg q(c_1, c_2) \vee \neg q(c_2, c_1)$.

Hence, $Skol(A, \emptyset)$ returns $p(x, f(x)) \wedge (\neg q(c_1, c_2) \vee \neg q(c_2, c_1))$. □

Theorem 5.4.14 (a) $Skol(A) \models A$, and (b) $Skol(A) \approx A$.

The proof of the above theorem is similar to that of Theorem 5.4.11 and is omitted here. As pointed out in Example 5.4.4, we have many different PNFs for the same formula and each PNF may give us a different result after Skolemization. All PNFs of the same formula are logically equivalent and all Skolemized formulas of the same formula are equisatisfiable.

In practice, we wish to produce Skolemized formulas where Skolem functions have minimal arities, so that the ground terms of these formulas will be simpler to deal with. Note that the equivalence rules for obtaining PNFs try to extend the scope of each quantified variable. To obtain Skolem functions with minimal number of arguments, we may need to use these equivalence rules backward.

5.4.3 Clausal Form

By CNF, we meant that the formula is a conjunction of clauses, and a clause is a disjunction of literals. To obtain CNF, a formula A goes through the following stages:

- Stage 1: Convert A into NNF A'.
- Stage 2: Skolemize A' into quantifier-free B.
- Stage 3: Convert B into CNF C.

Stage 3 converts Skolemized NNF B into clauses, as we did in propositional logic to convert NNF into clauses. That is, use the distribution law $A \vee (B \wedge C) \equiv (A \vee C) \wedge (A \vee C)$ repeatedly, until it no longer applies.

Stages 1 and 3 preserve the logical equivalence, $\equiv$, and stage 2 preserves only a weak equivalence, $\approx$ (equisatisfiability). Thus, we have $A \equiv A'$, $A' \approx B$, and $B \equiv C$. As a result, $A \approx C$. So, A is satisfiable iff C is satisfiable.

Example 5.4.15 Consider the following formulas and their intended meanings:

- $A = \forall x\, (human(x) \rightarrow mortal(x))$: Every human is mortal.
- $B = \forall x\, (iowan(x) \rightarrow human(x))$: Every Iowan is human.
- $C = \forall x\, (iowan(x) \rightarrow mortal(x))$: Every Iowan is mortal.

5.4 Conjunctive Normal Form (CNF)

To show $A \wedge B \models C$, we convert $A \wedge B \wedge \neg C$ into a set S of four clauses, where c_1 comes from A, c_2 from B, c_3 and c_4 from $\neg C$:

$$c_1 : (\neg human(x) \mid mortal(x))$$
$$c_2 : (\neg iowan(y) \mid human(y))$$
$$c_3 : (iowan(c))$$
$$c_4 : (\neg mortal(c))$$

An instance of c_1 is $c_1' = (\neg human(c) \mid mortal(c))$, and an instance of c_2 is $c_2' = (\neg iowan(c) \mid human(c))$. Resolution from c_3 and c_2', we obtain $c_5 : (human(c))$. From c_5 and c_1', we obtain $c_6 : (mortal(c))$. From c_6 and c_4, we obtain the empty clause ( ). □

Now, we know how to convert a formula into clausal form, which is required by the resolution rules. Keep in mind the variables in a clause are assumed free and have the same semantics as the universally quantified variables. These variables are different from clauses to clauses, even though they use the same name. A common practice is to give different names for variables in different clauses. For example, in clause c_i, we use x_i, y_i, z_i, etc. This way, we ensure that no variable appears in more than one clause.

Stage 3 of converting a formula into CNF may blow up the sizes of the output formulas, while stage 2 outputs a formula of the linear size of the input formula. In propositional logic, to obtain a set of clauses of linear size of the original formula, we may introduce new propositional variables for each occurrence of binary operators and the resulting clauses are 3CNF, which are not equivalent but equisatisfiable to the original formula. We may do the same in stage 3 so that the final clauses are of linear size of the original formula.

Clausal form might seem a highly constrained and unnatural representation for mathematical knowledge, and it has been criticized on those grounds. We will see that it is surprisingly natural in many cases. In fact, a problem can be expressed as a set of constraints over some variables, such as a constraint satisfaction problem (CSP). A constraint can be expressed in CNF and a set of constraints gives us a set of clauses.

We have seen that any predicate calculus formula can be converted into an equisatisfiable formula in CNF, which is useful for proving theoretical results by resolution.

Herbrand Models for CNF

When we consider interpretations for a set of formulas, there are simply too many ways to choose a set as the universe, too many ways to choose a relation for a predicate symbol, and too many ways to choose a function for a function symbol. We can improve on this considerably because there is a class of interpretations, invented by French mathematician Jacques Herbrand (1908–1931), called *Herbrand interpretations*, which can stand for all the others. The Herbrand interpretations share the same universe, called the *Herbrand universe*. What does this universe contain? For each n-ary function symbol f in the formula (including constants) and n elements, say $t_1, t_2, \ldots, t_n$, in the universe, it must contain the result of applying f to these n elements. This can be achieved by a simple syntactic device; we put all the ground terms in the universe and let each term denote the result of applying its function to its parameters. This way, function symbols are bound to a unique interpretation over the universe.

Definition 5.4.16 Given a CNF formula A in a first-order language $L = (P, F, X, Op)$, the *Herbrand universe* of A consists of all the ground terms, i.e., the set $T(F)$; the *Herbrand base* is the set of all ground atoms, denoted by $B(P, F)$. A *Herbrand interpretation* is a mapping $I : B(P, F) \mapsto \{1, 0\}$.

Note that $T(F)$ is the set of variable-free terms that we can make from the constants and functions in the formula. If there are no constants in F, $T(F)$ would be empty. To avoid this case, we assume that there must exist at least one constant in F, say a. From now on it will be useful to distinguish the constants from non-constant functions, so by "function" we mean non-nullary function. If F does not contain function symbols, then $T(F)$ will be a finite set of constants.

The Herbrand base $B(P, F)$ consists of all the atoms like $p(t_1, t_2, \ldots, t_n)$, where p is an n-ary predicate symbol in P and $t_i \in T(F)$. If $n = 0$ for all predicates, then $B(P, F) = P$, reduced to propositional logic. Otherwise, $B(P, F)$ is finite iff $T(F)$ is finite. For convenience, we may represent an interpretation I by a set H of ground literals such that for any $g \in B(P, F)$, $g \in H$ iff $I(g) = 1$ and $\overline{g} \in H$ iff $I(g) = 0$, where $\overline{g}$ stands for $\neg g$.

Example 5.4.17 In Example 5.2.12, we showed that the set S has no finite models:

$$S = \{\forall x \exists y\, p(x, y),\ \forall x\, \neg p(x, x),\ \forall x \forall y \forall z\, (p(x, y) \wedge p(y, z) \rightarrow p(x, z))\}$$

Note that S has no function symbols. Converting S to CNF, we obtain the following set S' of clauses:

$$S' = \{c_1 : (p(x, f(x))),\ c_2 : (\overline{p(x, x)}),\ c_3 : (\overline{p(x, y)} \mid \overline{p(y, z)} \mid p(x, z))\}$$

Adding a constant a, $F = \{f, a\}$ and $P = \{p\}$, then the Herbrand universe of S' is

$$T(F) = \{a, f(a), f(f(a)), \ldots, f^i(a), \ldots\},$$

where $f^0(a) = a$ and $f^{i+1}(a) = f(f^i(a))$. The Herbrand base of S' is

$$B(P, F) = \{p(a, a), p(a, f(a)), p(f(a), a),$$
$$p(a, f(f(a))), \ldots, p(f^i(a), f^j(a)), \ldots\}$$

A Herbrand interpretation of S' is

$$H = \{p(f^i(a), f^j(a)) \mid 0 \le i < j\} \cup \{\overline{p(f^i(a), f^j(a))} \mid 0 \le j \le i\}$$

If $f^i(a)$ denotes the natural number i, then H is identical to $R_< = \{(i, j) \mid i, j \in \mathcal{N}, i < j\}$. □

Proposition 5.4.18 *A Herbrand interpretation defines a unique interpretation for a formula in CNF.*

Proof Let A be a formula in CNF and H is a Herbrand interpretation. Define interpretation $I = (D, R, G)$, where $D = T(F)$ (the Herbrand universe),

$$R = \{r_p \mid p \in P, r_p = \{\langle t_1, \ldots, t_n \rangle \mid H(p(t_1, \ldots, t_n)) = 1\}\}$$

and $G = \{f^H \mid f \in F, f^H(t_1, \ldots, t_n) = f(t_1, \ldots, t_n)\}$, then I is an interpretation as defined in Definition 5.2.1 and is uniquely defined by H. □

Since D and G are fixed in I, H is the only part which affects I. That is, the only way in which Herbrand interpretations differ from one to another is over the meanings of predicate symbols. The above proposition gives us the justification for calling H "an interpretation."

Once we have an interpretation, we can compute the truth value of a formula A in this interpretation using a procedure like *eval* (Procedure 5.2.3). If A is true under a Herbrand interpretation H, we say H is a *Herbrand model* of A. Since a CNF A can be represented by a set S of clauses, and the variables in each clause have the same meaning as universally quantified variables, for any interpretation I, $I(A) = I(S) = \bigwedge_{C \in S} I(C)$, and for each clause C, $I(C) = I(\forall x_1 \ldots \forall x_k C)$, where $x_1, \ldots, x_k$ are the free variables of C.

The definition of Herbrand interpretation gives us a convenient way of expressing these meanings by assigning the truth value to each ground atom in the Herbrand base. Of course, the truth values of some atoms are irrelevant if these atoms cannot be instances of any predicate in the formula, as the truth of the formula does not depend on these atoms. From now on, when we say "a set of clauses has a model" we mean "it has a Herbrand model."

This chapter introduced the basic concepts of first-order logic, which is also called *predicate calculus* because it is an extension of propositional logic with predicates along with variables, functions, and quantifiers. In the next three chapters, we will introduce reasoning tools within first-order logic.

Exercises

1. Using a first-order language with variables ranging over people and predicates $trusts(x, y)$, $politician(x)$, $crazy(x)$, $know(x, y)$, and $related(x, y)$, and $rich(x)$, write down first-order sentences asserting the following:

    (a) Nobody trusts a politician.
    (b) Anyone who trusts a politician is crazy.
    (c) Everyone knows someone who is related to a politician.
    (d) Everyone who is rich is either a politician or knows a politician.

2. Let the meaning of $taken(x, y)$ be "student x has taken CS course y." Please write the meaning of the following formulas:

    (a) $\exists x \exists y\, taken(x, y)$
    (b) $\exists x \forall y\, taken(x, y)$
    (c) $\forall x \exists y\, taken(x, y)$
    (d) $\exists y \forall x\, taken(x, y)$
    (e) $\forall y \exists x\, taken(x, y)$

3. Let $c(x)$ mean "x is a criminal" and $s(x)$ mean "x is sane." Using them to write a first-order formula for each of the following sentences.

    (a) All criminals are sane.
    (b) Every criminal is sane.
    (c) Only criminals are sane.
    (d) Some criminals are sane.
    (e) There is a criminal that is sane.
    (f) No criminal is sane.
    (g) Not all criminals are sane.

4. Given the predicates $male(x)$, $female(x)$, and $parent(x, y)$ (x is a parent of y), define the following predicates in terms of *male, female, parent,* or other predicates defined from these three predicates, in first-order logic:

    (a) $father(x, y)$: x is the father of y.
    (b) $mother(x, y)$: x is the mother of y.
    (c) $ancestor(x, y)$: x is an ancestor of y.
    (d) $descendant(x, y)$: x is a descendant of y.
    (e) $son(x, y)$: x is a son of y.
    (f) $daughter(x, y)$: x is a daughter of y.
    (g) $sibling(x, y)$: x is a sibling of y.

5. Let $+, -, *, \%, >, <, \geq, \leq, =$ be usual operations on the natural numbers $\mathcal{N}$, where $(x \% y)$ returns the remainder of x divided by y for $y > 0$. Please define the following sets using first-order formulas:

Exercises

(a) The set of natural numbers divisible by 5, but not divisible by 3.
(b) The set of natural numbers whose square is the sum of two square numbers.
(c) The set of primes.

6. Provide proof of Theorem 5.2.25.
7. Let *Set* be defined by the following BNF:

$$\langle Set \rangle ::= \emptyset \mid adjoin(\langle Any \rangle, \langle Set \rangle)$$

where $adjoin(e, s)$ adjoins an element e into the set s. Please define the following predicates in terms of $\emptyset$ and $adjoin$ in first-order logic:

(a) $isSet(x)$: x is an object of *Set*.
(b) $member(x, s)$: x is a member of set s.
(c) $subset(s, t)$: s is a subset of set t.
(d) $equal(s, t)$: s and t contain the same elements.

8. Decide if the following formulas are valid, not valid but satisfiable, or unsatisfiable, using the definition of interpretations:

(a) $p(a, a) \vee \neg(\exists x\, p(x, a))$
(b) $\neg p(a, a) \vee (\exists x\, p(x, a))$
(c) $\neg q(a) \vee (\forall x\, q(x))$
(d) $\neg(\exists x\, q(x)) \vee (\forall x\, q(x))$
(e) $\exists x\, (q(x) \vee \neg \forall x\, q(x))$

9. Prove the following equivalence relations:

(a) $\neg \exists x\, p(x) \equiv \forall x\, \neg p(x)$
(b) $\exists x\, (p(x) \vee q(x)) \equiv (\exists x\, p(x)) \vee (\exists x\, q(x))$
(c) $(\exists x\, (p(x)) \wedge q) \equiv (\exists x\, (p(x) \wedge q))$
(d) $\exists x \exists y\, r(x, y) \equiv \exists y \exists x\, r(x, y)$
(e) $\exists x\, (p(x) \rightarrow q(x)) \equiv (\forall x\, p(x)) \rightarrow (\exists x\, q(x))$

10. Prove that the following entailment relations are true, but the converse of the relation is not true.

(a) $(\forall x\, p(x)) \vee (\forall x\, q(x)) \models \forall x\, (p(x) \vee q(x))$
(b) $\exists x\, (p(x) \wedge q(x)) \models (\exists x\, p(x)) \wedge (\exists x\, q(x))$
(c) $(\forall x\, p(x)) \rightarrow (\forall x\, q(x)) \models \forall x\, (p(x) \rightarrow q(x))$
(d) $(\exists x\, p(x)) \rightarrow (\exists x\, q(x)) \models \exists x\, (p(x) \rightarrow q(x))$
(e) $(\exists x\, p(x)) \rightarrow (\forall x\, q(x)) \models \forall x\, (p(x) \rightarrow q(x))$
(f) $\forall x\, (p(x) \rightarrow q(x)) \models (\exists x\, p(x)) \rightarrow (\exists x\, q(x))$
(g) $\forall x\, (p(x) \rightarrow q(x)) \models (\forall x\, p(x)) \rightarrow (\exists x\, q(x))$
(h) $\forall x\, (p(x) \leftrightarrow q(x)) \models (\forall x\, p(x)) \leftrightarrow (\forall x\, q(x))$
(i) $\forall x\, (p(x) \leftrightarrow q(x)) \models (\exists x\, p(x)) \leftrightarrow (\exists x\, q(x))$

11. Use the semantic tableau method to show the entailment relations are true in the previous problem.
12. Provide the pseudo-code for the algorithm *PrenexNF(A)* and show that it runs in $O(n)$ time when A is in NNF, where n is the size of A.

13. Show that the algorithm $Sko(A)$ runs in $O(n^2)$ time, where A is in NNF, and n is the size of A.
14. Show that the algorithm $Skol(A)$ runs in $O(n^2)$ time, where A is in NNF, and n is the size of A.
15. Show that formula $\forall x \exists y ((p(x) \land p(y)) \lor (q(x) \land q(y)))$ can be converted into an equivalent prenex normal form such that $\exists$ goes before $\forall$ in the prenex normal form.
16. Prove that if every predicate symbol is monodic (i.e., unary) and every function symbol is monodic or a constant, then any first-order formula can be converted into a prenex normal form where $\exists$ goes before any $\forall$.
17. Convert each one of the following formulas into a set of clauses:

    (a) $\neg((\forall x\, p(x)) \lor (\forall x\, q(x)) \to \forall x\, (p(x) \lor q(x)))$
    (b) $\neg((\exists x\, (p(x) \land q(x))) \to ((\exists x\, p(x)) \land (\exists x\, q(x))))$
    (c) $\neg(((\forall x\, p(x)) \to (\forall x\, q(x))) \to (\forall x\, (p(x) \to q(x))))$
    (d) $\neg(((\exists x\, (p(x) \to q(x))) \to (\exists x\, p(x))) \to (\exists x\, q(x)))$
    (e) $\neg((\exists x\, p(x)) \to ((\forall x\, q(x)) \to (\forall x\, (p(x) \to q(x)))))$
    (f) $\neg(((\forall x\, (p(x) \to q(x))) \to (\exists x\, p(x))) \to (\exists x\, q(x)))$
    (g) $\neg((\forall x\, (p(x) \to q(x))) \to ((\forall x\, p(x)) \to (\exists x\, q(x))))$
    (h) $\neg(((\forall x\, (p(x) \leftrightarrow q(x))) \to (\forall x\, p(x))) \leftrightarrow (\forall x\, q(x)))$
    (i) $\neg(((\forall x\, (p(x) \leftrightarrow q(x))) \to (\exists x\, p(x))) \leftrightarrow (\exists x\, q(x)))$

18. Given a set S of clauses, $S = \{(p(f(x), c)), (\neg p(f(w), z) \mid p(f(f(w)), f(z))), (\neg p(y, f(y)))\}$, please answer the following questions: (a) What is the Herbrand universe of S? (b) What is the Herbrand base of S? (c) What is the minimum Herbrand model of S (no subset of this model is a model)? (d) What is the maximum Herbrand model of S (no superset of this model is a model)?
19. Let c be $(\overline{p(x, y)} \mid \overline{p(y, z)} \mid q(z) \mid p(x, z))$ and $S = \{(\overline{p(x, y)} \mid \overline{p(y, z)} \mid r(x, z)), (\overline{r(x, z)} \mid q(x, z) \mid p(x, z))$. Show that $S \models c$ and $S \approx c$. In general, show that a clause c of n literals can be converted into a set S of $n - 2$ clauses of 3 literals each (3CNF) such that $S \models c$ and $S \approx c$.

References

1. Melvin Fitting, *First-Order Logic and Automated Theorem Proving*, Springer Science & Business Media. 1990 ISBN 978-1-4612-2360-3.
2. J. P. E. Hodgson, *First Order Logic*, Saint Joseph's University, Philadelphia, 1995.
3. William Ewald, "The Emergence of First-Order Logic", The Stanford Encyclopedia of Philosophy (Spring 2019 Edition), Edward N. Zalta (ed.),
   plato.stanford.edu/archives/spr2019/entries/logic-firstorder-emergence/, retrieved Oct. 2023.
4. Christoph Benzmüller and Peter Andrews, "Church's Type Theory", The Stanford Encyclopedia of Philosophy, Edward N. Zalta and Uri Nodelman (eds.),
   plato.stanford.edu/archives/win2023/entries/type-theory-church/. retrieved Oct. 2023.

Chapter 6
Unification and Resolution

Resolution is known to be a refutational proof method, which can show that the input formula is unsatisfiable. If we would like to show that formula A is valid, we need to convert $\neg A$ into clauses and show that the clauses are unsatisfiable. To show that $A \models B$, we need to convert $A \wedge \neg B$ into CNF and show that the clauses are unsatisfiable. Note that A and B must be closed formulas.

For example, if we want to express that $p(x, y)$ is a commutative relation, we may use the formula $p(x, y) \rightarrow p(y, x)$, where x and y are assumed to be arbitrary values. The negation of this formula is $\neg(\forall x \forall y\, (p(x, y) \rightarrow p(y, x)))$, not $\neg(p(x, y) \rightarrow p(y, x))$. The CNF of the former is $p(a, b) \wedge \neg p(b, a)$, where a and b are Skolem constants; the CNF of the latter is $p(x, y) \wedge \neg p(y, x)$, which is equivalent to the clause set $\{(p(x, y)), (\neg p(y, x))\}$. These two unit clauses will generate the empty clause by resolution when x is substituted by y. The process of finding a substitution to make two atomic formulas (i.e., atoms) identical is called "unification," which is needed in the definition of the resolution rule for first-order formulas.

6.1 Unification

In Example 5.4.15, when we apply the resolution rule

$$(A \mid \alpha),\ (\neg A \mid \beta) \vdash (\alpha \mid \beta)$$

on two clauses, we need to find the instances of the two clauses such that one instance contains A and the other instance contains $\neg A$, where A is an atom. A clause has infinite many instances, and the unification process helps us to find the right instances for resolution.

6.1.1 Substitutions and Unifiers

Given a first-order language $L = (P, F, X, Op)$, a *substitution* is a mapping $\sigma : X \mapsto T(F, X)$ which maps every variable of X to a term of $T(F, X)$. If $\sigma(x) \neq x$, we say x is *affected* by σ. If $\sigma(x)$ is a distinct variable for every affected x, σ is said to be a *renaming*.

When displaying σ, we use $[x \leftarrow \sigma(x)]$, where x is typically an affected variable. For example, if $X = \{x, y\}$, then x is affected by $\sigma_1 = [x \leftarrow f(y)]$ and $\sigma_2 = [x \leftarrow f(x)]$; both x and y are affected by $\sigma_3 = [x \leftarrow f(y), y \leftarrow a]$.

A substitution $\sigma : X \mapsto T(F, X)$ can be extended to be a mapping σ from terms to terms such that if $t = x$, then $\sigma(t) = \sigma(x)$; if $t = f(t_1, t_2, \ldots, t_k)$, then $\sigma(t) = f(\sigma(t_1), \sigma(t_2), \ldots, \sigma(t_k))$. In a similar way, σ can be extended to be a mapping from formulas to formulas.

We use $p(t)$ to denote an instance of $p(x)$, i.e., $p(t) = p(x)[x \leftarrow t]$, where every occurrence of x is replaced by term t. Here, $\sigma = [x \leftarrow t]$ is a substitution and $p(t) = \sigma(p(x)) = p(x)\sigma$, as we are free to write either $p(x)\sigma$ or $\sigma(p(x))$.

Definition 6.1.1 (Idempotent Substitution) Substitution σ is said to be *idempotent* if for any term t, $t\sigma = (t\sigma)\sigma$.

For example, $\sigma_1 = [x \leftarrow f(y)]$ is idempotent; $\sigma_2 = [x \leftarrow f(y), y \leftarrow a]$ is not idempotent, because $x\sigma_2 = f(y)$ and $(x\sigma_2)\sigma_2 = f(a)$. Some non-idempotent substitutions can be made into idempotent ones. For instance, σ_2 can be replaced by $\sigma_3 = [x \leftarrow f(a), y \leftarrow a]$, which is idempotent. Some substitutions like $\sigma_4 = [x \leftarrow f(x)]$ can never be made into idempotent ones.

For idempotent substitutions, an affected variable is never brought back by another affected variable.

Proposition 6.1.2 *A substitution σ is* idempotent *iff for every $x \in X$, either $\sigma(x) = x$ or x does not appear in $\sigma(y)$ for any $y \in X$.*

Proof Given any σ, if an affected x appears in $\sigma(y) = t(x)$, suppose $\sigma(x) = s$, then $y\sigma = t(x)$ and $(y\sigma)\sigma = t(s)$, so σ is not idempotent. Otherwise, for any term s, $s\sigma$ does not contain any variable affected by σ, so $(s\sigma)\sigma = s\sigma$ and σ is idempotent. □

Definition 6.1.3 (Unifier) Let s, t be either two terms of $T(F, X)$ or two atoms. We say s and t are *unifiable*, if there exists an idempotent substitution $\sigma : X \mapsto T(F, X)$ such that $s\sigma = t\sigma$; σ is said to be a *unifier* of s and t.

A set S of pairs of terms or atoms is *unifiable* if there exists a substitution σ such that for every pair $(s_i, t_i) \in S$. $s_i\sigma = t_i\sigma$, and σ is said to be a *unifier* of S.

Example 6.1.4 Let $s = f(x, g(y))$ and $t = f(g(z), x)$, then s and t are unifiable with unifier $\sigma = [x \leftarrow g(z), y \leftarrow z]$, since

$$s\sigma = f(x, g(y))[x \leftarrow g(z), y \leftarrow z] = f(g(z), g(z))$$
$$= f(g(z), x)[x \leftarrow g(z), y \leftarrow z] = t\sigma$$

In fact, s and t have infinite many unifiers like

$$[x \leftarrow g(w), y \leftarrow w, z \leftarrow w]$$

for any $w \in T(F, X)$ and $w \neq x$. □

In the above example, we see that a pair of terms may have multiple unifiers. How do you compare these unifiers? We need some relation over unifiers and this relation is based on the composition of substitutions.

6.1.2 Combining Substitutions

Definition 6.1.5 (Composition of Substitution) Given two substitutions σ and θ, the composition of σ and θ is the substitution γ, written $\gamma = \sigma \cdot \theta$, or simply $\sigma\theta$, where

$$\gamma = [x \leftarrow t\theta \mid (x \leftarrow t) \in \sigma]$$

The above definition assumes that σ contains every variable y, including $y = y\sigma$. If not, we need to add explicitly $[y \leftarrow s \mid y = y\sigma, (y \leftarrow s) \in \theta]$ into γ.

We will assume that $t\sigma\theta = (t\sigma)\theta = \theta(\sigma(t))$. To apply the composition $\sigma\theta$ to t, we write $t(\sigma\theta)$.

Example 6.1.6 Given $\sigma = [x \leftarrow f(y), z \leftarrow f(u), v \leftarrow u]$ and $\theta = [y \leftarrow g(a), u \leftarrow z, v \leftarrow f(f(a))]$, then

$$\sigma\theta = [x \leftarrow f(g(a)), y \leftarrow g(a), z \leftarrow f(z), u \leftarrow z, v \leftarrow z]$$
$$p(x, y, z, u, v)\sigma\theta = p(f(y), y, f(u), u, u)\theta$$
$$= p(f(g(a)), g(a), f(z), z, z)$$
$$= p(x, y, z, u, v)(\sigma\theta)$$

Note that the composition does not necessarily preserve the idempotent property as both σ and θ are idempotent but $z \leftarrow f(z)$ appears in $\sigma\theta$. □

Proposition 6.1.7 For any substitutions σ and θ, and any term t, $t\sigma\theta = t(\sigma\theta)$.

Proof For any $x \in X$, if x is affected by σ, say $\sigma(x) = s$, then $x\sigma\theta = s\theta = x(\sigma\theta)$. If x is unaffected by σ, then $x\sigma\theta = x\theta = x(\sigma\theta)$. □

The above proposition states the property that applying two substitutions to a term in succession produces the same result as applying the composition of two substitutions to the term.

The composition operation is not commutative in general. For example, if $\sigma = [x \leftarrow a]$ and $\theta = [x \leftarrow b]$, then $\sigma\theta = \sigma$ and $\theta\sigma = \theta$. Thus, $\sigma\theta \neq \theta\sigma$. By definition, if x is affected by both σ and θ, which are idempotent, then the pair $x \leftarrow \theta(x)$ is ignored in $\sigma\theta$ and the pair $x \leftarrow \sigma(x)$ is ignored in $\theta\sigma$.

Proposition 6.1.8 *The composition is associative, i.e., for any substitutions σ, θ, and γ, $(\sigma\theta)\gamma = \sigma(\theta\gamma)$.*

The proof is left as an exercise.

6.1.3 Rule-Based Unification

Definition 6.1.9 (Most General Unifier) Let σ and θ be unifiers of s and t. We say σ is *more general* than θ, if there exists a substitution γ such that $\theta = \sigma\gamma$. σ is the *most general unifier* (mgu) of s and t, if σ is more general than any unifier of s and t.

Note that mgu is unique upon renaming of variables. For instance, $[x \leftarrow y]$, $[y \leftarrow x]$, and $[x \leftarrow z, y \leftarrow z]$ are mgus of $f(x)$ and $f(y)$. They can be converted into each other by renaming substitutions.

Example 6.1.10 In Example 6.1.4, $\sigma = [x \leftarrow g(z), y \leftarrow z]$ is a mgu of $s = f(x, g(y))$ and $t = f(g(z), x)$. For any unifier $\theta = [x \leftarrow g(t'), y \leftarrow t', z \leftarrow t']$, $\theta = \sigma\gamma$, where $\gamma = [z \leftarrow t']$. Another mgu of s and t is $\sigma' = [x \leftarrow g(y), z \leftarrow y]$, $\sigma = \sigma'[y \leftarrow z]$ and $\sigma' = \sigma[z \leftarrow y]$. □

As illustrated by the above example, there are more than one mgu for a pair of terms; they differ by the renaming of variables. We can say *the* most general unifier if the renaming of variables is considered.

Given a pair s and t of terms, how to decide if s and t are unifiable? If they are, how to compute their mgu? These questions will be answered by unification algorithms.

Let $s \doteq t$ denote that "s and t are unifiable," and a unification algorithm will decide if $s \doteq t$ is true or false. Suppose $s = f(s_1, s_2, \ldots, s_m)$ and $t = g(t_1, t_2, \ldots, t_n)$. If $f \neq g$ or $m \neq n$, then $s \doteq t$ is false, i.e., s and t are not unifiable. This is called *clash* failure. If $s \doteq t$ is true, so are $s_i \doteq t_i$ for $1 \leq i \leq m = n$. In other words, if $s_i \doteq t_i$ is false for some i, then $s \doteq t$ is false. If $s = x$ and $t = f(x)$ (x occurs in t), then there exists no substitution σ such that $x\sigma = f(x)\sigma$. Hence, $x \doteq f(x)$ must be false. This is called *occur-check* failure, as x occurs in $f(x)$. If we accept $x \leftarrow f(x)$, then the unifier cannot be idempotent.

Rule-based unification algorithm uses a set of rules which transform $s \doteq t$ into a substitution σ through a set of pairs $\{s_i \doteq t_i\}$ and each rule preserves the unifiability of $s \doteq t$. These rules apply to $\langle S, \sigma \rangle$, where S is a set of pairs $s_i \doteq t_i$ and σ is a unifier to be computed. Initially, $S = \{s \doteq t\}$ and σ is empty.

6.1 Unification

Definition 6.1.11 The rule-based unification algorithm will try to transform $\langle \{s \doteq t\}, [\,] \rangle$ to $\langle \emptyset, \sigma \rangle$, where σ is a mgu of s and t, by the following unification rules:

- **Decompose**: $\langle S \cup \{f(s_1, s_2, \ldots, s_m) \doteq f(t_1, t_2, \ldots, t_m)\}, \sigma \rangle \mapsto \langle S \cup \{s_i \doteq t_i \mid 1 \le i \le m\}, \sigma \rangle$.
- **Clash**: $\langle S \cup \{f(s_1, s_2, \ldots, s_m) \doteq g(t_1, t_2, \ldots, t_n)\}, \sigma \rangle \mapsto \langle \emptyset, fail \rangle$.
- **Occur-check**: $\langle S \cup \{x \doteq t \mid x \ne t, x \text{ occurs in } t\}, \sigma \rangle \mapsto \langle \emptyset, fail \rangle$.
- **Redundant**: $\langle S \cup \{t \doteq t\}, \sigma \rangle \mapsto \langle S, \sigma \rangle$.
- **Orient**: $\langle S \cup \{t \doteq x \mid t \text{ is not variable}\}, \sigma \rangle \mapsto \langle S \cup \{x \doteq t\}, \sigma \rangle$.
- **Substitute**: $\langle S \cup \{x \doteq t \mid x \text{ does not occur in } t\}, \sigma \rangle \mapsto \langle S[x \leftarrow t], \sigma \cdot [x \leftarrow t] \rangle$.

Algorithm 6.1.12 The algorithm $unify(s, t)$ takes two terms s and t and returns either $fail$ or σ, a mgu of s and t.

proc $unify(s, t)$
1    $S := \{s \doteq t\};\ \sigma = [\,]$
2   **while** (true) **do**
3        $\langle S, \sigma \rangle :=$ apply one unification rule on $\langle S, \sigma \rangle$
4   **if** ($\sigma = fail$) **return** $fail$
5   **if** ($S = \emptyset$) **return** σ

Example 6.1.13 Given $s = p(f(x, h(x), y), g(y))$ and $t = p(f(g(z), w, z), x)$, $unify(s, t)$ will transform $\langle \{s \doteq t\}, [\,] \rangle$ into $\langle \emptyset, \sigma \rangle$ as follows:

$$\{s \doteq t\}, [\,]$$
$$\textbf{decompose} \mapsto \{f(x, h(x), y) \doteq f(g(z), w, z), g(y) \doteq x\}, [\,]$$
$$\textbf{orient} \mapsto \{f(x, h(x), y) \doteq f(g(z), w, z), x \doteq g(y)\}, [\,]$$
$$\textbf{substitute} \mapsto \{f(g(y), h(g(y)), y) \doteq f(g(z), w, z)\}, [x \leftarrow g(y)]$$
$$\textbf{decompose} \mapsto \{g(y) \doteq g(z), h(g(y)) \doteq w, y \doteq z\}, [x \leftarrow g(y)]$$
$$\textbf{substitute} \mapsto \{g(z) \doteq g(z), h(g(z)) \doteq w\}, [x \leftarrow g(z), y \leftarrow z]$$
$$\textbf{redundant} \mapsto \{h(g(z)) \doteq w\}, [x \leftarrow g(z), y \leftarrow z]$$
$$\textbf{orient} \mapsto \{w \doteq h(g(z))\}, [x \leftarrow g(z), y \leftarrow z]$$
$$\textbf{substitute} \mapsto \emptyset, [x \leftarrow g(z), y \leftarrow z, w \leftarrow h(g(z))]$$

□

The algorithm $unify(s, t)$ does not specify the order of rules, so we may use these rules in any order when they are applicable. However, none of these rules can be applied forever. Hence, $unify(s, t)$ will terminate eventually.

Lemma 6.1.14 *Given any pair s and t, $unify(s, t)$ will terminate with either $fail$ or σ.*

Proof First, we point out that the **substitute** rule may increase the sizes of terms in S but it moves variable x from S into σ, that is, x disappears from S, so this rule can be applied no more than the number of variables.

For the rules used between two applications of the **substitute** rule, if the **clash** rule or the **occur-check** rule apply, the process ends with failure immediately. Otherwise, the **decompose** rule removes two occurrences of f. The **orient** rule switches $t \doteq x$ to $x \doteq t$, where t is not a variable. The **redundant** rule removes $t \doteq t$. None of these rules can be used forever. In the end, either $fail$ is returned, or S becomes empty and σ is returned. □

Lemma 6.1.15 *If $\langle S', \sigma' \rangle$ is the result of applying the **substitute** rule on $\langle S, \sigma \rangle$ and θ' is a mgu of S', then $\theta = [x \leftarrow t] \cdot \theta'$ is a mgu of S.*

Proof By the definition of the **substitute** rule, $S' = (S - \{x \doteq t\})[x \leftarrow t]$, where x does not appear in t. By the assumption, S' is unifiable with a mgu θ'. Since x does not appear in S', $\theta = [x \leftarrow t] \cdot \theta'$ is a mgu of $S' \cup [x \doteq t]$. It is easy to check that θ is also a mgu of S because $S\theta = S([x \leftarrow t] \cdot \theta') = (S[x \leftarrow t])\theta' = S'\theta' \cup \{t \doteq t\}$. □

Theorem 6.1.16 *The algorithm $unify(s, t)$ will return a mgu of s and t iff s and t are unifiable.*

Proof Lemma 6.1.14 ensures that $unify(s, t)$ will terminate. Inside $unify(s, t)$, when $\langle S, \sigma \rangle$ is transformed to $\langle S', \sigma' \rangle$, for each rule, we can check that S is unifiable (σ does not affect the unifiability of S) iff $\sigma' \neq fail$ and S' is unifiable. That is, when σ' is $fail$, S cannot be unifiable. When $S' = \emptyset$, S' is trivially unifiable, so S must be unifiable.

Before the **substitute** rule moves $x \doteq t$ from S, x is not affected by σ, $t\sigma = t$ and t does not contain x. Let $\sigma' = \sigma \cdot [x \leftarrow t] = \sigma[x \leftarrow t] \cup [x \leftarrow t]$. Since $S[x \leftarrow t]$ and $\sigma[x \leftarrow t]$ remove all the occurrences of x in S and σ, x has a unique occurrence in σ' and that is also true for all affected variables in σ. Thus, σ and σ' are idempotent by Proposition 6.1.2.

Now assume s and t are unifiable. Let $S_0 = \{s \doteq t\}$, $\sigma_0 = [\,]$, and $\sigma_i = \sigma_{i-1} \cdot [x_i \leftarrow t_i]$ records every application of the **substitute** rule for $1 \leq i \leq m$. Let the corresponding S be $S_1, S_2, \ldots, S_m$, respectively, where $S_m = \emptyset$. Since the mgu of S_m is $\theta_m = [\,]$. So, by Lemma 6.1.15,

$$\theta_{m-1} = [x_m \leftarrow t_m]\theta_m = [x_m \leftarrow t_m]$$

is a mgu of S_{m-1}. Again, by Lemma 6.1.15,

$$\theta_{m-2} = [x_{m-1} \leftarrow t_{m-1}]\theta_{m-1} = [x_{m-1} \leftarrow t_{m-1}, x_m \leftarrow t_m]$$

is a mgu of S_{m-2}, and so on. Finally,

$$\theta_0 = [x_1 \leftarrow t_1]\theta_1 = [x_1 \leftarrow t_1, \ldots, x_{m-1} \leftarrow t_{m-1}, x_m \leftarrow t_m] = \sigma_m$$

is a mgu of $S_0 = \{s \doteq t\}$. □

6.1 Unification

Example 6.1.17 Consider the following pair of terms:

$$s = f(f(f(f(a, x_1), x_2), x_3), x_4), \quad t = f(x_4, f(x_3, f(x_2, f(x_1, a))))$$

$$\{s \doteq t\}, [\,]$$
decompose $\mapsto \{f(f(f(a, x_1), x_2), x_3) \doteq x_4, x_4 \doteq f(x_3, f(x_2, f(x_1, a)))\}, [\,]$
substitute $\mapsto \{f(f(f(a, x_1), x_2), x_3) \doteq f(x_3, f(x_2, f(x_1, a)))\},$
$\qquad [x_4 \leftarrow f(x_3, f(x_2, f(x_1, a)))]$
decompose $\mapsto \{f(f(a, x_1), x_2) \doteq x_3, x_3 \doteq f(x_2, f(x_1, a))\},$
$\qquad [x_4 \leftarrow f(x_3, f(x_2, f(x_1, a)))]$
substitute $\mapsto \{f(f(a, x_1), x_2) \doteq f(x_2, f(x_1, a))\},$
$\qquad [x_4 \leftarrow f(f(x_2, f(x_1, a)), f(x_2, f(x_1, a))), x_3 \leftarrow f(x_2, f(x_1, a))]$
decompose $\mapsto \{f(a, x_1) \doteq x_2, x_2 \doteq f(x_1, a)\}, [x_4 \leftarrow \ldots, x_3 \leftarrow f(x_2, f(x_1, a))]$
substitute $\mapsto \{f(a, x_1) \doteq f(x_1, a)\}, [x_4 \leftarrow \ldots, x_3 \leftarrow \ldots, x_2 \leftarrow f(x_1, a)]$
decompose $\mapsto \{a \doteq x_1, x_1 \doteq a\}, [x_4 \leftarrow \ldots, x_3 \leftarrow \ldots, x_2 \leftarrow f(x_1, a)]$
substitute $\mapsto \{a \doteq a\}, [x_4 \leftarrow \ldots, x_3 \leftarrow \ldots, x_2 \leftarrow f(a, a), x_1 \leftarrow a]$
redundant $\mapsto \emptyset, [x_4 \leftarrow \ldots, x_3 \leftarrow \ldots, x_2 \leftarrow f(a, a), x_1 \leftarrow a]$

Let $t_1 = a$, $t_2 = f(t_1, t_1)$, $t_3 = f(t_2, t_2)$, $t_4 = f(t_3, t_3)$, then the mgu of $s \doteq t$ is $[x_i \leftarrow t_i \mid 1 \leq i \leq 4]$.

In general, if s and t are the following pair of terms,

$$s = f(f(f(\ldots f(f(a, x_1), x_2), \ldots), x_{n-1}), x_n)$$
$$t = f(x_n, f(x_{n-1}, f(\ldots, f(x_2, f(x_1, a)) \ldots)))$$

then the mgu of s and t is $\sigma = [x_i \leftarrow t_i \mid 1 \leq i \leq n]$, where $t_1 = a$ and $t_{i+1} = f(t_i, t_i)$. We have $s\sigma = t\sigma = t_{n+1}$. The sizes of s and t are the same, i.e., $2n + 1$. However, the size of t_n is $2^n - 1$. □

The above example illustrates that $unify(s, t)$ will take $O(2^n)$ time and space for the input terms of size $O(n)$. This high cost cannot be avoided if we use trees to represent terms. Using the data structure of directed graphs for terms, identical subterms can be shared, so we do not need a space of exponential size.

6.1.4 Practically Linear Time Unification Algorithm

Corbin and Bidoit (1983) improved Alan Robinson's unification algorithm to a quadratic one by using directed graphs. Ruzicka and Privara (1988) [1] further improved it to an almost linear one, by using the union/find data structures and postponing the occur-check at the end of unification. The union/find data structure stores a collection of disjoint sets, and each of the sets represents an equivalence class. Thus, the disjoint sets represent an equivalence relation and are a partition of the involved elements. The *Union* operation will merge two sets into one, and the *Find* operation will return a representative member of a set, so that it allows us to find out efficiently if two elements are in the same set or not.

In the following, we introduce an almost-linear time unification algorithm slightly different from that of Ruzicka and Privara [1].

Given two terms s and t, the *term graph* for s and t is the directed acyclic graph $G = (V, E)$ obtained by combining the formula trees of s and t with the following operation: All the nodes labeled with the same variable in s and t are merged into a single node of G.

The graph G has the following features:

- Each node v of G is labeled with a symbol in s or t, denoted by $label(v)$.
- The roots of the formula trees are the two nodes in G with zero in-degree, representing s and t, respectively. The other non-variable nodes' in-degrees are one. The in-degree of a variable node is the number of occurrences of that variable in s and t.
- The leaf nodes of the formula trees are the nodes with zero out-degree in G, which are labeled with variables or constants.
- If a node is labeled with a function symbol f whose arity is k, then this node has k ordered successors. We will use $child(v, i)$ to refer to the ith successor of v.

The term associated with each node n of G can be recursively defined as follows:

1. If $label(n)$ is a variable or constant, $term(n) = label(n)$.
2. If $label(n)$ is a function f other than a constant, $term(n) = f(s_1, s_2, \ldots, s_k)$, where $s_i = term(child(n, i))$ for $1 \le i \le arity(f)$.

We say $term(n)$ is the term represented by n in G.

The unification algorithm will work on $G = (V, E)$ and is divided into two stages: In stage 1, $unify_1$ checks if s and t have no clash failure; in stage 2, $postOccurCheck$ returns true iff they have no occur-check failure.

In stage 1, algorithm $unify_1$ takes the two nodes representing s and t and visits G in tandem by depth-first search. During the search, we will possibly assign another node u to node v as the *representative* of v, denoted by $up(v) = u$. Initially, $up(v) = v$ for every node of G. If $up(v) = u$ and $v \ne u$, it means the terms represented by v and u, respectively, are likely unifiable (no clash failure) and we use u as the representative of v in the rest of the search. At the end of the search, $unify_1$ returns true if no clash failure was found. In this case, the relation R defined by up, i.e., $R = \{(u, up(u)) \mid u \in V\}$, can generate an equivalence relation E (i.e., E is the reflexive, symmetric, and transitive closure of R). The relation E partitions V into equivalence classes and the nodes in the same equivalent class represent the terms which must be unifiable if s and t are unifiable.

Example 6.1.18 Given $t = f(f(f(a, x_1), x_2), x_3)$ and $s = f(x_3, f(x_2, f(x_1, a)))$, their term graph is shown in Fig. 6.1. The up relation is initialized as $up(v) = v$.

6.1 Unification

Fig. 6.1 The term graphs for
$t = f(f(f(a, x_1), x_2), x_3)$
and
$s = f(x_3, f(x_2, f(x_1, a)))$

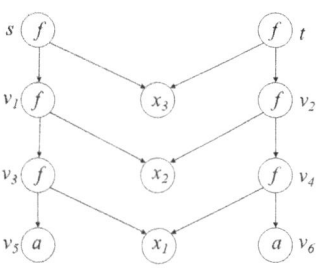

The algorithm $unify_1(t, s)$ starts the depth-first search in tandem and goes to the first child of (t, s), i.e., (v_1, x_3). Since x_3 is a variable node, $unify_1(v_1, x_3)$ will do $up(x_3) := v_1$ and backtrack. The second child of (t, s) is (x_3, v_2). Since $up(x_3) = v_1$, we use v_1 for x_3 and call $unify_1(v_1, v_2)$ recursively. Since v_1 and v_2 have the same label, the search goes to their children. The first child of (v_1, v_2) is (v_3, x_2). Since x_2 is a variable node, we do $up(x_2) := v_3$ and backtrack. The second child of (v_1, v_2) is (x_2, v_4). Use v_3 for x_2 and call $unify_1(v_3, v_4)$ recursively. Since v_3 and v_4 have the same label, the search goes to their children. The first child of (v_3, v_4) is (v_5, x_1). We do $up(x_1) := v_5$ and backtrack. The second child of (v_3, v_4) is (x_1, v_6). Use v_5 for x_1 and call $unify_1(v_5, v_6)$ recursively. Since (v_5, v_6) have the same constant label, the search backtracks after doing $up(v_6) := v_5$. Now, all children of (v_3, v_4) are done. After $up(v_4) := v_3$, the search backtracks. Similarly, all children of (v_1, v_2) are done, and we backtrack after doing $up(v_2) := v_1$. Finally, all children of (t, s) are done and $unify_1(t, s)$ finishes with $up(s) := t$.

The content of $up(v)$ after the termination of $unify_1(t, s)$ is shown below:

node n	s	t	v_1	v_2	v_3	v_4	v_5	v_6	x_1	x_2	x_3
$up(n)$	t	t	v_1	v_1	v_3	v_3	v_5	v_5	v_5	v_3	v_1

The equivalence relation E generated by $(v, up(v))$ can be represented by the following partition of V:

$$\{s, t\}, \{v_1, v_2, x_3\}, \{v_3, v_4, x_2\}, \{v_5, v_6, x_1\}$$

The next step is occur-check. For instance, whether x_3 occurs in the term represented by v_1 or v_2? The procedure *postOccurCheck* in the second stage will confirm that the answer is "no." □

It is clear from the above example that each node of V is in a singleton set initially. When the terms represented by the two nodes are unifiable, the sets containing the two nodes are merged into one through the assignment $up(v) := u$, i.e., we obtain "union" of the two sets containing u and v, respectively. To know which node is represented by which node, we need the "find" operation. Thus, we may employ the well-known union/find data structure to support the implementation of $unify_1$.

In view of the union/find data structure, each disjoint set is represented by a tree: $up(v) = u$ means the parent of v is u in the tree if $v \neq u$. If $up(u) = u$, then u is the root of the tree, which also serves as the name (or representative) of the set. $Find(u)$ will find the name of the set containing u and will compress the tree if necessary. $Union(u, v)$ will merge the two sets whose names are u and v, respectively. A typical implementation of $Union$ will let the root of shorter/smaller tree point to the root of a taller/larger tree and can be done in constant time.

As stated earlier, the algorithm $unify_1$ takes the two nodes representing s and t, respectively, and visits G in tandem by depth-first search. Algorithm $unify_1$ will first replace the current two nodes by their set names, which are also the nodes in G. If the two set names are the same, backtrack; If one of them is a variable node, make the other as its parent in the union/find tree, and backtrack. If both are function nodes and their labels are different, report *clash failure*; otherwise, recursively visit their children. If no failure is found from the children, merge the two sets named by the current two nodes and backtrack. During the depth-first search, we also use a marking mechanism to detect if a cycle is created because a variable is unified with a term; if yes, an occur-check failure happened. Only a portion of occur-check failures can be found this way. The pseudo-code of algorithm *lunify* is given below.

Algorithm 6.1.19 Algorithm $lunify(s, t)$ takes terms s and t, as input, creates the formula graph $G = (V, E)$ from s and t, merging all the nodes of the same variable into one, and calls $unify_1(v_s, v_t)$, where v_s and v_t are the nodes representing s and t, respectively. If $unify_1(v_s, v_t)$ returns *true*, $postOccurCheck(v_s, v_t)$ is called, which returns *true* iff there exist no cycles in $G = (V, E)$ by the depth-first search.

proc $lunify(s, t)$: Boolean
1   $G := createGraph(s, t)$ // $up(v) = v$ for each v in G
2   **if** $unify_1(v_s, v_t)$ // v_s, v_t represent s, t in G
3   **return** $postOccurCheck(v_s, v_t)$
4   **else return** *false*

proc $unify_1(v_1, v_2)$: Boolean
    // Use the union/find data structure on the term graph G.
1   $s_1 := Find(v_1); s_2 := Find(v_2)$
2   **if** $s_1 = s_2$ **return** *true* // already unified.
3   **if** ($label(s_1)$ and $label(s_2)$ are variables) $Union(s_1, s_2)$ **return** *true*
4   **if** ($label(s_1)$ is variable) $up(s_1) := s_2$; **return** *true*
5   **if** ($label(s_2)$ is variable) $up(s_2) := s_1$; **return** *true*
6   **if** $label(s_1) \neq label(s_2)$ **return** *false* // clash failure
    // use $mark(s)$ to detect if G has a cycle
7   **for** $i := 1$ **to** $arity(label(s_1))$ **do**
8      $c_1 := child(s_1, i); c_2 := child(s_2, i)$
9   **if** ($marked(c_1)$ **or** $marked(c_2)$) **return** *false* // occur-failure
10     $mark(c_1); mark(c_2)$
11   **if** ($unify_1(c_1, c_2) = false$) **return** *false*

6.1 Unification

12      $unmark(c_1)$; $unmark(c_2)$
13      $Union(s_1, s_2)$
14   **return** *true*

proc $Union(s_1, s_2)$
     // Initially, for any node s, $h(s) = 0$, the height of a singleton tree.
1   **if** $h(s_1) < h(s_2)$ $up(s_1) := s_2$; **return**
2      $up(s_2) := s_1$ // the parent of s_2 is s_1
3   **if** $h(s_1) = h(s_2)$ $h(s_1) := h(s_1) + 1$

proc $Find(v)$
1      $r := up(v)$ // look for the root of the tree containing v
2   **while** $(r \neq up(r))$ $r := up(r)$ // r is root iff $r = up(r)$
3   **if** $(r \neq v)$ // compress the path from v to r
4   **while** $(r \neq v)$ $up(v) := r$; $v := up(v)$
5   **return** r

As said earlier, in the union-find data structure, each set of the nodes is represented by a tree (defined by $up(v) = u$, where the parent of v is u) and the root of the tree is the name of the set. The union operation implements the "union by height," making the root of the higher tree as the parent of the root of the shorter tree. $Find(v)$ will return the name of the set, i.e., the root of the tree containing v; this is done by the first while loop. For the efficiency of future calls to $Find$, the path from v to the root is compressed after the root is found; this is completed by the second while loop.

There are two parent–child relations used in the above algorithm. The procedure $unify_1$ uses the depth-first search based on the parent–child relation defined by the term graph, not the parent–child relation defined by $up(v) = u$. The latter induces an equivalence relation and is used by the union/find algorithm.

Example 6.1.20 Let $s = f(x, x)$ and $t = f(y, a)$. A graph $G = (V, E)$ will be created with $V = \{v_s, v_t, v_x, v_y, v_a\}$ and $E = \{(v_s, v_x)_1, (v_s, v_x)_2, (v_t, v_y)_1, (v_t, v_a)_2\}$, where the subscripts are the order of children. Algorithm $unify_1(v_s, v_t)$ will call $unify_1(v_x, v_y)$, which changes $up(v_x) = v_y$, and $unify_1(v_x, v_a)$, which changes $up(v_y) = v_a$. After the two recursive calls, $unify_1(v_s, v_t)$ will change $up(v_s) = v_t$ and return true. V is partitioned into two sets: $\{v_s, v_t\}$ and $\{v_x, v_y, v_a\}$. □

In algorithm $unify_1$, the graph $G = (V, E)$ never changes; only $up(v)$ changed. The value of $up(v)$ defines the equivalence relation in the union/find data structure: v and $up(v)$ are in the same disjoint set. The value of $up(v)$ is changed indirectly either by $Find$ (called at line 1 of $unify_1$; $up(v)$ is changed at line 4 of $Find$), or by $Union$ (called at lines 3 and 13; $up(v)$ is modified at lines 1 and 2 of $Union$), or directly at lines 4 and 5 of $unify_1$, which are special cases of $Union$: a variable node's parent is set to be a function node (not the other direction). Thus, the algorithm uses only linear space. The algorithm uses almost linear time because of the following observations:

- Let $G = (V, E)$ be the term graph of s and t, and $n = |V|$ is bound by the total size of s and t.
- The number of $Find$ operations performed in $unify_1$ is twice of the number of recursive calls of $unify_1$.
- Each $Union$ operation reduces the number of node sets by one; thus, the number of union operations is bound by n.
- Once two subterms are shown to be unifiable, the two nodes representing them are put into the same set; thus, we never try to unify again the same pair of subterms and the number of recursive calls of $unify_1$ is bound n.
- The total cost of $Find$ and $Union$ performed in $unify_1$ is $O(n\alpha(n))$, where $\alpha(n)$ is the inverse of Ackermann function and grows very slowly. In fact, $\alpha(n)$ is practically a constant function, e.g., $\alpha(10^9) \leq 4$.
- The special unions performed at lines 4 and 5 of $unify_1$ do not affect the almost linear complexity, because the total cost of such unions between a variable node and a non-variable is bounded by the number of variable nodes.
- The procedure $postOccurCheck$ will check if $G' = (V, E')$, where $E' = E \cup \{(v, up(v)) | v \in V, v \neq up(v)\}$, contains a cycle and can be implemented in $O(n)$ time by the depth-first search.

Theorem 6.1.21 *The algorithm lunify(s, t) takes $O(n\alpha(n))$ time, where n is the total sizes of s and t, and $\alpha(n)$ is the inverse of Ackermann function; lunify(s, t) returns* **true** *iff s and t are unifiable.*

The algorithm *lunify* tells us if two terms are unifiable or not. However, this is different from knowing what their unifiers actually are. We leave the following problem as an exercise: finding the mgu of s and t after *lunify(s, t)* returns true.

Since *lunify* does not change the term graph of s and t, we may wonder how to get the term $t\sigma$ from G if σ is the mgu of s and t. The term $t\sigma$, which is equal to $s\sigma$, can be obtained by $term'(v_s)$ (or $term'(v_t)$), where $term'$ can be recursively defined as follows: For any node v of G:

1. If $up(v) \neq v$, $term'(v) = term'(up(v))$.
2. If $label(v)$ is a constant or a variable and $up(v) = v$, $term'(v) = label(v)$.
3. If $label(v)$ is a function f other than a constant, $term'(v) = f(s_1, s_2, \ldots, s_k)$, where $s_i = term'(child(v, i))$ for $1 \leq i \leq arity(f)$.

It is easy to check that $term'(v) = term(v)$ before the execution of *lunify* as at that time, $up(v) = v$ for any v. We may view $\{(v, up(v)) \mid up(v) \neq v\}$ as a new set of edges added to $G = (V, E)$ during the execution of *lunify*. This set of edges defines not only the equivalence relation represented by the union/find data structure, but also the mgu σ when s and t are unifiable.

The contrived examples in Table 6.1 illustrate the behavior of *lunify*. The size of a term is the total number of symbols in the term, excluding parentheses and commas. The column **visited** gives the number of nodes visited before a result is returned.

6.2 Resolution

Table 6.1 The last column gives the number of visited nodes by $unify_1$ in the term graph for each example

	Pair of terms	Size	Failure	Visited
P1	$f(f(\ldots f(f(a, x_1), x_2), \ldots), x_n)$ $f(x_n, f(x_{n-1}, f(\ldots, f(x_1, a) \ldots)))$	$4n+2$	None	$4n+2$
P2	$f(z, g^n(x))$ $f(g(z), g^n(y))$	$n+3$	Occur	$n+3$
P3	$h(x, f(x, f(x, \ldots, f(x, f(x, x)) \ldots)), x)$ $h(f(y, \ldots, f(y, f(y, y)) \ldots), y, y)$	$8n+3$	Occur	8
P4	$f(f(\ldots f(f(a, x_1), x_2), \ldots), x_n)$ $f(x_n, f(x_{n-1}, f(\ldots, f(x_1, b) \ldots)))$	$4n+2$	Clash	$4n+2$
P5	$h(x, g^n(x), x)$ $h(g^n(y), y, y)$	$2n+8$	Occur	$2n+8$
P6	$f(f(f(\ldots f(f(a, x_1), x_2), \ldots), x_n), z)$ $f(f(x_n, f(x_{n-1}, f(\ldots, f(x_1, a) \ldots))), g(z))$	$4n+7$	Occur	$4n+7$

In P1, the two terms are unifiable, and the occur-check can be awfully expensive if we do not postpone the check. In P2, the first occur-check (on $z \leftarrow g(z)$) will fail, though *lunify* will not know it until postOccurCheck is called. In P3, the last occur-check will fail. It is interesting to notice that $unify_1$ will report this failure by the marking mechanism in a constant number of steps, after visiting only eight nodes of the term graph. In P4, a clash failure takes place at the bottom of the depth-first search. P5 is similar to P3, though an occur-check failure can be found only after a thorough search. P6 is similar to P1, though the last occur-check will locate the failure.

6.2 Resolution

Resolution is an inference rule of first-order logic proposed by J. Alan Robinson in 1965.

6.2.1 Formal Definition

Definition 6.2.1 (Resolution) Supposing clause c_1 is $(A \mid \alpha)$ and c_2 is $(\overline{B} \mid \beta)$, where A and B are atoms and α and β are the rest literals in c_1 and c_2, respectively, *resolution* is the following inference rule:

$$\frac{(A \mid \alpha) \quad (\overline{B} \mid \beta)}{(\alpha \mid \beta)\sigma}$$

where σ is a mgu of A and B. The clause $(\alpha \mid \beta)\sigma$ produced by the resolution rule is called *resolvent* of the resolution; c_1 and c_2 are the *parents* of the resolvent.

This resolution rule is also called *binary resolution*, as it involves two clauses as premises.

Example 6.2.2 Given a pair of clauses $(p(f(x_1)) \mid q(x_1))$ and $(\overline{p(x_2)} \mid \overline{q(g(x_2))})$. The resolution can generate two resolvents from them: Resolve on p to get $(q(x_1) \mid \overline{q(g(f(x_1))))}$ with the mgu $[x_2 \leftarrow f(x_1)]$ and on q to get $(p(f(g(x_2))) \mid \overline{p(x_2)})$ with the mgu $[x_1 \leftarrow g(x_2)]$. □

The above example shows that more than one resolvent can be generated from two clauses. We will use $resolve(c_1, c_2)$ to denote the set of all resolvents from clauses c_1 and c_2. In propositional logic, if two clauses do not have duplicated literals and can generate more than one resolvent, then these resolvents are tautology.

Proposition 6.2.3 *The resolution rule is sound, that is,* $c_1 \wedge c_2 \models resolve(c_1, c_2)$.

Proof Suppose c_1 is $(A \mid \alpha)$, c_2 is $(\overline{B} \mid \beta)$, and $resolve(c_1, c_2)$ contains $(\alpha \vee \beta)\sigma$ and σ is a mgu of A and B. For any model I of $c_1 \wedge c_2$, I is also a model of $c_1\sigma = A\sigma \vee \alpha\sigma$ and $c_2\sigma = \overline{B}\sigma \vee \beta\sigma$, because free variables are treated as universally quantified variables. Since σ is a unifier of A and B, $A\sigma = B\sigma$. Thus, the truth values of $A\sigma$ and $\overline{B}\sigma$ are different in I. If $I(A\sigma) = 0$, then $I(\alpha\sigma) = 1$; otherwise $I(c_1\sigma)$ would be 0. If $I(A\sigma) = 1$, then $I(\overline{B}\sigma) = 0$, so $I(\beta\sigma) = 1$. In both cases, either $\alpha\sigma$ or $\beta\sigma$ or both are true in I, so $(\alpha \mid \beta)\sigma$ is true in I. □

By the above proposition, every resolvent is a logical consequence of the input clauses, because the entailment relation is transitive. If the empty clause is generated, then the input clauses are unsatisfiable. In other words, we may use the resolution rule to design a refutation prover. Suppose we have some axioms, say A, and we want to see if some conjecture, say B, is a logical consequence of A. By the refutational strategy, we want to show that $A \wedge \overline{B}$ is unsatisfiable, by putting $A \wedge \overline{B}$ into clausal form and finding a contradiction through resolution. We wish the following logical equivalences hold.

- B is a logical consequence of A, i.e., $A \models B$, iff
- $A \wedge \overline{B}$ is unsatisfiable, iff
- S, the clause set derived from $A \wedge \overline{B}$, is unsatisfiable, iff
- The empty clause can be generated by resolution from S.

Unfortunately, the last "iff" does not hold in first-order logic. Indeed, when S is satisfiable, resolution will never generate the empty clause. If the empty clause can be generated, we can claim that S is unsatisfiable, because resolution is sound. However, when S is unsatisfiable, resolution may not generate the empty clause.

Example 6.2.4 Let $S = \{(\underline{p(x_1)} \mid p(y_1)), (\overline{p(x_2)} \mid \overline{p(y_2)})\}$. One resolvent of the two clauses in S is $(p(x_3) \mid \overline{p(y_3)})$, and we may add it into S. Any other resolvent from S will be a renaming of the three clauses. In other words, resolution alone cannot generate the empty clause from S. On the other hand, an instance of the first clause is $(p(a) \mid p(a))$, which can be simplified to $(p(a))$. Adding this unit clause to S, we can easily obtain the empty clause from S. □

In propositional logic, we defined a proof system which uses the resolution rule as the only inference rule and showed that this proof system is a decision procedure for propositional clauses. In first-order logic, we do not have such a luck.

6.2.2 Factoring

To get rid of the problem illustrated in the previous example, we need the following inference rule:

Definition 6.2.5 (Factoring) Let clause C be $(A \mid B \mid \alpha)$, where A and B are literals and α are the rest literals in C. *Factoring* is the following inference rule:

$$\frac{(A \mid B \mid \alpha)}{(A \mid \alpha)\sigma}$$

where σ is the mgu of A and B.

For the clause $(p(x_1) \mid p(y_1))$ in the previous example, factoring will generate $(p(x_1))$; for $(\overline{p(x_2)} \mid \overline{p(y_2)})$, factoring will generate $(\overline{p(x_2)})$. Now resolution will generate the empty clause from $(p(x_1))$ and $(\overline{p(x_2)})$.

Proposition 6.2.6 *The factoring rule is sound, that is, $C \models C'$, where C' is the clause generated by factoring from C.*

Proof C' is an instance of C, i.e., $C' = C\sigma$, where σ is the unifier used in factoring. Any instance of a formula is a logical consequence of the formula. □

Armed with resolution and factoring, now we can claim the following result whose proof can be found in Sect. 6.3.3.

Theorem 6.2.7 (Refutational Completeness of Resolution) *A set S of clauses is unsatisfiable iff the empty clause can be generated by resolution and factoring.*

Though factoring is needed for the completeness of resolution, in practice, factoring is rarely needed.

6.2.3 A Refutational Proof Procedure

For propositional logic, we introduced three deletion strategies for deleting clauses without worrying about missing a proof:

- **Pure literal deletion**: Clauses containing pure literals are discarded.
- **Tautology deletion**: Tautology clauses are discarded.
- **Subsumption deletion**: Subsumed clauses are discarded.

Pure literal deletion and tautology deletion can be used without modification. For the subsumption deletion, the rule is modified as follows:

Definition 6.2.8 (Subsumption) Clause c_1 subsumes clause c_2 if there exists a substitution σ such that every literal of $c_1\sigma$ appears in c_2.

For example, $(p(x, y) \mid \overline{p(y, x)})$ subsumes $(p(a, b) \mid \overline{p(b, a)} \mid q(a))$ with $\sigma = [x \leftarrow a, y \leftarrow b]$.

Deletion strategies can be integrated into the algorithm *resolution* as follows.

Procedure 6.2.9 The procedure *resolution(C)* takes a set C of clauses and returns false iff C is unsatisfiable. It uses *preprocessing(C)* to simplify the input clauses C. Procedure *resolve*(α, β) returns the set of resolvents of α and β. Procedure *tautology*(α) checks if clause α is a tautology. Procedure *subsumedBy*(α, S) checks if α is subsumed by a clause in S. Procedure *factoring*(α) takes clause α as input and generates the set of results from the factoring rule on α, including α itself.

proc *resolution(C)*
1    $G := preprocessing(C)$ // G: given clauses
2    $K := \emptyset$ // K: kept clauses
3   **while** $G \neq \emptyset$ **do**
4        $\alpha := pickClause(G)$ // heuristic for picking a clause
5        $G := G - \{\alpha\}$
6        $N := \emptyset$ // new clauses from α and K by resolution
7   **for** $\beta \in K$ **if** *resolvable*(α, β) **do**
8   **for** $\gamma \in resolve(\alpha, \beta)$ **do**
9   **if** $\gamma = ()$ **return** *false* // the empty clause is found
10   **if** *tautology*(γ) $\vee$ *subsumedBy*($\gamma, G \cup K$) **continue**
11               $N := N \cup factoring(\gamma)$
12       $K := K \cup \{\alpha\}$
13   **for** $\alpha \in G$ **if** *subsumedBy*(α, N) **do** $G := G - \{\alpha\}$
14   **for** $\beta \in K$ **if** *subsumedBy*(β, N) **do** $K := K - \{\beta\}$
15       $G := G \cup N$
16   **return** *true* // K is saturated by resolution.

The above procedure is almost identical to Algorithm 3.3.27 with three differences: (*a*) *resolve*(α, β) returns a set of resolvents instead of a single resolvent (line 8); (*b*) *factoring* is used (line 11); (*c*) the termination of the procedure is not guaranteed as it may generate an infinite number of new clauses. Thus, it is not an algorithm. Despite this, this procedure is a base for many resolution theorem provers, including McCune's Otter and its successor Prover9, which will be introduced shortly. In Prover9, the subsumption check at line 10 is called *forward subsumption* and the subsumption check at lines 13 and 14 is called *backward subsumption*.

Prover9 also uses another simplification rule called *unit deletion*.

Definition 6.2.10 (Unit Deletion) Unit deletion is the simplification rule that deletes literal B from clause α if there exist a unit clause (A) and a substitution σ such that $A\sigma = \overline{B}$ (or $\overline{A}\sigma = B$).

Unit deletion is sound because by the requirement, there is a resolution between (A) and α and the resolvent of this resolution will subsume α. For example, if we have unit clause $(\overline{p(x,b)})$, then the literal $p(a,b)$ in $(p(a,b) \mid q(a))$ can be deleted to get a new clause $(q(a))$, which subsumes $(p(a,b) \mid q(a))$.

Unit deletion can be used at line 10, so that the new clause γ is simplified by the unit clauses in $G \cup K$. This is called *forward unit deletion*. *Backward unit deletion* can be inserted at lines 13 and 14 when the unit clauses in N are used to simplify clauses in $G \cup K$.

6.3 Simplification Orders and Ordered Resolution

In Sect. 3.3.2, we listed some well-known resolution strategies which add restrictions to the use of the resolution rule: unit resolution, input resolution, ordered resolution, positive resolution, negative resolution, set of support, and linear resolution. They can be carried over to first-order logic without modification, with the exception of ordered resolution. The procedure $resolvable(A, B)$ (line 7 of *resolution*) is the place where various restrictions are implemented so we can use these restricted resolution strategies.

For unit resolution and input resolution, they are still incomplete in general, but are equivalent in the sense that there is a resolution proof for one strategy, then there exists a resolution proof for the other strategy.

To use ordered resolution in first-order logic, we need the concept of simplification orders, which are well-founded partial orders over terms.

6.3.1 Well-Founded Partial Orders

Recall that a partial order $\succeq$ is an antisymmetric and transitive binary relation over a set S of elements, and $\succeq$ is well-founded if there is no infinite sequence of distinct elements $x_1, x_2, \ldots, x_i, \ldots$, such that

$$x_1 \succeq x_2 \succeq \cdots \succeq x_i \succeq \cdots$$

We will use $\succ$ and $\succeq$ as needed, where $x \succ y$ iff $x \succeq y$ and $x \neq y$. We write $x \prec y$ if $y \succ x$.

The reason we stick to well-founded orders is that well-founded orders allow us to use mathematical induction. Let S be a set with a well-founded order $\succ$. To show a property $P(x)$ holds for every $x \in S$, we need to do two things:

- For every minimal element $m \in S$, $P(m)$ is true.
- Prove that $P(x)$ is true, assuming $P(x')$ is true for every $x' \prec x$.

The above is called the *well-founded induction principle*.

There are many well-founded orders over different sets. For example, $\geq$ over the set of natural numbers is well-founded; $\geq$ over the set of integers is not well-founded. Any partial order over a finite set is well-founded.

Multiset Extension

There are many ways to obtain well-found orders. A *multiset* or *bag* over a set S of elements is a modification of the concept of a subset of S, that, unlike a set, allows for multiple instances for each of its elements from S. The union, intersection, and subtraction operations can be extended over multisets. If there exists a strict partial order $\succ$ over the set S, we may extend $\succ$ to $\succ^{mul}$ to compare multisets of S:

Definition 6.3.1 For any finite multisets S_1, S_2 over S, $S_1 \succ^{mul} S_2$ if $S_1 - S_2 \neq \emptyset$, and for any $y \in S_2 - S_1$, there exists $x \in S_1 - S_2, x \succ y$.

The "finite" condition is necessary. For instance, if S_1 and S_2 are the sets of odd and even natural numbers, respectively, then we have $S_1 \succ^{mul} S_2$ and $S_2 \succ^{mul} S_1$. In other words, $\succ^{mul}$ is not antisymmetric over infinite multisets.

Example 6.3.2 Let $S_1 = \{a, b, b, c\}$ and $S_2 = \{a, a, b, c\}$, then $S_1 \cup S_2 = \{a, a, a, b, b, b, c, c\}$, $S_1 \cap S_2 = \{a, b, c\}$, $S_1 - S_2 = \{b\}$, and $S_2 - S_1 = \{a\}$. If $c \succ b \succ a$, then $S_1 \succ^{mul} S_2$ because $b \succ a$. □

Recall that a well-order is total and well-founded.

Proposition 6.3.3 *If $\succ$ is a well-order over a set S, then $\succ^{mul}$ over the multisets of S is a well-order.*

Lexicographic Extension

One popular way is to use lexicographic combination over Cartesian products (Sect. 1.3.1) from well-founded orders of individual sets.

Definition 6.3.4 Let $\succ_i$ be a partial order over set S_i, $1 \leq i \leq n$. The lexicographic order $\succ^{lex}$ over $S_1 \times S_2 \times \cdots \times S_n$ is defined as follows: For any $\langle x_1, x_2, \ldots, x_n \rangle$, $\langle y_1, y_2, \ldots, y_n \rangle \in S_1 \times S_2 \times \cdots \times S_n$, $\langle x_1, x_2, \ldots, x_n \rangle \succ^{lex} \langle y_1, y_2, \ldots, y_n \rangle$ iff there exists k, $1 \leq k \leq n$, such that $x_i \succeq_i y_i$ for $1 \leq i < k$ and $x_k \succ_k y_k$.

6.3 Simplification Orders and Ordered Resolution

Proposition 6.3.5 *If $\succ_i$ is a well-order over S_i, $1 \leq i \leq n$, then the lexicographic order $\succ^{lex}$ is a well-order over $S_1 \times S_2 \times \cdots \times S_n$.*

The lexicographic combination can be extended to compare two finite sequences of different lengths, assuming a non-empty sequence is greater than the empty sequence, like the dictionary order which compares two words of various lengths.

Precedence

When using ordered resolution, the user is often required to provide a partial order, called *precedence*, over function or predicate symbols. Given a first-order language $L = (P, F, X, Op)$, let $\Sigma = P \cup F \cup Op$. We say $\succ$ is a *precedence* if $\succ$ is a well-order over Σ. Note that if Σ is finite, $\succ$ is trivially well-founded.

For some applications, we may need to define an equivalence $\approx$ over a set. For instance, we may allow $\vee \approx \wedge$ in a precedence. By definition, a partial order is an antisymmetric and transitive relation. Allowing $\approx$ more than $=$ violates the antisymmetry property, which requires $s = t$ if $s \succeq t$ and $t \succeq s$. In literature, if $s \succeq t$ and $t \succeq s$ imply $s \approx t$, $\succeq$ is called "quasi-order." A quasi-orders $\succeq'$ can be obtained from a partial order $\succeq$ by adding $\approx$, that is, let $\succeq' = \succeq \cup \approx$. In other words, $a \succeq' b$ iff $a \succeq b$ or $a \approx b$.

Proposition 6.3.6 *If partial order $\succeq$ is well-founded and any equivalent class of elements under $\approx$ is finite, then quasi-order $\succeq' = \succeq \cup \approx$ is well-founded.*

Since quasi-orders inherit most properties of partial orders, we will treat partial orders and quasi-orders with finite equivalent classes indifferently.

6.3.2 Simplification Orders

There are several requirements on the orders used for ordered resolution and we call it "simplification order."

Definition 6.3.7 (Stable and Monotonic Relation) A binary relation R over S, where S denotes the set of terms, atoms, and clauses, is said to be *stable* if for any $s, t \in S$, $R(s, t)$ implies $R(s\sigma, t\sigma)$ for any substitution $\sigma : X \mapsto T(F, X)$. R is said to be *monotonic* if $R(s, t)$ implies $R(f(\ldots, s, \ldots), f(\ldots, t, \ldots))$ for any symbol f, where f is either a function or predicate symbol or a Boolean operator.

For a stable order $\succ$ over terms, we cannot have $f(y) \succ x$ because $f(y)\sigma \succ x\sigma$ does not hold for $\sigma = [x \leftarrow f(y)]$.

Definition 6.3.8 (Simplification Order) A partial order $\succeq$ over terms or clauses is called *simplification order* if it is well-founded, stable, and monotonic.

In practice, simplification orders are often constructed from a *precedence* over function and predicate symbols. We may extend any precedence over Σ to terms, atoms, and clauses, by several different ways.

- LPO (Lexicographic Path Order). The term order is determined entirely by the symbol precedence.
- RPO (Recursive Path Order). Like LPO, it is induced from the precedence.
- KBO (Knuth–Bendix Order). This order uses a weighting function on symbols as well as the symbol precedence. The weighting function is used first, and the symbol precedence breaks ties.

LPO: Lexicographic Path Order

For any term or clause t, let $var(t)$ denote the set of variables appearing in t.

Definition 6.3.9 ($\succ_{lpo}$) The *lexicographical path order* $\succ_{lpo}$ is defined recursively as follows: $s \succ_{lpo} t$ if $s \neq t$ and either (a) $t \in var(s)$ or (b) $s = f(s_1, \ldots, s_m) \succ_{lpo} t = g(t_1, \ldots, t_n)$ by one of the following conditions:

1. $f \succ g$ and $s \succ_{lpo} t_i$ for $1 \leq i \leq n$, or
2. $f \approx g$, $[s_1, \ldots, s_m] \succ^{lex}_{lpo} [t_1, \ldots, t_n]$, and $s \succ_{lpo} t_i$ for $1 \leq i \leq n$, or
3. $f \prec g$ and there exists j, $1 \leq j \leq m$, $s_j \succeq_{lpo} t$.

Example 6.3.10 Let $* \succ +$, s be $x*(y+z)$ and t be $(x*y)+(x*z)$. Then $s \succ_{lpo} t$ by condition (b)1, because $* \succ +$, $s \succ_{lpo} x*y$ and $s \succ_{lpo} x*z$, both by condition (b)2. That is, from $[x, (y+z)] \succ^{lex}_{lpo} [x, y]$ and $[x, (y+z)] \succ^{lex}_{lpo} [x, z]$. □

Proposition 6.3.11 $\succ_{lpo}$ *is a simplification order and is total on ground terms, if the precedence $\succ$ is a well-order.*

The proof is left as exercise.

RPO: Recursive Path Order

Definition 6.3.12 ($\succ_{rpo}$) The *recursive path order* $\succ_{rpo}$ over $T(F, X)$ is defined recursively as follows: $s \succ_{rpo} t$ if $s \neq t$ and either (a) $t \in var(s)$ or (b) $s = f(s_1, \ldots, s_m) \succ_{rpo} t = g(t_1, \ldots, t_n)$ by one of the following conditions:

1. $f \succ g$ and $s \succ_{rpo} t_i$ for $1 \leq i \leq n$, or
2. $f \approx g$ and $\{s_1, \ldots, s_m\} \succ^{mul}_{rpo} \{t_1, \ldots, t_n\}$, or
3. $f \prec g$ and there exists j, $1 \leq j \leq m$, $s_j \succeq_{rpo} t$.

Example 6.3.13 Let s be $i(x*y)$ and t be $i(y)*i(x)$ with $i \succ *$. To show $s \succ_{rpo} t$, condition (b)1 applies since $i \succ *$, and we need to show $s \succ_{rpo} i(x)$ and $s \succ_{rpo} i(y)$. To show $s \succ_{rpo} i(x)$, condition (b)2 applies and we need $x*y \succ^{mul}_{rpo} x$, which holds by (a). The proof of $s \succ_{rpo} i(y)$ is identical to that of $s \succ_{rpo} i(x)$. In comparison, $s \succ_{lpo} t$ holds by the same proof.

6.3 Simplification Orders and Ordered Resolution

To show $t \succ_{rpo} s$, we need a new precedence in which $* \succ i$ and use $\{i(x), i(y)\} \succ_{rpo}^{mul} \{x, y\}$ (which is true). In comparison, we do not have $t \succ_{lpo} s$ with $* \succ i$, because $[i(y), i(x)] \succ_{lpo}^{lex} [x, y]$ is not true. □

Proposition 6.3.14 $\succ_{rpo}$ *is a simplification order, if the precedence $\succ$ is a well-founded total order.*

Example 6.3.15 Let $s = (x * y) * z$ and $t = x * (y * z)$, then $s \succ_{lpo} t$ because of condition (b)2: $[x * y, z] \succ_{lpo}^{lex} [x, y * z]$. However, $\succ_{rpo}$ cannot compare these two terms, because $\{x * y, z\}$ and $\{x, y * z\}$ are not comparable by $\succ_{rpo}^{mul}$. In fact, $\succ_{rpo}$ cannot compare $(a * a) * a$ and $a * (a * a)$, where a is a constant. Thus, in general, $\succ_{rpo}$ is not a total order on ground terms. □

Let us define $s \approx_{rpo} t$ if $s = t$ or $s = f(s_1, \ldots, s_m), t = f(t_1, \ldots, t_n), f \approx g$ and $\{s_1, \ldots, s_m\} \approx_{rpo}^{mul} \{t_1, \ldots, t_n\}$, and $s \succeq_{rpo} t$ iff $s \succ_{rpo} t$ or $s \approx_{rpo} t$. For example, $(a * a) * a \approx_{rpo} a * (a * a)$. Obviously, this extension of $\succeq_{rpo}$ gives us only finite equivalence classes and preserves well-foundedness. This extension will allow us to compare more pairs of terms than before and make $\succ_{rpo}$ more useful in practice. For instance, it will allow us to show $f(b, (a*a)*a) \succ_{rpo} f(a, a*(a*a))$ when $b \succ a$.

Relation $\approx_{rpo}$ also allows us to combine $\succ_{rpo}$ with another simplification order, say $\succ_{lpo}$: When $s \approx_{rpo} t$ and $s \neq t$, we use the result of $s \succ_{lpo} t$ for $s \succ_{rpo} t$. This idea makes $\succ_{rpo}$ total on ground terms.

The difference of $\succ_{lpo}$ and $\succ_{rpo}$ lies on condition (b)2 of LPO and RPO when $f \approx g$: The former uses the lexcographical extension and the latter uses the multiset extension.

Definition 6.3.16 LPO and RPO are combined into one simplification order $\succ_{lrpo}$ by the *status* of function symbols: For every function f, arity(f) > 1: status(f) ∈ $\{l, r, m\}$. If $f \approx g$, then status(f) = status(g). Condition (b)2 of LPO (or RPO) is replaced by the following conditions:

- status(f) = l: When $f \approx g$, condition (b)2 of LPO is used.
- status(f) = r: When $f \approx g$, condition (b)2 of LPO is used and the lexcographical extension applies on the arguments of f and g from right to left.
- status(f) = m: When $f \approx g$, condition (b)2 of RPO is used.

The use of status allows us to compare more pairs of terms in the desired direction. For instance, to have $sub(x, s(y)) \succ_{lrpo} sub(pre(x), y)$, we let status($sub$) = r; thus, $[x, s(y)] \succ_{lpo}^{lex} [pre(x), y]$.

KBO: Knuth–Bendix Order

The *Knuth–Bendix order* ($\succ_{kbo}$), invented by Donald Knuth and Peter Bendix in 1970, assigns a number, called *weight*, to each function, predicate, or variable symbol, and uses the sum of weights to compare terms. When two terms have the same weight, the precedence relation is used to break ties. Using weights gives us flexibility as well as complication.

Definition 6.3.17 (Weight Function) A function $w : X \cup \Sigma \mapsto \mathcal{N}$ is said be a *weight function* if it satisfies (i) $w(c) > 0$ for every constant $t \in \Sigma$, and (ii) there exists at most one unary function symbol $f \in \Sigma$ such that $w(f) = 0$ and f must be maximal in the precedence if $w(f) = 0$.

The weight function w can be extended a function over terms or formulas as follows:

$$w(x) = w_0 \text{ for any variable } x;$$
$$w(f(t_1, \ldots, t_n)) = w(f) + w(t_1) + \cdots + w(t_n)$$

where $w_0 > 0$ denotes the minimal weight of all constants.

Definition 6.3.18 ($\succ_{kbo}$) Let w be a weight function and $mv(t)$ be the multiset of variables appearing in t. The *Knuth–Bendix order* $\succ_{kbo}$ is defined recursively as follows: $s \succeq_{kbo} t$ if $w(s) \geq w(t)$ and $mv(s) \supseteq mv(t)$; $s \succ_{kbo} t$ if $s \succeq_{kbo} t$ and either (a) $w(s) > w(t)$ or (b) $s \neq t$ and t is a variable, or (c) $s = f(s_1, \ldots, s_m)$, $t = g(t_1, \ldots, t_n)$, and one of the two conditions is true:

1. $f \succ g$, or
2. $f \approx g$ and $[s_1, \ldots, s_m] \succ_{kbo}^{lex} [t_1, \ldots, t_n]$.

We like to point out that the condition $mv(s) \supseteq mv(t)$ is necessary. Let $s = f(g(x), y)$ and $t = f(y, y)$. If $w(g) > 0$, then $w(s) > w(t)$, but $mv(s) = \{x, y\} \supseteq \{y, y\} = mv(t)$ does not hold. If we ignore the condition $mv(s) \supseteq mv(t)$, then $s \succ_{kbo} t$ would imply $s\theta \succ_{kbo} t\theta$ for $\theta = [y \leftarrow g(x)]$, where $s\theta = t\theta = f(g(x), g(x))$.

If $w(s) = w(t)$, then condition (b), i.e., $s \neq t$ and t is a variable, implies that $s = f^i(x)$ and $t = x$, where $i > 0$, $w(f) = 0$. One reason we need a function of 0 weight is that we want $i(x * y) \succ_{kbo} i(x) * i(y)$ (see Example 6.3.13) and it is doable with $w(i) = 0$ and $i \succ *$.

Example 6.3.19 Let $w(f) = w(a) = 1$, $w(b) = 2$, then $f(x, b, a) \succ_{kbo} f(a, a, b)$ by condition $(c)2$, because $x \succeq_{kbo} a$ and $b \succ_{kbo} a$. Similarly, $p(x, b, y) \succ_{kbo} p(a, a, y)$. Note that neither $\succ_{lpo}$ nor $\succ_{rpo}$ can compare $p(x, b, a)$ and $p(a, a, b)$, or $p(x, b, y)$ and $p(a, a, y)$. □

Proposition 6.3.20 $\succ_{kbo}$ *is a simplification order and is total on ground terms, if the precedence $\succ$ is a well-order.*

Example 6.3.21 Let the precedence relation be $b/1 \succ a/1 \succ e/0$, where a and b are unary function symbols and e is a constant. We will write $a(a(b(a(e))))$ as $aaba$ for brevity. Let $w(e) = 1$ and the weights of a and b, as well as the least 13 terms, with the exception of the last row, are listed (in the order of $\prec_{kbo}$) in the following table:

6.3 Simplification Orders and Ordered Resolution

$w(a)$	$w(b)$	The least 13 terms in the order of $\prec_{kbo}$
1	1	$e, a, b, aa, ab, ba, bb, aaa, aab, aba, abb, baa, bab, \ldots$
2	1	$e, b, a, bb, ab, ba, bbb, aa, abb, bab, bba, bbbb, abb, \ldots$
2	3	$e, a, b, aa, ab, ba, aaa, bb, aab, aba, baa, aaaa, abb, \ldots$
2	0	$e, b, bb, \ldots, b^i, \ldots, a, ab, \ldots, ab^i, \ldots, ba, bab, babb, \ldots$

On the other hand, the list of terms in the increasing order of $\succ_{lpo}$ (or $\succ_{rpo}$) is

$$e, a, \ldots, a^i, \ldots, b, ab, \ldots, a^i b, \ldots, ba, aba, \ldots, a^i ba, \ldots,$$
$$baa, abaa, \ldots, a^i baa, \ldots$$

Note that all of these orders provided by KBO, LPO, or RPO are well-founded. □

The above example shows the flexibility of $\succ_{kbo}$, which provides various ways of comparing two terms by changing weight functions as needed. However, $\succ_{kbo}$ lacks the flexibility of comparing terms like $x * (y + z)$ and $x * y + x * z$. We cannot make $x * (y + z) \succ_{kbo} x * y + x * z$, while $x * (y + z) \succ_{lpo} x * y + x * z$ if $* \succ +$ and $x * y + x * z \succ_{lpo} x * (y + z)$ if $+ \succ *$.

Now when we say to use a simplification order $\succ$, we could use either $\succ_{lpo}$, $\succ_{rpo}$, or $\succ_{kbo}$.

Example 6.3.22 Let the operators $\leftrightarrow, \oplus, \rightarrow, \neg, \vee, \wedge$ be listed in the descending order of the precedence $\succ$, then the right side of each following equivalence relation is less than the corresponding left side by either $\succ_{lpo}$ or $\succ_{rpo}$.

$$A \oplus B \equiv (A \vee B) \wedge (\neg A \vee \neg B);$$
$$A \leftrightarrow B \equiv (A \vee \neg B) \wedge (\neg A \vee B);$$
$$A \rightarrow B \equiv \neg A \vee B;$$
$$\neg \neg A \equiv A;$$
$$\neg (A \vee B) \equiv \neg A \wedge \neg B;$$
$$\neg (A \wedge B) \equiv \neg A \vee \neg B;$$
$$A \vee (B \wedge C) \equiv (A \vee B) \wedge (A \vee C)$$

These equivalence relations are used to translate a formula into CNF. Thus, the term orderings like $\succ_{rpo}$ and $\succ_{lpo}$ can be used to show the termination of the process of converting a formula into CNF. □

If we want to convert formulas into DNF, then the precedence relation should be $\wedge \succ \vee$, so that the termination of converting a formula into DNF can be shown by $\succ_{rpo}$ or $\succ_{lpo}$.

6.3.3 Completeness of Ordered Resolution

Let $\succ$ be a simplification order, which is a well-founded, stable partial order over the set of atoms and total over the set of ground atoms, and we will use $\succ$ in ordered resolution. Given $\succ$, an atom $A \in S$ is said to be *maximal* if there is no $B \in S$ such that $B \succ A$. A is said to be *minimal* if there is no $B \in S$ such that $A \succ B$. S may have several minimal or maximal atoms. In fact, an atom A can be both minimal and maximal in S if A cannot be compared by $\succ$ with any other atom of S. Literals are compared by comparing their atoms. Since a clause is represented by a set of literals, we may say a literal is maximal or minimal in a clause. For example, the three atoms in $(\overline{p(x,y)} \mid \overline{p(y,z)} \mid p(x,z))$ are not comparable under any stable $\succ$, so each atom is both minimal and maximal.

Definition 6.3.23 (Ordered Resolution) Given a simplification order $\succ$ over the set of atoms, the ordered resolution is the resolution with the condition that A is a maximal atom in $(A \mid \alpha)$ and B is a maximal atom in $(\overline{B} \mid \beta)$, where σ is a mgu of A and B:

$$\frac{(A \mid \alpha) \quad (\overline{B} \mid \beta)}{(\alpha \mid \beta)\sigma}$$

Given a set S of clauses, let S^* denote the set of clauses *saturated* by ordered resolution and factoring. That is, any resolvent from ordered resolution on any two clauses of S^* is in S^* and any clause from factoring on any clause in S^* is also in S^*. Let GC be the set of all ground instances of S^*.

Lemma 6.3.24 *GC is saturated by ordered resolution.*

Proof Suppose $(g \mid \alpha)$ and $(\overline{g} \mid \beta)$ are in GC, where g is the maximal atom in both clauses. The resolvent of the two clauses is $(\alpha \mid \beta)$. We need to show that $(\alpha \mid \beta) \in GC$.

Let $(g \mid \alpha)$ be an instance of $(A \mid \alpha') \in S^*$, i.e., $g = A\theta$ and $\alpha = \alpha'\theta$ for some substitution θ. We assume that A is the only literal in $(A \mid \alpha')$ such that $A\alpha = g$; otherwise, we use factoring to achieve this condition. Similarly, let $(\overline{g} \mid \beta)$ be an instance of $(\overline{B} \mid \beta') \in S^*$, i.e., $\overline{g} = \overline{B}\gamma$ and $\beta = \beta'\gamma$ for some γ, and B is the only literal in $(\overline{B} \mid \beta')$ such that $B\gamma = g$.

Since $g = A\theta = B\gamma$ is maximal, A and B must be unifiable and maximal in $(A \mid \alpha')$ and $(\overline{B} \mid \beta)$, respectively. Let λ be the mgu of A and B, then the resolvent of $(A \mid \alpha')$ and $(\overline{B} \mid \beta')$ by ordered resolution is $(\alpha' \mid \beta')\lambda$, which must be in S^*. Since $\theta = \lambda\theta'$ and $\gamma = \lambda\gamma'$ for some θ' and γ', we must have $(\alpha \mid \beta) = (\alpha' \mid \beta')\lambda\theta'\gamma'$, which is an instance of $(\alpha' \mid \beta')\lambda$. Hence, $(\alpha \mid \beta) \in GC$. □

Theorem 6.3.25 (Refutational Completeness of Ordered Resolution) *S is unsatisfiable iff the empty clause is in S^*.*

Proof If the empty clause is in S^*, since $S \models \bot$, S must be unsatisfiable.

6.3 Simplification Orders and Ordered Resolution

If the empty clause is not in S^*, we will construct a Herbrand model for S from GC, which is the set of all ground instances of S^*.

Let GA be the set of ground atoms appearing in GC. The empty clause is not in GC because any instance of an non-empty clause cannot be empty. By the assumption, $\succ$ is a well-order over GA.

We start with an empty model H in which no atoms have a truth value. We then add ground literals into H one by one, starting from the minimal atom, which exists because $\succ$ is well-founded. Let g be any atom in GA, and

$$pc(g) = \{(g \mid \alpha) : g \text{ is maximal in } (g \mid \alpha) \in G\}$$
$$nc(g) = \{(\overline{g} \mid \beta) : g \text{ is maximal in } (\overline{g} \mid \beta) \in G\}$$
$$upto(g) = \bigcup\nolimits_{g' \in GA, g \succeq g'} (pc(g') \cup nc(g'))$$

That is, $pc(g)$ contains all the clauses which contains g as its maximal literal; $nc(g)$ contains all the clauses which contains $\overline{g}$ as its maximal literal; and $upto(g)$ contains all the clauses from GC whose maximal literals are less or equal to g.

We claim that after g or $\overline{g}$ is added into H, the following property is true:

Claim: H is a model of $upto(g)$ after g or $\overline{g}$ is added into H.

Let g_1 be the minimal atom of GA. We cannot have both (g_1) and $(\overline{g_1})$ in GC; otherwise, the empty clause must have been generated from (g_1) and $(\overline{g_1})$ because GC is saturated. If $(g_1) \in GC$, we add g_1 into H (it means $H(g_1) = 1$); otherwise, we add $\overline{g_1}$ into H (it means $H(g_1) = 0$). It is trivial to check that H is a model of $upto(g_1)$ (which may be empty), so the claim is true for g_1. This is the base case of the induction based on $\succ$.

Let g be the minimal atom in GA which has no truth value in H. Assume as the induction hypothesis that the claim is true for all $g' \prec g$. If there exists a clause $(g \mid \alpha) \in pc(g)$ such that $H(\alpha) = 0$, we add g into H; otherwise, add $\overline{g}$ into H.

Suppose g is added into H, then every clause in $pc(g)$ will be true under H. For any clause $(\overline{g} \mid \beta) \in nc(g)$, then there exists an ordered resolution between $(g \mid \alpha)$ and $(\overline{g} \mid \beta)$, and their resolvent is $(\alpha \mid \beta)$. By Lemma 6.3.24, $(\alpha \mid \beta) \in GC$. Since every literal g' of $(\alpha \mid \beta)$ is less than g, by the induction hypothesis, $(\alpha \mid \beta)$ is true in H. Since $H(\alpha) = 0$, we must have $H(\beta) = 1$. That means no clause in $nc(g)$ will be false in H.

Now suppose $\overline{g}$ is added into H. The above analysis also holds by changing the role of g and $\overline{g}$. In both cases, all clauses in $pc(g)$ and $nc(g)$ are true in H after adding either g or $\overline{g}$ into H. Thus, the claim is true for every $g \in GA$. Once every atom in GA is processed, we have found a model H of GC by the claim. Since GC is the set of all ground instances of S^*, H is a Herbrand model of S^* or S. In other words, S is satisfiable. □

Note that the model construction in the above proof is not algorithmic, because GC is infinite in general. The induction is sound because it is based on the well-founded order $\succ$.

Now it is easy to prove Theorem 6.2.7: If the empty clause is generated from S, then S must be unsatisfiable; if S is unsatisfiable, then the empty clause will be generated by ordered resolution, which is a special case of resolution.

A traditional proof of Theorem 6.2.7 uses Herbrand's theorem:

Theorem 6.3.26 (Herbrand's Theorem) *A set S of clauses is unsatisfiable iff there exists a finite unsatisfiable set of ground instances of S.*

From Herbrand' theorem, a proof of Theorem 6.2.7 goes as follows:

1. If S is unsatisfiable, then by Herbrand's theorem, there exists a finite unsatisfiable set G of ground instances of S.
2. Resolution for propositional logic will find a proof from G, treating each ground atom in G as a propositional variable.
3. This ground resolution proof can be lifted to a general resolution proof in S, using a lemma similar to Lemma 6.3.24.

The completeness proof based on simplification orders comes from [4] and does not use Herbrand's theorem.

6.4 Prover9: A Resolution Theorem Prover

Prover9 is a full-fledged automated theorem prover for first-order logic based on resolution. Prover9 is the successor of the Otter prover; both Otter and Prover9 were created by William McCune (1953–2011).

6.4.1 Input Formulas to Prover9

Prover9 has a fully automatic mode in which the user simply gives it formulas representing the problem. A good way to learn about Prover9 is to browse and study the example input and output files that are available with the distribution of Prover9. Let us look at an example.

Example 6.4.1 Once we can specify the following puzzle in first-order logic, it is trivial to find a solution by Prover9:

> Jack owns a dog. Every dog owner is an animal lover. No animal lover kills an animal. Either Jack or Curiosity killed the cat, who is named Tuna. Did Curiosity kill the cat?

The following first-order formulas come from the statements of the puzzle:

6.4 Prover9: A Resolution Theorem Prover

(1) $\exists x(dog(x) \wedge owns(Jack, x))$
(2) $\forall x(\exists y(dog(y) \wedge owns(x, y)) \rightarrow animalLover(x))$
(3) $\forall x(animalLover(x) \rightarrow (\forall y(animal(y) \rightarrow \neg kills(x, y))))$
(4) $kills(Jack, Tuna) \vee kills(Curiosity, Tuna)$
(5) $cat(Tuna)$
(6) $\forall x(cat(x) \rightarrow animal(x))$

The goal is to show that $kills(Curiosity, Tuna)$. The above formulas can be written in Prover9's syntax as follows:

```
exists x (dog(x) & owns(Jack, x)).
all x (exists y (dog(y) & owns(x, y)) -> animalLover(x)).
all x (animalLover(x) -> (all y (animal(y) -> -kills(x, y)))).
kills(Jack, Tuna) | kills(Curiosity, Tuna).
cat(Tuna).
all x (cat(x) -> animal(x)).
```

We may either use kills(Curiosity, Tuna) as the goal or add -kills (Curiosity, Tuna) (the negation of the goal) into the clause set so that Prove9 will find a resolution proof. □

From this example, we can see that each formula in Prover9 is ended with ".", and we use "->" for $\rightarrow$, "|" for $\vee$, "&" for $\wedge$, "−" for $\neg$, "all" for $\forall$, and " exists" for $\exists$. In fact, the topmost "all" is optional as free variables are assumed to be universally quantified. Prover9 will convert formulas in the "assumptions" list and the negation of the formulas in the "goals" list into clauses before resolution is called.

Note that Prover9 uses the same precedence for the Boolean operators and the quantifiers, "all" and "exists," as the one in this book for omitting some parentheses. For instance,

```
exists x dog(x) & owns(Jack, x)
```

represents the formula $(\exists x\, dog(x)) \wedge owns(Jack, x)$, not $\exists x(dog(x) \wedge owns(Jack, x))$. To avoid confusion, insert a pair of parentheses when you are uncertain about the scope of a quantified variable.

Prover9 implements *resolution* (Procedure 6.2.9) with bells and whistles. The given list is called the "sos" (set of support) list and the kept list is called the "usable" list. You may use the "usable" list and the "sos" list as follows:

```
formulas(usable). % the kept list
...
end_of_list.

formulas(sos). % the given list
...
end_of_list.
```

The resolution is always done between a clause from the sos list (the given list) and some clauses from the usable list (the kept list). Once all the resolvents between this clause and the usable list have been computed, this clause moves from the sos list to the usable list. In other words, resolutions between the clauses in the original usable list are omitted.

A basic Prover9 command on a Linux machine will look like

```
prover9 -f Tuna.in > Tuna.out
```

or

```
prover9 < Tuna.in > Tuna.out
```

where the file named "Tuna.in" contains the input to Prover9 and the file named "Tuna.out" contains the output of Prover9. If the file "Tuna.in" contains the formulas in Example 6.4.1, then the file "Tuna.out" will contain the following proof:

```
=========================== PROOF ==============================

% Proof 1 at 0.03 (+ 0.05) seconds.
% Length of proof is 19.
% Level of proof is 6.
% Maximum clause weight is 6.
% Given clauses 0.

1 (exists x (dog(x) & owns(Jack,x))). [assumption].
2 (all x ((exists y (dog(y) & owns(x,y))) -> animalLover(x))).
       [assumption].
3 (all x (animalLover(x) -> (all y (animal(y) -> -kills(x,y))))) 
       [assumption].
4 (all x (cat(x) -> animal(x))). [assumption].
5 kills(Curiosity,Tuna). [goal].
6 -dog(x) | -owns(y,x) | animalLover(y). [clausify(2)].
7 dog(c1). [clausify(1)].
8 -owns(x,c1) | animalLover(x). [resolve(6,a,7,a)].
9 owns(Jack,c1). [clausify(1)].
10 animalLover(Jack). [resolve(8,a,9,a)].
11 -animalLover(x) | -animal(y) | -kills(x,y). [clausify(3)].
12 -cat(x) | animal(x). [clausify(4)].
13 cat(Tuna). [assumption].
14 animal(Tuna). [resolve(12,a,13,a)].
15 -animal(x) | -kills(Jack,x). [resolve(10,a,11,a)].
16 kills(Jack,Tuna) | kills(Curiosity,Tuna). [assumption].
17 -kills(Curiosity,Tuna). [deny(5)].
18 -kills(Jack,Tuna). [resolve(14,a,15,a)].
19 $F. [back_unit_del(16),unit_del(a,18),unit_del(b,17)].

=========================== end of proof =======================
```

In Prover9, the empty clause is denoted by $F. In each clause, the literals are numbered by a, b, c, ... For example, to generate clause 18, -kills(Jack, Tuna), from

```
14 animal(Tuna). [resolve(12,a,13,a)].
15 -animal(x) | -kills(Jack,x). [resolve(10,a,11,a)].
```

Prove9 performed a resolution on the first literal (a) of clause 14 and the first literal (a) of clause 15, with the mgu x ⟵ Tuna, and the resolvent is

    18 -kills(Jack,Tuna). [resolve(14,a,15,a)].

The message followed by each clause can be checked by either human or machine, to ensure the correctness of the proof.

6.4.2 Inference Rules and Options

There are two types of parameters in Prover9: Boolean flags and numeric parameters. To change the default values of the former type, use set(flag) or clear(flag). The latter can be changed by assign(parameter, value). For example, the default automatic mode is set by

    set(auto).

Turning "auto" on will cause a list of other flags to turn on, including setting up "hyper-resolution" as the main inference rule for resolution. Hyper-resolution reduces the number of intermediate resolvents by combining several resolution steps into a single inference step.

Definition 6.4.2 (Hyper-Resolution) *Positive hyper-resolution* consists of a sequence of positive resolutions between one non-positive clause (called *nucleus*) and a set of positive clauses (called *satellites*), until a positive clause or the empty clause is produced.

The number of positive resolutions in hyper-resolution is equal to the number of negative literals in the nucleus clause.

Example 6.4.3 Let the nucleus clause be $(\overline{p(x)} \mid \overline{q(x)} \mid r(x))$, and the satellite clauses be $(q(a) \mid r(b))$ and $(p(a) \mid r(c))$. The first resolution between the nucleus and the first satellite produces $(\overline{p(a)} \mid r(a) \mid r(b))$. The second resolution between the resolvent and the second satellite produces the second resolvent $(r(a) \mid r(b) \mid r(c))$, which is the result of positive hyper-resolution between the nucleus and the satellites. □

Negative hyper-resolution can be defined similarly by replacing "positive" by "negative." Hyper-resolution means both positive and negative hyper-resolutions.

Definition 6.4.4 (ur-Resolution) *Unit-resulting (ur)-resolution* consists of a sequence of unit resolutions between one non-unit clause (called *nucleus*) and a set of unit clauses (called *satellites*), until a unit clause or the empty clause is produced.

Example 6.4.5 In Sect. 3.2.2, we discussed the Hilbert system which consists of three axioms and one inference rule, i.e., *modus ponens*. To prove $x \to x$ in the Hilbert system, we may use the following input: □

```
op(400, infix_right, ["->"]).   % infix operator

formulas(usable).
P((n(y) -> n(x)) -> (x -> y)).
-P(x) | -P(x -> y) | P(y).
end_of_list.

formulas(sos).
P(x -> (y -> x)).
P((x -> (y -> z)) -> ((x -> y) -> (x -> z))).
-P(a -> a).
end_of_list.
```

Prover9 will produce the following proof:

```
1  -P(x) | -P(x -> y) | P(y).  [assumption].
2  P(x -> y -> x).  [assumption].
3  P((x -> y -> z) -> (x -> y) -> x -> z).  [assumption].
4  -P(a -> a).  [assumption].
5  P(x -> y -> z -> y).  [hyper(1,a,2,a,b,2,a)].
7  P((x -> y) -> x -> x).  [hyper(1,a,2,a,b,3,a)].
15 P(x -> x).  [hyper(1,a,5,a,b,7,a)].
16 $F.  [resolve(15,a,4,a)].
```

For the above example, ur-resolution will produce the same proof as hyper-resolution does. Hyper-resolution is the default inference rule for the automatic mode. If you like to see a binary resolution proof, you have to turn off "auto" and turn on "binary_resolution":

```
clear(auto).
set(binary_resolution).
```

Turning off `hyper-resolution`, Prover9 will produce the following proof:

```
1  -P(x) | -P(x -> y) | P(y).  [assumption].
2  P(x -> y -> x).  [assumption].
3  P((x -> y -> z) -> (x -> y) -> x -> z).  [assumption].
4  -P(a -> a).  [assumption].
5  -P(x) | P(y -> x).  [resolve(2,a,1,b)].
6  -P(x -> y -> z) | P((x -> y) -> x -> z).  [resolve(3,a,1,b)].
10 P((x -> y) -> x -> x).  [resolve(6,a,2,a)].
19 P(x -> (y -> z) -> y -> y).  [resolve(10,a,5,a)].
20 -P(x -> y) | P(x -> x).  [resolve(10,a,1,b)].
45 P(x -> x).  [resolve(20,a,19,a)].
46 $F.  [resolve(45,a,4,a)].
```

6.4 Prover9: A Resolution Theorem Prover

Since positive resolution is complete and unit resolution is not complete, hyper-resolution is complete and unit-resulting resolution is not complete. To use "unit-resulting resolution," use the command

set(ur_resolution).

To use "hyper-resolution" in Prover9, use the command

set(hyper_resolution).

This option will cause the flags pos_hyper_resolution and neg_hyper_resolution to be true.

The resolution inference rules provided by Prover9 include "binary resolution," "hyper resolution," "ordered resolution," and "unit-resulting resolution." To use "ordered resolution," use the command

set(ordered_res).

This option puts restrictions on the binary and hyper-resolution inference rules. It says that resolved literals in one or more of the parents must be maximal in the clause. Continuing from the previous example, if we turn on the ordered resolution flag, the proof generated by Prover9 will be the following.

```
1 -P(x) | -P(x -> y) | P(y). [assumption].
2 P(x -> y -> x). [assumption].
3 P((x -> y -> z) -> (x -> y) -> x -> z). [assumption].
4 -P(a -> a). [assumption].
5 -P(x) | P(y -> x). [resolve(2,a,1,b)].
6 -P(x -> y -> z) | P((x -> y) -> x -> z). [resolve(3,a,1,b)].
8 P(x -> y -> z -> y). [resolve(5,a,2,a)].
9 P(x -> y -> z -> u -> z). [resolve(8,a,5,a)].
10 P(x -> y -> z -> u -> w -> u). [resolve(9,a,5,a)].
14 P((x -> y) -> x -> x). [resolve(6,a,2,a)].
19 -P(x -> y) | P(x -> x). [resolve(14,a,1,b)].
21 P(x -> x). [resolve(19,a,10,a)].
22 $F. [resolve(21,a,4,a)].
```

6.4.3 Simplification Orders in Prover9

Prover9 has several methods available for comparing terms or literals. The term orders are partial orders (and sometimes total on ground terms), and they are used to decide which literals in clauses are admissible for application of ordered resolution. Several of the resolution rules require that some of the literals be maximal in their clause.

The symbol precedence is a total order on function and predicate symbols (including constants). The symbol weighting function maps symbols to non-negative integers. Prover9 supports three simplification orders: LPO (lexicographic path order), RPO (recursive path order), and KBO (Knuth–Bendix order), which are introduced in Sect. 6.3.2.

Here is the command for choosing a simplification order.

```
assign(order, string).  % default string=lpo, range [lpo,rpo,kbo]
```

This option is used to select the primary order to be used for determining maximal literals in clauses. The choices are "lpo" (lexicographic path order), "rpo" (recursive path order), and "kbo" (Knuth–Bendix order).

The default symbol precedence (for LPO, RPO, and KBO) is given by the following rules (in order).

- Function symbols < non-equality predicate symbols.
- For function symbols: $c/0 < f/2 < g/1 < h/3 < i/4 < \ldots$, where c is any constant, f is any function of arity 2, g is any function of arity 1, ...(note the position of $g/1$).
- For predicate symbols: lower arity < higher arity.
- Non-Skolem symbols < Skolem symbols.
- For Skolem symbols, the lower index is the lesser.
- For non-Skolem symbols, more occurrences < fewer occurrences.
- the lexical ASCII order (UNIX strcmp() function).

The function_order and predicate_order commands can be used to change the default symbol precedence. They contain lists of symbols ordered by the precedence from the smallest to the largest. For example,

```
predicate_order([=, <=, P, Q]).         % = < <= < P < Q
function_order([a, b, c, +, *, h, g]).  % a < b < c < + < * < h < g
```

We need two separate commands for defining the precedence, because predicate symbols are assumed to be always greater than function symbols in the precedence. The used symbol precedence for a problem is always printed in the output file (in the section PROCESS INPUT).

6.4.4 The TPTP Library

Prover9 has proved automatically many theorems in the TPTP library, where TPTP stands for Thousands of Problems for Theorem Provers. TPTP is maintained by Geoff Sutcliffe at University of Miami. According to the website of TPTP, tptp.org, TPTP contains thousands of test problems over more than 50 domains and supports input formats for more than 50 automated theorem proving (ATP) systems. TPTP has been used to support many ATP competitions. TPTP supplies the ATP community with the following functions:

- A comprehensive library of the test problems that are available today, in order to provide an overview and a simple, unambiguous reference mechanism.
- A comprehensive list of references and other interesting information for each problem.

- Arbitrary size instances of generic problems (e.g., the N-queens problem).
- A utility to convert the problems to existing ATP systems' formats.
- General guidelines outlining the requirements for ATP system evaluation.
- Standards for input and output for ATP systems.

For people who are interested in automated theorem proving, TPTP is a rich place to explore.

Exercises

1. Let $\theta = [x \leftarrow u, y \leftarrow f(g(u)), z \leftarrow f(u)]$ and $\sigma = [u \leftarrow x, x \leftarrow f(u), y \leftarrow a, z \leftarrow f(u)]$. Compute $\theta\sigma$ and $\sigma\theta$. Among $\theta, \sigma, \theta\sigma$, and $\sigma\theta$, which substitutions are idempotent?
2. Use the rule-based unification algorithm *unify* to decide if the following pairs of terms are unifiable or not and provide each state in *unify* after applying a transformation rule.

    (a) $p(x, a) \doteq p(f(y), y)$
    (b) $p(x, f(x)) \doteq p(f(y), y)$
    (c) $p(f(a), g(x)) \doteq p(y, y)$
    (d) $q(x, y, f(y)) \doteq q(u, h(v, v), u)$
    (e) $q(a, x, f(g(y))) \doteq q(z, f(z), f(u))$

3. How to modify *unify*(s, t) so that it can be used to decide if a set S of terms are unifiable? That is, find an idempotent, most general substitution σ and a term t such that such that $s\sigma = t$ for every $s \in S$.
4. Find the mgu of $s = f(f(\ldots f(f(a, x_1), x_2), \ldots), x_n)$ and $t = f(x_n, f(x_{n-1}, f(\ldots, f(x_1, a)\ldots)))$ for $n = 5$.
5. Create the term graph $G = (V, E)$ for each of the following pairs of terms:

    (a) $s = p(x, a)$ and $t = p(f(y), y)$
    (b) $s = p(x, f(x))$ and $t = p(f(y), y)$
    (c) $s = q(x, y, f(y))$ and $t = q(u, h(v, v), u)$
    (d) $s = q(a, x, f(g(y)))$ and $t = q(z, f(z), f(u))$
    (e) $s = f(f(f(f(a, x_1), x_2), x_3), x_4)$ and $t = f(x_4, f(x_3, f(x_2, f(x_1, a))))$

    and show the equivalence classes of V after calling *unify*$_1$ with a success.
6. Provide the pseudo-code for the procedure *postOccurCheck* used in *lunify*. What is the complexity of your algorithm?
7. Design an efficient algorithm and provide its pseudo-code for creating the mgu when the procedure *lunify* returns true. What is the complexity of your algorithm?

8. Find counterexamples to show that KBO is not well-founded when the following condition is false: "there exists at most one unary function symbol $f \in \Sigma$ such that $w(f) = 0$ and f must maximal in the precedence if $w(f) = 0$."
9. Show that all the left side s of each equivalence relation in Example 6.3.22 is greater than their right side t by $\succ_{lpo}$, and explain what precedence is used and what conditions of $\succ_{lpo}$ are used for $s \succ_{lpo} t$.
10. Provide a precedence for $\{\times, +\}$ and show that all the left side s of each following equation is greater than their right side t by $\succ_{lpo}$. Please explain what conditions of $\succ_{lpo}$ are used for $s \succ_{lpo} t$.

$$(x \times y) \times z = x \times (y \times z)$$
$$x \times (y + z) = (x \times y) + (x \times z)$$
$$(y + z) \times x = (y \times x) + (z \times x)$$

Can you use $\succ_{rpo}$ or $\succ_{kbo}$ to achieve the same result? Why?
11. Choose with justification a simplification order from $\succ_{lpo}$, $\succ_{rpo}$ and $\succ_{kbo}$ such that the left side s is greater than the right side t for each of the following equations (which defines the Ackermann's function):

$$A(0, y) = s(y)$$
$$A(s(x), 0) = A(x, s(0))$$
$$A(s(x), s(0)) = A(x, A(s(x), y))$$

12. The Hoofers Club problem assumes the following statements:

> Tony, Tom, and Liz belong to the Hoofers Club. Every member of the Hoofers Club is either a skier or a mountain climber or both. No mountain climber likes rain, and all skiers like snow. Liz dislikes whatever Tony likes and likes whatever Tony dislikes. Tony likes rain and snow.

And it asks for the solution of the following question: "Is there a member of the Hoofers Club who is a mountain climber but not a skier?" Please find the following resolution proofs for the Hoofers Club problem: (*a*) unit, (*b*) input, (*c*) positive, (*d*) negative, and (*e*) linear resolutions.
13. Use Prover9 to answer the question in the Hoofers Club problem of the previous problem. Please prepare the input to Prover9 and turn in the output file of Prover9.
14. Use binary resolution, hyper-resolution, and ur-resolution, respectively, of Prover9 to answer the question "Is John happy?" from the following statements:

> Anyone passing his logic exams and winning the lottery is happy. But anyone who studies or is lucky can pass all his exams. John did not study but he is lucky. Anyone who is lucky wins the lottery.

15. Use hyper-resolution and ur-resolution, respectively, of Prover9 to show that "the sprinklers are on" from the following statements:

> Someone living in your house is wet. If a person is wet, it is because of the rain, the sprinklers, or both. If a person is wet because of the sprinklers, the sprinklers must be on. If a person is wet because of rain, that person must not be carrying an umbrella. There is an umbrella in your house, which is not in the closet. An umbrella that is not in the closet must be carried by some person who lives in that house. Nobody is carrying an umbrella.

Please prepare the input to Prover9 and turn in the output file of Prover9.

16. Following the approach illustrated in Example 6.4.5, use Prover9 to prove that the following properties are true in the Hilbert system:

(a) $(x \rightarrow y) \rightarrow ((y \rightarrow z) \rightarrow (x \rightarrow z))$
(b) $(x \rightarrow (y \rightarrow z)) \rightarrow (y \rightarrow (x \rightarrow z))$
(c) $\neg x \rightarrow (x \rightarrow y)$
(d) $x \rightarrow (\neg x \rightarrow y))$
(e) $\neg\neg x \rightarrow x$
(f) $x \rightarrow \neg\neg x$
(g) $(x \rightarrow \neg x) \rightarrow \neg x$
(h) $(\neg x \rightarrow x) \rightarrow x$
(i) $(x \rightarrow y) \rightarrow (\neg y \rightarrow \neg x)$

References

1. Peter Ruzicka and Igor Privara, *An almost linear Robinson unification algorithm*, Acta Informatica, 1988, v27, pp. 61–71
2. Alexander Leitsch, *The Resolution Calculus*. Texts in Theoretical Computer Science. An EATCS Series. Springer, 1997. ISBN 978-3642606052
3. Williams McCune, *Prover9 and Mace4*, www.cs.unm.edu/mccune/prover9/, retrieved Nov. 11, 2023.
4. Hantao Zhang, *Reduction, superposition and induction: automated reasoning in an equational logic*, Ph.D. Thesis, Rensselaer Polytechnic institute, New York, 1988

Chapter 7
First-Order Logic with Equality

There are many equivalence relations in logic. When looking at a semantic level, we have logical equivalence and equisatisfiability, which are denoted by the symbols $\equiv$ and $\approx$, respectively, for both propositional logic and first-order logic. When looking at a syntactic level, we have the logical operator $\leftrightarrow$ to express the equality between two statements. A first-order language has two types of objects, i.e., formulas and terms, and we use $\leftrightarrow$ for the formulas. The conventional symbol for the equality of terms is "=", which is a predicate symbol. In the previous chapter, we studied first-order logic without equality and have been using "=" as an identity relation outside of first-order logic. In this chapter, we study first-order logic with equality, that is, we study "=" as a predicate symbol in first-order logic.

7.1 Equality of Terms

Equality is an omnipresent and important relation in every field of mathematics. For example, how would you specify that "for every x, there exists a unique y such that the relation $p(x, y)$ holds"? If the meaning of $p(x, y)$ is "x is the mother of y," then how would you state in first-order logic that "everybody's mother is unique"? If the meaning of $p(x, y)$ is $f(x) = y$, then how would you state that "for every x, $f(x)$ has a unique value"? Using "=", the answer is easy:

$$\forall x \exists y \, (p(x, y) \land \forall z \, (p(x, z) \rightarrow y = z))$$

The above formula is often denoted by $\forall x \exists! y \, p(x, y)$ in mathematics. Using "=" gives us higher expressive power to specify various statements:

- We can express "there are at least two elements for x such that $A(x)$ holds" as

$$\exists x \exists y \, x \neq y \land A(x) \land A(y)$$

- We can express "there are at most two elements for x such that $A(x)$ holds" as

$$\forall x \forall y \forall z \, ((A(x) \wedge A(y) \wedge A(z)) \rightarrow (x = y \vee y = z \vee x = z))$$

  This states that if we have three elements satisfying $A(x)$, then two of them must be equal.
- We can express "there are exactly two elements for x such that $A(x)$ holds" as the conjunction of the above two statements.

7.1.1 Axioms of Equality

The axioms of equality consist of five types of formulas; each is sufficiently famous to have earned itself a name [1]. In the following, the variables appearing in the formulas are assumed universally quantified:

- **Reflexivity**:
  $x = x$. An object is always equal to itself.
- **Commutativity**:
  $(x = y) \rightarrow (y = x)$. It does not matter how the parameters of $=$ are ordered. That is, $=$ is commutative.
- **Transitivity**:
  $(x = y) \wedge (y = z) \rightarrow (x = z)$. Equality can be propagated. That is, $=$ is transitive.
- **Function monotonicity**:
  For $1 \leq i \leq k$, $(x_i = y_i) \rightarrow (f(x_1, \ldots, x_k) = f(y_1, \ldots, y_k))$. If the arguments of a function are pair-wisely equal, the composed terms will be equal. That is, $=$ is a monotonic relation over terms.
- **Predicate monotonicity**:
  For $1 \leq i \leq k$, $(x_i = y_i) \wedge p(x_1, \ldots, x_k) \rightarrow p(y_1, \ldots, y_k)$. If the arguments of a predicate are pair-wisely equal, then the resulting atoms are logically equivalent.

The first three axioms together say that "=" is an equivalence relation. The fourth and fifth are very similar and share the name *monotonicity*. The first monotonicity rule governs terms and the second monotonicity rule governs formulas. Both of them assert that having equal arguments ensures equal results. In fact, these are axiom schemata as we need a version for each function f and each predicate p. In other words, we will need different versions of the fourth and fifth axioms in different first-order languages. Sometimes, we may split the monotonicity axioms into a set of axioms for each argument. For example, if the arity of predicate p is 2, we obtain a set of two axioms from the predicate monotonicity:

$$(x_1 = y_1) \wedge p(x_1, x_2) \rightarrow p(y_1, x_2)$$
$$(x_2 = y_2) \wedge p(x_1, x_2) \rightarrow p(x_1, y_2)$$

7.1 Equality of Terms

These monotonicity axioms are not to be confused with the substitution rule, which allows the monotonicity of terms for variables only.

All the five types of axioms can be easily converted into clauses so that we can use resolution to prove theorems involving $=$.

Example 7.1.1 Let $f(a) = b$ and $f(b) = a$, if we are not allowed to use "equal-by-equal substitution," how could you prove that $f(f(a)) = a$? The answer is resolution. Since resolution is a refutational prover, we need to add the negation of $f(f(a)) = a$, i.e., $f(f(a)) \neq a$, into the set of clauses from the equality axioms and the premises and show the clause set is unsatisfiable:

1	$(f(a) = b)$	premise
2	$(f(b) = a)$	premise
3	$(f(f(a)) \neq a)$	negation of the goal
4	$(x \neq y \mid y \neq z \mid x = z)$	transitivity
5	$(x \neq y \mid f(x) = f(y))$	monotonicity
6	$(f(f(a)) = f(b))$	resolvent from 1 and 5
7	$(f(b) \neq z \mid f(f(a)) = z)$	resolvent from 6 and 4
8	$(f(f(a)) = a)$	resolvent from 2 and 7
9	$()$	resolvent from 3 and 8

It shows clearly that without equality axioms, we cannot have a resolution proof of $f(f(a)) = a$, and the resolution proof is cumbersome if we cannot use "equal-by-equal substitution." □

7.1.2 Semantics of "="

The equality symbol is meant to model the equality of objects in a domain. For example, when we say "2 + 0 = 2" or $add(s(s(0)), 0) = s(s(0))$, we meant that "2+0" and "2" represent the same object in a domain. That is, $s = t$ is true if s and t represent "equal" or "identical" objects in any domain. We are asserting that two different descriptions refer to the same object. Because the notion of identity can be applied to virtually any domain of objects, equality is often assumed to be omnipresent in every logic. However, talk of "equality" or "identity" raises messy philosophical questions. "Am I the same person I was three days ago?"

We need a simple and clear way to define the meaning of $=$. For a first-order language $L = (P, F, X, Op)$, a Herbrand model for a CNF A can be simply represented by a set H of ground atoms such that only those atoms in H are interpreted to be true. Because of the existence of the equality axioms, the set H must hold some properties.

Definition 7.1.2 (Congruence) A relation $=_m$ over the set of terms $T(F, X)$ is called a *congruence* if $=_m$ is an equivalence relation closed under instantiation:

$$s\sigma =_m t\sigma \text{ if } s =_m t$$

for every substitution $\sigma : X \mapsto T(F, X)$ and closed under monotonicity

$$f(s_1, \ldots, s_k) =_m f(t_1, \ldots, t_k) \text{ if for any } 1 \leq i \leq k, s_i =_m t_i$$

for every function f/k.

Recall that a binary relation R is *stable* if $R(s, t)$ implies $R(s\sigma, t\sigma)$ for any substitution σ (Definition 6.3.7). By definition, R is stable iff R is closed under instantiation.

By definition, $=$ is a congruence as it is an equivalence relation closed under instantiation and monotonicity. Given a set A of clauses which contain all the equality axioms, if A is satisfiable, then any model of A induces a congruence as shown by the following proposition.

Proposition 7.1.3 *Let H be a Herbrand model of A which is in CNF and contains all the equality axioms. Define $s =_H t$ iff $(s = t) \in H$ for every $s, t \in T(F)$, the set of ground terms built from F. Then, $=_H$ is a congruence over $T(F)$.*

Proof The first three equality axioms ensure that $=_H$ is an equivalence relation, and the monotonicity axioms ensure that $=_H$ is monotonic. $=_H$ is closed under instantiation because H does not contain variables. □

Since a congruence relation $=_m$ is an equivalence relation, we may partition $T(F)$ by $=_m$ into equivalence classes, also called *congruence classes*, such that each class contains equivalent objects. For every $s \in T(F)$, let

$$[s] = \{t \mid s =_m t, t \in T(F)\},$$

denote the *congruence class* of s. Let

$$T(F)/=_m \ = \{[s] \mid s \in T(F)\},$$

the set of all congruence classes, which is also called the *Herbrand base modulo* $=_m$.

Example 7.1.4 Let $F = \{0/0, s/1\}$, and A contains $s(s(s(x))) = x$ and the equality axioms. Let H be a Herbrand model of A in which $(0 = s(0)) \notin H$, then

$$T(F) = \{0, s(0), s(s(0)), s(s(s(0))), s^4(0), \ldots\}$$
$$[0] = \{0, s(s(s(0))), s^6(0), s^9(0), \ldots\},$$
$$[s(0)] = \{s(0), s^4(0), s^7(0), s^{10}(0), \ldots\},$$
$$[s(s(0))] = \{s(s(0)), s^5(0), s^8(0), s^{11}(0), \ldots\},$$
$$T(F)/=_H \ = \{[0], [s(0)], [s(s(0))]\}$$

In this example, the Herbrand base modulo $=_H$, $T(F)/=_H$, happens to be finite. □

7.1 Equality of Terms

Now, it is easy to check if two ground terms s and t are equal or not under H: $(s = t) \in H$ iff $t \in [s]$ (or $s \in [t]$), i.e., $s =_H t$. In fact, when we define a Herbrand model for a formula with equality, at first we can define a congruence $=_m$ and then regard $T(F)/=_m$ as the domain of the Herbrand model with equality: $(s = t) \in H$ iff $s =_m t$. This way, we do not have to concern about the equality axioms explicitly.

Example 7.1.5 Let $F = \{0/0, s/1, -/1\}$, and A contains $-(0) = 0$, $-(-(x)) = x$, $s(-(s(x))) = -x$, and the equality axioms:

$$T(F) = \{0, s(0), -(0), s(s(0)), s(-(0)), -(-(0)), s(s(s(0))), \ldots\}$$
$$[0] = \{0, -(0), -(-(0)), -(-(-(0))), s(-(s(0))), -^4(0), -^5(0), \ldots\},$$
$$[s(0)] = \{s(0), s(-(0)), s(-(-(0))), -(-(s(0))), s(s(-(s(0)))), \ldots\},$$
$$[-(s(0))] = \{-(s(0)), -(s(-(0))), s(-(s(s(0)))),$$
$$-(-(-(s(0)))), -(s(-(-(0)))), \ldots\},$$
$$[s(s(0))] = \{s(s(0)), s(s(-(0))), s(s(-(-(0)))), s(s(s(-(s(0))))), \ldots\}$$
$$T(F)/=_A = \{[0], [s(0)], [-(s(0))], [s(s(0))], [-(s(s(0)))], \ldots\}$$

In this example, each congruence class $[t]$ is infinite, and the collection of all congruence classes, i.e., $T(F)/=_A$, is infinite, too. A natural model of A is the set $\mathcal{Z}$ of integers, where a nonnegative integer n is represented by $s^n(0)$ and a negative integer $-n$ by $-(s^n(0))$ for $n > 0$. The meaning of s is "to add one" and $-$ is the minus sign. We may use $T(F)/=_A$ as the Herbrand base to define a Herbrand model. However, it would be cumbersome to define usual operations like $+$, $-$, $*$, etc., in this setting. An alternative solution is to use $\{0, s, p\}$ with $\{s(p(x)) = x, p(s(x)) = x\}$ as the building blocks for the set of integers: negative integer $-n$ is represented by $p^n(0)$ and positive integer n by $s^n(0)$. □

7.1.3 Theory of Equations

A set of equations is simply a set of positive unit clauses where the only predicate symbol is "=".

Example 7.1.6 In modern algebra, a *group* is $G = (S, *)$, where S is a set of elements and $*$ is a binary operator satisfying the properties that *(a)* $*$ is associative and closed on S; *(b)* there exists an identity element in S; and *(c)* every element of S has an inverse in S. Using the first-order logic approach, the closure property is implicit; the identity is denoted by e; for each $x \in S$, the inverse of x is denoted by $i(x)$. We just need the following three equations as axioms:

(1) $e * x = x$ e is the left identity element
(2) $i(x) * x = e$ $i(x)$ is the left inverse of x
(3) $(x * y) * z = x * (y * z)$ $*$ is associative

From these three equations, we may prove many interesting properties of the group theory, such as $i(i(x)) = x$, $x * e = x$, $x * i(x) = e$, etc. □

Definition 7.1.7 Given a set E of equations, the relation $=_E$ is recursively defined as follows:

1. $s =_E t$ if $s = t \in E$
2. $t =_E t$ if $t \in T(F, X)$
3. $s =_E t$ if $t =_E s$
4. $s =_E t$ if $s =_E r, r =_E t$
5. $f(\ldots, s_i, \ldots) =_E f(\ldots, t_i, \ldots)$ if $s_i =_E t_i$ and $f \in F$
6. $s\sigma =_E t\sigma$ if $s =_E t$ and $\sigma : X \mapsto T(F, X)$ is any substitution

The *congruence* problem of firt-order logic with equality is to decide if $s =_E t$ or not for any terms s and t.

Recall that every binary relation R, including $=_E$, can be regarded as a set of pairs of items, i.e., $\{(s, t) \mid R(s, t)\}$. Similarly, both E and $=_E$ can be also regarded as a set of pairs of terms and $E \subseteq =_E$.

Definition 7.1.8 Given a binary relation R over $T(F, X)$, R is said to be *closed* under the following:

- *Reflexivity* if $(t, t) \in R$ for any $t \in T(F, X)$
- *Commutativity* if $(t, s) \in R$ whenever $(s, t) \in R$
- *Transitivity* if $(r, t) \in R$ whenever $(r, s), (s, t) \in R$
- *Monotonicity* if $(f(s_1, \ldots, s_k), f(t_1, \ldots, t_k)) \in R$ whenever $(s_i, t_i) \in R$ for $1 \leq i \leq k$
- *Instantiation* if $(s\sigma, t\sigma) \in R$ whenever $(s, t) \in R$ for any $\sigma : X \mapsto T(F, X)$

The *transitive closure* is the minimum superset of R that is closed under transitivity; the *equivalence closure* is the minimum superset of R that is closed under reflexivity, commutativity, and transitivity; the *congruence closure* is the minimum equivalence closure of R that is also closed under monotonicity and instantiation.

Note that monotonicity and instantiation are two different concepts. If $(f(a, x), f(b, x)) \in R$, monotonicity adds $(g(f(a, x)), g(f(b, x)))$ into R and instantiation adds $(f(a, g(x)), f(b, g(x)))$ into R.

Proposition 7.1.9 *For any set E of equations, $=_E$ is the minimum congruence containing E.*

Proof Let us check the conditions for $s =_E t$ to be true: condition 1 says $E \subset =_E$; conditions 2–4 say $=_E$ is an equivalence relation. Conditions 5 and 6 say $=_E$ is closed under monotonicity and instantiation. □

7.2 Rewrite Systems

Example 7.1.10 Using F from Example 7.1.4, let $E = \{s(s(s(x))) = x\}$ and $X = \{x\}$. Since $=_E$ is a congruence, we can check that

$$T(F)/=_E \ = \ T(F)/=_H \ = \ \{[0], [s(0)], [s(s(0))]\},$$

where $[0] = \{0, s(s(s(0))), s^6(x), \ldots\}$, $[s(0)] = \{s(0), s^4(x), s^7(x), \ldots\}$, and $[s(s(0))] = \{s(s(0)), s^5(0), s^8(0), \ldots\}$.

$$T(F, X)/=_E \ = \ \{[0], [s(0)], [s(s(0))], [x], [s(x)], [s(s(x))]\},$$

where $[x] = \{x, s(s(s(x))), s^6(x), \ldots\}$, $[s(x)] = \{s(x), s^4(x), s^7(x), \ldots\}$, and $[s(s(x))] = \{s(s(x)), s^5(x), s^8(x), \ldots\}$. □

Proposition 7.1.11 *Given E, let E_{ax} be the equality axioms associated with E, then for every $(s, t) \in =_E$, $E \cup E_{ax} \models (s = t)$.*

Proof (Sketch) Check all the six conditions of Definition 7.1.7 for $s =_E t$: condition 1 holds because $A \models A$. Conditions 2–5 hold because of E_{ax}. Condition 6 holds because the free variables are assumed to be universally quantified. □

Since $=_E$ contains all the equations that are logical consequence of E and the equality axioms, we call $=_E$ the *theory* of E.

Example 7.1.12 Let E be the three equations in Example 7.1.6, we may show that $x * e =_E x$, $x * i(x) =_E e$, $i(e) =_E e$, $i(i(x)) =_E x$, $i(x * y) =_E i(y) * i(x)$, etc. A proof will be provided in the next section. □

Given any two terms, say s and t, does $s =_E t$? This is an important decision problem with many applications in mathematics. In computation theory, all computable functions can be constructed by equations. Unfortunately, it is undecidable in general to answer $(s, t) \in =_E$; in other words, we do not have an algorithm which takes E, s, and t as input and returns yes if $s =_E t$ (see Chap. 11).

In computer science, the *congruence closure problem* refers to the problem of deciding $s =_E t$, when E is a set of ground equations. This is a decidable problem and there exist efficient algorithms. There are a number of applications using congruence closures. The detailed discussion on this topic can be found in Chap. 12.

7.2 Rewrite Systems

In Example 7.1.1, we showed how to prove $f(f(a)) = a$ from $f(a) = b$ and $f(b) = a$, using resolution and the equality axioms. A straightforward proof is to replace $f(a)$ in $f(f(a)) = a$ by b to get $f(b) = a$ and then replace $f(b)$ by a to get $a = a$. In this simple example, we treat $f(a) = b$ and $f(b) = a$ as two rewrite rules, and each replacement is one application of a rewrite rule, called *rewriting*.

To replace "equal" by "equal" is a common practice in mathematics and logic. For example, we used equivalence relations as rewrite rules to simplify formulas to normal forms. Rewrite rules are not only a common inference system in everyday mathematics but also a very efficient computational technique, as they can involve very little search. We will devote this section to a brief study of rewrite systems, including a discussion of some interesting theoretical results [2].

7.2.1 Rewrite Rules

From now on, we will write the predicate symbol "=" as "$\doteq$", so that "=" is no longer a predicate symbol of first-order logic. Hence, equation $s = t$ is written as $s \doteq t$. However, $\neg(s \doteq t)$ will be written as $s \neq t$. Recall that a *position* of a term is a sequence of positive integers to locate the subterm at that position (Definition 5.1.4).

Definition 7.2.1 A *rewrite rule* is an ordered pair of terms (l, r), often denoted by $l \to r$, where l is the left side and r is the right side of the rule. A term t is *rewritten* to t' by $l \to r$, denoted by $t \Rightarrow t'$, if there exists a substitution σ and a position p of t such that $l\sigma = t/p$ and $t' = t[p \leftarrow r\sigma]$, where t/p denotes the subterm of t at position p.

Mathematically, both an equation and a rewrite rule are a pair of terms. They are often stored as a pair of terms in a computer.

To rewrite t by $l \to r$, we have to find a subterm of t at position p and a substitution σ such that $t/p = l\sigma$. In this case, we say l matches t/p and σ is the *matching*. Matching is one-way unification, and finding matching is easier than finding a unifier as the variables in t cannot be affected by σ. If we treat the variables of t as constants, then finding a matching is the same as finding a unifier of t/p and l.

Once the matching is found, we need to replace the matched subterm in t by (an instance of) the right side of the rewrite rule. Note that $t[p \leftarrow s]$ denote the term where the subterm at position p, t/p, (ref. Definition 5.1.4) is replaced by s. For example, if $g(x) \to h(x)$ is a rewrite rule, then term $t = f(a, g(b))$ can be rewritten to $t' = f(a, h(b))$, because $t/1 = g(b) = g(x)\sigma$ and $t'/1 = h(x)\sigma = h(b)$, where $\sigma = [x \leftarrow b]$.

Definition 7.2.2 A rewrite system R is a set of rewrite rules. A term t is rewritten to t' by R, denoted by $t \Rightarrow_R t'$, or simply $t \Rightarrow t'$, if there exists a rewrite rule in R which rewrites t to t'.

By convention, we will use $\Rightarrow_R^+$ and $\Rightarrow_R^*$ to denote the transitive closure and the reflexive and transitive closure of $\Rightarrow$, respectively. That is, $s \Rightarrow_R^+ t$ means that s is rewritten to t by R using one or more rewritings; $s \Rightarrow_R^* t$ means that s is rewritten to t by R using 0 or more rewritings.

7.2 Rewrite Systems

While the major use of R is to represent a congruence relation, R can be used for representing any transitive, stable, and monotonic relation. This might be inequality (like $>$ or $\geq$) or implication.

Example 7.2.3 In Chap. 1 (Sect. 1.3.2), we have seen that the functions can be constructed by equations. For instance, we used equations to define the predecessor, addition, subtraction, and multiplication functions over the natural numbers. We may treat these equations as rewrite rules:

$$pre(0) \rightarrow 0$$
$$pre(s(x)) \rightarrow x$$
$$add(0, y) \rightarrow y$$
$$add(s(x), y) \rightarrow s(add(x, y))$$
$$sub(x, 0) \rightarrow x$$
$$sub(x, s(y)) \rightarrow sub(pre(x), y)$$
$$mul(0, y) \rightarrow 0$$
$$mul(s(x), y) \rightarrow add(mul(x, y), y)$$
$$exp(x, 0) \rightarrow s(0)$$
$$exp(x, s(y)) \rightarrow mul(x, exp(x, y))$$

To compute $2 \times 2 = 4$, we rewrite $mul(s(s(0)), s(s(0)))$ to $s(s(s(s(0))))$. We give below also the positions for each rewriting, and you may find out which rule is used for each rewriting:

$$mul(s(s(0)), s(s(0))) \quad \text{at } \epsilon$$
$$\Rightarrow add(mul(s(0), s(s(0))), s(s(0))) \quad \text{at } 1$$
$$\Rightarrow add(add(mul(0, s(s(0))), s(s(0))), s(s(0))) \quad \text{at } 1.1$$
$$\Rightarrow add(add(0, s(s(0))), s(s(0))) \quad \text{at } 1$$
$$\Rightarrow add(s(s(0)), s(s(0))) \quad \text{at } \epsilon$$
$$\Rightarrow s(add(s(0), s(s(0)))) \quad \text{at } 1$$
$$\Rightarrow s(s(add(0, s(s(0))))) \quad \text{at } 1.1$$
$$\Rightarrow s(s(s(s(0))))$$

In this language, if a term is ground, it is not hard to prove by induction that the rewriting process will eventually terminate with a number, i.e., a term containing only 0 and s. Using this rewrite system, we can do all the addition, subtraction, and multiplication over the natural numbers by rewriting, if we do not care about the speed. □

7.2.2 Termination of Rewriting

A nice property of many rewrite systems, including the previous example, is that the application of rules to terms, ground or not, cannot go on forever; it will eventually terminate.

Definition 7.2.4 A rewrite system R is said to be *terminating* if there exists no infinite sequences of terms $t_0, t_1, \ldots, t_i, \ldots$ such that $t_i \Rightarrow t_{i+1}$.

To prove that a rewrite system R is terminating, we may use a simplification order $\succ$ (ref. Sect. 6.3.2).

Proposition 7.2.5 *Let $\succ$ be a simplification order and R a rewrite system. If $l \succ r$ for all every $l \to r$ in R, then R is terminating.*

Proof By definition, $\succ$ is a well-founded, stable, and monotonic order on terms. If $t_i \Rightarrow t_{i+1}$ by $l \to r$, then $t_i \succ t_{i+1}$, because $l \succ r$, and $\succ$ is stable and monotonic. If there exists an infinite sequence, $t_0, t_1, \ldots, t_i, \ldots$ such that $t_i \succ t_{i+1}$, then $\succ$ cannot be well-founded. □

The above proposition provides a sufficient condition for the termination of a rewrite system. In Sect. 6.3.2, we discussed lexicographic path order (LPO), recursive path order (RPO), and Knuth-Bendix order (KBO). These orders are simplification orders and can be used to prove the termination of R.

Example 7.2.6 To show the termination of R in Example 7.2.3, we may use the lexicographic path order $\succ_{lpo}$ with the precedence $s \prec pre \prec add \prec sub \prec mul$ and every binary operator has the left-to-right status, with the exception of sub: sub must have the right-to-left status (Definition 6.3.16), as we want $[x, s(y)] \succ^{lex} [pre(x), y]$. □

In general, how can we show that the rewriting process using R will terminate? Unfortunately, this is an undecidable problem. That is, we do not have an algorithm to answer this question for general R.

Definition 7.2.7 Given a rewrite system R, if $s \Rightarrow_R^* t$ and there exists no t' such that $t \Rightarrow_R t'$, we say t is a *normal form* of s in R.

Why is termination an important property of a rewrite system? The termination of R ensures that there exists a normal form for every term. Without termination, we will have hard time to control rewriting. In computation theory, termination is a divider for telling whether a decision problem is decidable or not. Some equations, such as the commutativity of $+$ or $\vee$, can never be made into a terminating rule. It is undecidable to tell if a rewrite system is terminating or not.

7.2.3 Confluence of Rewriting

Termination is one important property of a set of rewrite rules; the other important property is confluence.

Example 7.2.8 In Example 7.1.6, we used three equations to specify a group $(S, *)$. These equations can be made into a rewrite system R:

$$
\begin{aligned}
&(1) & e * x &\to x \\
&(2) & i(x) * x &\to e \\
&(3) & (x * y) * z &\to x * (y * z)
\end{aligned}
$$

There are two ways for R to rewrite $(i(x) * x) * y$:

$$\underline{(i(x) * x) * y} \Rightarrow \begin{cases} (3) \ i(x) * (x * y) \\ (2) \ \underline{e * y} \Rightarrow (1) \ y \end{cases}$$

That is, there is more than one way to rewrite a term. For $(i(x) * x) * y$, we have two normal forms: $i(x) * (x * y)$ and y. □

Definition 7.2.9 A rewrite system R is said to be *confluent* if, for any term s, if $s \Rightarrow_R^* t_1$ and $s \Rightarrow_R^* t_2$, then there exists a term t' such that $t_1 \Rightarrow_R^* t'$ and $t_2 \Rightarrow_R^* t'$. R is said to be *canonical* if R is both terminating and confluent.

Figure 7.1 illustrates the concepts of "terminating," "confluent," and "canonical." That is, if R is confluent, it does not matter how you go about applying the rules to a term, you will get the same result. There are several concepts of "confluence" proposed in literature. They are equivalent for terminating rewrite systems. The termination of R ensures the existence of a normal form for every term; the confluence of R ensures the uniqueness of normal forms. When R is canonical, you do not need to worry about which positions and choose which rule to rewrite a term. Any choice can be made, and the end result will be the same. Let us denote the unique normal form of t in a canonical rewrite system by $t\downarrow_R$, the canonical form of t.

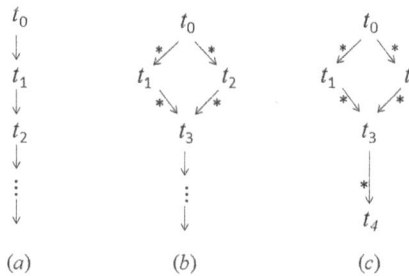

Fig. 7.1 Illustration of (**a**) a nonterminating rewriting sequence. (**b**) confluent: if $t_0 \Rightarrow^* t_1$ and $t_0 \Rightarrow^* t_2$, then there exists t_3 such that $t_1 \Rightarrow^* t_3$ and $t_2 \Rightarrow^* t_3$. (**c**) canonical: if $t_0 \Rightarrow^* t_1$ and $t_0 \Rightarrow^* t_2$, then there exists a unique normal form t_4 such that $t_1 \Rightarrow^* t_4$ and $t_2 \Rightarrow^* t_4$

Definition 7.2.10 A rewrite system R is a decision procedure for a set E of equations if *(a)* R is terminating and *(b)* for any two terms s and t, $s =_E t$ iff $s\downarrow_R = t\downarrow_R$.

We say R is a decision procedure for E, i.e., R is a decision procedure for the congruence problem (Definition 7.1.7), because the computation of normal forms in R is terminating and the equality checking of $s =_E t$ is reduced to the comparison of the normal forms of s and t.

Theorem 7.2.11 *Let R be a canonical rewrite system satisfying the following conditions:*

1. *For each rewrite rule $l \to r$ of R, $l =_E r$.*
2. *For each equation $s_1 = s_2$ of E, $s_1\downarrow_R = s_2\downarrow_R$.*

Then, R is a decision procedure for E.

Proof Since R is canonical, R must be terminating. We just need to show that the given conditions imply that $s =_E t$ iff $s\downarrow_R = t\downarrow_R$ for any two terms s and t.

For each rule $l \to r$ of R, since $l =_E r$, if $s \Rightarrow t$ by $l \to r$, then $s =_E t$ by applying the instantiation and monotonicity rules on $l =_E r$. Applying this result to each rewriting, we have $s =_E s\downarrow_R$ and $t =_E t\downarrow_R$. If $s\downarrow_R = t\downarrow_R$, then $s =_E t$.

If $s =_E t$, check all the six conditions for obtaining $s =_E t$: if $s =_E t$ because $s = t \in E$, then $s \downarrow_R = t \downarrow_R$ by the second condition. If $s =_E t$ because $t =_E s$, then $t \downarrow_R = s \downarrow_R$ implies $s \downarrow_R = t \downarrow_R$. If $s =_E t$ because $s =_E r$ and $r =_E t$, then $s \downarrow_R = r \downarrow_R = t \downarrow_R$. If $s = f(\ldots, s_i, \ldots) =_E f(\ldots, t_i, \ldots) = t$ because $s_i =_E t_i$, then $s_i \downarrow_R = t_i \downarrow_R$ implies $s \downarrow_R = t \downarrow_R$. If $s = s'\sigma =_E t'\sigma = t$ because $s' =_E t'$, then $s' \downarrow_R = t' \downarrow_R$ implies $s \downarrow_R = t \downarrow_R$. In all cases, we have $s \downarrow_R = t \downarrow_R$ if $s =_E t$. □

Even when a rewrite system is canonical, we may still want to exercise carefully our choices of rules and positions in a term to apply rules, in order to minimize the amount of computational effort required to find the canonical form. We may want to explore only the shortest rewriting steps. For instance, choosing a position for rewriting, we may use an outermost-position-first strategy or an innermost-position-first strategy. A careful inspection of the rules to be applied may reveal which strategy works better.

7.2.4 The Knuth-Bendix Completion Procedure

In Example 7.2.8, we have seen that term $t = (i(x) * x) * y$ has two normal forms, i.e., $i(x) * (x * y)$ and y. The rewrite system in the example is not confluent. In 1970, Donald Knuth and Peter Bendix introduced a procedure which takes a set of equations and a simplification order as input and makes rewrite rules from the equations according to the simplification order. So-called *critical pairs* are then computed from the rewrite rules. If all the critical pairs have the same normal form,

7.2 Rewrite Systems

we obtain a canonical rewrite system. If not, such critical pairs are added into the set of equations and continue.

Definition 7.2.12 Given two rewrite rules which share no variables and may differ only by variable names, say $l_1 \to r_1$ and $l_2 \to r_2$, a *critical pair* of the two rules is a pair of terms (s, t), where $s = r_1\sigma, t = l_1[p \leftarrow r_2]\sigma$, p is a non-variable position of l_1, and σ is the mgu of l_1/p and l_2. We say l_2 is *superposed* at p of l_1 to produce the critical pair and this process is called *superposition*.

In Example 7.2.8, we are given three rewrite rules. If we rename the variables in rule 3 as $(x_3 * y_3) * z_3 \to x_3 * (y_3 * z_3)$, we can compute a critical pair from rule 3 and rule 2, by superposing $l_2 = i(x) * x$ at position 1 of l_3, that is, unifying l_2 and $l_3/1 = x_3 * y_3$ with the mgu $\sigma = [x_3 \leftarrow i(x), y_3 \leftarrow x, z_3 \leftarrow y]$ and generate the critical pair $(i(x) * (x * y), e * y)$ from $l_3\sigma = (i(x) * x) * y$ as illustrated in Example 7.2.8.

During the computation of a critical pair, applying the mgu σ to l_1, $l_1\sigma$ can be rewritten to $r_1\sigma$ by $l_1 \to r_1$. At the same time, since $l_1/p\sigma = l_2\sigma$, $l_1\sigma$ contains an instance of l_2 at position p, so $l_1\sigma$ can be also rewritten to $l_1[p \leftarrow l_2]\sigma$ by $l_2 \to r_2$. In other words, the critical pair is obtained by two rewritings on $l_1\sigma$ using the two rules, respectively:

$$l_1\sigma \Rightarrow \begin{cases} r_1\sigma & \text{by } l_1 \to r_1 \\ l_1[p \leftarrow r_2]\sigma & \text{by } l_2 \to r_2 \end{cases}$$

Lemma 7.2.13 *If $l_1 =_E r_1$, $l_2 =_E r_2$, and (s, t) is a critical pair from $l_1 \to r_1$ and $l_2 \to r_2$, then $s =_E t$.*

Proof From $l_1 =_E r_1$, we have $l_1\sigma =_E r_1\sigma$. From $l_2 =_E r_2$ and $l_1/p\sigma = l_2\sigma$, we have $l_1\sigma =_E l_1[p \leftarrow r_2]\sigma$. Hence, $s = r_1\sigma =_E l_1[p \leftarrow r_2]\sigma = t$. □

Example 7.2.14 We will show how to obtain a canonical rewrite system from the three rewrite rules in Example 7.2.8:

(1) $e * x \to x$
(2) $i(x) * x \to e$
(3) $(x * y) * z \to x * (y * z)$

The lexicographical path order is used with the precedence $e \prec * \prec i$. When computing a critical pair of the two rules, if a variable name appears in both rules, we assume that the variable name in one of the rules is implicitly subscribed by the rule number. For example, if both (2) and (3) contain x, then x in (3) is assumed to be x_3 (y and z are not renamed because they do not appear in (2)). If the two rules come from the same rewrite rule, then one rule is a copy of the other rule with all the variables renamed.

The first critical pair is $(i(x) * (x * y), e * y)$, which comes from (2) and (3) and is regarded as an equation $i(x) * (x * y) \doteq e * y$. After rewriting $e * y$ to y, this equation is oriented into the fourth rewrite rule:

$$(4)\ i(x) * (x * y) \to y$$

A critical pair comes from rule (4) (i.e., $i(x_4) * (x_4 * y_4) \to y_4$ is $l'_4 \to r'_4$) and itself (i.e., $i(x) * (x * y) \to y$ is $l_4 \to r_4$), and the mgu of $l'_4/2$ and l_4 is $\sigma = [x_4 \leftarrow i(x), y_4 \leftarrow x * y]$:

$$i(i(x)) * \underline{(i(x) * (x * y))} \Rightarrow \begin{cases} (4)\ x * y \\ (4)\ i(i(x)) * \underline{y} \end{cases}$$

The critical pair is $x * y \doteq i(i(x)) * y$, which produces the fifth rule:

$$(5)\ i(i(x)) * y \to x * y$$

A critical pair comes from rules (4) and (2), and the mgu of $l_4/2$ and l_2 is $\sigma = [x_4 \leftarrow i(x), y_4 \leftarrow x]$:

$$i(i(x)) * \underline{(i(x) * x)} \Rightarrow \begin{cases} (4)\ x \\ (2)\ i(i(x)) * \underline{e} \to (5)\ x * e \end{cases}$$

The critical pair is $x \doteq x * e$, which produces the sixth rule:

$$(6)\ x * e \to x$$

A critical pair comes from rules (6) and (2) with $\sigma = [x_6 \leftarrow i(e)]$ and generates the seventh rule:

$$(7)\ i(e) \to e$$

A critical pair comes from rule (5) and (2) with $\sigma = [x_2 \leftarrow i(x), y \leftarrow i(x)]$:

$$i(i(x)) * i(x) \Rightarrow \begin{cases} (5)\ x * i(x) \\ (2)\ e \end{cases}$$

and generates the eighth rule

$$(8)\ x * i(x) \to e$$

A critical pair comes from rules (6) and (5) with $\sigma = [x_6 \leftarrow i(i(x)), y \leftarrow e]$:

$$i(i(x)) * e \Rightarrow \begin{cases} (6)\ i(i(x)) \\ (5)\ x * e \to (6)\ x \end{cases}$$

7.2 Rewrite Systems

and generates the ninth rule

$$(9) \ i(i(x)) \to x$$

A critical pair comes from rules (4) and (3), and the mgu of $l_4/2$ and l_3 is $\sigma = [x_4 \leftarrow x * y, y_4 \leftarrow z]$:

$$i(x * y) * \underline{((x * y) * z)} \Rightarrow \Big\langle \begin{array}{l} (4) \ z \\ (3) \ i(x * y) * (x * (y * z)) \end{array}$$

and generates the tenth rule

$$(10) \ i(x * y) * (x * (y * z)) \to z$$

A critical pair comes from rules (10) and (8), and the mgu of $l_{10}/2.2$ and l_8 is $\sigma = [x_8 \leftarrow y, z \leftarrow i(y)]$:

$$i(x * y) * (x * \underline{(y * i(y))}) \Rightarrow \Big\langle \begin{array}{l} (10) \ i(y) \\ (8) \ i(x * y) * (x * \underline{e}) \Rightarrow (6) \ i(x * y) * x \end{array}$$

and generates the eleventh rule

$$(11) \ i(x * y) * x \to i(y)$$

A critical pair comes from rules (11) and (4), and the mgu of $l_{11}/1.1$ and l_4 is $\sigma = [x_{11} \leftarrow i(x), y_{11} \leftarrow x * y]$:

$$\underline{i(i(x) * (x * y))} * i(x) \Rightarrow \Big\langle \begin{array}{l} (11) \ i(x * y) \\ (4) \ i(\underline{y}) * i(x) \end{array}$$

and generates the twelfth rule

$$(12) \ i(x * y) \to i(y) * i(x)$$

The final critical pair comes from rules (4) and (5) and generates the thirteenth rule:

$$(13) \ x * (i(x) * y) \to y$$

Since rules (5), (10), and (11) can be rewritten to the identity, i.e., $t = t$, by rules (9) and (12) (and then by other rules), the final set of rewrite rules is

(1)	$e * x \to x$		(7)	$i(e) \to e$
(2)	$i(x) * x \to e$		(8)	$x * i(x) \to e$
(3)	$(x * y) * z \to x * (y * z)$		(9)	$i(i(x)) \to x$
(4)	$i(x) * (x * y) \to y$		(12)	$i(x * y) \to i(y) * i(x)$
(6)	$x * e \to x$		(13)	$x * (i(x) * y) \to y$

All the critical pairs from this set of ten rules can be reduced to the identity by this rewrite system. □

The above example illustrates the execution of the Knuth-Bendix completion procedure, invented by Donald Knuth and Peter Bendix, which is described by the following pseudo-code.

Procedure 7.2.15 The procedure $KnuthBendix(E, \succ)$ takes a set E of equations and a simplification order $\succ$ as input and generates a canonical rewrite system from E when it succeeds. It calls four procedures:

- $pickEquation(E)$ picks an equation from E.
- $NF(t, R)$ returns a normal form of term t by R.
- $rewritable(t, r)$ checks if rewrite rule r can rewrite term t.
- $criticalPairs(r, R)$ returns the set of all critical pairs (in the form of equation) between rule r and every rule in R.

proc $KnuthBendix(E, \succ)$
1   $R := \emptyset$
2   **while** $(E \neq \emptyset)$ **do**
3       $(s \doteq t) := pickEquation(E); E := E - \{(s \doteq t)\}$
4       $s := NF(s, R); t := NF(t, R)$ // normalize s and t by R.
5   **if** $(s = t)$ **continue** // identity is discarded.
6   **else if** $(s \succ t)$ $r := (s \to t)$
7   **else if** $(t \succ s)$ $r := (t \to s)$
8   **else return** "failure" // s and t are not comparable by $\succ$.
9       $R := R \cup \{r\}$ // add new rule into R
10   **for** $(d \to e) \in R - \{r\}$ **do** // inter-reduction
11   **if** $rewritable(d, r)$ $R := R - \{d \to e\}; E := E \cup \{d \doteq e\}$
12   **if** $rewritable(e, r)$ $R := R - \{d \to e\} \cup \{(d \to NF(e, R))\}$
13      $E := E \cup criticalPairs(r, R)$
14   **return** R

Example 7.2.14 is a successful example of the above procedure which takes three equations and produces a rewrite system of ten rules. This rewrite system is sufficient to solve the *word problem* for a free group with no other properties. The *word problem* in abstract algebra is to find a decision procedure to decide whether

7.2 Rewrite Systems

any two terms are equal according to the axioms of group theory. The rewrite system returned by the above procedure can serve as a decision procedure by checking if the two terms have the same canonical form.

The Knuth-Bendix completion is not an algorithm because it may go on forever due to an ever-growing set of rules and critical pairs. Note that the procedure may fail if an equation cannot be oriented into a rewrite rule. There are many attempts to reduce the cases of failures. For instance, if an equation like $f(x, g(x)) \doteq h(y, g(y))$ cannot be oriented, we may introduce a new constant c to make two equations $f(x, g(x)) \doteq c$ and $h(y, g(y)) \doteq c$. In general, if we have an equation $s(x, z) \doteq t(y, z)$, where x does not appear in t and y does not appear in s, we may introduce a new function $h(z)$ to create two new equations: $s(x, z) \doteq h(z)$ and $t(y, z) \doteq h(z)$. For the axioms $AC = \{(x * y) * y \doteq x * (y * z), x * y \doteq y * x\}$, we may consider the rewriting over $T(F, X)/{=}_{AC}$, the so-called rewriting modulo AC (associativity and commutativity).

Definition 7.2.16 A rewrite system R is said to be *inter-reduced* if for any rule $l \to r \in R$, l is not rewritable by $R - \{l \to r\}$; R is *reduced* if both l and r are not rewritable by $R - \{l \to r\}$.

Lemma 7.2.17 *If KnuthBendix($E, \succ$) returns R, then R is reduced.*

Proof Line 11 (resp. 12) ensures the left (resp. right) side of each rule in R is not rewritable by any other rule. □

When *KnuthBendix* terminates successfully, we obtain a reduced canonical rewrite system.

Theorem 7.2.18 *If KnuthBendix($E, \succ$) returns R, then R is reduced and canonical.*

Proof R is reduced by Lemma 7.2.17. For canonicality, at first, R is terminating, because for every $l \to r$ of R, $l \succ r$, where $\succ$ is a simplification order. Secondly, all the critical pairs of R are computed, and each pair is rewritten to the same term by R (if not, a new rule will be added into R).

To show that R is confluent, we do induction on t_0 based on $\succ$. As a base case, if t_0 is not rewritable by R, then R is confluent trivially at t. Suppose, for $i = 1, 2$, $t_0 \Rightarrow t_i$ by $l_i \to r_i \in R$ at position p_i of t, i.e., $t_i = t[p_i \leftarrow r_i \sigma_i]$.

Case 1: If p_1 and p_2 are disjoint positions, then both t_1 and t_2 can be written to a common term $t_3 = t_0[p_1 \leftarrow r_1\sigma_1, p_2 \leftarrow r_2\sigma_2]$.

Case 2: If p_1 and p_2 are not disjoint positions, without loss of generality, assume $p_1 = \epsilon$, then $t_0/p_1 = t_0 = l_1\sigma_1$, and $t_0/p_2 = (l_1\sigma_1)/p_2 = l_2\sigma_2$. There are two subcases to consider:

Case 2.1: If p_2 is a non-variable position of l_1, then $(l_1\sigma_1)/p_2 = (l_1/p_2)\sigma_1 = l_2\sigma_2$, that is, l_1/p_2 and l_2 are unifiable. Thus, there exists a critical pair (s_1, s_2) between the two rules such that $t_1 = s_1\theta$ and $t_2 = s_2\theta$ for some substitution θ. Since all critical pairs are rewritten to the same term by R, so is (s_1, s_2), as well as (t_1, t_2).

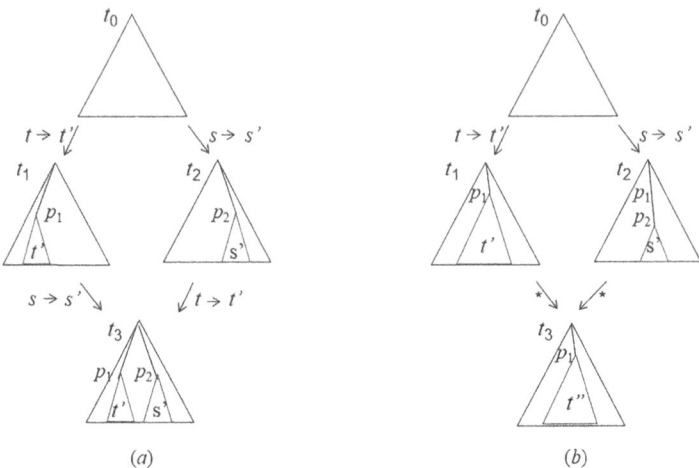

Fig. 7.2 Illustration of the proof of Theorem 7.2.18: (**a**) two disjoint positions; (**b**) two overlapping positions

Case 2.2: When p_2 is not a non-variable position of l_1, that is, either p_2 or a prefix of p_2 is a variable position of l_1. A sketch of the proof goes as follows: we may find a common successor of t_1 and t_2 by first applying $l_2 \to r_2$ to t_2 $n-1$ times if the variable at that position of l_1 occurs n times and then apply $l_1 \to r_1$ once. An example showing this case is given after the proof.

By induction hypotheses, R is confluent at t_1 and t_2, respectively, because $t_0 \succ t_1$ and $t_0 \succ t_2$. Now, we have shown that t_1 and t_2 are rewritten to a common term in all cases; R must be confluent at t_0.

The two cases in this proof are illustrated in Fig. 7.2, where $t \to t'$ is $l_1 \to r_1$ and $s \to s'$ is $l_2 \to r_2$.

□

Example 7.2.19 To illustrate case 2.2 in the above proof, let $t = i(i(i(a))) * i(i(a))$ and R be the ten rules in Example 7.2.14. Then, t can be rewritten to e by rule 2 and to $i(a) * i(i(a))$ by rule 9 at position 1.1. The left side of rule 2 is $i(x) * x$, where x occurs twice ($n = 2$).

To obtain a common term, we apply rule 9 one more time ($n - 1 = 1$) to $i(a) * i(i(a))$ at position 2 to obtain $i(a) * a$, which can be rewritten to e by rule 2.

$$i(i(i(a))) * i(i(a)) \Rightarrow \begin{cases} (2) \ e \\ (9) \ i(a) * i(i(a)) \ \to \ (9) \ i(a) * a \ \to \ (2) \ e \end{cases}$$

Note that both 1.1 and 2 of t are variable positions of $i(x) * x$, the left side of rule 2.

□

Theorem 7.2.20 *If KnuthBendix($E, \succ$) returns R, then R is a decision procedure for E.*

7.2 Rewrite Systems

Proof (Sketch) The previous theorem tells us that R is canonical. In light of Theorem 7.2.11, condition 1 is ensured by Lemma 7.2.13; condition 2 holds because every equation of E is either made into a rule or simplified to identity. □

In conclusion, the desirable properties of rewrite systems are termination and confluence. Rewrite systems can be shown to terminate with the aid of simplification orders and be confluent by looking at critical pairs formed from pairs of rules. In Chap. 11, we will see that it is undecidable to answer if a rewrite system is terminating or not in general. In practice, the Knuth-Bendix completion often runs forever or stops with failure for many inputs.

7.2.5 Special Rewrite Systems

Rewrite systems can be used to put terms in normal form, to prove terms equal or equivalent, or to serve as a decision procedure for some transitive monotonic relations. Their application represents a powerful method of mathematical reasoning, because they can sometimes overcome the combinatorial explosions caused by other proof procedures, e.g., resolution. In the following, we will briefly introduce several special cases of rewrite systems.

Rewrite Systems for Ground Equations

A *ground equation* is a pair of ground terms, i.e., no variables appear in the equation. The Knuth-Bendix completion can be greatly simplified on ground equations because of the following:

- There exists a total simplification order for ground terms. This can be achieved by defining a total precedence over the function symbols and then applying one of the known simplification orderings, such as the Knuth-Bendix order (KBO), to terms.
- The critical pair computation is not needed if the rewrite system is inter-reduced. That is, if there exists a critical pair between two rewrite rules, then one rule can rewrite the left side of the other rule.

Example 7.2.21 Let $f^3 a$ and $f^5 a$ denote $f(f(f(a)))$ and $f(f(f(f(f(a)))))$, respectively. Feeding $E = \{f^5 a \doteq a, f^3 a \doteq a\}$ to the Knuth-Bendix completion, the rewrite rules made from E should be $\{(1)\ f^5 a \to a,\ (2)\ f^3 a \to a\}$. Since $f^5 a = f^2 f^3 a$, the critical pair from (1) and (2) is $f^2 a \doteq a$, and this is the same as we rewrite (1) to $f^2 a \doteq a$ by (2). In fact, rewriting happens first (line 11) in the Knuth-Bendix procedure. From $f^2 a \doteq a$, we have (3) $f^2 a \to a$, which rewrites (2) to $f^1 a \doteq a$. The Knuth-Bendix completion halts with $\{f^1 a \to a\}$. □

Theorem 7.2.22 *For any set E of ground equations, there exists a canonical rewrite system R serving as a decision procedure of E.*

We leave the proof of this theorem as an exercise.

String Rewrite Systems

In theoretical computer science and mathematical logic, a *string rewrite system* (SRS), historically called a *semi-Thue system*, is a rewrite system over strings from an alphabet.

Example 7.2.23 $R = \{(1)\ ba \rightarrow ab,\ (2)\ ca \rightarrow ac,\ (3)\ cb \rightarrow bc\}$ is an SRS over the alphabet $\Sigma = \{a, b, c\}$. We can rewrite cba by R to abc by the following sequence of rewriting:

$$cba \Rightarrow (1)\ cab \Rightarrow (2)\ acb \Rightarrow (3)\ abc$$

We may also apply (3) first to cba and achieve the same result. □

Formally, an SRS is defined as $P = (\Sigma, R)$, where Σ is the alphabet and R is a finite subset of $\Sigma^* \times \Sigma^*$. Each member (s, t) of R is called *rewrite rule* and written as $s \rightarrow t$.

A term rewrite system (TRS) is more expressive than SRS as an SRS is easily converted into a TRS, where each symbol s in the SRS is replaced by a unary function $s/1$ in the TRS. For example, rewrite rule $ba \rightarrow ab$ in R of the above example is replaced by $b(a(x))) \rightarrow a(b(x))$. All the important concepts of TRS, such as termination and confluence, can be carried over to SRS. For instance, it is easy to show that R in the above example is a canonical rewrite system.

For TRS, a rewrite system defines a transitive, monotonic, and stable binary relation over terms. For SRS, a rewrite system defines a transitive and monotonic binary relation over Σ^*. That is, given a set R of rewirte rules, which is binary relation $\rightarrow$ between fixed strings over Σ^*, an SRS extends the rewriting relation to all strings in which the left and right sides of the rules appear as substrings, that is, $usv \Rightarrow utv$ from $s \rightarrow t$, where s, t, u, and v are strings of Σ^*. If we consider the symmetric closure of $\Rightarrow$, then R is called *Thue system*, whose name comes from the Norwegian mathematician Axel Thue, who introduced systematic treatment of SRS in 1914.

Rewriting with Polynomial Equations

Every reader of this book has seen polynomial equations in middle school algebra. For example, the following is a system of polynomial equations:

$$(1)\quad x^3 + 4x^2 - 4xz - 8 = 0$$
$$(2)\ x^2y + 4xy - 4yz + 4 = 0$$

We will use this example to illustrate all the concepts about rewriting with polynomials.

In abstract algebra, a *field* is a set of elements closed under arithmetic operations $+$, $-$, $*$, and $/$. The set of rational numbers is the best-known field. Given a set

7.2 Rewrite Systems

of n variables, say $x_1, \ldots, x_n$, a *monomial* is a product $cx_1^{a_1} x_2^{a_2} \cdots x_n^{a_n}$, where $a_i \in \mathcal{N}$ and c, called *coefficient*, is an element of a given field. This monomial can be represented by the list $[c, a_1, a_2, \ldots, a_n]$. A *polynomial* is a sum of one or more monomials. A *system of polynomial equations*, or simply a *polynomial system*, is a set of simultaneous equations $f_1 = 0, \ldots, f_m = 0$, where each f_i, $1 \leq i \leq m$, is a polynomial. Polynomial systems are usually classified by the number of variables and the maximal number of variables in any monomial. For instance, *linear programming* considers only polynomial systems with at most one variable in each monomial.

Given two monomials represented by $M = [c_1, a_1, a_2, \ldots, a_n]$ and $N = [c_2, b_1, b_2, \ldots, b_n]$, the multiplication of M and N is $MN = [c_1 c_2, a_1 + b_1, a_2 + b_2, \ldots, a_n + b_n]$. Two monomials differing only by coefficients can be merged into one by the addition of the coefficients.

If we define a total order $\succ$ on the variables appearing in a polynomial system, this order can be extended in several ways to a total order over monomials. Without loss of generality, let us assume $x_1 \succ x_2 \succ \cdots \succ x_n$. Suppose $M = [c_1, a_1, a_2, \ldots, a_n]$ and $N = [c_2, b_1, b_2, \ldots, b_n]$, where $c_i \neq 0$. Define $M \succ N$ by the *lexicographical order* on lists of natural numbers: $M \succ N$ iff $[a_1, a_2, \ldots, a_n] \succ^{lex} [b_1, b_2, \ldots, b_n]$. It is easy to show that $\succ$ is a well-order over monomials (Proposition 6.3.5). Suppose $x \succ y \succ z$ in the above example, then the monomials in each polynomial are sorted by $\succ$ from left to right

Given a total ordering over monomials, a polynomial equation can be viewed as a rewrite rule whose left side is the maximal monomial and whose right side contains the negation of the rest monomials. We will display rewrite rules as before and the two rewrite rules in our example look like the following:

$$(1) \quad x^3 \rightarrow -4x^2 + 4xz + 8$$
$$(2) \quad x^2 y \rightarrow -4xy + 4yz - 4$$

Here, the variables are treated as constants of first-order logic and the coefficient of the left side of any rewrite rule is always simplified to one. We can use these rewrite rules to simplify polynomials and compute *critical pairs* by superposition.

Given two monomials $M = [1, a_1, a_2, \ldots, a_n]$ and $N = [1, b_1, b_2, \ldots, b_n]$, let the *least common multiple* (lcm) of M and N be

$$lcm(M, N) = [1, max(a_1, b_1), max(a_2, b_2), \ldots, max(a_n, b_n)].$$

To compute a critical pair, we consider two rewrite rules of which the left sides share a variable and rewrite the lcm of the left sides by the two rules, respectively. For our example, the lcm from (1) and (2) is $lcm(x^3, x^2 y) = x^3 y$:

$$x^3 y \Rightarrow \begin{cases} (1) \ (-4x^2 + 4xz + 8)y = -4x^2 y + 4xyz + 8y \\ (2) \ x(-4xy + 4yz - 4) = -4x^2 y + 4xyz - 4x \end{cases}$$

Thus, the critical pair of (1) and (2) is the equation:

$$-4x^2y + 4xyz + 8y = -4x^2y + 4xyz - 4x$$

After the simplification, we obtain $x + 2y = 0$, which generates a new rule:

$$(3) \quad x \rightarrow -2y$$

Formally, let $M \rightarrow P$ and $N \rightarrow Q$ denote two polynomial rewrite rules, M and N share a variable, and $lcm(M, N) = MM' = N'N$ for some monomials M' and N', the *critical pair* of these two rules is the equation $PM' = N'Q$. As before, this process is called *superposition*. For the above example, $M = x^3$, $N = x^2y$, $M' = y$, and $N' = x$.

Now, we use (3) to simplify (1) and obtain a new rule:

$$(4) \quad y^3 \rightarrow 2y^2 + yz - 1$$

Using (3) and (4), (2) is simplified to *true*, and we obtain the following set of rewrite rules:

$$\{ (3) \ x \rightarrow -2y, \quad (4) \ y^3 \rightarrow 2y^2 + yz - 1 \}$$

This is a canonical rewrite system for the original polynomial system.

The above description illustrates an extension of the Knuth-Bendix completion (Procedure 7.2.15) for polynomial equations. In theory, since polynomial variables are considered as constants by the completion procedure, it will terminate with a canonical rewrite system as in the case of ground equations. In practice, the completion procedure may run out of memory before termination.

In literature, a canonical rewrite system for a polynomial system is called *Gröbner basis* and the procedure for obtaining it is called *Buchberger's algorithm*, which was introduced by Bruno Buchberger in 1965. Buchberger's algorithm can be seen as a generalization of both Gaussian elimination for linear programming and Euclid's algorithm for computing polynomial greatest common divisors. Rewriting as well as superposition is a generalization of both row reduction occurring in Gaussian elimination and Euclidean division of univariate polynomials. Over the years, Gröbner bases have been used successfully in many areas of science and engineering, including algebraic geometry, coding theory, cryptography, invariant theory, integer programming, graph theory, and computer algebra.

Wenjun Wu's method for polynomial systems is based on the concept of *characteristic set*, introduced in the late 1940s by Joseph Ritt. Fully independent of Buchberger's work, Wu's method uses division for converting a polynomial system into a triangular form, called *Ritt-Wu characteristic set*. It is easy to describe Wu's method as an extension of the Knuth-Bendix completion as illustrated above, because division can be implemented by rewriting and superposition. A canonical rewrite system from polynomial equations can serve as a Ritt-Wu characteristic set for many problems in geometry. In addition to automated proofs in geometry, Wu's method has been successfully applied in various scientific fields, such as biology, computer vision, and robot kinematics.

7.3 Inductive Theorem Proving

In Example 7.2.3, we converted a set E of equations into rewrite rules, where add is defined using two equations: $E = \{add(0, y) \doteq y, add(s(x), y) \doteq s(add(x, y))\}$. It is easy to prove by induction on i that $add(s^i(0), s^j(0)) = s^{i+j}(0) = add(s^j(0), s^i(0))$. Since $add(s^i(0), s^j(0)) = add(s^j(0), s^i(0))$, it is natural to state that $add(x, y) \doteq add(y, x)$ is true for all ground terms x and y. Obviously, $add(x, y) \doteq add(y, x)$ is not a logical consequence of E and E_{ax} (the equality axioms), because $E \cup E_{ax}$ may have models in which $add(x, y) \doteq add(y, x)$ does not hold. It is known that some properties of a first-order formula cannot be proved by deduction or contradiction; they can be proved only by induction. That is why we call these properties *inductive theorems*, and $add(x, y) \doteq add(y, x)$ is just one of them.

Mathematical induction is a fundamental and powerful rule of inference. Inductive theorem proving has important applications in formal verification and has been used to prove the correctness of computer architecture or algorithms. Like the previous chapters, we are particularly interested in inductive proving methods which can be automated in a computer.

7.3.1 Inductive Theorems

Definition 7.3.1 Given a first-order language $L = (P, F, X, Op)$, a quantifier-free formula B is said to be an *inductive theorem* of A, if for all substitution $\sigma : X \mapsto T(F)$, $A \models B\sigma$.

In other words, a formula is an inductive theorem of A if all its ground instances are logical consequence of A. We can show that $add(x, y) \doteq add(y, x)$ and $add(add(x, y), z) \doteq add(x, add(y, z))$ are inductive theorems of $E = \{add(0, y) \doteq y, add(s(x), y) \doteq s(add(x, y))\}$. Typically, A is quantifier-free, with the understanding that all free variables act as universally quantified variables.

To prove an inductive theorem, we need an inference rule for induction, which has many different versions. One induction rule requires that function symbols are divided into two parts: *constructors* and *defined functions*.

Definition 7.3.2 Given a set E of equations over a set F of function symbols, a subset C of F called *constructors* if C contains at least one constant and $D = F - C$ called *defined functions*.

A function $f \in D$ of arity k is said to be *C-complete* if for $t_1, \ldots, t_k \in T(C)$, there exists a term $t \in T(C)$, $f(t_1, \ldots, t_k) =_E t$. D is *C-complete* if every $f \in D$ is C-complete.

E is *C-consistent* if $t_1 \neq_E t_2$ for $t_1, t_2 \in T(C)$ and $t_1 \neq t_2$.

If $f \in D$ is C-complete, it means f is completely defined for the terms in $T(C)$. E will be called *equational definitions* of D over C. If $T(C)$ represents a domain, then f represents a total function over the domain. In the literature, the concept of "C-complete" is also called "sufficiently complete."

Example 7.3.3 In Example 7.2.3, we convert a set E of equations into a rewrite system so that the computation can be carried out by rewriting:

$$
\begin{aligned}
&(1) \quad pre(0) \to 0 \\
&(2) \quad pre(s(x)) \to x \\
&(3) \quad add(0, y) \to y \\
&(4) \; add(s(x), y) \to s(add(x, y)) \\
&(5) \quad sub(x, 0) \to x \\
&(6) \; sub(x, s(y)) \to sub(pre(x), y) \\
&(7) \quad mul(0, y) \to 0 \\
&(8) \; mul(s(x), y) \to add(mul(x, y), y) \\
&(9) \quad exp(x, 0) \to s(0) \\
&(10) \; exp(x, s(y)) \to mul(x, exp(x, y))
\end{aligned}
$$

In this example, the function symbols are $F = \{0, s, pre, add, sub, mul\}$, which can be divided into two parts: $C = \{0, s\}$ and $D = \{pre, add, sub, mul\}$. $T(C) = \{0, s(0), s^2(0), \ldots\}$ is a representation of the natural numbers. D represents a set of functions over the natural numbers.

For each function f in D, f is C-complete:

- $pre(0) = 0$ and $pre(s^i(0)) = s^{i-1}(0)$ for $i > 0$
- $add(s^i(0), s^j(0)) = s^{i+j}(0)$ for any $i, j \geq 0$
- $sub(s^i(0), s^j(0)) = 0$ if $i \leq j$ and $pre(s^i(0), s^j(0)) = s^{i-j}(0)$ for $i > j$
- $mul(s^i(0), s^j(0)) = s^{i*j}(0)$ for any $i, j \geq 0$

It is easy to check that $T(F)/=_E = \{[0], [s(0)], \ldots, [s^i(0)], \ldots\}$, which is the domain of the Herbrand model with equality and has a bijection h from $T(C)$ to $T(F)/=_E$ by $h(t) = [t]$. □

If $C = \{0, 1\}$, then the concept of C-consistency coincides with the consistency of Boolean logic. The C-consistency of E does not allow equations over C. For some applications, we may need such equations. For example, to model the set of integers, we may use three constructors: 0, s (plus one), and p (minus one), so that a negative integer $-i$ is represented by $p^i(0)$. In this case, we need $E_C = \{p(s(x)) \doteq$

7.3 Inductive Theorem Proving

x, $s(p(x)) \doteq x\}$, and the definition of C-consistency can be modified by replacing $T(C)$ by $T(C)/{=}_{E_C}$.

We leave the proof of the following lemma as an exercise.

Lemma 7.3.4 *Let E be the equational definitions of $D = F - C$. D is C-complete iff for every ground term $s \in T(F)$, there exists $t \in T(C)$ such that $s =_E t$.*

Proposition 7.3.5 *Let E be a set of equations over F, $C \subset F$ be the set of constructors and $D = F - C$. Let $h : T(C) \mapsto T(F)/{=}_E$ be defined by $h(t) = [t]$. The following statements are true:*

(a) h is injective iff E is C-consistent.
(b) h is surjective iff D is C-complete.
(c) h is bijective iff E is C-consistent and D is C-complete.

Proof

(a) $h : T(C) \mapsto T(F)/{=}_E$ is injective iff $h(t_1) \neq h(t_2)$ whenever $t_1 \neq t_2$ for $t_1, t_2 \in T(C)$. Thus, $h(t_1) = h(t_2)$ iff $[t_1] = h(t_1) = h(t_2) = [t_2]$, iff $t_1 =_E t_2$, iff E is C-inconsistent.
(b) $h : T(C) \mapsto T(F)/{=}_E$ is surjective iff $\forall s \in T(F) \exists t \in T(C) h(t) = [t] = [s]$, iff $\forall s \in T(F) \exists t \in T(C) s =_E t$, iff D is C-complete (Lemma 7.3.4).
(c) Comes from (a) and (b).

□

It is not always an easy task to show that every defined function is C-complete or E is C-consistent. The above proposition does not provide an effective method for this task but may help us understand these concepts.

To prove an inductive theorem in the presence of constructors and C-complete functions, we need to consider only the ground terms in $T(C)$, not $T(F)$, thus making the inductive proof simpler. In general, inductive theorems are properties of defined functions.

7.3.2 Structural Induction

One induction rule is called the *structural induction rule*, which is based on the structure of terms built up by the constructors. In the current example, the constructors are $C = \{0, s\}$.

Structural Induction Rule for $C = \{0, s\}$

$$\frac{B(0) \quad B(y) \to B(s(y))}{B(x)}$$

where $B(x)$ stands for an inductive theorem to be proved by induction on x.

The above rule states that to prove an inductive theorem $B(x)$, we need to prove two formulas: $B(0)$ and $B(y) \to B(s(y))$. Its soundness is based on the condition that all the functions in $F - C$ are C-complete. The C-completeness ensures that every term in $T(F)$ is equivalent to a term in $T(C)$ (Lemma 7.3.4). Since every term $x \in T(C)$ can be represented by 0 or $s(y)$, if $B(0)$ and $B(s(y))$ are true, so is $B(x)$. $B(0)$ is the base case of the induction. To prove $B(s(y))$, we use $B(y)$ as "induction hypothesis" because y is less than $s(y)$ in terms of the term structure of $T(C)$.

Example 7.3.6 Let $B(x)$ be $add(x, 0) \doteq x$. Applying the structural induction on this equation, we need to prove (a) $add(0, 0) \doteq 0$ and (b) $add(x, 0) \doteq x \to add(s(x), 0) \doteq s(x)$. (a) is trivial; (b) is simplified by rule (4), the definition of add, to: $add(x, 0) \doteq x \to s(add(x, 0)) \doteq s(x)$, which is true by *equality crossing*, as described below. □

Equality Crossing

$$\frac{(t_1 \neq t_2 \mid A(t_1))}{(t_1 \neq t_2 \mid A(t_2))}$$

Since $(t_1 \neq t_2 \mid A(t_1))$ is equivalent to $(t_1 \doteq t_2) \to A(t_1)$, the result $(t_1 \neq t_2 \mid A(t_2))$ can be regarded as "rewrite $A(t_1)$ to $A(t_2)$ by the context $t_1 \doteq t_2$" [4]. For instance, $add(x, 0) \doteq x \to s(add(x, 0)) \doteq s(x)$ is written to $add(x, 0) \doteq x \to s(x) \doteq s(x)$ by equality crossing, where $t_1 = add(x, 0)$ and $t_2 = x$. In other words, we used the induction hypothesis $add(x, 0) \doteq x$ to show that $s(add(x, 0)) \doteq s(x)$ is true.

Equality crossing, called *crossing fertilization* by Boyer and Moore, is a simplification rule in the presence of equality axioms, due to the following result.

Proposition 7.3.7 (Soundness of Equality Crossing) $E_{ax} \models (t_1 \neq t_2 \mid A(t_1)) \leftrightarrow (t_1 \neq t_2 \mid A(t_2))$, where E_{ax} is the equality axioms.

Proof From the monotonic axiom of equality, i.e., $(x_i = y_i) \land p(\ldots, x_i, \ldots) \to p(\ldots, y_i, \ldots)$, we can obtain $(t_1 \doteq t_2) \land A(t_1) \to A(t_2)$, whose clausal form is

$$(t_1 \neq t_2 \mid \overline{A(t_1)} \mid A(t_2))$$

The resolvent of this clause and $(t_1 \neq t_2 \mid A(t_1))$ on $A(t_1)$ is $(t_1 \neq t_2 \mid A(t_2))$. Switching t_1 and t_2, the other implication also holds. □

The fundamental difference between using an inference rule like resolution and using an induction rule is not that a well-founded order, e.g., $s(x) \succ x$, is needed when applying induction. When using resolution for theorem proving by refutation, the empty clause is the goal, and the resolution rule is used to derive this goal from the input clauses. When using an induction rule for theorem proving, the goal is the formula to be proved. However, we never try (and it is almost impossible) to

7.3 Inductive Theorem Proving

derive this goal directly from the input formulas by the induction rule. Instead, we try to reduce the goal to subgoals by using the induction rule backward. That is, to prove $B(x)$ by induction, we generate two subgoals $B(0)$ and $B(y) \to B(s(y))$ from $B(x)$ and try to establish the validity of these subgoals. Remember that when every defined function is C-complete, this condition ensures the soundness of the induction rule.

Example 7.3.8 Continuing from Example 7.3.3, let us prove that $sub(add(x, y), y) \doteq x$. If we do induction on x, the proof will not go through. On the other hand, we can do induction on y. By inspecting the definitions of add and sub, we see that x is the induction position for $add(x, y)$ and y is the induction position for $sub(x, y)$.

The base case is $sub(add(x, 0), 0) \doteq x$, which is simplified to true by rules (3) and (5). The inductive case is

$$sub(add(x, y), y) \doteq x \to sub(add(x, s(y)), s(y)) \doteq x$$

To simplify $add(x, s(y))$, we need $add(x, s(y)) \doteq s(add(x, y))$, which is also an inductive theorem, and its proof is left as an exercise. Suppose this theorem is available to us, then

$$
\begin{aligned}
& sub(add(x, s(y)), s(y)) \\
={} & sub(s(add(x, y)), s(y)) && \text{by } add(x, s(y)) \doteq s(add(x, y)), \\
={} & sub(pre(s(add(x, y)), y) && \text{by rule (6)}, \\
={} & sub(add(x, y), y) && \text{by rule (2)}, \\
={} & x && \text{by equality crossing.}
\end{aligned}
$$

Equality crossing can be used as a simplification rule in resolution proof with equality. □

7.3.3 Induction on Two Variables

The previous example shows that we need to choose the right variable to do the induction. Sometimes, we need to do induction on multiple variables.

Example 7.3.9 Adding the following two predicates to Example 7.3.3:

(11) $\quad y < 0 \to \bot$
(12) $\quad 0 < s(x) \to \top$
(13) $\quad s(y) < s(x) \to y < x$
(14) $\quad 0 \leq y \to \top$
(15) $\quad s(x) \leq 0 \to \bot$
(16) $\quad s(x) \leq s(y) \to x \leq y$

To prove $(x \leq y) \doteq \neg(y < x)$ by structural induction, neither on x alone nor on y alone will work. We need an induction on two variables. □

Induction Rule on Two Variables for $C = \{0, s\}$

$$\frac{B(0, y) \quad B(s(x), 0) \quad B(x, y) \to B(s(x), s(y))}{B(x, y)}$$

where $B(x, y)$ stands for an inductive theorem to be inducted on both x and y.

This is not the only induction rule for the two variables. Why should we use this rule for $(x \leq y) \doteq \neg(y < x)$? The clue is in the definitions of $\leq$ and $<$: for $(x \leq y)$, we defined $\leq$ on $\langle 0, y \rangle$, $\langle s(x), 0 \rangle$, and $\langle s(x), s(y) \rangle$ with a recursive call on $\langle x, y \rangle$. For $(y < x)$, we also defined $<$ on $\langle 0, y \rangle$, $\langle s(x), 0 \rangle$, and $\langle s(x), s(y) \rangle$ with a recursive call on $\langle x, y \rangle$ (the positions of x and y are switched here). In other words, the definition equations of these two predicates tell us what cases needed to consider: the base cases come from the definition equations without recursion; the inductive cases come from the definition equations with recursion. This is the major induction heuristic in Boyer and Moore's computational logic [3], where lisp functions are used instead of equational definitions. An induction inference rule is always obtained automatically from the definition of a function appearing in the theorem to be proved.

Now, back on the proof of $(x \leq y) \doteq \neg(y < x)$, by the above induction rule, we need to prove three subgoals:

1. $(0 \leq y) \doteq \neg(y < 0)$, which can be easily proved by showing $0 \leq y = \top$ by rule (14) and $\neg(y < 0) = \neg(\bot) = \top$ by rule 11.
2. $(s(x) \leq 0) \doteq \neg(0 < s(x))$, which can be proved by showing $(s(x) \leq 0) = \bot$ by rule 15 and $\neg(0 < s(x)) = \neg(\top)) = \bot$ by rules 12.
3. $(x \leq y) \doteq \neg(y < x) \to (s(x) \leq s(y)) \doteq \neg(s(y) < s(x))$. Proof of 3. is also easy: $s(x) \leq s(y)$ is simplified to $x \leq y$ by rule 16 and $\neg(s(y) < s(x))$ to $\neg(y < x)$ by rule 13, then 3. becomes a trivial implication: $(x \leq y) \doteq \neg(y < x) \to (x \leq y) \doteq \neg(y < x)$.

7.3.4 Many-Sorted Algebraic Specification

By far, we have focused on the natural number functions based on the constructors 0 and s. In many applications, we need other data structures, such as lists, trees, etc., which have their own constructors. In Sect. 5.2.4, we discussed briefly *many-sorted logic*, which is a small extension of first-order logic. A *type* or *sort* corresponds to an unary (or monadic) predicate $p(x)$ and defines a set $S = \{x \mid p(x)\}$. Many-sorted algebraic specification is a language in which we define functions by equations

7.3 Inductive Theorem Proving

in first-order logic with multiple types. Many-sorted algebraic specification will provide a convenient tool for applications with multiple types. For example, to work on the lists of natural numbers, we need two types: natural numbers and lists, and their domains can be specified by BNF as follows:

$$\langle nat \rangle ::= 0 \mid s(\langle nat \rangle)$$
$$\langle list \rangle ::= nil \mid cons(\langle nat \rangle, \langle list \rangle)$$

where the constructors for *nat* are $0 : nat$ and $s : nat \mapsto nat$ and the constructors for *list* are $nil : list$ and $cons : nat, list \mapsto list$. We have seen the functions on *nat*; the functions over *list* can be also defined using equations. To prove inductive theorems involving *nil* and *cons*, we need the following rule:

Structural Induction Rule for $C = \{nil, cons\}$

$$\frac{B(nil) \qquad B(y) \rightarrow B(cons(x, y))}{B(y)}$$

where $B(y)$ stands for an inductive theorem to be inducted on y.

Example 7.3.10 The *append* and *reverse* operations of lists can be defined as follows:

$$app : list, list \mapsto list$$
(1) $\quad app(nil, y) \rightarrow y$
(2) $app(cons(x, y), z) \rightarrow cons(x, app(y, z))$
$$rev : list \mapsto list$$
(3) $\quad rev(nil) \rightarrow nil$
(4) $\quad rev(cons(x, y)) \rightarrow app(rev(y), cons(x, nil))$

To prove $rev(app(y, z)) \doteq app(rev(z), rev(y))$ for all lists y and z, the structural induction rule reduces $rev(app(y, z)) \doteq app(rev(z), rev(y))$ to two subgoals:

$G_1 : rev(app(nil, z)) \doteq app(rev(z), rev(nil))$
$G_2 : rev(app(y, z)) \doteq app(rev(z), rev(y)) \rightarrow$
$\quad\quad rev(app(cons(x, y), z)) \doteq app(rev(z), rev(cons(x, y)))$

We can simplify G_1 by rules 1 and 3 to $rev(z) \doteq app(rev(z), nil)$, which suggests a more general property: $app(y, nil) \doteq y$. This heuristic is called *generalization* and can be implemented automatically. Indeed, $app(y, nil) \doteq y$ is another inductive theorem, and its proof is left as an exercise. Using this lemma, $rev(z) \doteq app(rev(z), nil)$ becomes true.

For G_2, we can simplify

$$rev(app(cons(x, y), z)) \doteq app(rev(z), rev(cons(x, y)))$$

by rule 2 to $rev(cons(x, app(y, z)) \doteq app(rev(z), rev(cons(x, y)))$, then to

$$app(rev(app(y, z)), cons(x, nil)) \doteq app(rev(z), app(rev(y), cons(x, nil)))$$

by rule 4. Using equality crossing, $rev(app(y, z))$ is rewritten to $app(rev(z), rev(y))$, and we are facing

$$app(app(rev(z), rev(y)), cons(x, nil)) \doteq app(rev(z), app(rev(y), cons(x, nil)))$$

A generalization of the above formula is the associativity of app:

$$app(app(x', y'), z') \doteq app(x', app(y', z'))$$

where x' comes from $rev(z)$, y' from $rev(y)$, and z' from $cons(x, nil)$. The associativity of app can be shown to be an inductive theorem separately. Once this is done, the proof of G_2 is complete. Generalization is a heuristic technique originally used in Boyer-Moore's theorem prover. □

The above example illustrates an interesting feature of inductive theorem proving: in order to prove $rev(app(y, z)) \doteq app(rev(z), rev(y))$, we need to prove extra lemmas: $app(y, nil) \doteq y$ and $app(app(x, y), z) \doteq app(x, app(y, z))$. These lemmas can sometimes be generated automatically and sometimes need to be provided by the user. For rigorousness, they are needed to be proved before they are used in a proof.

Example 7.3.11 Let us prove another inductive theorem, $rev(rev(y)) \doteq y$, by the same induction rule. The induction rule reduces $rev(rev(y)) \doteq y$ to two subgoals: $G_3 : rev(rev(nil)) \doteq nil$, which is easy to prove by rule 2, and

$$G_4 : rev(rev(y)) \doteq y \to rev(rev(cons(x, y))) \doteq cons(x, y).$$

Applying rule 4, we rewrite $rev(rev(cons(x, y)))$ to $rev(app(rev(y), cons(x, nil)))$, which is then simplified to $app(rev(cons(x, nil)), rev(rev(y)))$. Since $rev(cons(x, nil))$ is rewritten to x by rules 4, 3, and 1, so $app(rev(cons(x, nil)), rev(rev(y)))$ is rewritten $cons(x, rev(rev(y)))$ by rules 2 and 1. Thus, G_4 becomes

$$G_5 : rev(rev(y)) \doteq y \to cons(x, rev(rev(y))) \doteq cons(x, y)$$

Applying equality crossing, $rev(rev(y))$ is replaced by y and G_5 becomes true. □

7.3 Inductive Theorem Proving

We have shown how to use the language of algebraic specification of abstract data types to write axioms (definitions of functions and predicates), theorems or lemmas (intermediate inferences), etc. for inductive reasoning. In particular, we require that the axioms as well as theorems in an algebraic specification be expressed as a set of (conditional) equations. In other words, we are interested in automatic methods which can prove equations from equations by mathematical induction.

Example 7.3.12 We may define the insertion sort algorithm using equations:

$$isort : list \mapsto list$$
(5) $\quad isort(nil) \to nil$
(6) $\quad isort(cons(x, y)) \to insert(x, isort(y))$
$$insert : nat, list \mapsto list$$
(7) $\quad insert(x, nil) \to cons(x, nil)$
(8) $\quad insert(x, cons(y, z)) \to cons(x, cons(y, z)) \text{ if } x \leq y$
(9) $\quad insert(x, cons(y, z)) \to cons(y, insert(x, z)) \text{ if } \neg(x \leq y)$
$$sorted : list \mapsto boolean$$
(10) $\quad sorted(nil) \to \top$
(11) $\quad sorted(cons(x, nil)) \to \top$
(12) $sorted(cons(x, cons(y, z))) \to (x \leq y) \land sorted(y, z)$

Note that $\leq$ used here is defined in Example 7.3.9.

For the correctness of $isort(x)$, we may prove that $sorted(isort(x))$ is an inductive theorem (an exercise problem). If we define $perm(x, y)$ to be true iff x is a permutation of y, then we may also prove that $perm(x, isort(x))$ is an inductive theorem. Both theorems ensure the correctness of the insertion sort algorithm. □

The inductive theorems shown in the above examples can be easily proved by today's inductive theorem prover. Many algorithms can be specified by equations in this way and can be formally verified automatically using this approach.

The notable success of inductive theorem proving comes from the field of functional programming (see Sect. 9.1.2). Boyer-Moore's theorem prover NQTHM, its successor ACL2, Coq, and Isabelle, are well-known software tools that use functional language to specify digital circuits, operational systems, compilers, algorithms, etc. and prove that these specifications have desired properties by automated induction proofs and other automated reasoning techniques. An understanding of induction proofs in the context of rewrite systems will help us to understand these software tools because they share the same principles and techniques.

7.3.5 Recursion, Induction, and Self-Reference

Recursion, induction, and self-reference are sometimes treated as synonyms. For instance, the following sentences can be used to explain one another:

- A function is recursively defined.
- A function is inductively defined.
- A function is defined by self-reference.

Induction as an inference rule cannot be replaced by recursion or self-reference. The soundness of the induction rule is based on a well-founded ordering as we have seen in this section. This order is implicit when we define a function (or a set of objects) by recursion. Without this order, we may have an undefined function value or inconsistency.

Example 7.3.13 The unsolved Collatz conjecture asks if the following recursive function $f(x)$ is totally defined, i.e., $f(x) = 1$, for any positive integer x:

$$f(x) = \textbf{if } (x = 1) \textbf{ then } 1 \textbf{ else if } even(x) \textbf{ then } f(x/2) \textbf{ else } f((3x+1)/2)$$

If we can find an odd integer x such that $f(x) = f((3x+1)/2) = \cdots = f(x)$, then x will be a counterexample to the Collatz conjecture and $f(x)$ is undefined. A procedure for computing $f(x)$ will loop forever on this x.

As of 2020, the conjecture has been checked by computer for all $x \leq 268^2 \approx 2.95 \times 10^{20}$. This experimental evidence is still not rigorous proof that the conjecture is true for all positive integers, as counterexamples may be found when considering very large (or possibly immense) values. The renowned mathematician Paul Erdös once said about the Collatz conjecture: "Mathematics is not yet ready for such problems." Erdös offered $500 for its solution. □

In the case of using recursion incorrectly, we often use self-reference. The *circular reasoning* fallacy is a typical error of self-reference, where two statements (or concepts) of the same fact are used to support each other. If the meaning of a propositional variable is defined by self-reference, it may lead to a contradiction. For example, the statement "I am lying" leads to a contradiction if we assume that "if someone is lying, then what he said is false." If the statement "I am lying" is true, since it is uttered by someone who is lying, then the statement is false. If the statement is false, then "I am not lying" is true, so what I said (i.e., "I am lying") should be true. The above contradiction is called the "liar paradox" and can be found in first-order logic, too. For instance, the Epimenides paradox, "All Cretans are liars" when uttered by an ancient Greek Cretan, is one of the first recorded instances.

By interpreting "lying" as "false," the liar paradox can be expressed in logic as a self-reference: let p denote "p is false," assigning any truth value to p will result in a contradiction. This paradox can be extended to multiple sentences:

- p: The next statement, i.e., q, is false.
- q: The next statement, i.e., r, is false.
- r: The first statement, i.e., p, is false.

From $p \equiv \neg q$, $q \equiv \neg r$, and $r \equiv \neg r$, it is easy to deduce $p \equiv \neg p$.

Arthur Prior claims that there is nothing paradoxical about the liar paradox. His argument is that every statement includes an implicit assertion of its own truth. For example, the statement "It is true that "2+2 = 4" contains no more information than the statement "$2 + 2 = 4$," because the phrase "it is true that..." is always implicitly there. That is, the phrase "it is true that..." is equivalent to "this whole statement is true and ...." Thus in the liar paradox, the following two statements are equivalent:

- This statement is false.
- This statement is true and this statement is false.

The latter is a simple contradiction of the form "p and not p," and hence, there is no paradox.

Curry's paradox, named after logician Haskell Curry, is another paradox of self-reference in logic. The paradox can be described as follows: let p denote "if p then q," or equivalently, $\neg p \vee q$. If p is true, then q is true from $\neg p \vee q$. Since q is arbitrary and can be $\bot$, we have a contradiction. If p is false, then the negation of $\neg p \vee q$ is true. Since $\neg(\neg p \vee q) \equiv p \wedge \neg q$ is false when p is false, we have a contradiction, too. If we use Prior's argument in Curry's paradox, then p will denote "p and if p then q." This expression is equivalent to $p \wedge q$, and there is no apparent paradox. Curry's paradox may be expressed in natural language and in various logics, including certain naive set theory, lambda calculus, and combinatory logic.

Self-reference is blamed as the source of Russell's paradox in naive set theory. However, self-reference is indispensable in mathematics and computer science. As seen in Chap. 1, recursion is a powerful tool to construct objects, functions, and procedures. In Chap. 11, we will see that recursion is the cornerstone of computability theory.

7.4 Resolution with Equality

Up to this point in this chapter, our focus was on equations, which are positive unit clauses with "$\doteq$" as the only predicate. Equations are expressive enough to describe all the computations. For instance, all Turing machines can be expressed as a set of equations. On the other hand, it is more convenient to express axioms and theorems in general clauses. The Knuth-Bendix completion suggests that we do not need the equality axioms explicitly if we have an inference like superposition which computes critical pairs from two rewrite rules. This is also true in resolution-based proving as Alan Robinson and Larry Wos suggested the *paramodulation* rule [5], shortly after the invention of the resolution rule.

7.4.1 Paramodulation

Definition 7.4.1 (Paramodulation) Suppose clause c_1 is $(s \doteq t \mid \alpha)$ and c_2 is $(A \mid \beta)$, where A is a literal and p is a non-variable position of A, α and β are the rest literals in c_1 and c_2, respectively, the *paramodulation* rule is defined as follows:

$$\frac{(s \doteq t \mid \alpha) \quad (A \mid \beta)}{(A[p \leftarrow t] \mid \alpha \mid \beta)\sigma}$$

where σ is the mgu of s and A/p. The clause $(A[p \leftarrow t] \mid \alpha \mid \beta)\sigma$ produced by the paramodulation rule is called *paramodulant* of the paramodulation; c_1 and c_2 are the *parents* of the paramodulant. We use $paramod(c_1, c_2)$ to denote the set of all clauses like $(A[p \leftarrow t] \mid \alpha \mid \beta)\sigma$ from c_1 and c_2.

Superposition is a special case of paramodulation where α and β are empty and A is also an equation.

Example 7.4.2 Let c_1 be $(f(g(y)) \doteq a \mid r(y))$ and c_2 be $(p(g(f(x))) \mid q(x))$, then $f(g(y))$ and $p(g(f(x)))/1.1 = f(x)$ are unifiable with $\sigma = [x \leftarrow g(y)]$. The paramodulant is $(p(g(a)) \mid q(g(y)) \mid r(y))$. □

Proposition 7.4.3 (Soundness of Paramodulation) *For any clauses c_1 and c_2, $E_{ax} \cup \{c_1, c_2\} \models paramod(c_1, c_2)$, where E_{ax} is the equality axioms.*

Proof If $paramod(c_1, c_2)$ is empty, then $paramod(c_1, c_2) \equiv \top$ by our assumption and the proposition is trivially true. If $paramod(c_1, c_2)$ is not empty, let σ be a mgu used in the paramodulation, $c_1\sigma$ is $(s' \doteq t' \mid \alpha')$ and $c_2\sigma$ is $(A'[s'] \mid \beta')$, then $paramod(c_1, c_2)$ contains $c_3 : (A'[t'] \mid \alpha' \mid \beta')$.

From the monotonic axiom in E_{ax}, we can deduce $(s' \doteq t') \wedge A'(s') \rightarrow A'(t')$, whose clausal form is $c_4 : (s' \neq t' \mid \overline{A'(s')} \mid A'(t'))$. Apply resolution on $(s' \doteq t')$ of $c_1\sigma$ and c_4, we get $c_5 : (\overline{A'(s')} \mid A'(t') \mid \alpha')$. Finally, apply resolution on $A'(s')$ of $c_2\sigma$ and c_5, we get $c_6 : (A'(t') \mid \alpha' \mid \beta')$, which is the same as c_3. Since resolution is sound, c_3 is a logical consequence of c_1, c_2, and E_{ax}. □

The above proof shows that one paramodulation is equivalent to two resolutions with an instance of the monotonic axiom.

Using paramodulation, we do not need most of the equality axioms. It can be shown that any consequence from the equality axioms can be deduced from paramodulation, with the only exception of the reflexivity axiom: $x \doteq x$. $(x \doteq x)$

7.4 Resolution with Equality

is indispensable in showing that $A = \{f(x) \neq f(y)\}$ is unsatisfiable. We may introduce the following inference rule which makes the axiom $(x \doteq x)$ redundant.

Definition 7.4.4 (Reflexing) Suppose clause C is $(s \neq t \mid \alpha)$, *reflexing* is an inference rule defined by the following schema:

$$\frac{(s \neq t \mid \alpha)}{(\alpha)\sigma}$$

where σ is the mgu of s and t.

Reflexing works in the same way as factoring: factoring removes one literal by unifying two literals; reflexing removes $(s \neq t)$ by unifying s and t.

Paramodulation combined with resolution, factoring, and reflexing is refutationally complete for theorem proving with equality. That is, if $A \cup E_{ax}$ is unsatisfiable, where E_{ax} is the equality axioms, then the empty clause can be derived from A by paramodulation, resolution, factoring, and reflexing. It remains complete with any complete resolution strategies, such as positive resolution or ordered resolution. We may add restrictions on paramodulation as we put restrictions on resolution, without sacrificing the refutational completeness.

Definition 7.4.5 (Ordered Paramodulation) Given a simplification order $\succ$, suppose $s \doteq t$ is maximal in clause c_1 $(s \doteq t \mid \alpha)$, $s \not\succ t$, and literal A is maximal in c_2 $(A \mid \beta)$, p is a non-variable position of A, α and β are the rest literals in c_1 and c_2, respectively. The *ordered paramodulation* rule will generate the paramodulant $(A[p \leftarrow t] \mid \alpha \mid \beta)\sigma$, where σ is the mgu of s and A/p.

The paramodulation in Example 7.4.2 is an ordered paramodulation if $f(g(y)) \doteq a$ is maximal in c_1 and $p(g(f(x)))$ is maximal in c_2.

Using the same simplification order, ordered paramodulation can be combined with ordered resolution without losing the refutational completeness. When a set of clauses is just a set of equations (and regarded as a set of rewrite rules), ordered paramodulation is reduced to superposition. Robinson and Wos's work on paramodulation preceded, but apparently did not influence, the work of Knuth and Bendix, who independently formulated the concepts of superposition and rewriting in 1970. In comparison, superposition is a special case of Robinson and Wos' paramodulation inference rule, and rewriting is the same as Robinson and Wos' *demodulation* simplfication rule.

7.4.2 Simplification Rules

For a theorem prover based on resolution and paramodulation, huge number of new clauses are generated, and we need to control redundancy in large search spaces. Restricted strategies for using resolution and paramodulation are effective to reduce redundant clauses, deleting unnecessary clauses in the search for the empty clause is also important to control the search space. As shown before, we may apply the following deletion strategies in a resolution-based theorem prover:

- **Pure literal deletion**: Clauses containing pure literals are discarded.
- **Tautology deletion**: Tautology clauses are discarded.
- **Subsumption deletion**: Subsumed clauses are discarded.
- **Unit deletion**: Those literals which are the negation of instances of some unit clauses are discarded from clauses.

Simplification strategies allow us to keep clauses in simplified forms so that multiple copies of equivalent clauses are not needed. For instance, if a clause contains a term s, say $(p[s] \mid q)$, and we have a rewrite rule $s \to t$ or simply $s \doteq t$, then this clause can be rewritten to $(p[t] \mid q)$ by $s \doteq t$. Rewriting is not the only way to obtain $(p[t] \mid q)$, because paramodulation can generate it from $(p[s] \mid q)$ and $s \doteq t$. However, there is no need to keep both $(p[s] \mid q)$ and $(p[t] \mid q)$, because both are logically equivalent under equality. In other words, when $(p[s] \mid q)$ is rewritten to $(p[t] \mid q)$, we keep the latter and throw the former away. That is why we call this kind of strategy a *simplification strategy*.

To work with ordered resolution and paramodulation, we may enforce a restriction on equality crossing as a simplification rule:

Ordered Equality Crossing

$$\frac{(s \neq t) \mid A(s)}{(s \neq t) \mid A(t)}$$

if $s \succ t$ for the simplification order $\succ$ used in the ordered paramodulation.

Its soundness follows from Proposition 7.3.7. A more general form of equality crossing is the following simplification rule [4]:

Definition 7.4.6 (Contextual Rewriting) Given a clause $c_1 : (s \doteq t \mid \alpha)$, where $s \succ t$, clause $c_2 : (A \mid \beta)$ can be rewritten by c_1 if there exists a substitution σ such that $A/p = s\sigma$ for a position p of A and $\alpha\sigma \subseteq \beta$. The result of this rewriting is $c_2' : (A[p \leftarrow t\sigma] \mid \beta)$.

Note that if α is empty, then c_1 is a positive unit clause and contextual rewriting is the same as ordinary rewriting. In [4], $\neg\alpha$ is called the *condition* of the rewriting (because c_1 is equivalent to $\neg\alpha \to s \doteq t$) and $\neg\beta$ is called the *context* of the rewriting (which implies that the condition $\neg\alpha$ is true).

Example 7.4.7 Let c_1 be $(f(g(y)) \doteq a) \mid q(y)$ and c_2 be $(p(f(g(b))) \mid q(b))$, then c_1 can rewrite c_2 to $(p(a) \mid q(b))$ because $p(f(g(b)))/1 = f(g(b)) = f(g(y))\sigma$, where $\sigma = [y \leftarrow b]$ and $q(b) = q(y)\sigma$ appears in c_2. □

Proposition 7.4.8 (Soundness of Contextual Rewriting) *If c_1 rewrites c_2 to c_2' by contextual rewriting, then $E_{ax} \cup \{c_1\} \models c_2 \leftrightarrow c_2'$, where E_{ax} is the equality axioms.*

7.4 Resolution with Equality

Proof Suppose c_1 is $(s \doteq t \mid \alpha)$ and σ is the substitution used in contextual rewriting, let $s' = s\sigma$ and $t' = t\sigma$, then c_2 is $(A[s'] \mid \beta)$, and c_2' is $(A[t'] \mid \beta)$. Using paramodulation, we can obtain c_2' from $c_1\sigma$ and c_2 as well as c_2 from $c_1\sigma$ and c_2' (by switching the role of s and t). The soundness of paramodulation ensures the equivalence of c_2 and c_2' under c_1 and E_{ax}. □

There are two possible extensions to contextual rewriting. The first extension is to replace the condition $\alpha\sigma \subseteq \beta$ by $Ax \cup \{\neg\beta\} \models \neg\alpha\sigma$, where Ax is the set of axioms used in the current problem. Note that $\alpha\sigma \subseteq \beta$ implies $\neg\beta \models \neg\alpha\sigma$.

The second extension of contextual rewriting is to treat non-equality literals as rewrite rules. That is, if c_1 is $(A \mid \alpha)$, where A is a maximal non-equality literal in c_1, then we regard c_1 as a rewrite rule "$A \doteq \top$ **if** $\neg\alpha$" if A is positive; or "$B \doteq \bot$ **if** $\neg\alpha$" if A is $\neg B$. Then, c_1 can rewrite atoms in other clauses to $\top$ or $\bot$. If the literals in the other clauses are rewritten to $\top$, this is equivalent to subsumption. If they are rewritten to false, the other clauses will have one less literal. In particular, if α is empty, this is equivalent to unit deletion (Definition 6.2.10).

The soundness of contextual rewriting remains true with the two extensions; thus, contextual rewriting can serve as a good simplification rule which covers ordinary rewriting, subsumption, and unit deletion [4].

Theorem 7.4.9 *Ordered paramodulation combined with ordered resolution, factoring, reflexing, ordered equality crossing, and ordered contextual rewriting is refutationally complete for first-order logic with equality.*

The proof of this theorem follows the same proof line of Theorem 3.3.13: if the empty clause is generated, then the input clauses are unsatisfiable because all inference and simplification rules are sound. If the empty clause can never be generated and the clause set is saturated by the inference rules, then a Herbrand model can be constructed using the simplification order on $T(F)$.

7.4.3 Equality in Prover9

All the inference and simplification rules discussed in this section have been implemented in Prover9, with the exception of contextual rewriting. Rewriting using equations, called *demodulation*, is supported in Prover9. In fact, most examples provided in the distribution of Prover9 contain equality.

Example 7.4.10 In Example 7.1.6, we are given three equations which specify a free group $(S, *)$. Can we reduce the number of equations to one? What is the minimal size of such a single equation? These are examples of questions interested by mathematicians. Here is an answer to the first question:

$$y * i(z * (((u * i(u)) * i(x * z)) * y)) \doteq x$$

which involves four variables.

Prover9's input file to this problem is simply the single axiom and the goals will be the three equations in Example 7.1.6, with the only exception that we use here $y * i(y)$ for the identity e.

```
formulas(sos).
y * i(z * (((u * i(u)) * i(x * z)) * y)) = x # label(oneAxiom).
end_of_list.

formulas(goals).
(x * y) * z = x * (y * z) # label(associativity).
x * i(x) = y * i(y)       # label(inverse).
x * (y * i(y)) = x        # label(identity).
end_of_list.
```

It took Prover9 less than a second to find proofs for all the three equations after generating 1587 clauses. The proof of the associativity law is the hardest, which consists of 48 steps after the proofs of the other equations. □

The equality symbol "=" is built-in; so is "!=", which stands for the negation of equality. The major options involving paramodulation are the following:

```
clear(ordered_para).    % default set
set(ordered_par).

assign(para_lit_limit, n). % default n=-1, range [-1 .. INT_MAX]
```

If $n \neq -1$, each parent in paramodulation can have at most n literals. This option may cause incompleteness of the inference system.

```
set(para_units_only).
clear(para_units_only). % default clear
```

This flag says that both parents for paramodulation must be unit clauses. The only effect of this flag is the same as to assign 1 to the parameter `para_lit_limit`.

The major options involving rewriting are the following:

```
assign(demod_step_limit, n). % default n=1000, range [-1 .. INT_MAX]
```

This parameter limits the number of rewrite steps that are applied to a clause during demodulation. If $n = -1$, there is no limit.

```
assign(demod_size_limit, n). % default n=1000, range [-1 .. INT_MAX]
```

This parameter limits the size (measured as symbol count) of terms as they are demodulated. If any term being demodulated has more than n symbols, demodulation of the clause stops. If $n = -1$, there is no limit.

```
set(back_demod).
clear(back_demod). % default clear
```

If this flag is set, a new rewrite rule will try to rewrite equations in the usable and sos lists. Prover9 also provides options for non-orientable equations to be rewrite rules under certain conditions, and rewriting using these rules is restricted to ensure termination.

7.5 Finite Model Finding in First-Order Logic

In theory, the problem of finding a model of the fixed size in first-order logic can be formulated as a SAT problem. In practice, converting into a SAT instance is not always the most effective approach even though we have advanced SAT solvers. There are many finite model finders proposed in the literature. In the following, we will introduce Mace4 because it is available and has the same specification language as Prover9.

7.5.1 Mace4: A Finite Model Finding Tool

Mace4 is a software tool for finite model finding in a first-order language. Mace4 is distributed together with Prover9 and was created by William McCune. Prover9 and Mace4 share a first-order language and there is a GUI interface for both of them. Mace4 searches for finite models satisfying the formulas in a first-order language. For a given domain size, all instances of the formulas over the domain are constructed. The result is a set of ground clauses with equality. Then, a decision procedure based on ground equational rewriting is applied. If satisfiability is detected, one or more models are printed. If the formula is the denial of some conjecture, any model found by Mace4 is a counterexample to the conjecture. Mace4 is a valuable complement to Prover9, looking for counterexamples before (or at the same time as) Prover9 is used to search for proof. It can also be used to help debugging input clauses and formulas for Prover9.

For the most part, Mace4 accepts the same input files as Prover9. If the input file contains commands that Mace4 does not understand, then the argument "-c" must be given to tell Mace4 to ignore those commands. For example, we can run the following two jobs in parallel on a Linux machine, with Prover9 looking for proof and Mace4 looking for a counterexample.

```
prover9 -f x2.in > x2.prover9.out
mace4 -c -f x2.in > x2.mace4.out
```

Most of the options accepted by Mace4 can be given either on the command line or in the input file. The following command lists the command-line options accepted by Mace4.

```
mace4 -help
```

Mace4 searches for unsorted finite models only. That is, a model has one underlying finite set, called the domain (or the universe), and the members are always $0, 1, \ldots, n - 1$ for a set of size n. The models are the structures which define functions and relations over the domain, as an interpretation to the function and predicate symbols in the formula (using the same symbols). By default, Mace4 starts searching for a structure of domain size 2, and then, it increments the size until it succeeds or reaches some limit. The size of the initial domain or the incremental size can be specified by the user.

If a formula contains constants that are natural numbers, $\{0, 1, \ldots\}$, Mace4 assumes they are members of the domain of some structure, that is, they are distinct objects; in effect, Mace4 operates under the assumptions $0 \neq 1$, $0 \neq 2$, and so on. To Prover9, natural numbers are just ordinary constants. This is a subtle difference between Prover9 and Mace4. Because Mace4 assumes that natural numbers are members of the domain, if a formula contains a natural number that is out of range ($\geq n$, when searching for a structure of size n), Mace4 will terminate with a fatal error.

Mace4 and Prover9 have the same restrictions on the goal formulas it accepts. Mace4 negates the goals and translates them to clauses in the same way as Prover9. The term "goal" is not particularly intuitive for Mace4 users, because Mace4 does not prove things. It makes more sense, however, when one thinks of Mace4 as searching for a counterexample to the goal.

Mace4 uses the following commands to specify the initial domain size, the maximal domain size, and the increment of the domain size.

```
assign(domain_size, n).    % default n=2, range [2 .. 200]
                           % command-line -n n
assign(iterate_up_to, n).  % default n=10, range [-1 .. 200]
                           % command-line -N n
assign(increment, n).      % default n=1, range [1 .. 200]
                           % command-line -i n
```

These three parameters work together to determine the domain sizes to be searched. The search starts for structures of size `domain_size`; if that search fails, the size is incremented, and another search starts. This continues up through the value `iterate_up_to` (or until some other limit terminates the process).

Example 7.5.1 In Sect. 4.3.3, we showed how to find a Latin square with special constraints by a SAT solver. We used the propositional variable $p_{x,y,z}$ to represent the meaning that the cell at row x and column y contains value z. The same relation between x, y, and z can be equally specified by the predicate $p(x, y, z)$. Thus, the five sets of propositional clauses can be specified by the five first-order formulas as follows:

1. $\forall x \forall y \exists z\, p(x, y, z)$.
2. $\forall x \forall y \forall z \forall w\, ((z \doteq w) \mid \overline{p(x, y, z)} \mid \overline{p(x, y, w)})$.
3. $\forall x \forall y \forall z \forall w\, ((y \doteq w) \mid \overline{p(x, y, z)} \mid \overline{p(x, w, z)})$.
4. $\forall x \forall y \forall z \forall w\, ((x \doteq w) \mid \overline{p(x, y, z)} \mid p(w, y, z))$.
5. $\forall x \forall y \forall z \forall w\, (\overline{p(y, x, z)} \mid \overline{p(z, y, w)} \mid p(w, y, x))$.

These formulas can be easily converted into clauses and can be specified in Mace4 as the following: □

```
assign(domain_size, 5).
formulas(assumptions).
  p(x, y, f(x, y)).
```

7.5 Finite Model Finding in First-Order Logic

```
    (z = w) | -p(x, y, z) | -p(x, y, w).
    (y = w) | -p(x, y, z) | -p(x, w, z).
    (x = w) | -p(x, y, z) | -p(w, y, z).
    -p(y, x, z) | -p(z, y, w) | p(w, y, x).
end_of_list.
```

Note that the function f in the first clause is a Skolem function from $\exists z$. Mace4 will find a model of size 5 for this input, and it displays the value of p(x, y, z) as a Boolean matrix of 25 rows and 5 columns, not a very readable format for a Latin square. From the meaning of $p(x, y, z)$, we may reconstruct the Latin square as follows:

```
    0 2 1 4 3
    4 1 3 2 0
    3 4 2 0 1
    1 0 4 3 2
    2 3 0 1 4
```

If we specify the Latin square using $x * y \doteq z$ instead of $p(x, y, z)$, the input to Mace4 is the following:

```
assign(domain_size, 5).

formulas(assumptions).
    x * z != y * z | x = y.     % "!=" means "not equal"
    x * y != x * z | y = z.
    x * x = x.                  % idempotent law
    ((y * x) * y) * y = x.      % special constraint
end_of_list.
```

Note that "$x * z \neq y * z \mid x \doteq y$" specifies that if rows x and y contain the same value at column z, then $x = y$. Similarly, "$x * y \neq x * z \mid y \doteq z$" specifies that if columns y and z contain the same value at row x, then $y = z$. An alternative specification will use \ and / as the Skolem functions for u and v, respectively, in

$$\forall x, y \exists! u, v \, (x * u \doteq y) \wedge (v * y \doteq x)$$

where $\exists!$ means "there exists uniquely." Now, Mace4's input will look like

```
assign(domain_size, 5).

formulas(assumptions).
  % quasigroup axioms (equational)
    x * (x \ y) = y.
    x \ (x * y) = y.
    (x / y) * y = x.
    (x * y) / y = x.

    ((y * x) * y) * y = x. % special constraint
end_of_list.
```

Mace4 will find the following model:

```
interpretation( 5, [number=1, seconds=0], [
    function(*( , ), [      function(/(_,_),[      function($_,_), [
        0, 2, 1, 4, 3,          0, 4, 3, 1, 2,         0, 2, 1, 4, 3,
        3, 1, 4, 0, 2,          4, 1, 0, 2, 3,         3, 1, 4, 0, 2,
        4, 3, 2, 1, 0,          3, 0, 2, 4, 1,         4, 3, 2, 1, 0,
        2, 4, 0, 3, 1,          1, 2, 4, 3, 0,         2, 4, 0, 3, 1,
        1, 0, 3, 2, 4 ]),       2, 3, 1, 0, 4 ]),      1, 0, 3, 2, 4 ])
]).
```

The Latin square defined by $*$ is different from the one obtained by $p(x, y, z)$. If we use the command `assign(max_models, -1)`, Mace4 will find all the six models when the domain size is 5; it will find 120 models when the domain size is 7 (you need to turn off the `auto` command).

7.5.2 Finite Model Finding by SAT Solvers

Earlier versions of Mace (before Mace3) are based on the fact that a set C of first-order clauses has a finite model if the propositional formula translated from C is satisfiable. Let $D = \{0, 1, \ldots, n-1\}$ be the domain of the finite model I, we will do the following to obtain a set of propositional clauses from C:

- For each predicate symbol p/k, define a set of propositional variables $q^p_{x_1,\ldots,x_k}$, $x_i \in D$, such that $q^p_{x_1,\ldots,x_k}$ is true iff $p^I(x_1, \ldots, x_k)$ is true.
- For each function symbol f/k, define a set of propositional variables $q^f_{x_1,\ldots,x_k,y}$, $x_i, y \in D$, such that $q^f_{x_1,\ldots,x_k,y}$ is true iff $f^I(x_1, \ldots, x_k) = y$, and add the following clause into C:

$$(f(x_1, \ldots, x_k) \neq y_1 \mid f(x_1, \ldots, x_k) \neq y_2 \mid y_1 \doteq y_2)$$

  where $s \neq t$ stands for $\neg(s \doteq t)$.
- For each clause A of C, if $f(t_1, \ldots, t_k)$ appears in A and not in the format of $f(x_1, \ldots, x_k) \doteq y$ (or $y \doteq f(x_1, \ldots, x_k)$), then replace A by $(A[f(t_1, \ldots, t_k) \leftarrow z] \mid f(t_1, \ldots, t_k) \neq z)$, where z is a new variable. This step is called *flattening*. Once this process stops, every atom in C will be either $p(x_1, \ldots, x_k)$ or $f(x_1, \ldots, x_k) \doteq y$, where $x_1, \ldots, x_k, y$ are free variables.
- For each flattened clause A of C, for each variable x of A, for each value d of D, create all the ground instances of A. Let the set of all ground clauses be G.
- Finally, for each ground clause in G, replace $p(d_1, \ldots, d_k)$ by $q^p_{d_1,\ldots,d_k}$ and $f(d_1, \ldots, d_k) \doteq d$ by $q^f_{d_1,\ldots,d_k,d}$. Let the set of all propositional clauses be $propo(C)$.

Example 7.5.2 For the equational specification of the Latin square problem, we used three function symbols: $*$, $\backslash$, and $/$. The propositional variables will be $p_{x,y,z}$

7.5 Finite Model Finding in First-Order Logic

for $x * y \doteq z$, $q_{x,y,z}$ for $x \backslash y \doteq z$, and $r_{x,y,z}$ for $x/y \doteq z$; if $|D| = 5$, there will be 375 propositional variables. The flattened clauses from the axioms plus the functional constraints are

$$(x \backslash y \neq z \mid x * z \doteq y) \quad // \; x * (x \backslash y) \doteq y.$$
$$(x * y \neq z \mid x \backslash z \doteq y) \quad // \; x \backslash (x * y) \doteq y.$$
$$(x/y \neq z \mid z * y \doteq x) \quad // \; (x/y) * y \doteq x.$$
$$(x * y \neq z \mid z/y \doteq x) \quad // \; (x * y)/y \doteq x.$$
$$(y * z \neq z \mid z * y \neq w \mid w * y \doteq x) \quad // \; ((y * x) * y) * y \doteq x.$$
$$(x * y \neq z) \mid x * y \neq w \mid z \doteq w) \quad // \text{``}*\text{''} \text{ is a function}$$
$$(x \backslash y \neq z) \mid x \backslash y \neq w \mid z \doteq w) \quad // \text{``}\backslash\text{''} \text{ is a function}$$
$$(x/y \neq z) \mid x/y \neq w \mid z \doteq w) \quad // \text{``}/\text{''} \text{ is a function}$$

If $|D| = 5$, a clause with three free variables will generate $5^3 = 125$ ground clauses; a clause with four free variables will generate $5^4 = 625$ ground clauses. The ground instances of the above clauses are easy to convert to propositional clauses. For the literal $(z \doteq w)$, its ground instance becomes $\top$ if we have $z\sigma = w\sigma = d$ and $\bot$ if $z\sigma = d_1 \neq d_2 = w\sigma$. □

While the conversion from a first-order formula to a set of propositional clauses uses the equality "=", the equality axioms (introduced in the first section of this chapter) are not needed for the conversion because they produce only trivial propositional clauses. Only the function axiom is needed in the conversion for each function f:

$$(f(x_1, \ldots, x_k) \neq y_1 \mid f(x_1, \ldots, x_k) \neq y_2 \mid y_1 \doteq y_2)$$

Proposition 7.5.3 *A first-order formula C in CNF has a finite model iff the propositional formula $propo(C)$ is satisfiable.*

Proof Let C be a CNF formula of (P, F, X, Op) and $D = \{0, 1, \ldots, n-1\}$ be the finite domain.

If $propo(C)$ is satisfiable, then $propo(C)$ has a model σ and σ defines a relation R^p over D^k for each predicate symbol p/k:

$$R^p = \{\langle d_1, \ldots, d_k \rangle \mid \sigma(q^p_{d_1,\ldots,d_k}) = 1\}$$

and a function $f^\sigma : D^k \mapsto D$ for each function f/k

$$f^\sigma(d_1, \ldots, d_k) \doteq d \text{ iff } \sigma(q^f_{d_1,\ldots,d_k,d}) = 1.$$

Then, it is ready to check that $I = (D, \{R^p \mid p \in P\}, \{f^m \mid f \in F\})$ is a model of C.

On the other hand, if C has a finite model $I = (D, R, G)$, where D is finite, then we can create the interpretation σ_I for $propo(C)$ as follows:

$$\sigma_I(q^p_{d_1,\ldots,d_k}) = 1 \text{ iff } p^I(d_1, \ldots, d_k) = 1$$

and

$$\sigma_I(q^f_{d_1,\ldots,d_k,d}) = 1 \text{ iff } f^I(d_1, \ldots, d_k) = d$$

It is ready to check that σ is a model of $propo(C)$. □

To use SAT provers to solve a problem of modest difficulty, we need to write an *encoder* which generates propositional clauses and then feed these clauses into a SAT solver. We also need a *decoder* which translates SAT solver's solution into a solution to the problem. Mace4 releases the burden of writing encoders/decoders by converting automatically the first-order clauses into propositional clauses with the free variables being substituted by all the values in the finite domain and then call a SAT solver to find a model; it then converts the propositional model into a first-order model.

Later versions of Mace, including Mace4, maintain a set of ground clauses with equality and work in the same way as SEM, an early model generator developed by the authors of this book before Mace was created. This is because the method based on the DPLL method works well in many cases, but if the original problem has clauses with many variables or deeply nested terms, the transformation process to SAT clauses blows up, and only very small models can be sought. The interested reader may consult Mace4's user manual or an introduction to SEM for details.

Exercises

1. Show that $(f(x_1, \ldots, x_k) \neq y_1 \mid f(x_1, \ldots, x_k) \neq y_2 \mid y_1 \doteq y_2)$ from the equality axioms.
2. Write out in first-order logic the statements that there are at least, at most, and exactly three elements x such that $A(x)$ holds.
3. How many critical pairs can be computed between the rewrite rules in the final set of Example 7.2.14? Please count them by the pair of rules.
4. Adding the equation $a * b \doteq b * a$ to the rewrite system of Example 7.2.14, please use the Knuth-Bendix completion to obtain a canonical rewrite system with LPO and the precedence $a \succ b \succ i \succ * \succ e$.
5. Given the following set of ground equations

$$E = \{I \doteq J, \ K \doteq L, \ A[I] \doteq B[K], \ J \doteq A[J], \ M \doteq B[L]\},$$

run the Knuth-Bendix completion on E with KBO, assuming $B \succ A \succ M \succ L \succ K \succ J \succ I$, and show how each rewrite rule is generated by the procedure.

6. Let $W = \{aaa \doteq \epsilon, bbb \doteq \epsilon, ababab \doteq \epsilon\}$ be equations over $\{a, b\}^*$. Convert W into equations of terms and run the Knuth-Bendix completion on it with the simplification order LPO (or RPO) and the precedence $b \succ a$.

7. Define formally the concept of *critical pair* for string rewrite systems and illustrate the concept by $R = \{ab \to ba, bc \to cb\}$. Assuming $a \succ b \succ c$ and strings are compared lexicographically. Can you obtain a canonical rewrite system from R? Why?

8. Assuming $x \succ y \succ z$, please run the Knuth-Bendix completion on the following polynomial systems:

(a) (1) $x^7 y^2 + x^3 y^2 - y + 1 \doteq 0$
    (2) $\qquad\qquad xy^2 - x \doteq 0$

(b) (1) $x^3 - x^2 y - x^2 z + x \doteq 0$
    (2) $\qquad\qquad x^2 y - z \doteq 0$

(c) (1) $\qquad\qquad x^4 y^2 - z \doteq 0$
    (2) $\qquad\qquad x^3 y^3 - 1 \doteq 0$
    (3) $\qquad\qquad x^2 y^4 - 2z \doteq 0$

9. Prove Lemma 7.3.4.

10. Given the definitions of *add* and *mul* in Example 7.3.3, prove by structural induction the following inductive theorems:

(a) $\quad add(x, s(y)) \doteq s(add(x, y))$
(b) $\qquad add(x, y) \doteq add(y, x)$
(c) $\qquad mul(x, y) \doteq mul(y, x)$
(d) $add(add(x, y), z) \doteq add(x, add(y, z))$
(e) $mul(mul(x, y), z) \doteq mul(x, mul(y, z))$

11. Prove that the following rules from Example 7.3.3 are terminating by a simplification order:

(a) $\quad exp(x, 0) \to s(0)$
(b) $exp(x, s(y)) \to mul(x, exp(x, y))$

where $exp(x, y)$ computes x^y, the exponential function. Prove by structural induction that $exp(x, add(y, z)) \doteq mul(exp(x, y), exp(x, z))$.

12. Prove that the following rules are terminating by a simplification order:

    (a) $fib(0) \to 0$
    (b) $fib(s(0)) \to s(0)$
    (c) $fib(s(s(x))) \to add(fib(s(x)), fib(x))$
    (d) $fsum(0) \to 0$
    (e) $fsum(s(x)) \to add(fsum(x), fib(s(x)))$

    where $fib(x)$ stands for the Fibonacci number and $fsum(x)$ stands for the summation of the first x Fibonacci numbers. Add these rules to Example 7.3.3 and prove by the structural induction that $fsum(x) \doteq pre(fib(s(s(x))))$ is true.

13. Prove that the following rules are terminating by a simplification order:

    (a) $A(0, y) \to s(y)$
    (b) $A(s(x), 0) \to A(x, s(0))$
    (c) $A(s(x), s(y)) \to A(x, A(s(x), y))$

    where $A(x, y)$ stands for the Ackermann's function. Add these rules to Example 7.3.3 and prove by the structural induction that $A(s(0), y) \doteq s(s(y))$ and $A(s(s(0)), y) \doteq s(s(s(add(y, y))))$.

14. Let $C = \{0, s, p\}$ be the constructors to model the set of integers with $E_C = \{p(s(x)) \doteq x, \ s(p(x)) \doteq x\}$:

    (a) Please provide a C-complete definition for add (integer addition) and sub (integer subtraction).
    (b) Prove that add is commutative and associative.
    (c) Prove $sub(add(x, y), y) \doteq x$ and $add(sub(x, y), y) \doteq x$.

15. Given the definitions of app and rev in Example 7.3.10, prove that

    (a) $app(x, nil) \doteq x$
    (b) $app(x, app(y, z)) \doteq app(app(x, y), z)$
    (c) $rev(app(x, y)) \doteq app(rev(y), rev(x))$

    are inductive theorems by structural induction.

16. Given the definition of $isort$ in Example 7.3.12, prove that $sorted(isort(y)) \doteq \top$ is an inductive theorem by structural induction.

17. Use Prover9 to run the Knuth-Bendix completion on the following:

    (a) The three equations in Example 7.2.8
    (b) The equation $((x \uparrow y) \uparrow z) \uparrow (x \uparrow ((x \uparrow z) \uparrow x)) \doteq z$ in Proposition 2.4.5.

18. Use Mace4 to decide if there exist Latin squares of size between 4 and 9 for each of the following constraints (called "short conjugate-orthogonal identities"), respectively.

Code	Constraint			Name
QG3	$(x * y) * (y * x)$	$\doteq$	x	Schröder quasigroup
QG4	$(y * x) * (x * y)$	$\doteq$	x	Stein's third law
QG5	$((y * x) * y) * y$	$\doteq$	x	
QG6	$(x * y) * y$	$\doteq$	$x * (x * y)$	Schröder's first law
QG7	$(y * x) * y$	$\doteq$	$x * (y * x)$	Stein's second law
QG8	$x * (x * y)$	$\doteq$	$y * x$	Stein's first law
QG9	$((x * y) * y) * y$	$\doteq$	x	C_3-quasigroup

19. Repeat the previous problem when an additional constraint, $x * x \doteq x$, is added into the axiom set.

References

1. Alex Sakharov, *Equational Logic*, in Eric W. Weisstein (ed.) MathWorld–A Wolfram Web Resource, *mathworld.wolfram.com/EquationalLogic.html*, retrieved 2023-06-20.
2. Nachum Dershowitz and Jean-Pierre Jouannaud, *Rewrite Systems*, in Jan van Leeuwen (ed) Handbook of Theoretical Computer Science. Vol. B. Elsevier. pp. 243–320, 1990
3. Robert S. Boyer and J. Strother Moore, *A computational logic*, Academic Press, New York, 1979
4. Hantao Zhang, *Reduction, superposition and induction: automated reasoning in an equational logic*, Ph.D. Thesis, Rensselaer Polytechnic Institute, New York, 1988
5. Larry Wos and Gail Pieper, *Collected Works of Larry Wos*, World Scientific, January 2000, ISBN: 9810240015

Part III
Logic in Programming

Chapter 8
Prolog: Programming in Logic

Prolog is a programming language whose name comes from programming in logic. Unlike many other programming languages, Prolog comes from first-order logic and is intended primarily as a declarative programming language: the program is expressed in terms of relations, represented as Prolog's facts and rules. A computation is initiated by running a query over these relations.

Based upon Robinson's resolution principle, Prolog was first developed in 1972 by Alain Colmerauer with Philippe Roussel, called "Marseille Prolog." The Japanese Fifth-Generation Computer Project, announced in 1981, adopted Prolog as the default language of a computer and thereby focused considerable attention on the language and its capabilities. Prolog is an example of the fourth-generation programming languages supporting the declarative programming paradigm and is widely regarded as an excellent language for "exploratory" and "prototype programming" [1].

There are several free implementations of Prolog available today, such as GNU's gprolog [2] and SWI-prolog, and they are the descendants of Marseille Prolog. The language has been used for theorem proving, expert systems, automated planning, and natural language processing and is a popular tool in artificial intelligence and computational linguistics. Prolog is well-suited for specific tasks that benefit from rule-based logical queries such as databases and voice control systems.

8.1 Prolog's Working Principle

In logic programming, program logic is expressed in terms of a set of facts and rules, which define relations. A computation is initiated by running a query over these relations.

8.1.1 Horn Clauses in Prolog

In logic programming, relations and queries are usually constructed by Horn clauses. In Sect. 3.4.2, we have seen that Horn clauses are clauses with at most one positive literal. In Prolog, Horn clauses are divided into three types:

- **fact**: a unit positive clause
- **rule**: a clause with one positive literal and one or more negative literals
- **query**: a negative clause

A Prolog program consists of facts and rules, which contain exactly one positive literal and are collectively called *definite clauses*. A rule clause is $(A \mid \overline{B_1} \mid \ldots \mid \overline{B_n})$, where $A, B_1, \ldots, B_n, n \geq 1$, are atomic formulas (atoms). A rule clause is written in Prolog as

$$A :- B_1, \ldots, B_n.$$

This format resembles closely the formula:

$$B_1 \wedge \cdots \wedge B_n \rightarrow A.$$

In other words, the symbol :− is the inverse of → and the commas are ∧ in Prolog. For the above rule, A is called the *head* and "$B_1, \ldots, B_n$" the *body*. It is read as "A is true if "$B_1, \ldots, B_n$" are all true." Prolog's facts are clauses with an empty body and do not need the symbol ":-". Equivalently, we may write a fact in Prolog as "A :- true."

Prolog uses negative clauses $(\overline{B_1} \mid \ldots \mid \overline{B_n})$ for queries. Each Prolog query is written as ?- $B_1, \ldots, B_n$, which resembles the equivalent formula $\neg(B_1 \wedge \cdots \wedge B_n)$.

Example 8.1.1 Suppose we want to build the family tree of a big family. We may state the father-child and mother-child relations by facts and define other relations by rules. Here is a Prolog program for the family tree.

```
mom(rose, john).     % Rose is the mother of John
mom(rose, bob).
mom(rose, ted).
papa(joe, john).     % Joe is the father of John
papa(joe, bob).
papa(joe, ted).
papa(ted, patrick).
papa(john, caroline).
papa(bob, kathleen).
papa(bob, joseph).
papa(bob, mary).
parent(X, Y) :- papa(X, Y).   % X is parent of Y if X is papa of Y.
parent(X, Y) :- mom(X, Y).    % X is parent of Y if X is mom of Y.
grandpa(X, Y) :- papa(X, Z), parent(Z, Y).
grandmo(X, Y) :- mom(X, Z), parent(Z, Y).
```

From this program, we see that variables in Prolog are identifiers whose first character is a capital letter. Once the program is loaded into Prolog, we may issue queries after the symbol ?-.

To query who are the parents of ted, after the symbol ?-, we type parent(X, ted). After we hit a "return," Prolog will answer with X = joe. If you type ";", which means for "more solutions," Prolog will give you another answer: X = rose. If you hit ";" again, Prolog will terminate with "no," because it runs out of answers.

To query the sibling relation, we define the sibling relation as a rule

```
sibling(X, Y) :- parent(Z, X), parent(Z, Y), X =\= Y.
              % X and Y share parent Z.
```

where =\= stands for "not equal." Now, the result can be obtained by the query

```
?- sibling(X, Y).
```

On the other hand, we may issue the query

```
?- parent(Z, X), parent(Z, Y), X =\= Y.
```

This query prints out the common parent Z of X and Y. For instance, the first answer found by Prolog is

```
X = john, Y = bob, Z = joe.
```

□

8.1.2 Resolution Proof in Prolog

In Prolog, associated with every successful query, there exists a resolution proof. For the Prolog program in Example 8.1.1, we have an answer X = joe for the query "?- parent(X, ted)." The corresponding resolution proof is given below, where p stands for parent and f for papa.

Prolog format	clausal format	
1 papa(joe, ted).	$(f(joe, ted))$	// input
2 parent(X, Y) :- papa(X, Y).	$(p(X,Y) \mid \overline{f(X,Y)})$	// input
3 ?- parent(X, ted).	$(\overline{p(X, ted)})$	// query
4 ?- papa(X, ted).	$(\overline{f(X, ted)})$	// resolvent of 2 and 3
5 ⊥	⊥	// resolvent of 1 and 4.

In the last resolution, the mgu is $\{X \mapsto joe\}$, which is the answer to the query. The query comes from the negation of the formula $\exists X\, p(X, ted)$, which is a logical consequence of the input clauses.

As another example, we list below the corresponding resolution proof for the query "parent(Z, X), parent(Z, Y), X =\= Y." Note again that "?- A, B, C" denotes the clause $(\neg A \mid \neg B \mid \neg C)$.

1 papa(joe, john).		// input
2 papa(joe, bob).		// input
3 parent(X, Y) :- papa(X, Y).		// input

```
4 ?- parent(Z, X), parent(Z, Y), X =\= Y.  // query
5 ?- papa(Z, X), parent(Z, Y), X =\= Y.    // resolvent of 3 and 4
6 ?- parent(joe, Y), john =\= Y.           // resolvent of 1 and 5
7 ?- papa(joe, Y), john =\= Y.             // resolvent of 3 and 6
8 ?- john =\= bob.                         // resolvent of 2 and 7
9 ⊥                                        // ⊥ denotes the empty clause,
```

The mgus used in the resolutions provide the solution of the query: the resolution between 1 and 5 provides X = john, Z = joe and the resolution between 2 and 7 provides Y = bob.

Both above resolution proofs are a negative, input and linear resolution proof. That is, every resolution includes a negative clause and an input clause as parents, and the latest resolvent (which is negative) is used in the next resolution. This is true for every resolution proof found by Prolog. Corollary 3.4.11 claims that, for propositional Horn clauses, if there exists only one negative clause, then the negative, input, and linear resolution is complete for Horn clauses. This result can be extended to first-order clauses. In other words, the negative, input, and linear resolution strategy is complete for first-order Horn clauses. However, the depth-first search strategy used by Prolog is not a *fair* strategy. Thus, even if there exists a negative, input, and linear resolution proof for a set of Horn clauses, Prolog may fail to find a proof. We will discuss this issue when talking about recursions in Prolog.

In short, given a query, the Prolog engine attempts to find a resolution proof from the program and the query. If the empty clause can be derived, it follows that the query, along with the found substitution being applied, is a logical consequence of the program. This makes Prolog particularly useful for symbolic computation and language parsing applications.

Example 8.1.2 Suppose we are asked to write a program to convert a propositional formula into negation normal form. This would be a day-long exercise using a conventional programming language. Suppose we use only three propositional variables, say p, q, r, and the logical operators are "n" for ¬, "a" for ∧, "o" for ∨, "i" for →, and "e" for ↔, the Prolog program is given below, where $nnf(X, Y)$ is true iff Y is an NNF of X.

```
pvar(p).    % definition of propositional variables
pvar(q).
pvar(r).

literal(n(L)) :- pvar(L). % definition of literals
literal(L) :- pvar(L).

nnf(n(n(X)), NNF) :- nnf(X, NNF). % double negation

nnf(a(A, B), a(N1, N2)) :- nnf(A, N1), nnf(B, N2). % and
nnf(n(a(A, B)), o(N1, N2)) :- nnf(n(A), N1), nnf(n(B), N2).

nnf(o(A, B), o(N1, N2)) :- nnf(A, N1), nnf(B, N2). % or
nnf(n(o(A, B)), a(N1, N2)) :- nnf(n(A), N1), nnf(n(B), N2).
```

```
nnf(i(A, B), o(N1, N2)) :- nnf(n(A), N1), nnf(B, N2).  % implication
nnf(n(i(A, B)), a(N1, N2)) :- nnf(A, N1), nnf(n(B), N2).

nnf(e(A, B), o(a(NA, NB), a(NnA, NnB))) :-    % equivalence
      nnf(A, NA), nnf(n(A), NnA), nnf(B, NB), nnf(n(B), NnB).
nnf(n(e(A, B)), o(a(NA, NnB), a(NnA, NB))) :-
      nnf(A, NA), nnf(n(A), NnA), nnf(B, NB), nnf(n(B), NnB).

nnf(X, X) :- literal(X).        % NNF of literal is itself
```

The following is the result of a query to the above program:

```
?- nnf(n(o(n(a(p,q)),e(p,q))), X).
X = a(a(p,q),o(a(p,n(q)),a(n(p),q))).
```

The above Prolog program could be completed in a few hours by a student. □

8.1.3 A Goal-Reduction Procedure

The Prolog engine can be regarded as a goal-reduction process: for each atom G in the original query, the engine will search for a rule in the Prolog program whose head unifies with G and recursively apply the engine to the subgoals, which are the atoms in the body of the rule. If the body is empty, then this rule is a fact. When all the subgoals of G are solved, or when the body is empty, goal G is considered as *solved* and a substitution for the variables of G is returned; otherwise, the label "fail" is returned.

Procedure 8.1.3 Assuming P is a Prolog program, procedure $engine_1$ takes a single goal G as input and returns a substitution as output if one rule of P can solve G; otherwise, $fail$ is returned. Procedure $engine$ takes a list Gs of goals as input and returns a substitution as output if every goal of Gs is solved by P.

proc $engine_1(G)$
1   **for** $(H :- B) \in P$ **do**
2   **if** (H and G are unifiable with mgu σ)
3           $\theta := engine(B\sigma)$
4   **if** $\theta \neq fail$ **return** $\sigma\theta$
5   **return** $fail$

proc $engine(Gs)$
1   $\theta := [\,]$
2   **for** $A \in Gs$ **do**
3       $\sigma := engine_1(A\theta)$
4   **if** $\sigma = fail$ **break else** $\theta := \theta\sigma$
5   **if** $\sigma = fail$ **return** $fail$ **else return** θ

The following are assumed in Prolog: $H := B$ is a fact if B is empty. $(H := B) \in P$ selects $H := B$ in P (and $A \in Gs$ selects A in Gs) from first to last. The first call is $engine(Qs)$, where Qs are the goals in a query.

Obviously, $engine_1$ and $engine$ are mutually recursive procedures. If we examine the recursion tree of $engine_1$ and $engine$, the links going out of the $engine_1$ node are labeled by rules of P; and the links going out of the $engine$ node are labeled by a goal in Gs. The $engine_1$ node is an ∨-node: if one of its children succeeds, it will succeed. The $engine$ node is an ∧-node: if one of its children fails, it will fail. In other words, the recursion tree of $engine$ and $engine_1$ is an ∧-∨ tree.

Procedure $engine_1$ as given intends to find one answer; if more answers are desired, then line 4 of $engine_1$ needs to be modified to remember which rule is used to get σ and will use the next rule when more solutions are desired. In this case, the $engine_1$ node may have more than one successful child.

A leaf node in the ∧-∨ tree is said to be *ok* if either *(a)* it is an $engine$ node and Gs is empty or *(b)* it is an $engine_1$ node and G is a built-in atom evaluated to be true. An internal node in the ∧-∨ tree is said to be *ok* if either *(c)* it is an $engine$ node and every child node is ok or *(d)* it is an $engine_1$ node and one of its children is ok.

In this case, the ∧-∨ tree has a solution iff its root node is *ok*, and it represents all the solutions dictated by the ∨ nodes.

Example 8.1.4 The following program implements a depth-first search procedure for a directed graph:

```
% s(u, v): directed edge (u, v)
1: s(a,b).
2: s(a,c).
3: s(b,d).
4: s(c,g).
5: s(d,g).
6: goal(e).   % goal(x): x is a goal
7: goal(g).
8: dfs(N, N) :- goal(N).  % dfs(startNode, solution)
9: dfs(N, l(N, R)) :- s(N, X), dfs(X, R).
```

There are two solutions to the query "?- dfs(a, N).". That is, N = l(a, l(b, l(d, g))) and N = l(a, l(c, g)), where a term like $l(a, l(c, g))$ represents a path (a, c, g) in the graph. Later, after introducing the list, the same path can be represented by [a, c, g]. Figure 8.1 shows the ∨-nodes in the search tree for the query ":- dfs(a, N)." The ∧-nodes are implicit, the shaded nodes are the *failed* nodes, and the other nodes are *ok* nodes. □

Note that the two recursive procedures $engine$ and $engine_1$ just illustrate the goal-reduction process; the actual Prolog engine is an interactive procedure and uses a stack to save all the information in the ∧-∨ tree.

Procedure $engine_1$ shows that the order of rules in P is important. Procedure $engine$ shows that the order of atoms in the body of a rule is important. In other words, the links in the ∧-∨ tree are ordered, and it implements a special case of the negative, input, and linear resolution strategy: the literals in a rule are handled

8.1 Prolog's Working Principle

Fig. 8.1 ∨-nodes in the ∧-∨ tree for query ?- dfs(a, N). Clause numbers are given as the labels of the links between the nodes

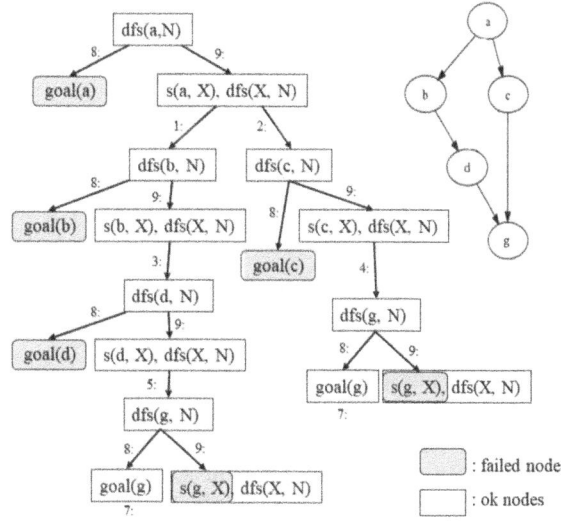

from left to right, and the commas, which represent ∧ in logic, are not commutative. The input clauses are tried to unify with a given goal from first to last. From the viewpoint of search strategies, this is the depth-first search in the ∧-∨ tree.

Example 8.1.5 Adding the following rules to the Prolog program in Example 8.1.1:

```
descen(X, Y) :- parent(Y, X).  % X is a descendant of Y
descen(X, Z) :- parent(Y, X), descen(Y, Z).
```

The last rule is different from the previous rules in that descen is recursively defined. If we ask, "who are the descendants of Joe," the query is "?- descen(X, joe)." Prolog will find that X = john, bob, ted, patrick, caroline, kathleen, joseph, and mary, in the given order. For the first answer, i.e., X = john, the recursive calls of *engine* and *engine*$_1$ go as follows:

$engine_1$(descen(X, joe))
   $engine$([parent(joe, X)])
      $engine_1$(parent(joe, X)])
         $engine$([papa(joe, X)])
            $engine_1$(papa(joe, X)) **return** X = john
. . .
   $engine_1$([descen(X, joe)]) **return** X = john

For the last answer, i.e., X = mary, the recursive calls go as follows. (the calls to *engine* are ignored if the list contains a single goal):

$engine_1$(descen(X, joe))
   $engine$([parent(Y, X), descen(Y, joe)])
      $engine_1$(parent(Y, X))

$engine_1(\text{papa}(Y, X))$ **return** $\{X = \text{mary}, Y = \text{bob}\}$
$engine_1(\text{descen}(\text{bob}, \text{joe}))$
$engine_1(\text{parent}(\text{joe}, \text{bob}))$
$engine_1(\text{papa}(\text{joe}, \text{bob}))$ **return** $\emptyset$
...
$engine([\text{parent}(Y, X), \text{descen}(Y, \text{joe})])$ **return** $\{X = \text{mary}, Y = \text{bob}\}$
$engine_1(\text{descen}(X, \text{joe}))$ **return** $\{X = \text{mary}, Y = \text{bob}\}$

However, if we change the last rule to a logically equivalent one

```
descen(X, Z) :- descen(Y, Z), parent(Y, X).
```

The same query, "?- descen(X, joe)," will produce X = john, bob, ted, caroline, kathleen, joseph, mary, and patrick, the same set of eight answers in different orders, and will cause Prolog to crash if you ask for more answers. If we switch the order of the last two clauses, i.e., the last two clauses are

```
descen(X, Z) :- descen(Y, Z), parent(Y, X).
descen(X, Y) :- parent(Y, X).
```

The same query will cause Prolog to crash without generating any answer. The first-order logic cannot explain the difference because these two clauses are logically equivalent. The procedures *engine* and *engine*$_1$ can explain the difference perfectly:

The initial goal list is [descen(X, joe)] in all the three cases. In the last case, the rule descen(X, Z) :- descen(Y, Z), parent(Y, X) is applied first, and the recursive calls of *engine* and *engine*$_1$ go as follows:

$engine_1(\text{descen}(X, \text{joe}))$
  $engine([\text{descen}(Y, \text{joe}), \text{parent}(Y, X)])$
    $engine_1(\text{descen}(Y, \text{joe}))$
      $engine([\text{descen}(Y1, \text{joe}), \text{parent}(Y1, Y)])$
        $engine_1(\text{descen}(Y1, \text{joe}))$
          $engine([\text{descen}(Y2, \text{joe}), \text{parent}(Y2, Y1)])$
...

The unsolved goal list contains a long sequence of goals like

..., parent(Y3, Y2), parent(Y2, Y1), parent(Y1, Y), parent(Y, X)

until Prolog crashes when it runs out of memory. Here, we changed the variable names because each clause is assumed to have different variables from others. □

The above example shows that Prolog uses a depth-first strategy, which is not a fair strategy, to implement resolution, and becomes an incomplete theorem prover for Horn clauses. As a Prolog programmer, one needs to be careful to avoid the pitfall of infinite loops associated with the depth-first strategy. We will address this issue in the next section.

Example 8.1.6 Hanoi's tower, introduced in the introduction of this book, asks to move n disks of different sizes from peg X to peg Y using peg Z as an auxiliary holding peg. At no time can a larger disk be placed upon a smaller disk. The Prolog code contains two predicate definitions: move writes out the move information; hanoi(N,X,Y,Z) checks if $N = 1$, then moves disk 1 from X to Y; otherwise, it tries to move the top N-1 disks from X to Z; move the bottom disk N from X to Y; and then move the top N-1 disks from Z to Y.

```
move(N,X,Y) :-        % print a move info
   write('Move disk '), write(N), write(' from '),
   write(X), write(' to '), write(Y), nl. % 'nl' = newline

hanoi(1,X,Y,Z) :-     % disk 1 can move freely from X to Y
   move(1,X,Y).
hanoi(N,X,Y,Z) :-     % To move N disks from X to Y,
   N>1, M is N-1,     % where N > 1,
   hanoi(M,X,Z,Y),    % move the top N-1 disks from X to Z;
   move(N,X,Y),       % then move disk N, the largest, from X to Y;
   hanoi(M,Z,Y,X).    % then move the top N-1 disks from Z to Y.

?- hanoi(3, a, b, c).
Move disk 1 from a to b
Move disk 2 from a to c
Move disk 1 from b to c
Move disk 3 from a to b
Move disk 1 from c to a
Move disk 2 from c to b
Move disk 1 from a to b
```

The code comes directly from the description of the solution. The body of the second rule of hanoi contains two recursive calls of hanoi. The recursive calls will terminate because the first argument of hanoi is reduced by one in each successive call. □

8.2 Prolog's Data Types

As a programming language, Prolog has several built-in data types, such as numbers (floats or integers), strings, lists, and associated functions. These data types are subtypes of a supertype, called *term*. In this view, *term* is the only data type in Prolog. Term is either atoms, numbers, variables, or compound terms. Atomic formulas in Prolog are called *callable term*.

8.2.1 Atoms, Numbers, and Variables

We have been using "atom" for "atomic formula"; in this section, we will use "atomic formula" explicitly because "atom" has special meaning in Prolog: an *atom* is either an identifier other than variables or a single-quoted string of characters. Examples of atoms include x, red, "Taco," and "some atom." An atom is a general-purpose name with no inherent meaning.

Prolog provides the predicate atom/1, where atom(X) is true iff X is an atom. For example, atom("some atom") is true; atom(123) is false.

Numbers can be floats or integers. The range of numbers depends on the number of bits (amount of computer memory) used to represent the number. Most of the major Prolog systems support arbitrary length integer numbers. The treatment of float numbers in Prolog varies from implementation to implementation; float numbers are not all that heavily used in Prolog programs, as the emphasis in Prolog is on symbol manipulation.

Prolog provides all the usually predicates on numbers, such as >/2, </2, =</2, >=/2, etc., and arithmetic operations, such as -/1, +/1, +/2, -/2, */2, //2, mod/2, rem/2, gcd/2, lcm/2, abs/1, sign/1, max/2, min/2, floor/1, ceiling/1, truncate/1, sqrt/1, sin/1, cos/1, etc. You use =:=/2 to check if two numbers are equal and =\=/2 for not equal. The symbol =/2 is reserved for unification: s = t returns true if s and t are unifiable.

Prolog also provides predicate number/1 to check if a term is a number or not; integer/1 for being an integer; and float/1 for being a float number with "." For example, float(1.0) returns true; float(1) returns false. Note that integer(1+1) returns false as 1+1 is a compound term. If we check the value of 1+1<3, Prolog will return true because $1 + 1$ will be evaluated to 2 before < is called on (2, 3).

Variables are denoted by an identifier consisting of letters, numbers, and underscore characters and beginning with an uppercase letter or underscore. The underscore alone, "_", denotes a nameless distinct variable. A nameless variable is used when we do not care about the value of the variable. For example, if we define the multiplication relation mul(X, Y, Z), where Z = X*Y, the first rule can be mul(0, _, 0), where the second position is a nameless variable.

Prolog variables closely resemble variables in logic in that they can be instantiated by arbitrary terms through unification. If a variable is instantiated, we say "the variable is bounded." Some operations require the involved variables to be bounded. For instance, to show

    descen(X, Z) :- parent(Y, X), descen(Y, Z).

is a terminating rule in Example 8.1.5, we need to check the bindings of X, Y, and Z. The predicate parent(Y, X) ensures that X and Y will be bounded to atoms when we reach descen(Y, Z). When these variables are bounded by atoms, the atom bounded to Y is one step closer than that of X to Z in the descendant relation, this meaning defines a well-founded ordering $\succ$ such that $X \succ Y$ when we compare descen(X, Z) and descen(Y, Z).

8.2 Prolog's Data Types

In Prolog, there is a special infix predicate called is/2, which is used in the Hanoi example. It resembles the assignment statement of a variable by a numeric expression in a conventional programming language. For example, the value of X is 1+1 is true with X is instantiated with 2. That is, Prolog will evaluate the right side of is, if it is a number, that number is bounded to the variable on the left side of is; otherwise, it will abort with error. In contrast, X = 1+1 will also return true, but it will bound the variable X to the compound term 1+1. As another example, "X = a" will return true and "X is a" will be aborted, because a is not a number.

8.2.2 Compound Terms and Lists

A *compound term* is composed of a predicate or function symbol, called a "functor" and a number of "arguments," which are terms. Compound terms are ordinarily written as a functor followed by a comma-separated list of argument terms surrounded by parentheses. The number of arguments is called the term's arity.

There is a built-in predicate called functor/3 which allows us to extract the functor and its arity from a compound term:

```
?- functor(f(a, b), F, N).
F = f, N = 2
```

Compound terms are indispensable for representing various data structures of conventional languages. For example, to represent a binary tree where each non-empty tree node contains a key, we may use the compound term node(key, leftChild, rightChild). We can define easily predicate tree/1, which checks if its argument is a tree, and predicate dfs/2, which tries to find a key by depth-first search in a tree.

```
tree(nil).                            % nil denotes the empty tree
tree(node(_, Left, Right)) :- tree(Left), tree(Right).

dfs(node(K, _, _), K) :- write('found '), write(K).
dfs(node(_, L, _), K) :- dfs(L, K).   % go to Left branch
dfs(node(_, _, R), K) :- dfs(R, K).   % go to Right branch
```

In the definition of dfs, we used "_" twice in each rule; each "_" represents a unique variable.

One special member of compound terms is *list*, which is an ordered collection of terms. It is denoted by square brackets with the terms separated by commas or in the case of the empty list, by []. Here are some lists of three elements, [1,2,3], [red,green,blue], or [34,tom,[2,3]]. Lists are the replacement of arrays of conventional programming languages.

For a non-empty list, the first member of a list is separated from the rest members by the symbol |. For example, the result of [X | Y] = [1,2,3] is X = 1 and Y = [2, 3] and the result of [X | Y] = [1] is X = 1 and Y = [].

Two lists are unifiable if they are the same length, and all their elements are pairwise unifiable.

```
?-[a,b,c,d]=[Head|Tail].          ?- [(a+X),(Y+b)]=[(W+c),(d+b)].
Head = a,                         W = a,
Tail = [b,c,d]?                   X = c,
yes                               Y = d?
                                  yes
?- [a,b,c,d]=[X,Y|Z].
X = a,
Y = b,                            ?- [(a+X),(Y+b)]=[(W+c),(X+Y)].
Z = [c,d]?                        no
yes
```

8.2.3 Popular Predicates over Lists

Most Prolog implementations provide efficient implementations of the following predicates over lists. We also provide Prolog codes of these predicates; they are not actual implementations.

- list/1: Check if its argument is a list.
    ```
    list([]).
    list([_ | X]) :- list(X).
    ```
- member/2: Check if the first argument is a member of the second argument.
    ```
    member(X, [X | _]).
    member(X, [_ | Y]) :- member(X, Y).
    ```
- append/3: Check if the third list is the concatenation of the first two lists.
    ```
    append([], Y, Y).
    append([F|X], Y, [F|Z]) :- append(X, Y, Z).
    ```
- select/3: select(E, List1, List2) is true when List2 is the result of removing E from List1.
    ```
    select(E, [E|Y], Y).
    select(E, [F|Y], [F|Z]) :- select(E, Y, Z).
    ```
    For instance, select(E, [a,b], Z) has two solutions: (1) E = a and Z = [b]; (2) E = b and Z = [a]. An alternative implementation of select is to use append:
    ```
    select(E, L, R) :- append(X, [E|Y], L), append(X, Y, R).
    ```
- delete/3: delete(E, List1, List2) is true when List2 is the result of removing from List1 all its members that unify with E.
    ```
    delete(_, [], []).
    delete(E, [E|Y], Z) :- delete(E, Y, Z).
    delete(E, [F|Y], [F|Z]) :- E \= F, delete(E, Y, Z).
    ```

where \= means "not unifiable." For instance, delete(E, [a,b], Z) has only one solution: E = a, Z = [b]; the second answer, E = b, Z=[a], could not be generated by the above code, because Prolog treats E \= a as false (i.e., E = a is true).[1] This shows another difference of select and delete.

- reverse/2: reverse(List1, List2) is true when the elements of List2 are in reverse order compared to List1.

    ```
    reverse([], []).
    reverse([F|X], Z) :- reverse(X, Y), append(Y, [F], Z).
    ```

- permutation/2: permutation(List1, List2) is true when List2 is a permutation of List1.

    ```
    permutation([], []).
    permutation([F|X], Y) :- permutation(X, Z), select(F, Y, Z).
    ```

- length/2: length(List, N) is true when the length of List is N.

    ```
    length([], 0).
    length([_|X], N) :- length(X, M), N is M+1.
    ```

One common feature of these predicates is that they are all defined recursively. The termination of these definitions is easy to establish as one of the arguments to the recursive calls goes smaller, or we may show that the head of each rule is strictly greater than its body under a simplification ordering.

Due to the relational nature of many built-in predicates, they can be used in several directions. For example, append/3 can be used either to append two lists into one (the first two arguments are input and the third is the output) or to split a given list into two parts (the third argument is the input, and the first two arguments are the output). For instance, the query "append(X, Y, [1,2,3])." will produce the following four answers:

```
X = [], Y = [1,2,3]
X = [1], Y = [2,3]
X = [1,2], Y = [3]
X = [1,2,3], Y = []
```

For this reason, a comparatively small set of library predicates suffices for many Prolog programs. Note that the suggested implementation for length can compute the answer N = 3 from length([a,b,c], N) but cannot generate a list of 3 placeholders, L = [_1, _2, _3], from length(L, 3). Some versions of Prolog, e.g., SWI Prolog, can do so.

For some applications, we need to obtain a set $S = \{x \mid A(x)\}$, that is, S is a set of elements satisfying the condition $A(x)$. It is not easy to write a Prolog program which generates such a set. Fortunately, in most implementations of Prolog, such a set can be obtained by calling bagof(X, Goal(X), S). For example, in Example 8.1.1, we may define children(F) = { C | papa(F, C) }. A query like the following will give us a satisfactory answer:

---

[1] For the same query, the built-in delete in gprolog gives an infinite number of solutions.

```
?- bagof(C, papa(Father, C), Children).
Father = bob, Children = [kathleen,joseph,mary] ? ;
Father = joe, Children = [john,bob,ted] ? ;
Father = john, Children = [caroline] ? ;
Father = ted, Children = [patrick]
```

Some Prolog implementations also provide the set operations, such as intersection/3, union/3, subtract/3, subset/2, when a list is viewed as a set.

In Prolog, a string is a sequence of characters surrounded by double quotes and is equivalent to a list of numeric character codes. One common mistake is to use strings in the write predicate instead of a single-quoted sequence of characters (an atom).

Prolog also provides some debugging tools. For instance, to trace the execution, you may use trace to turn on the tracing and notrace to turn off it.

8.2.4 Sorting Algorithms in Prolog

To illustrate how an algorithm is implemented in Prolog, we show below the Prolog code of insertion sort and quicksort.

Example 8.2.1 The following program defines four predicates for the implementation of insertion sort and quicksort: isort/2, insert/3, split/4, and qsort/2.

```
% isort(L, R): R is the output of L by insert sort
isort([], []).
isort([X|L], R) :-
   isort(L, T),           % sort L into T
   insert(X, T, R).       % insert X into T to obtain R.

% insert(X, L, R): insert X into sorted list L and R is the result.
insert(X, [], [X]).
insert(X, [Y|L], [X, Y|L]) :- X =< Y.
insert(X, [Y|L], [Y|R]) :- X > Y, insert(X, L, R).

% split(P, I, S, B): split the input list I into S and B by pivot P.
split(_, [], [], []).
split(P, [H|I], [H|S], B) :- H =< P, split(P, I, S, B).
split(P, [H|I], S, [H|B]) :- H > P, split(P, I, S, B).

% qsort(L, R): L is the input list and R is the output by quicksort
qsort([], []).
qsort([X|Y], Z) :- split(X, Y, S, B),
    qsort(S, S1), qsort(B, B1), append(S1, [X|B1], Z).
```

Other sorting algorithms can be defined in a similar way. □

The above code is lucid, neat and easy to understand and maintain. This is a typical feature of a declarative programming language. Of course, its execution is not as efficient as conventional programming languages. That is why Prolog is used mostly for prototype programming. To compensate this weakness, most Prolog implementations provide efficient built-in implementations for popular functions like arithmetic operations or sorting.

Another feature of declarative programming is that the correctness of the algorithm can be verified. For sorting, the correctness is ensured by the two properties: (1) the output is sorted; (2) the output is a permutation of the input list. In Prolog, `permutation` is a built-in function and we can define `sorted` with ease.

Example 8.2.2 The following Prolog program defines the predicate sorted(*L*), which returns true iff *L* is a sorted list, and some utility predicates for the verification purpose.

```
sorted([]).
sorted([_]).
sorted([X,Y|L]) :- X=<Y, sorted([Y|L]).

verifyIsort(X, Y) :- isort(X, Y), permutation(X, Y), sorted(Y).
verifyQsort(X, Y) :- qsort(X, Y), permutation(X, Y), sorted(Y).
slowsort(X, Y) :- permutation(X, Y), sorted(Y).
```

In the above code, the first three clauses define `sorted/1`. The next two clauses illustrate the verification of `isort` and `qsort`. The last clause shows how the requirement can be the code itself. The code of `permutation/2` can be found in Sect. 8.2.3. □

Most Prolog implementations provide two versions of predicates for sorting: `sort/2`, which removes duplicate elements, and `msort/2`, which does not remove duplicates. For instance, the query "`sort([b,a,c,a], X).`" will produce X = [a,b,c], and "`msort([b,a,c,a], X).`" will produce X = [a,a,b,c]. These built-in predicates run faster than the user-defined sorting algorithms.

8.3 Recursion in Prolog

In most programming tasks, we often need to repeatedly perform some operation either over a whole collection of data or until a certain point is reached. Conventional programming languages provide various loop controls for performing such tasks. The only way we can do this in Prolog is by recursion. In Sect. 8.2.3, we have seen many Prolog programs which contain recursive definitions. Recursion allows us to write clear and elegant code. Data in such formulas or lists are often recursively defined and best processed recursively. Recursion allows Prolog to perform complex searches of a problem space without any dedicated algorithms. Without recursion, Prolog goes nowhere. To master Prolog, we have to learn how to use recursion correctly and efficiently.

Any recursive definition, whether in Prolog or some other languages, needs two things: *base cases* and *recursive cases*. The base cases specify a condition of when the recursion terminates. Without this, the recursion would never stop! For example, the base case of append(X, Y, Z) is append([], Y, Y) when X=[]. In a Prolog program, the clauses for the base cases almost always come before those for the recursive cases, which ask a program to handle a similar problem of smaller size. For example, append([U|X], Y, [U|Z]) :- append(X, Y, Z).

8.3.1 Program Termination

The danger of recursion is that the program may loop forever and never terminate. The termination problem is more complicated in Prolog than in conventional programs because the input to a Prolog program may contain variables, and we may ask for multiple solutions from a single query.

Definition 8.3.1 A Prolog program P is said to have *first termination* on an input x if P terminates with either no solution or the first solution on x. P is said to have *last termination* on x if P terminates after all solutions of x are generated. If P has first (last) termination on any ground input x, then P is said to have *first (last) ground termination*.

Example 8.3.2 Given the following program P, defining ancestor(X, Y) as "X is an ancestor of Y."

    ancestor(tom, jack).
    ancestor(X, Y) :- ancestor(X, Z), ancestor(Z, Y).

P has first termination on the query "?- ancestor(A, B)" with the solution A=tom, B=jack, but not on ?- ancestor(jerry, tom), as the second clause will be used forever during the search. Since the second query is ground, P has no first ground termination. Since last termination implies first termination, P has no last ground termination. □

Prolog is a relational programming language, and relations can hold between multiple entities that are reported on backtracking. Therefore, first termination does not fully characterize the procedural behavior of a Prolog query, and we need the concept of last termination. We may use the termination of the query "?- Q, false." to check if Q has last termination, because all the solutions of Q must be generated before the query terminates.

Consider the member/2 program in Sect. 8.2.3, the program has ground termination when the second argument of *member* is ground. However, the query "?- member(X, Y), false." will fail with a stack overflow error, thus showing that member(X, Y) has no last termination.

Example 8.3.3 Given a propositional formula A in negation normal form (NNF, see Sect. 2.3.1), we may use the following program to decide if A is satisfiable,

8.3 Recursion in Prolog

assuming the formula uses Prolog variables for propositional variables, and, or, not for $\wedge$, $\vee$, $\neg$, respectively.

```
sat(true).                    % positive literal
sat(not(false)).              % negative literal
sat(or(X, _)) :- sat(X).
sat(or(_, Y)) :- sat(Y).
sat(and(X, Y)) :- sat(X), sat(Y).
```

The first two clauses are the base cases of sat which handle literals, and the next three clauses are the recursive cases which handle or and and.

It is easy to see that the argument to the recursive call is strictly less than the argument in the head of each rule. Thus, we expect that the program will terminate with the test examples. Now, we create three queries to test the above program:

```
test1(X, Y) :- sat(and(not(X), X)).     % test examples
test2(X, Y) :- sat(and(X, not(Y))).
test3(X, Y) :- sat(and(X, or(not(X), Y))).
```

The query "?- test1(X, Y)." will say no; thus, sat has last termination on and(not(X), X). For the second query, "?- test2(X, Y).", Prolog will give the first solution X=true, Y=false. The third query "?- test3(X, Y)." will produce the first solution X=true, Y=true. Thus, sat has first termination on the last two queries.

However, if we look for more answers from ?- test2(X, Y), we will see the following answers:

```
X = not(false), Y = false
X = or(true,_), Y = false
X = or(not(false),_), Y = false
X = or(or(true,_),_), Y = false
X = or(or(not(false),_),_), Y = false
X = or(or(or(true,_),_),_), Y = false
X = or(or(or(not(false),_),_),_), Y = false
...
```

These answers are correct in Prolog, but not correct for propositional satisfiability (which asks for the models of a formula). If you ask for more answers from ?- test3(X, Y), you will see an infinite number of answers, too. Thus, sat has last termination neither on and(X, not(Y)) nor on and(X, or(not(X), Y)). It is easy to check that the program has both first and last ground terminations by an induction on the term structure of the input. □

The problem illustrated by the above example is that when variables appearing in the arguments of a recursive program are not bounded, they can be substituted by terms containing other unbound variables, thus creating an infinite number of possibilities. This is the same reason why member(X, Y) has no last termination.

8.3.2 Focused Recursion

A poor definition of a recursive predicate may cause the program to loop forever. In Chap. 11, it is shown that checking if a Prolog program terminates on a given input is an undecidable problem. Thus, people are interested in sufficient conditions which ensure the termination of a Prolog program. Focused recursion is one of such conditions.

Definition 8.3.4 A recursive predicate is said to be *focused* on an input t_0 if there exists a well-founded order $\succ$ such that $t_1 \succ t_2 \succ \cdots \succ t_n$, where t_i are the values of the same argument appearing in the predicate's consecutive recursive calls.

The argument in the above definition is called the *focused position* of the predicate. Given the definition of append/3 in Sect. 8.2.3, the predicate append is focused on the first argument when the first argument is bound to a list of items (which can be variables) of finite length. Each recursive call reduces the number of items in the list by one. On the other hand, if the first argument is bound to a variable, the call may go on forever. The same can be said about member/2 or sat in Example 8.3.3, where one sequence of calls of sat in the query "test2(X, Y)." may contain the following values as the argument of sat:

```
and(X, not(Y)), X, or(X1, _), X1, or(X2, _), X2, or(X3, _), X3, ...
```

Example 8.3.5 In Example 8.3.2, the transitivity of the ancestor relation is expressed by the rule ancestor(X, Y) :- ancestor(X, Z), ancestor(Z, Y). This rule cannot be used in Prolog as it causes an infinite loop. To define ancestor(X, Y) as "X is an ancestor of Y," we may use the parent/2 relation, which is defined by a set of ground facts, and can be represented by a directed acyclic graph (DAG) among individuals. Then, we can show that ancestor/2 defined by the following rules is a focused recursion on the first argument.

```
ancestor(X, Y) :- parent(X, Y).
ancestor(X, Y) :- parent(X, Z), ancestor(Z, Y).
```

Let $X_0, X_1, \ldots, X_n$ be the values of the first argument in the recursive calls of ancestor. Under the assumption that parent defines a DAG among the individuals, we can see that the distance from X_i to Y in the DAG is reduced by one when i increases by one. Thus, the length of longest paths of the DAG provides an upper bound for the number of recursive calls of ancestor. It is easy to see that the ancestor relation is the transitive closure of the parent relation. □

Proposition 8.3.6 *If a recursive predicate P is focused on a ground input, then P has last termination.*

The proof is left as an exercise. Note that the other arguments of P may contain variables.

8.3.3 Tail Recursion

In Prolog, due to the power of logic variables, many predicates can be naturally written in a tail recursive way, where the recursion happens in the last position of the body. For example, the predicates `list, member, append, select`, and `delete` are tail recursive as given in Sect. 8.2.3. On the other hand, `reverse, permutation`, and *length* are not tail recursive.

In many cases, tail recursion is good for performance. For conventional programs, a smart compiler can change tail recursion to an interactive one, thus saving the time and space of using a stack to store the environments of a recursive call. In Prolog, the same technique can be used so that a tail recursive call means that the Prolog system can automatically reuse the allocated space of the environment on the local stack. In typical cases, this measurably reduces memory consumption of your programs, from $O(N)$ in the number of recursive calls to $O(1)$. Since decreased memory consumption also reduces the stress on memory allocation and garbage collection, writing tail recursive predicates often improves both space and time efficiency of your Prolog programs.

Example 8.3.7 The `reverse/2` predicate defined in Sect. 8.2.3 is copied here:

```
reverse([], []).
reverse([F|X], Z) :- reverse(X, Y), append(Y, [F], Z).
```

Suppose the query is "?- `reverse([a, b, c, d], L).`" The subgoal list sees ?- `reverse([], []), append([], [d], L3), append(L3, [c], L2), append(L2, [b], L1), append(L1, [a], L)`, before it is shrinking. To avoid this long list of subgoals, we may introduce a tail recursive `rev/3`:

```
reverse(X, Y) :- rev(X, [], Y).
rev([], R, R).
rev([U|X], Y, R) :- rev(X, [U|Y], R).
```

Now, for query ?- `reverse([a, b, c, d], L)`, the subgoal list never contains more than one item at any time:

```
?- reverse([a, b, c, d], L)
?- rev([a, b, c, d], [], L)
?- rev([b, c, d], [a], L)
?- rev([c, d], [b, a], L)
?- rev([d], [c, b, a], L)
?- rev([], [d, c, b, a], L)
L = [d, c, b, a].
```

This example illustrates that tail recursion can save space and time. However, you should not overemphasize it, as for beginners, it is more important to understand termination and to focus on clear declarative descriptions. □

8.3.4 A Prolog Program for N-Queen Puzzle

Example 8.3.8 Below is a simple Prolog program for generating all legal positions for the N-queens problem. The predicate queen(N, R) takes N as the number of queens and generates a list R for the row numbers of N queens in the N columns.

□

```
% queen(N, R) generates at first a list R of N variables.
% Assuming one variable per column, then generate values for R,
% which gives the row number of the queen in each column.
queen(N, Res) :-
    length( Res, N),
    gen_list( 1, N, L),          % L = [1, 2, ..., N]
    solution( N, Res, L).        % Res is a permutation of L
                                 % Res = N variables for columns

% gen_list( X, N, R) is true iff R = [X, X+1, X+2, ..., N].
gen_list( X, N, []) :- X>N.
gen_list( X, N, [X | Res]) :- Y is X+1, gen_list( Y, N, Res).

% solution( N, RowValues, List) is true if
%     RowValues represent a list of N non-attacking queens
solution( 0, [], _).                    % run out of columns
solution( C, [R|Others], L) :-          % 1st queen at position [R,C]
    C1 is C-1,
    select( R, L, L2),                  % pick row R for column C
    solution( C1, Others, L2),          % find a partial solution
    noattack( R/C, C1, Others).         % 1st queen doesn't attack Others

% noattack(R/C, N, List) is true if the queen at R/C, where R is the row
%    number, C is the column number, does not attack the other queens at
%    R_i/Ci, where List = [R1, R2, ..., RN], Ci = i for 1<= i <= N.
%    Since R is different from any Ri and C is different from any Ci,
%    we check only if positions R/C and Ri/Ci share a diagonal.
noattack( _, _, []).                    % Nothing to attack
noattack( R/C, C1, [R1 | Others]) :-
    C1-C =\= R1-R,                      % R/C and R1/C1 on different diagonals
    C1-C =\= R-R1,                      % R/C and R1/C1 on different diagonals
    C2 is C1-1,                         % recursive on C2 = C1-1
    noattack( R/C, C2, Others).         % no attacks for Others
```

Given the query "?- queen(4, R).", the first answer is R = [2,4,1,3], which says that the queens are placed at row 2 in the first column, row 4 in the second column, etc., and the second answer is R = [3,1,4,2]. In the above program, each position is generated and checked against the positions generated previously, so that no two positions can attack each other by the move of queens. Using Prolog's depth-first search, the program is simple to write but highly inefficient. When $N > 12$, it will take quite a while to find one solution, much slower than a SAT solver. Sudoku puzzles can also be solved by Prolog using the same idea. However, a Sudoku puzzle cannot be solved efficiently using this generate-and-test approach, as the search space is huge. SAT solvers are a much better tool for Sudoku puzzles.

8.4 Beyond Clauses and Logic

As a general-purpose language, Prolog provides various built-in predicates to perform routine activities like input/output, using graphics and other medias to communicate with the user. These predicates are not given a relational meaning and are only useful for the side effects they exhibit on the system. For example, the predicate `write/1` displays a term and returns true when the display is successful. Besides these kinds of extensions, various kinds of extensions are proposed, some intended to improve the expressive power, and some speed up the execution. Some of these extensions follow the first-order logic; some go beyond the first-order logic.

For instance, since the occur-check in the unification algorithm is expensive, most prolog implementations choose to ignore the occur-check. This does not cause any problem in most applications. If you run into this problem and need a sound unification algorithm, you may call the built-in function `unify_with_occurs_check`.

```
?- X = f(Y,Y).
X = f(Y,Y)
yes

?- X = f(X,Y).
cannot display cyclic term for X

?- unify_with_occurs_check(X, f(X,Y)).
no
```

8.4.1 The Cut Operator "!"

The cut, in Prolog, is a predicate of arity 0, written as !, which always succeeds but cannot be backtracked. It is best used to prevent unwanted backtracking, including the finding of extra solutions by Prolog and to avoid unnecessary computations.

Example 8.4.1 In this contrived example, a cut operator is placed in the first rule:

```
a :- b, !, c.       % 1
a :- c.             % 2
b :- d.             % 3
b :- e.             % 4
c :- d.             % 5
c :- e.             % 6
d :- write('d').    % 7
e :- write('e').    % 8
```

For the query ":- a.", Prolog will try rule 1 first. The goal list now is [b, !, c]. To solve b, Prolog will try rule 3, reducing b to d. Prolog will succeed with rule 7. Now, the cut operator is passed, and Prolog will solve c by rules 5 and 7 with the first success. If we look for the next solution, Prolog will succeed for the second time by solving c of rule 1 with rules 6 and 8. If we ask Prolog for more solutions, Prolog will backtrack to the cut operator. However, the meaning of the cut operator is "no longer backtrack." Thus, Prolog will terminate with only two successes. Without the cut operator, Prolog should find four more solutions: two solutions from *b* of rule 1 with rules 4 and 5–8 and two solutions from rules 2, 5–8. □

The above example illustrates the use of the cut operator: when "!" is reached by $engine_1$, alternative rules for the head of this rule, such as rule 2 for a, are eliminated; alternative rules for all the left siblings of "!", such as rule 4 for b, are also eliminated. The right siblings of "!" are not affected by "!".

Some programmers call the cut a controversial control facility because it was added for efficiency only and is not a function of logic.

A *green cut* is a use of cut which only improves efficiency. Green cuts are used to make programs more efficient without changing program output. For example, in the definition of split predicate of qsort, we may insert a cut after H =< P of the second rule, that is, the three clauses become the following:

```
split(_, [], [], []).
split(P, [H|I], [H|S], B) :- H =< P, !, split(P, I, S, B).
split(P, [H|I], S, [H|B]) :- H > P, split(P, I, S, B).
```

If the current goal *G* is split(3, [2, 4, 1], L, R), then *G* unifies with the heads of the last two rules. Once the subgoal H =< P, i.e., 2 =< 3, succeeds, the last clause will never be tried due to the cut operator, even if *G* unifies with the head of this clause, and we want to see more answers. This cut is a green cut, because we know the subgoal H>P, i.e., 2 > 3, will fail and the last clause cannot be applied. The cut operator saves us the time for trying unifying *G* with the head of the last clause. With or without the cut, the program performs the same way as desired.

A *red cut* is a use of cut which changes the meaning of the program if the cut is missing. For the above split example, if H =< P fails, then H > P will succeed. To save the cost of H > P, we can write the split program as the following:

```
split(_, [], [], []).
split(P, [H|I], [H|S], B) :- H =< P, !, split(P, I, S, B).
split(P, [H|I], S, [H|B]) :- split(P, I, S, B).
```

This program will give us the same result as the previous one for any input. However, this program will perform differently from the following program, where the cut is gone:

```
split(_, [], [], []).
split(P, [H|I], [H|S], B) :- H =< P, split(P, I, S, B).
split(P, [H|I], S, [H|B]) :- split(P, I, S, B).
```

So, the above cut is a red cut. Red cuts are potential pitfalls for bugs. For example, if the order of the two rules is reversed, the green cut will work correctly and the red cut will produce wrong results. If the saving is significant when using a red cut, detailed documentation should be provided along the code. The cut should be used sparingly. Proper placement of the cut operator and the order of the rules is required to determine their logical meaning.

8.4.2 Negation as Failure

A non-Horn clause like $A \vee B \vee \neg C$ can be represented as $\neg B \wedge C \rightarrow A$. Can we write it in Prolog as $A : --\neg B, C$? The answer lies in how to solve the subgoal $\neg B$ by the goal-reduction process. One approach is called "negation as failure": try to solve B. If B succeeds, then $\neg B$ fails; if B fails, then $\neg B$ succeeds.

Using the cut operator together with the built-in predicate fail, we may define this kind of negation, so that properties that cannot be specified by Horn clauses can be specified.

For example, to express "Mary likes any animal except reptiles" in Prolog

```
likes(mary, X) :- reptile(X), !, fail.
likes(mary, X) :- animal(X).
```

We can use the same idea of "cut fail" to define the predicate not, which takes a term as an argument, and use the built-in predicate "call." not(G) will "call" the term G, evaluate G as though G is a goal. If G succeeds, so is call(G), and not(G) will fail. Otherwise, not(G) succeeds.

```
not(G) :- call(G), !, fail.
not(_).
```

Most Prolog systems have a built-in predicate like not. For instance, SWI-Prolog calls it \+. Remember, "not" is not the same as ¬, because it is based on the success/failure of goals. It can, however, be useful:

```
likes(mary, X) :- not(reptile(X)).    % Mary does not like reptiles.
different(X, Y) :- not(X = Y).         % Not equal means "different"
```

Negation as failure can be misleading. Suppose the database held the names of members of the public, marked by whether they are innocent or guilty of some offense, and expressed in Prolog:

```
innocent(peter_pan).
innocent(winnie_the_pooh).
innocent(X) :- occupation(X, nun).
guilty(joe_bloggs).
guilty(X) :- occupation(X, thief).
```

```
?- innocent(einstein).
no.
```

If we add one more rule into the above program

```
guilty(X) :- not(innocent(X)).
```

```
?- guilty(einstein).
yes.
```

It is one thing to show that "Einstein cannot be demonstrated to be innocent"; but it is quite another thing to incorrectly show that he is guilty. To justify the behavior of "negation as failure," people suggest the *closed world assumption*: every ground atomic formula is assumed to be false unless it can be shown to be true by the program. Under this assumption, since we cannot show that innocent(einstein) is true in the program, so innocent(einstein) is assumed to be false, thus it justifies the answer guilty(einstein).

Some disturbing behavior is even more subtle than the innocent/guilty problem and can lead to some extremely obscure programming errors. Here is a Prolog program about restaurants:

```
good_standard(godels).
good_standard(hilberts).
expensive(godels).
reasonable(R) :- not(expensive(R)).

?- good_standard(X), reasonable(X).
X = hilberts
yes

?- reasonable(X), good_standard(X).
no.
```

Logically, the query "good_standard(X), reasonable(X)." is equivalent to the query "reasonable(X), good_standard(X)." Why do we get different answers for what seem to be logically equivalent queries? The difference between the two queries is as follows. In the first query, the variable X is always instantiated when reasonable(X) is executed. In the second question, X is not instantiated when reasonable(X) is executed. The semantics of reasonable(X) differ depending on whether its argument is instantiated. To avoid this kind of problem, some implementations of Prolog ask that G is ground in $not(G)$.

Because of the problems associated with "negation as failure," extra care is needed when using it.

8.4.3 Beyond Clauses

Typically, a rule's body consists of atomic formulas, separated by commas, and denotes the conjunction of subgoals. If the commas are replaced by the semicolons (;/2), then it will denote the disjunction of subgoals. A rule like H :- (C; D), E is no longer a clause: it is equivalent to two clauses: H :- C, E and H :- D, E, because ";" is considered to be ∨. Conjunctions "," and disjunctions ";" can only appear in the body, not in the head of a rule.

Prolog also supports the if-then relation by using the infix symbol "−> /2". A rule like H :- C -> G has the same effect as H :- C, G: if C returns true, G will be called; if C returns false, C -> G returns false. When -> is combined with ";", we then have the if-then-else relation implemented as H :- C -> G1; G2: If C returns true, G1 will be called; otherwise, G2 will be called. Logically, H :- C

-> G1; G2, which is not a clause, is equivalent to two clauses: H :- C, G1 and H :- ¬C, G2 (which is not a Horn clause).

Once a predicate is declared as *dynamic*, Prolog allows the clauses of that predicate to be inserted or removed during the execution. For Example 8.1.1, if we have a file which contains the pair of father-children names, we can declare papa as dynamic: dynamic(papa). During the execution, we read names into Father and Child and call assertz(papa(Father, Child)), which adds the clause papa(Father, Child) at the end of all existing clauses of the predicate papa. If we want to add the new clause at the beginning of the existing clauses, the command is asserta/1. To remove a clause during the execution, the command is extract/1. This set of commands allows the user to dynamically change the program that occurs as the result of executing assertz/1 or retract/1, which removes the clause added by assertz/1. The change does not affect any activation that is currently being executed. Thus, the database is frozen during the execution of a goal, and the list of clauses defining a predicate is fixed at the moment of its execution.

8.4.4 Variations and Extensions

Since the birth of Prolog, other paradigms for logic programming have also been introduced. In this section, we briefly describe some of them.

Datalog

Datalog is a declarative programming language. Originally, it was mainly designed as a query language for deductive databases; but it can be used in other domains (like program analysis and big data processing). Syntactically, Datalog is a subset of Prolog and it uses Horn clauses (facts and rules) as its program code. For example, it does not have Prolog's cut operator. Unlike Prolog's depth-first restricted resolution strategy, Datalog typically adopts a bottom-up evaluation strategy, which starts with the facts in the program and repeatedly applies the rules to obtain new facts.

Constraint Logic Programming

In Prolog, the subgoals from the body of a Prolog rule are solved from left to right. Constraint logic programming (CLP) extends Prolog by relaxing this left-to-right order for certain subgoals. A *constraint* is a predefined relation on variables, e.g., $X + Y > 8$, which appears also in Prolog programs. The evaluation of a constraint is similar to that of Prolog. Unlike Prolog, when the CLP interpreter encounters a constraint which is not evaluable, it will put the constraint in a constraint store.

Later, if the constraint store is found to be unsatisfiable, the interpreter backtracks. If the constraints are satisfiable, a solution will be provided.

Answer Set Programming

Answer set programming (ASP) is another kind of declarative programming paradigm for solving constraint satisfaction problems. It is based on the stable model (answer set) semantics of logic programming. ASP follows the model generation approach rather than the theorem proving approach. Answer set solvers try to generate stable models using backtrack search procedures (enhanced with techniques like the conflict-driven mechanism in modern SAT solvers). And a solution is given by a model of the problem specification. Some ASP solvers are available: SMODELS (www.tcs.hut.fi/Software/smodels), CMODELS (www.cs.utexas.edu/~tag/cmodels) and CLASP (potassco.org/clasp).

Exercises

1. Given the following Prolog program
    ```
    append([], Y, Y).
    append([H | X], Y, [H | Z]) :- append(X, Y, Z).
    ```
    Find the four solutions for the query "?- append(X, Y, [a, b, c])." For each of the four solutions, find the corresponding resolution proof of each solution.
2. Given the program in Example 8.1.2, please answer the following questions:
    (a) What is the result of the program when converting $\neg(p \rightarrow (q \vee \neg p))$ to NNF? Please display your NNF using both Prolog and conventional formats.
    (b) Do you get the same result if you change the order of clauses? Please explain all possible cases.
    (c) What is the resolution proof associated with your result?
3. For the program in Example 8.3.3, place the clause for handling and before those for or, and run the query "?- test2(X, Y)." for three answers. Please produce the resolution proof for each of the three answers.
4. Implement in Prolog the selection sort, $\mathrm{ssort}(X, Y)$, where X is a list of integers and Y is the result of sorting X. You will need to implement maxNum(L, A), which returns true when A is maximal among the numbers in list L. Your selection sort will move repeatedly the maximal number from the input to the output.
5. Implement in Prolog the merge sort, $\mathrm{msort}(X, Y)$, where X is a list of integers and Y is the result of sorting X.

6. The predicate `slowsort` is defined in Example 8.2.2. Construct the resolution proof for the query "?- `slowsort([3,2,1], Y).`"
7. Assume `xor/2` is the logical operation $\oplus$ (exclusive-or). Add the rules for converting the formula containing `xor` into NNF in the Prolog program in Example 8.1.2.
8. Add a predicate `cnf/2` into the Prolog program in Example 8.1.2, where $cnf(F, X)$ takes a propositional formula F as input and converts F into an equivalent formula X in CNF.
9. Write a Prolog program which converts a first-order formula A into NNF (negation normal form), with the following assumptions: the predicate symbols are p/2, q/1, and r/0. The quantifiers are $all(X, A)$ and $ex(X, A)$, where X takes a value in {x, y, z}.
10. Decide with evidence if the following Prolog programs are terminating or not. Do they have first termination? last termination? ground termination? on what inputs? under what conditions?

    (a) ```
        goHome(no_12).
        goHome(X) :- get_next_house(X,Y), goHome(Y).
        ```
    (b) ```
        hasFlu(rebecca).
        hasFlu(john).
        hasFlu(X) :- kisses(X, Y), hasFlu(Y).
        kisses(janet,john).
        ```
    (c) ```
        search(end).
        search(X) :- path(X,Y), search(Y).
        ```

11. Introduce a tail recursive predicate for each of the following predicate: *(a)* `length/2` and *(b)* `permutation/2`, so that the same result can be obtained through the tail recursive calls.
12. Given the following Prolog program which gives reviews on restaurants

    ```
    not(P) :- call(P), !, fail.
    not(_).
    good_standard(godels).
    good_standard(hilberts).
    expensive(godels).
    reasonable(R) :- not(expensive(R)).
    ```

    Do the following two queries provide the same result? Why?

    ```
    ?- good_standard(X), reasonable(X).
    ?- reasonable(X), good_standard(X).
    ```

References

1. William F. Clocksin and Christopher S. Mellish: *Programming in Prolog*. Berlin; New York: Springer-Verlag, 2003
2. Daniel Diaz, *The GNU Prolog web site*, http://gprolog.org, retrieved 2023-06-20
    ```

Chapter 9
Hoare Logic

One of the most important research objectives in logic is to obtain means of reasoning about computer systems (be it a CPU, a small chip, a library function, a whole program, or operating system). Proving properties of computer systems is extremely important. A 2003 study commissioned by the Department of Commerce's National Institute of Standards and Technology found that software bugs cost the US economy $59.5 billion annually. In safety-critical systems, even small mistakes can provoke disasters. Since the amount of data which has to be handled in verification of computer systems is usually huge, computer support is necessary. Many approaches have been proposed to formally verify the properties of software and hardware systems. For software formal verification, Hoare logic is one of the most influential approaches for proving the correctness of software designs in popular programming languages [1]. By utilizing the methods of formal verification in software development, one gains the insurance that a software product will work correctly.

9.1 Formal Verification of Computer Systems

Formal verification is the process of proving the correctness of computer systems (hardware/software /algorithms) using mathematical methods. Essentially, it is used for ensuring that a computer system works as expected. For doing so, one formulates the expected properties mathematically and proves or disproves these theorems by reasoning tools. As we formally specify and prove the desired system behaviors, we strongly diminish the possibility of having logical errors in these systems.

In general, formal verification is time-consuming and technically complicated. For life-critical systems, the expense of formal verification is much less than that of potential disasters. Moreover, formal verification will reduce the expense of debugging and testing. For instance, when we write a program that operates on natural numbers (therefore should work for every natural number), we just do not have the ability to test the correct outcome for every natural number. Formal verification may prove the desired property for every natural number by mathematical tools.

9.1.1 Verification of Imperative Programs

Imperative programming languages like C, C++, Java, or Python are closely related to the *random-access stored-program* (RASP) computing model and is an example of the so-called *von Neumann architecture*. Algorithms written in imperative languages are called *imperative programs*, where a program describes subsequent steps that change a computer's memory. Many formal verification techniques are shared for imperative programs and functional programs.

We will use the word "state" to denote a valuation of all variables in an imperative program. A command in the program may change the value of variables, thus causing a transition of states. For example, suppose x holds the value 1 before the command "$x := x + 1$," the value of x will be 2 after the execution of $x := x + 1$. A state can be easily specified by a first-order formula, called *assertions*, such as "$x \doteq 1$," "$A[0] \doteq 2$," etc.

Executing an imperative program has the effect of changing the current state. To reason about such a program, one first establishes an initial state by specifying the initial values of variables. One then executes the program, and this transforms the initial state into a final one through a sequence of state transitions. One then retrieves the values of variables in the final state to get the desired results. For example, to compute "$z := exp(x, y)$," the initial state specifies the values of x and y, and the final state specifies the value of z satisfying "$z \doteq x^y$" to judge that the program $exp(x, y)$ is correct.

Hoare logic (also known as *Floyd-Hoare logic* or *Hoare rules*) is a formal system with a set of logical rules for reasoning rigorously about the correctness of imperative programs. It was proposed in 1969 by the British computer scientist and logician Tony Hoare and subsequently refined by Hoare and other researchers. The original ideas were seeded by the work of Robert Floyd, who had published a similar system for flowcharts.

Hoare logic is an axiomatic approach to formal verification, which consists of (1) a first-order logic language for making assertions about programs and (2) rules for establishing state transitions, that is, what assertions hold after a transition of states. From the assumption on the input, we create a precondition of a program, and from the requirement on the output, we create a postcondition of the program. Both the precondition and the postcondition are assertions, i.e., formulas of the first-order logic. We then use various methods, including theorem provers, to show that the postcondition is a logical consequence of a formula derived from the precondition and the program code by the rules of Hoare logic.

9.1.2 Verification of Functional Programs

In the previous chapter, we mentioned that general recursive functions and Lambda calculus are the foundations for today's functional programming languages, such as Lisp, ML, Clojure, or Haskell. In the imperative paradigm, we specify exactly how an algorithm should run; in the functional paradigm, we specify what the result of the algorithm looks like.

Example 9.1.1 Consider the following imperative program which computes the sum of all numbers in an integer array. After the execution of the program, we expect that "$s \doteq sum(n, A)$" is true.

$$s := 0 \qquad\qquad sum(0, A) \doteq 0$$
$$\textbf{for } i := 0 \textbf{ to } n - 1 \textbf{ do} \qquad sum(i + 1, A) \doteq sum(i, A) + A[i]$$
$$\quad s := s + A[i]$$
$$\textbf{od}$$

The definition of *sum* is in fact a functional program, which specifies the expected outcome of the imperative program. □

In imperative programs, variables can be passed through parameters into a function, and there are in general two modes: "pass by value" and "pass by reference." In the "pass by value" mode, values of the variables are copied to the parameters, and then, the copied objects are used in the function. In the "pass by reference" mode, a variable itself is passed by the parameter into the function, and any change to the parameter will affect the value of the variable (outside of the function). Therefore, an assignment inside of a function may modify the global state. This modification is called a "side effect." Side effects often create unexpected outcomes, as well as hard-to-detect bugs. Because of this unsafe feature, some programming languages, such as Java, prohibit "call by reference." That is, Java passes everything by value, including references, though peple can simulate "call by reference" with Java's container objects such as *array* or *collection*.

In functional programs, variables are absent. Parameters of a function look like variables and serve as placeholders of values. We cannot create global variables that influence how our functions work. Functional programming intends to make all functions pure, with the exception of I/O functions that can be modeled using special constructs such as *monad*, which is a structure that combines program fragments and wraps their return values in a type with additional computation.

Pure functional programs do not have the burden of maintaining a global state and resemble mathematical functions. Thus, various mathematical tools can be used to reason about their properties. For instance, in Example 7.3.12, we defined the insert sorting algorithm, $isort$, by a set of equations, which can be regarded as a functional program. The correctness of $isort$ can be reduced to two theorems: *(a) $sorted(isort(x))$* (the output of $isort(x)$ is a sorted list), and *(b) $permu(x, isort(x))$* ($isort(x)$ is a *permutation* of x).

The Boyer-Moore theorem prover (NQTHM) [2], started in 1971, uses Lisp as a working logic to define total recursive functions. Over the decades, NQTHM evolved considerably and achieved significant progress in formal specification and verification:

- (1976) an expression compiler for a stack machine (Boyer and Moore)
- (1983) invertibility of the RSA encryption algorithm (Boyer and Moore)
- (1985) FM8501 microprocessor (Warren Hunt)
- (1988) an operating system kernel (William Bevier)
- (1989) Piton assembly language on FM9001 (Boyer and Moore)
- (1990) Gauss' law of quadratic reciprocity (David Russinoff)
- (1992) Byzantine Generals and clock synchronization (Bevier and Young)
- (1992) a compiler for the NQTHM language (A. Faltau)
- (1992) the 32-bit FM9001 microprocessor (Hunt and Brock)
- (1993) bi-phase mark asynchronous communications protocol (Boyer and Moore)
- (1993) a small compiler (Debora Weber-Wulff)
- (1993) Motorola MC68020 and Berkeley C String Library (Yuan Yu)

An outstanding successor to NQTHM is ACL2 (a shorthand for "A Computational Logic for Applicative Common Lisp"), developed by Matt Kaufmann and J.S. Moore. They received the 2005 *ACM Software System Award* for building ACL2. Like NQTHM, ACL2 is a logic and programming language in which you can model computer systems, together with a tool to help you prove properties of those models. For more information, the interested reader may visit ACL2's website:

www.cs.utexas.edu/users/moore/acl2/

Recent work using Coq and Isabelle for the verification of kernels of operating systems and compilers has also been encouraging and promising. For people who are doubtful about the performance of functional language, you may use them at least as a logic language in which other designs could be conveniently formalized and verified. For instance, verification of operating systems such as sel4 is generally about imperative C programs, using functional programs for specification. You then use verified compilers to generate efficient and secure code for added safety and correctness.

9.2 Hoare Triples

The central idea of Hoare logic is *Hoare triples*, a notation proposed first by Hoare. A postcondition specifies the desired behavior of the program. Given a postcondition, a Hoare triple specifies exactly what precondition that a program asks before the execution. That is, a Hoare triple describes how the execution of a piece of code changes the state of the computation, we regard precondition and

9.2 Hoare Triples

postcondition as description/constraint of states. A *Hoare triple* is of the form:

$$\{P\}\, C\, \{Q\}$$

where P and Q are assertions, which are formulas in a first-order language, and C is a program code. P is the *precondition* and Q the *postcondition*: assuming P is true before the execution of C, C finishes the execution, and Q is true after the execution of C, we say "$\{P\}\, C\, \{Q\}$" is true.

For instance, $\{x \doteq 1\}\, x := x + 1\, \{x \doteq 2\}$ is true, because the value of x changes from 1 to 2 after the execution of $x := x + 1$. On the other hand, $\{x \doteq 1\}\, x := x + 1\, \{x \doteq 3\}$ is false. Since a triple may be true or false, we call it a *statement* in the general sense.

Example 9.2.1 Consider the program from Example 9.1.1, which computes the sum of all numbers in an integer array with precondition and postcondition.

$\{\ P\colon A \text{ is an array of } n \text{ integers, and } n \geq 0\ \}$
$s := 0$
for $i := 0$ **to** $n - 1$ **do**
$\quad s := s + A[i]$
$\{\ Q\colon s \doteq \sum_{i=0}^{n-1} A[i]\ \}$

At first glance, P and Q are not the formulas we saw in the previous chapters. An array is not different from a list in logic. In Prolog, we have seen that a list is a compound term. Thus, it is not strange to represent an array by a term in first-order logic, and we may simply treat an array as a term in the logic. Similarly, we can also treat $\sum_{i=0}^{n-1} A[i]$ as another term. In fact, we may define a function $sum : N, Array \mapsto N$ recursively as follows:

$$sum(0, A) \doteq 0$$
$$sum(i + 1, A) \doteq sum(i, A) + A[i]$$

Then, $\sum_{i=0}^{n-1} A[i]$ is equivalent to term $sum(n, A)$. □

Now, suppose we can specify preconditions and postconditions in first-order logic, how do we prove that a postcondition is a theorem from the precondition and the program? This is handled by the Hoare rules for each type of command in a small programming language. In this language, for simplicity, we consider only three types of commands:

1. Assignment, i.e., $x := E$, where x is a variable and E is an expression as understood in a programming language
2. Conditional, i.e., **if** B **then** S_1 **else** S_2 **fi**, where B is a Boolean expression, S_1 and S_2 are sequences of commands
3. While loop, i.e., **while** B **do** S **od**, where B is a Boolean expression and S is a sequence of commands

Commands are separated by semicolons. We assume that the computation of all expressions, such as E and B above, has no side effect. That is, E or B cannot contain assignments like in the C language, and the computation of E or B does not change the value of any variable.

This language is small but covers the essential part of most imperative programming languages. There are many variations of the loop construct in imperative programming languages: the "while" form, which uses a continuation condition; the "do-until" form, which always executes the loop body at least once, testing for the condition at the end rather than on entry; and the "for" or "do" forms, which iterate over an integer interval or a data structure. They can all be expressed in a straightforward way as a while loop, on which we will rely throughout this chapter. For instance, a *for-loop* can be expressed by a *while-loop*. The for-loop in Example 9.2.1 can be replaced by the equivalent while loop.

$s := 0$
for $i := 0$ **to** $n - 1$ **do**
$\quad s := s + A[i]$
od

$s := 0; i := 0$
while $(i < n)$ **do**
$\quad s := s + A[i]; i := i + 1$
od

9.2.1 Hoare Rules

Hoare rules cover the three basic commands in the small programming language, plus one rule for the composition of commands and one rule for logical implications.

Assignment Rule

The *assignment rule* states that, after the assignment, any assertion that was previously true for the right side of the assignment now holds for the variable. Formally, let P be an assertion in which the variable x is free, then

$$\frac{\top}{\{P[x \leftarrow E]\} \, x := E \, \{P\}}$$

where $P[x \leftarrow E]$ denotes the assertion obtained from P in which each occurrence of x is replaced by the expression E.

The assignment rule means that the truth of P after $x := E$ is equivalent to the truth of $P[x \leftarrow E]$. Thus, if $P[x \leftarrow E]$ is true prior to the assignment, by the assignment rule, P would be true after the assignment because the value of x is E. All preconditions that are not modified by the assignment can be carried over to the postcondition.

9.2 Hoare Triples

Examples of valid triples include $\{x \doteq 4\}\, y := x + 1\, \{y \doteq 5\}$ and $\{x \leq N - 1\}\, x := x + 1\, \{x \leq N\}$. In the first example, to show that $y \doteq 5$ is true, the assignment rule produces the precondition $x + 1 \doteq 5$, which is equivalent to $x \doteq 4$, from $y \doteq 5$. In the second example, the assignment rule produces the precondition $x + 1 \leq N$, which is equivalent to $x \leq N - 1$, from $x \leq N$.

Since we have not given a formal definition of "state," we cannot prove rigorously the correctness of the assignment rule. However, we may argue informally. To show that $\{P[x \leftarrow E]\}\, x := E\, \{P\}$ is sound, let s be the state before $x := E$ and s' the state after. So, $s' = s[x \mapsto E]$ (assuming E has no side effect). $P[x \leftarrow E]$ holds in s iff P holds in $s' = s[x \mapsto E]$, because

1. Every variable, except x, has the same value in s and s'.
2. $P[x \leftarrow E]$ has every x in P replaced by E.
3. P has every x evaluated to E in $s' = s[x \mapsto E]$.

The major application of the assignment rule is to find the precondition, i.e., $P[x \leftarrow E]$, of $x := E$ from the desired postcondition P, not forward. Be careful not to try to do this "forward" by following this incorrect way of thinking: $\{P\}\, x := E\, \{P[x \leftarrow E]\}$; this rule leads to examples like $\{x \doteq 4\}\, x := 5\, \{5 \doteq 4\}$. Another incorrect rule looking tempting at first glance is $\{P\}\, x := E\, \{P \wedge (x \doteq E)\}$; it leads to illogical examples like $\{x \doteq 5\}\, x := x + 1\, \{x \doteq 5 \wedge x \doteq x + 1\}$.

While a given postcondition P uniquely determines the precondition $P[x \leftarrow E]$ by the assignment rule, the converse is not true. That is, different postconditions are available for the same precondition and the same code. For example,

- $\{0 \leq y \wedge y \leq 4\}\, x := y\, \{0 \leq x \wedge x \leq 4\}$
- $\{0 \leq y \wedge y \leq 4\}\, x := y\, \{0 \leq y \wedge x \leq 4\}$
- $\{0 \leq y \wedge y \leq 4\}\, x := y\, \{0 \leq y \wedge y \leq 4\}$

are valid instances of the assignment rule.

The assignment rule proposed by Hoare can apply to multiple mutual assignments. For example, to swap two elements of an array, some programming languages support the command $A[i], A[j] := A[j], A[i]$. The assignment rule for this type of commands is

$$\frac{\top}{\{P[x \leftarrow E_1, y \leftarrow E_2]\}\, x, y := E_1, E_2\, \{P\}}$$

Implication Rule

The *implication rule*, also called the *consequence rule*, is purely logical as it does not involve any command of a programming language but concerns the implication relation of assertions:

$$\frac{P_1 \rightarrow P_2 \quad \{P_2\}\, S\, \{Q_1\} \quad Q_1 \rightarrow Q_2}{\{P_1\}\, S\, \{Q_2\}}$$

This rule can be split into two rules:

$$\frac{P_1 \to P_2 \quad \{P_2\} S \{Q\}}{\{P_1\} S \{Q\}} \qquad \frac{\{P\} S \{Q_1\} \quad Q_1 \to Q_2}{\{P\} S \{Q_2\}}$$

We say formula A is *weaker* than formula B if $B \models A$, but not $A \models B$. If we regard the entailment relation $\models$ as an order $\succ$ on formulas, then $\bot$ is the strongest (or the largest) and $\top$ is the weakest (or the least).

Definition 9.2.2 We say P is the *weakest precondition* for program S and the postcondition Q if *(a)* $\{P\}S\{Q\}$ is true; and *(b)* for any precondition P', if $\{P'\}S\{Q\}$ is true, then $P' \models P$.

The implication rule allows us to strengthen the precondition P_2 to P_1, by assuming more than we need, and to weaken the postcondition Q_1 to Q_2, by concluding less than we are allowed to.

For instance, if we have shown that $\{n > 0\} sort(A) \{n > 0 \wedge sorted(A)\}$, and the target assertion is $sorted(A)$, since $n > 0 \wedge sorted(A) \to sorted(A)$, the implication rule allows us to get $\{n > 0\} sort(A) \{sorted(A)\}$.

Conditional Rule

The *conditional rule* states that a postcondition Q common to the two branches of an if-then-else statement is also a postcondition of the whole if-then-else statement:

$$\frac{\{B \wedge P\} S \{Q\} \qquad \{\neg B \wedge P\} T \{Q\}}{\{P\} \textbf{if } B \textbf{ then } S \textbf{ else } T \textbf{ fi} \{Q\}}$$

In the "then" part and the "else" part, B and $\neg B$ are added to the precondition P, respectively.

Example 9.2.3 To show the following triple

$$\{P : \top\} \textbf{if } x < 10 \textbf{ then } x := x \textbf{ else } x := 10 \textbf{ fi} \{Q : x \leq 10\},$$

is true by the conditional rule, we need to show the following two statements:

(1) $\{P_1 : \top \wedge x < 10\} x := x \{Q : x \leq 10\}$,
(2) $\{P_2 : \top \wedge x \geq 10\} x := 10 \{Q : x \leq 10\}$.

To prove (1) and (2), we obtain the following by the assignment rule:

(1') $\{x \leq 10\} x := x \{Q : x \leq 10\}$,
(2') $\{10 \leq 10\} x := 10 \{Q : x \leq 10\}$.

9.2 Hoare Triples

It is trivial to show that (1) and (2) are true by the implication rule based on (1′) and (2′). □

If the "else" part is missing, the conditional rule becomes

$$\frac{\{B \wedge P\} S \{Q\} \qquad \neg B \wedge P \to Q}{\{P\} \text{ if } B \text{ then } S \text{ fi } \{Q\}}$$

Example 9.2.4 To show the following statement

$$\{P : x \doteq a \wedge y \doteq b\} \text{ if } x > y \text{ then } y := x \text{ fi} \{Q : y \doteq max(a, b)\}$$

is true, by the conditional rule, we need to show the following two statements:

(1) $\{P_1 : x \doteq a \wedge y \doteq b \wedge x > y\} \, y := x \, \{Q : y \doteq max(a, b)\}$;
(2) $(x \doteq a \wedge y \doteq b \wedge \neg(x > y)) \to y \doteq max(a, b)$.

From the assignment rule, we have

$$\{x \doteq max(a, b)\} \, y := x \, \{Q : y \doteq max(a, b)\}$$

To prove (1), by the implication rule, we need to show $P_1 \to x \doteq max(a, b)$, which can be simplified by equality crossing as $a > b \to a \doteq max(a, b)$ is true. (2) can also be simplified by equality crossing to $\neg(a > b) \to b \doteq max(a, b)$, which is true in arithmetic, too. □

We may add a *skip* command, which does nothing, so that it is easy to use the if-then-else command for the if-then command: **if** B **then** S **fi** becomes **if** B **then** S **else** *skip* **fi**.

Proposition 9.2.5 *If* $\{P_1\} S \{Q\}$ *and* $\{P_2\} T \{Q\}$ *are true and* P *is* $(B \to P_1) \wedge (\neg B \to P_2)$, *then* $\{P\}$ **if** B **then** S **else** T **fi** $\{Q\}$ *is true.*

Proof It is easy to show that $(P \wedge B) \to P_1$ and $(P \wedge \neg B) \to P_2$. From $(P \wedge B) \to P_1$ and $\{P_1\} S \{Q\}$, we have $\{P \wedge B\} S \{Q\}$ by the implication rule. Similarly, we have $\{P \wedge \neg B\} T \{Q\}$. The final result follows from the conditional rule. □

The above proposition allows to design a new conditional rule:

$$\frac{\{P_1\} S \{Q\} \qquad \{P_2\} T \{Q\}}{\{(B \to P_1) \wedge (\neg B \to P_2)\} \text{ if } B \text{ then } S \text{ else } T \text{ fi} \{Q\}}$$

This rule is useful when we want to deduce preconditions from postconditions.

Iteration Rule

$$\frac{\{I \wedge B\} S \{I\}}{\{I\} \text{ while } B \text{ do } S \text{ od } \{\neg B \wedge I\}}$$

Here, I is the so-called *loop invariant*, which is true at the beginning of the loop and is preserved by the loop body S, i.e., after the loop is finished, I still holds. The loop will terminate if B becomes false. The assertion $\neg B$, which is called the *exit condition* of the loop, is added to I as the postcondition of the loop.

Example 9.2.6 Assuming x is an integer, to show that the following statement is true

$$\{I : x \leq 5\} \text{ while } x < 5 \text{ do } x := x + 1 \text{ od } \{Q : x \doteq 5\}$$

we apply the iteration rule and show that

$$\{x \leq 5 \wedge x < 5\} x := x + 1 \{I : x \leq 5\}$$

By the assignment rule, we have $\{x + 1 \leq 5\} x := x + 1 \{I : x \leq 5\}$, so we just need to prove $(x \leq 5 \wedge x < 5) \to (x + 1 \leq 5)$ by the implication rule. When x is an integer, $(x < 5) \to (x + 1 \leq 5)$ is true. So, $(x \leq 5 \wedge x < 5) \to (x + 1 \leq 5)$ is true. Finally, we need to show $(I \wedge \neg B) \to Q$ by the implication rule. Since $x \leq 5 \wedge \neg(x < 5) \to x \doteq 5$ is true for all number x, the program is correct when x is an integer. □

The iteration rule is considered as the most difficult in Hoare logic because of the difficulty of finding good loop invariants. We will address this later in the chapter.

Sequence Rule

The assignment, conditional, and iteration rules cover the three basic commands of the small programming language. The sequence rule applies to sequentially executed programs S and T, where S executes prior to T and is written as $S; T$:

$$\frac{\{P\} S \{Q\} \quad \{Q\} T \{R\}}{\{P\} S; T \{R\}}$$

where Q is called the *middle assertion*. For example, consider the following two instances of the assignment axiom: $\{x \doteq 4\} y := x + 1 \{y \doteq 5\}$ and $\{y \doteq 5\} z := y \{z \doteq 5\}$. By the sequence rule, one concludes $\{x \doteq 4\} y := x + 1; z := y \{z \doteq 5\}$.

9.2.2 Examples of Formal Verification

Now, we have all the rules needed to show the correctness of any program in the small programming language.

Example 9.2.7 Consider the program in Example 9.1.1:

$\{n \geq 0\}$ $sum(0, A) \doteq 0$
$s := 0; i := 0$ $sum(i + 1, A) \doteq sum(i, A) + A[i]$
while $(i < n)$ **do**
    $s := s + A[i]; i := i + 1$
od
$\{s \doteq sum(n, A)\}$

Since the code contains a while loop, a crucial step in applying the Hoare logic is to find a loop invariant. Examining the body of the loop, we see that the values of s and i are changed inside the body. A good loop invariant must reflect these changes. Since the variable s accumulates the sum of the first i elements of A, a first guess for a suitable loop invariant would be $I : s \doteq sum(i, A)$.

By the sequence rule, we need to prove the following statements:

(1) $\{n > 0\} s := 0; i := 0 \{I\}$
(2) $\{I\}$ **while** $i < n$ **do** $s := s + A[i]; i := i + 1$ **od** $\{s \doteq sum(n, A)\}$

The proof of (1) is easy as we just need to prove $n > 0 \rightarrow I[i \leftarrow 0, s \leftarrow 0]$, which comes from $\{I[i \leftarrow 0, s \leftarrow 0]\} s := 0 \{I[i \leftarrow 0]\}$ and $\{I[i \leftarrow 0]\} i := 0 \{I\}$. Since $I[i \leftarrow 0, s \leftarrow 0]$ is $0 \doteq sum(0, A)$, which is true by the definition of sum, hence (1) is true.

The proof of (2) is reduced to the following two statements by the iteration rule and the implication rule:

(2.1) $\{I \wedge i < n\} s := s + A[i]; i := i + 1 \{I\}$.
(2.2) $I \wedge \neg(i < n) \rightarrow \{s \doteq sum(n, A)\}$.

The way to prove (2.1) is the same as that of (1), and we need to prove the following formula:

              (2.1.1) $(I \wedge i < n) \rightarrow I[i \leftarrow i + 1, s \leftarrow s + A[i]]$

Opening I in (2.1.1), we have

$$(s \doteq sum(i, A) \wedge i < n) \rightarrow (s + A[i] \doteq sum(i + 1, A))$$

By the definition of sum, $sum(i + 1, A) = sum(i, A) + A[i] = s + A[i]$, so (2.1.1) is true; thus, (2.1) is true.

Opening I in (2.2), we have $(s \doteq sum(i, A) \wedge i \geq n) \rightarrow (s \doteq sum(n, A))$. Since the condition $i \geq n$ cannot give us $i = n$, the proof cannot go through. From the failure, we see that we need to strengthen the loop invariant by adding $i \leq n$ to it: the new loop invariant I' should be $s \doteq sum(i, A) \wedge i \leq n$. Using I', the proofs of (1) and (2.1) will go through as before, and the proof of (2.2) becomes easy as $(i \leq n \wedge i \geq n) \rightarrow i \doteq n$ is true. □

The above example shows that the loop invariant should contain all the information about how the loop works:

- It reflects what has been done so far together with what remains to be done and should contain all the changing variables.
- It is true at both the beginning and the end of each iteration.
- Together with the negation of the loop condition, it gives the desired result when the loop terminates.
- If a proof cannot go through, we may need to strengthen the loop invariant by adding a condition needed in the proof.

Example 9.2.8 Consider the following program which finds the maximal element in an array:

$\{n > 0\}$ $arrMax(1, A) \doteq A[0]$
$m := A[0]; i := 1;$ $arrMax(i + 1, A) \doteq$
while $i < n$ **do** $max(arrMax(i, A), A[i])$
    if $(m < A[i])$ **then** $m := A[i]$
    $i := i + 1$ $max(x, y) \doteq x$ **if** $x > y$
od $max(x, y) \doteq y$ **if** $x \leq y$
$\{m \doteq arrMax(n, A)\}$

Examining the body of the loop, we see that the values of m and i are changed inside the body. A good loop invariant must reflect these changes. Since variable m contains the maximal value of the first i elements in A, a reasonable guess for a loop invariant would be

$$I : m \doteq arrMax(i, A) \wedge i \leq n.$$

By the sequence rule, we need to prove the following statements:

(1) $\{n > 0\} \, m := A[0]; i := 1 \, \{I\}$
(2) $\{I\}$ **while** $i < n$ **do** ... **od** $\{m \doteq arrMax(n, A)\}$

The proof of (1) is easy as we just need to prove

$$(1.1): \quad n > 0 \rightarrow I[i \leftarrow 1, m \leftarrow A[0]],$$

which comes from $\{I[i \leftarrow 1, m \leftarrow A[0]]\} \, m := A[0] \, \{I[i \leftarrow 1]\}$ and $\{I[i \leftarrow 1]\} \, i := 1 \, \{I\}$. Since $I[i \leftarrow 1, m \leftarrow A[0]]$ is $A[0] \doteq arrMax(1, A) \wedge 1 \leq n$, which is true by the definition of $arrMax$ and $n > 0$, hence (1) is true.

9.2 Hoare Triples

The proof of (2) is reduced to the following two statements by the iteration rule and the implication rule:

(2.1) $\{I \wedge i < n\}$ **if** $(m < A[i])$ **then** $m := A[i]; i := i + 1 \{I\}$
(2.2) $I \wedge \neg(i < n) \rightarrow \{m \doteq arrMax(n, A)\}$

The proof of (2.1) is focused on the proof of the following statement:

(2.1.1) $\{I \wedge i < n\}$ **if** $(m < A[i])$ **then** $m := A[i] \{I[i \leftarrow i + 1]\}$

Applying the conditional rule, we need to prove the following two statements:

(2.1.1.1) $\{I \wedge i < n \wedge m < A[i]\}\ m := A[i] \{I[i \leftarrow i + 1]\}$
(2.1.1.2) $(I \wedge i < n \wedge \neg(m < A[i])) \rightarrow I[i \leftarrow i + 1]$

Applying the assignment and implication rules to (2.1.1.1), we need to prove

(2.1.1.1.1) $(I \wedge i < n \wedge m < A[i]) \rightarrow I[i \leftarrow i + 1, m \leftarrow A[i]]$

Opening I in (2.1.1.1.1), we need to show the following statement:

$$(m \doteq arrMax(i, A) \wedge i \leq n \wedge i < n \wedge m < A[i])$$
$$\rightarrow (A[i] \doteq arrMax(i + 1, A) \wedge i + 1 \leq n)$$

Note that $i + 1 \leq n$ is true because $i < n$ and

$$\begin{aligned} arrMax(i + 1; A) &= max(arrMax(i, A), A[i]) \quad \text{by the definition of } arrMax \\ &= max(m, A[i]) \quad \text{by } m \doteq arrMax(i, A) \\ &= A[i] \quad \text{by } m < A[i] \end{aligned}$$

Hence, (2.1.1.1.1) is true. Using the same approach, (2.1.1.2) can be proved to be true. Hence, (2.1.1) is true.

Opening I in (2.2), we have

$$(m \doteq arrMax(i, A) \wedge i \leq n \wedge \neg(i < n)) \rightarrow (m \doteq arrMax(n, A)).$$

Since $i \leq n$ and $\neg(i < n)$ imply $i = n$, so $(m \doteq arrMax(i, A))$ implies $(m \doteq arrMax(n, A))$. Thus, the proofs of (1) and (2) are complete and I is indeed a good loop invariant for showing $m \doteq arrMax(n, A)$. □

Example 9.2.9 The following program will compute the integer quotient.

$\{x, y > 0\}$
$r := x; q := 0$
while $y \leq r$ **do**
$\quad r := r - y$
$\quad q := q + 1$

od
$\{q \doteq \lfloor x/y \rfloor\}$

Since both r and q are changed inside the loop, let the loop invariant I be

$$I: \quad y > 0 \land r \geq 0 \land x \doteq r + y \cdot q$$

By the sequence rule, we need to show the following statements:

(1) $\{x, y > 0\}\, r := x;\, q := 0 \,\{I\}$
(2) $\{I\}$ **while** $y \leq r$ **do** $r := r - y;\, q := q + 1$ **od** $\{q \doteq \lfloor x/y \rfloor\}$

(1) is easy to prove by using the assignment rule twice, the sequence rule once, and the implication rule once: $x, y > 0 \to I[r \leftarrow x, q \leftarrow 0]$, that is, $x, y > 0 \to (y > 0 \land x \geq 0 \land x \doteq x + y \cdot 0)$ is true.

(2) needs the iteration rule and the implication rule: We need to prove

(2.1) $\{I \land y \leq r\}\, r := r - y;\, q := q + 1 \,\{I\}$
(2.2) $(I \land \neg(y \leq r)) \to (q \doteq \lfloor x/y \rfloor)$

The way to prove (2.1) is the same as to prove (1), we need to show

$$(I \land y \leq r) \to I[r \leftarrow r - y, q \leftarrow q + 1]$$

which, when I is open up, is

$$(y > 0 \land r \geq 0 \land x \doteq r + y \cdot q \land y \leq r) \to (y > 0 \land r - y \geq 0 \land x \doteq (r - y) + y \cdot (q + 1))$$

Since $r - y \geq 0$ is a logical consequence of $y \leq r$, and $x \doteq (r - y) + y \cdot (q + 1)$ is a logical consequence of $x \doteq r + y \cdot q$, (2.1) is true.

The proof of (2.2) is easy as $x = r + y \cdot q$ iff $x/y = r/y + q$ when $y > 0$. Since $(y > r \land r \geq 0)$ implies $0 \leq r/y < 1$, so $\lfloor x/y \rfloor = \lfloor r/y + q \rfloor = \lfloor r/y \rfloor + q = 0 + q = q$. □

9.2.3 Partial and Total Correctness

Hoare rules presented in the previous section concern the *partial correctness* of programs: if the program stops, the postcondition will hold. If a program is both partially correct and terminating, we say the program is *totally correct*. The relationship between partial and total correctness can be informally expressed by the equation:

$$\text{Total correctness} = \text{Termination} + \text{Partial correctness}.$$

We will see in Chap. 11 that the termination of programs is an undecidable problem and there are no decision procedures to solve this problem in general.

Partial correctness can be proved for some Hoare triples and termination of the programs needs to be proved separately. Thus, the intuitive reading of a Hoare triple is whenever P is true in the state before the execution of C, then Q will be true after the execution of C or C does not terminate. In the latter case, there is no "after," so Q can be any statement at all. Indeed, one can choose Q to be false to express that C does not terminate.

Example 9.2.10 Consider the triple $\{I\}$ **while** $x \neq a$ **do** *skip* **od** $\{x \doteq a\}$. Let the loop invariant I be $\top$, then both $I \wedge x \neq a \rightarrow I$ and $I \wedge \neg(x \neq a) \rightarrow x \doteq a$ are trivially true. Thus, the program is partially correct: if the program stops, we must have $x \doteq a$. The program will loop forever unless it starts with $x \doteq a$; therefore, it is not totally correct. □

Total correctness can be proved with an extended version of the iteration rule, where the value of a function on the changing variables of the loop decreases under a well-founded order. Total correctness is what we are ultimately interested in, but it is usually easier to prove the total correctness by establishing partial correctness and termination separately. Termination is often straightforward to establish by the method of complexity analysis, which often provides finer results than the termination. For example, the program in Example 9.2.7 is terminating because the time complexity of this program is $O(n)$. For the program in Example 9.2.9, let $f(q) = \lfloor x/y \rfloor - q$, then $f(q)$ decreases by one after each iteration, and the minimum value of $f(q)$ is 0 when the looping condition is false. The time complexity of this program is $O(\lfloor x/y \rfloor)$.

9.3 Automated Generation of Assertions

After seeing only a few examples, the following two things are painfully clear:
- Correctness proofs in the program verification are typically long and boring, even if the program being verified is quite simple.
- There are lots of fiddly little details to get right, many of which are trivial, e.g., proving $(r \doteq x \wedge q \doteq 0) \rightarrow (x \doteq r + y \cdot q)$.

Many attempts have been made (and are still being made) to automate correctness proofs by designing systems to do boring and tricky bits of generating formal proofs in Hoare logic. Unfortunately, logicians have shown that it is impossible in principle to design a decision procedure to decide automatically the truth or falsehood of an arbitrary mathematical statement. However, this does not mean that one cannot have procedures that will prove many useful theorems. The nonexistence of a general decision procedure merely shows that one cannot hope to prove everything automatically. In practice, it is quite possible to build a system that will mechanize many of routine aspects of verification. This section describes one common approach to doing this.

In the previous section, it was shown how to prove $\{P\}\,S\,\{Q\}$ by reducing this goal to several subgoals and then putting them together using the Hoare rules to get the desired property of S itself. For example, to prove $\{P\}\,S;\,T\,\{Q\}$, we first prove $\{R\}\,T\,\{Q\}$ and $\{P\}\,S\,\{R\}$ for a middle assertion R and then deduce $\{P\}\,S;\,T\,\{Q\}$ by the sequence rule. This process is called *goal reduction*, or *backward reasoning*, because one works backward: starting from the goal of showing $\{P\}\,S\,\{Q\}$, one generates subgoals, sub-subgoals, etc., until the problem is solved. Prolog also uses this process for computation.

Example 9.3.1 Suppose one wants to show

$$\{x \doteq x_0 \land y \doteq y_0\}\, r := x;\, x := y;\, y := r\, \{y \doteq x_0 \land x \doteq y_0\},$$

then by the assignment and sequence rules, the above statement is reduced to the subgoal:

$$\{x \doteq x_0 \land y \doteq y_0\}\, r := x;\, x := y\, \{r \doteq x_0 \land x \doteq y_0\}$$

because $\{r \doteq x_0 \land x \doteq y_0\}\, y := r\, \{y \doteq x_0 \land x \doteq y_0\}$ is true by the assignment rule. By a similar argument, this subgoal can be reduced to

$$\{x \doteq x_0 \land y \doteq y_0\}\, r := x\, \{r \doteq x_0 \land y \doteq y_0\}$$

which is true by the assignment rule. The middle assertions for the sequence rule are automatically generated. □

We wish that the process illustrated by the above example can be extended to arbitrary programs so that all middle assertions can be generated automatically, releasing programmers from the burden of providing all assertions. A programmer just needs to provide precondition and postcondition of the program derived from the specification of the problem, and all the needed assertions can be generated automatically. Coupling with a powerful theorem prover, all the formulas derived from Hoare rules can be proved automatically, and we arrive at the automation of software verification.

9.3.1 Verification Conditions

The formal verification of a program can be described succinctly as follows:

1. The program S is annotated by inserting into it the assertions expressing conditions that are meant to be true at various intermediate points. This step needs a good understanding of how the program works. The user needs to provide

9.3 Automated Generation of Assertions

at least the pre- and postconditions and loop invariants; other assertions can be generated automatically.
2. A set of formulas called *verification conditions* is then generated from the annotated specification. This process is purely mechanical and easily done by a software tool. We will describe how this is done in this section.
3. The verification conditions are proved by a theorem prover. Automating this remains to be a big challenge.

Definition 9.3.2 Given a program S with postcondition Q, let $vc(S, Q)$ denote the formula P such that $\{P\} S \{Q\}$ is true. $vc(S, Q)$ is called a *verification condition* of S with respect to Q.

If $vc(S, Q)$ can be generated automatically from S and Q, then the process of generating all assertions becomes automated. By definition, $vc(S, Q)$ is not unique, and this property gives flexibility for the automated generation of $vc(S, Q)$. Besides being the basis for automated verification systems, verification conditions are a useful way of working with Hoare logic by hand.

We will define a recursive procedure $VC(S, Q)$ on input S and Q to generate $vc(S, Q)$, coinciding with the goal reduction process of Hoare triples. The definition of $VC(S, Q)$ follows closely with Hoare rules on the three constructs and the sequence rule of our small programming language. There are only five cases to consider in the definition of $VC(S, Q)$ to generate $vc(S, Q)$.

In fact, $VC(S, Q)$ will generate the *weakest preconditions* in most cases, with the exception of while loops, but we will neither make this more precise nor go into details here. An in-depth study of weakest preconditions can be found in [3].

For while loops, we do not have an algorithm which can generate $vc(S, Q)$ for every loop; only some heuristic methods are available. To help the verification tool to generate $vc(S, Q)$, a programmer needs to suggest a loop invariant. That is, instead of the code **while** B **do** S **od**, we write **while** B **do** $\{I\}$ S **od**, where I is the suggested loop invariant. The pre- and postconditions can be also inserted into a program code. This type of program codes with assertions is called "annotated program," where assertions are provided by the programmer to assist the verification.

Another exception with while loops is that, in addition to returning a verification condition for the loop, it will create another formula whose validity is needed to ensure the validity of the Hoare triple. This formula will be pushed into a stack denoted by s during the execution of VC. The stack s is a global variable and is initially empty; it will hold as many formulas as the number of while loops in the program.

Algorithm 9.3.3 The algorithm VC takes an annotated program C, a formula Q as a postcondition of C, and returns a verification condition P such that $\{P\}\,C\,\{Q\}$ is true.

proc $VC(C, Q)$
1   **if** C is "$x := E$" **return** $Q[x \leftarrow E]$
2   **else if** C is "$skip$" **return** Q
3   **else if** C is "**if** B **then** S **else** T **fi**"
4   **return** $(B \to VC(S, Q)) \land (\neg B \to VC(T, Q))$
5   **else if** C is "**while** B **do** $\{I\}$ S **od**"
6        $push(I \to ((B \to VC(S, I)) \land (\neg B \to Q)), s)$; **return** I
7   **else if** C is "$S; T$" **return** $VC(S, VC(T, Q))$
8   **else** print("unknown command"); **return** $\bot$

In the above algorithm, lines 1–2 handle the base cases, when the command is an assignment or skip command. Lines 3–4 handle the conditional (for simplicity, we ignore the conditional case when the else part is missing). Lines 5–6 handle the while loop; and line 7 handles the sequence of commands. The last three cases involve the recursive calls of VC.

Theorem 9.3.4 *Assume S is a legal program of our small programming language. The output of $VC(S, Q)$ is a formula P in first-order logic such that $\{P\}\,S\,\{Q\}$ is true, if all the formulas generated by $VC(S, Q)$ in the stack s are valid.*

By definition, $VC(S, Q)$ returns a verification condition $vc(S, Q)$ if the conditions of the theorem are met. The following subsection will be devoted to the proof of the theorem.

9.3.2 Proof of Theorem 9.3.4

First, we need to show that VC will terminate. This is an easy task because the first argument of VC, which is a piece of code, becomes strictly smaller for each recursive call, and it will come down to an assignment or a skip command; there cannot exist an infinite chain of recursive calls. So, the algorithm VC must terminate.

The proof of the theorem will be by induction on the structure of C. Such inductive arguments have two parts. First, it is shown that the result holds for the assignment and skip commands. Second, it is shown that when C is not an assignment command, if the result holds for the constituent commands of C (this is called the induction hypothesis), then it holds also for C. The first of these parts is called the base cases of the induction, and the second part is called the inductive cases. From the base and inductive cases, it follows that the result holds for all commands. For instance, to show that VC returns a formula, since all the base cases return a formula, assuming all the recursive calls will return a formula, then VC will return a formula by examining all the inductive cases.

9.3 Automated Generation of Assertions

The case of the skip command is trivial; the other cases will be discussed separately.

Assignments

The verification condition for $x := E$ and Q is $Q[x \leftarrow E]$, which is the weakest precondition for $x := E$ and Q. Thus, we have $VC(x := E, Q) = Q[x \leftarrow E]$, and it is easy to verify that $VC(x := E, Q)$ is a $vc(x := E, Q)$.

Sequences

For the code "$S; T$" and postcondition Q, $VC(S; T, Q)$ returns $VC(S, VC(T, Q))$. If both $VC(S, Q')$ and $Q' = VC(T, Q)$ are $vc(S, Q')$ and $vc(T, Q)$, respectively, then $VC(S, VC(T, Q))$ is a $vc(S; T, Q)$.

Example 9.3.5 A verification condition for the code and the postcondition in Example 9.3.1 is easily obtained as it involves only two cases, assignment and sequence:

$$\begin{aligned}
&VC(\{r := x; x := y; y := r\}, y \doteq x_0 \wedge x \doteq y_0) \\
&= VC(\{r := x; x := y\}, VC(y := r, y \doteq x_0 \wedge x \doteq y_0)) \\
&= VC(\{r := x; x := y\}, r \doteq x_0 \wedge x \doteq y_0) \\
&= VC(r := x, VC(x := y, r \doteq x_0 \wedge x \doteq y_0)) \\
&= VC(r := x, r \doteq x_0 \wedge y \doteq y_0) \\
&= (x \doteq x_0 \wedge y \doteq y_0)
\end{aligned}$$

which is the precondition for the swapping command of x and y. The three recursive calls of VC generate $vc(y := r, y \doteq x_0 \wedge x \doteq y_0)$, $vc(x := y, r \doteq x_0 \wedge x \doteq y_0)$, and $vc(r := x, r \doteq x_0 \wedge y \doteq y_0)$, respectively. □

Conditionals

By Proposition 9.2.5, $P = (B \rightarrow vc(S, Q)) \wedge (\neg B \rightarrow vc(T, Q))$ is a verification condition for the code **if** B **then** S **else** T **fi** and Q. Note that P is the weakest precondition for the conditional command, if both $vc(S, Q)$ and $vc(T, Q)$ are the weakest preconditions for S and T, respectively.

Hence, $VC(\textbf{if } B \textbf{ then } S \textbf{ else } T \textbf{ fi}, Q)$ returns $P = (B \rightarrow VC(S, Q)) \wedge (\neg B \rightarrow VC(T, Q))$. By the induction hypothesis, both $\{VC(S, Q)\} S \{Q\}$ and $\{VC(T, Q)\} T \{Q\}$ are true. By Proposition 9.2.5

$$\{P\} \textbf{ if } B \textbf{ then } S \textbf{ else } T \textbf{ fi} \{Q\}$$

is true.

Example 9.3.6 The postcondition of Example 9.2.3 is $x \leq 10$.

$$VC(\text{if } (x < 10) \text{ then } x := x \text{ else } x := 10 \text{ fi}, x \leq 10)$$
$$= (x < 10 \rightarrow VC(x := x, x \leq 10)) \wedge (x \geq 10 \rightarrow VC(x := 10, x \leq 10))$$
$$= (x < 10 \rightarrow x \leq 10) \wedge (x \geq 10 \rightarrow 10 \leq 10)$$
$$\equiv (\top) \wedge (x \geq 10 \rightarrow \top)$$
$$\equiv \top$$

Note that $\top$ is the weakest among all assertions. □

Iterations

$VC(\text{while } B \text{ do } \{I\} S \text{ od}, Q)$ returns I, which is provided by the user to ensure that I is a loop invariant. Moreover, the formula

$$I \rightarrow ((B \rightarrow VC(S, I)) \wedge (\neg B \rightarrow Q)),$$

which is equivalent to $((I \wedge B) \rightarrow VC(S, I)) \wedge ((I \wedge \neg B) \rightarrow Q)$, is created and stored in the stack s.

The validity of $I \wedge B \rightarrow VC(S, I)$ ensures that I holds after the execution of the loop body S. By the induction hypothesis, $\{VC(S, I)\} S \{I\}$ is true. Since $I \wedge B \rightarrow VC(S, I)$, by the implication rule, $\{I \wedge B\} S \{I\}$ is true, that is, I is indeed a loop invariant. Thus, by the iteration rule, $\{I\}$ while B do S od $\{I \wedge \neg B\}$ is true.

Finally, the validity of $(I \wedge \neg B) \rightarrow Q$ ensures that Q holds when the loop terminates. By the implication rule, $\{I\}$ while B do S od $\{Q\}$ holds.

It is clear that for the iteration command, the loop invariant I returned by VC can be a verification condition if the formula $I \rightarrow ((B \rightarrow VC(S, I)) \wedge (\neg B \rightarrow Q))$ is valid.

The following table summarizes the verification conditions returned by VC for various commands and postcondition Q:

code C	what $VC(C, Q)$ returns
skip	Q
$x := E$	$Q[x \leftarrow E]$
$S; T$	$VC(S, VC(T, Q))$
if B then S else T fi	$(B \rightarrow VC(S, Q)) \wedge (\neg B \rightarrow VC(T, Q))$
while B do $\{I\} S$ od	I

All verification conditions can be automatically generated except for the case of iteration command. In this case, the validity of $I \rightarrow ((B \rightarrow VC(S, I)) \wedge (\neg B \rightarrow Q))$ ensures that I is a verification condition. This completes the proof of Theorem 9.3.4.

9.3 Automated Generation of Assertions

Algorithm $VC(S, Q)$ shows how assertions needed for Hoare rules are derived from the postcondition and loop invariants automatically in a goal-reduction or backward reasoning process. Backward reasoning can be useful for nested loops, where the inner loop's postcondition can be derived from the outer loop's invariant.

Example 9.3.7 The precondition for the following program is $P : n \geq 0 \wedge x \doteq n \wedge y \doteq 1$ and the loop invariant is $I : x! * y \doteq n!$, where $x!$ is the factorial function: $0! \doteq 1$ and $(x + 1)! \doteq x! * (x + 1)$.

$VC(\textbf{while } (x \neq 0) \textbf{ do } \{I : x!y \doteq n!\} \ y := x*y; \ x := x-1 \textbf{ od}, \ y \doteq n!) = (x!y \doteq n!)$

The formula pushed into the stack and needs to be proved is

$x!y \doteq n! \to ((x! \doteq 1 \wedge x \neq 0 \to VC([y := y * x; x := x - 1], x!y \doteq n!))$
$\wedge (x \doteq 0 \to y \doteq n!))$
$= x!y \doteq n! \to ((x! \doteq 1 \wedge x \neq 0 \to VC(y := y * x, VC(x := x - 1, x!y \doteq n!)))$
$\wedge (x \doteq 0 \to y \doteq n!))$
$= x!y \doteq n! \to ((x! \doteq 1 \wedge x \neq 0 \to VC(y := y * x, (x - 1)!y \doteq n!))$
$\wedge (x \doteq 0 \to y \doteq n!))$
$= x!y \doteq n! \to ((x! \doteq 1 \wedge x \neq 0 \to (x - 1)!(yx) \doteq n!) \wedge (x \doteq 0 \to y \doteq n!))$
$= x!y \doteq n! \to ((x \doteq 1 \to x!y \doteq n!) \wedge (x \doteq 0 \to y \doteq n!))$
$= x!y \doteq n! \to (x \doteq 0 \to y \doteq n!)$

The second last step uses the property that $x! \doteq 1 \wedge x \neq 0$ is equivalent to $x \doteq 1$ because $x! \doteq 1$ is equivalent to $x \doteq 0 \vee x \doteq 1$. The last step can be obtained in two ways: (a): using the logical equivalence $X \to ((Y \to X) \wedge Z) \equiv X \to Z$, where X is $(x!y \doteq n!)$, Y is $(x! \doteq 1 \wedge x \neq 0)$, and Z is $(x \doteq 0 \to y \doteq n!)$. (b): using equality crossing to simplfy $x!y \doteq n! \to (\ldots \to x!y \doteq n!) \wedge \ldots$ to $x!y \doteq n! \to (\ldots \to n! \doteq n!) \wedge \ldots$.

Applying equality crossing twice, $x!y \doteq n! \to (x \doteq 0 \to y \doteq n!)$ is simplified to $x!y \doteq n! \to (x \doteq 0 \to y \doteq x!y)$, then to $x!y \doteq n! \to (x \doteq 0 \to y \doteq 0!y)$, which is easily simplified to true by $0! \doteq 1$ and $1 * y \doteq y$. □

Can we get rid of the stack used in VC? That is, instead of storing $I \to ((B \to VC(S, I)) \wedge (\neg B \to Q))$, can we let $VC(\textbf{while } B \textbf{ do } \{I\} S \textbf{ od}, Q)$ returns $I \to ((B \to VC(S, I)) \wedge (\neg B \to Q))$ as part of its output? The following counterexample shows that this approach is unsound.

Example 9.3.8 Let the Hoare triple be

$$\{x \doteq 0\} \textbf{ while } (x < 5) \textbf{ do } x := x + 1 \textbf{ od } \{x \doteq 5\}$$

which can be proved to be true by the Hoare rules with the loop invariant $x \leq 5$.
If we modify VC such that $VC(\textbf{while } B \textbf{ do } \{I\} \, S \textbf{ od}, Q)$ returns

$$I \wedge (I \rightarrow ((B \rightarrow VC(S, I)) \wedge (\neg B \rightarrow Q)))$$

which is equivalent to

$$I \wedge (B \rightarrow VC(S, I)) \wedge (\neg B \rightarrow Q)$$

Let I be $x \doteq 0 \vee x \doteq 1$ for the above program, then $VC(\textbf{while } (x < 5) \textbf{ do } \{I\} \, x := x + 1 \textbf{ od}, x \doteq 5)$ would return

$$I \wedge (x < 5 \rightarrow VC(x := x + 1, I)) \wedge (x \geq 5 \rightarrow x \doteq 5)$$

which can be simplified to

$$P: (x \doteq 0 \vee x \doteq 1) \wedge (x < 5 \rightarrow (x \doteq -1 \vee x \doteq 0)) \wedge (x \geq 5 \rightarrow x \doteq 5)$$

because $VC(x := x + 1, I)$ returns $(x + 1 \doteq 0 \vee x + 1 \doteq 1)$. P can be simplified further to $x \doteq 0 \vee (x \doteq 1 \wedge x \doteq -1)$. Now, it is easy to show that $x \doteq 0 \rightarrow P$ is true.

By the original Hoare rule, we cannot show that $x \doteq 0 \vee x \doteq 1$ is a valid loop invariant. Thus, the modification of VC is unsound. □

9.3.3 Implementing VC in Prolog

Since Prolog is good at pattern matching, it is easy to implement the algorithm $VC(S, Q)$ in Prolog (as $vc(S, Q)$ because VC is a Prolog variable). The first job is to let the Prolog program parses codes easily. For this purpose, we change the syntax of our programming language as follows:

command	before	now
assignment	$x := E$	$let(x, E)$
conditional	**if** B **then** S **else** T **fi**	$ite(B, S, T)$
iteration	**while** B **do** $\{I\}$ S **od**	$while(B, I, S)$
sequence	$S_1; S_2; \ldots; S_n$	$[S_1, S_2, \ldots, S_n]$

9.3 Automated Generation of Assertions

For the iteration, instead of using a stack to store formulas, we simply write the formulas on the screen. Below is the Prolog code for treating assignments, conditionals, iterations, and skip:

```
% vc1(S1, Post, VC): VC = vc(S1, Post), S1 is a single command

% assignment
vc1(let(X, E), Post, VC) :- substitute(X, E, Post, VC).
% VC is the result of Post by replacing X in Post by E.

% conditional
vc1(ite(B, S, T), Post, VC) :-
    vc(S, Post, VC1), vc(T, Post, VC2),
    VC = and((B -> VC1), (neg(B) -> VC2)).

% iteration
% use vc1(while(B, Inv, S), Post, Inv) as an alternative.
vc1(while(B, Inv, S), Post, or(and(neg(B), Post), Inv)) :-
    vc(S, Inv, VC),                          % VC is vc(S, Inv)
    write((and(Inv, B) -> VC)), nl,          % part 1 of a conjunction
    write((and(Inv, neg(B)) -> Post)), nl.   % part 2 of a conjunction
```

The handling of assignment uses a utility function `substitute`, which performs the substitution:

```
% substitute(X, V, E, R) succeeds when every occurrence of X in
% expression E is replaced by V and the modified E is R.
substitute(X, V, X, V):- !.
substitute(X, V, E, R) :-
    E =.. [F|Args0],
    maplist(substitute(X, V), Args0, Args),
    R =.. [F|Args].
```

For a sequence S of commands, we first reverse S into S_2 and work on S_2 from the beginning to the end, as illustrated by the following code:

```
% vc(S, Post, VC) succeeds when VC is generated
% from the program S and the postcondition Post.

vc(P, Post, VC) :- is_list(P), !, reverse(P, P2), vcR(P2, Post, VC).
vc(P, Post, VC) :- vc1(P, Post, VC).      % if S is not a list

% vcR(SR, Post, VC) generates VC from SR and Post, where
% SR is a list of instructions in reversed order.
vcR([], Post, Post).
vcR([Last | Rest], Post, VC) :-
    vc1(Last, Post, Post1), vcR(Rest, Post1, VC).
```

In the following examples, we define a triple $tri_i(P, S, Q)$, where i is a number, the query q_i for generating and displaying a verification condition for $tri_i(P, S, Q)$. The validity of all the displayed formulas ensures that the triple is true.

Example 9.3.9 The triple is specified as follows:
```
tri1(true,
     [ ite(a>b, let(x, b), let(x, a)), ite(x > c, let(x, c), skip)],
     x=min(a,min(b,c))).

q1 :- tri1(P, S, Q), vc(S, Q, VC), write((P -> VC)), nl.
```
The query "?- q1." produces one formula:
```
true -> and((a>b -> and((b>c -> c=min(a, min(b,c))),
                        (neg(b>c) -> b=min(a, min(b,c))))),
            (neg(a>b) -> and((a>c -> c=min(a, min(b,c))),
                             (neg(a>c) -> a=min(a, min(b,c))))))
```
which represents the formula

$$\top \to ((a > b \to ((b > c \to c = m) \land (\neg(b > c) \to b = m))) \land (\neg(a > b) \to$$
$$((a > c \to c = m) \land (\neg(a > c) \to a = m))))$$

where m is a shorthand for $min(a, min(b, c))$. □

Example 9.3.10 The triple is specified as follows:
```
tri2(and(n>=0, and(k=0, result=0)),
     while(k<n,
           and(0=<k, and(k=<n, result=sum(0, k, a))),
           [let(result, result+a(k)),
            let(k, k+1)]),
     result=sum(0, n, a)).

q2 :- tri2(P, S, Q), vc(S, Q, VC), write((P -> VC)), nl.
```
The query "?- q2" produces three formulas, the first two from the stack and the last from "write((P -> VC))":
```
and(and(0=<k, and(k=<n, result=sum(0,k,a))), k<n) ->
    and(0=<k+1, and(k+1=<n, result+a(k)=sum(0,k+1,a)))
and(and(0=<k, and(k=<n, result=sum(0,k,a))), neg(k<n)) ->
    result=sum(0,n,a)
and(n>=0, and(k=0, result=0)) -> and(0=<k,and(k=<n, result=sum(0,k,a)))
```
where $sum(b, e, A) = 0$ if $(b \geq e)$; otherwise, $sum(b, e, A) = sum(b, e - 1, A) + A[e - 1]$. The three formulas can be rewritten as

$(0 \leq k \land k \leq n \land result = sum(0, k, a) \land k < n) \to$
$\quad (0 \leq k + 1 \land k + 1 \leq n \land result + a(k) = sum(0, k + 1, a))$
$(0 \leq k \land k \leq n \land result = sum(0, k, a) \land \neg(k < n)) \to result = sum(0, n, a)$
$(n \geq 0 \land k = 0 \land result = 0) \to (0 \leq k \land k \leq n \land result = sum(0, k, a))$

and can be proved easily. □

9.3 Automated Generation of Assertions

Example 9.3.11 The triple is specified as follows:

```
tri3(and(n >= 0, and(x=n, y=1)),
     while(neg(x = 0),
           fac(x)*y = fac(n),
           [let(y, x*y),
            let(x, x-1)]),
     y=fac(n)).

q3 :- tri3(P, S, Q), vc(S, Q, VC), write((P -> VC)), nl.
```

The query "?- q3" produces three formulas:

```
and(fac(x)*y=fac(n), neg(x=0)) -> fac(x-1)*(x*y)=fac(n)
and(fac(x)*y=fac(n), neg(neg(x=0))) -> y=fac(n)
and(n>=0, and(x=n,y=1)) -> fac(x)*y=fac(n)
```

where $fac(n)$ is the factorial function: $fac(0) = 1, fac(i+1) = fac(i) * (i+1)$. The three formulas can be rewritten as

$$(fac(x) * y = fac(n) \land \neg(x = 0)) \rightarrow fac(x-1) * (x * y) = fac(n)$$
$$(fac(x) * y = fac(n) \land \neg\neg(x = 0)) \rightarrow y = fac(n)$$
$$(n \geq 0 \land x = n \land y = 1) \rightarrow fac(x) * y = fac(n)$$

and can be proved easily. □

The above three examples show how formulas can be generated from postconditions and loop invariants. All middle assertions, with the exception of loop invariants, are generated automatically. These formulas are then fed to an automated theorem prover. Once they are proved to be valid, the Hoare triple is shown to be true, and the program is partially correct.

More than often, a theorem prover cannot prove these formulas automatically; if it fails, advice is sought from the user. The analysis of the failure may suggest that we would need to add a conjunct to a loop invariant, to the precondition, or modify the postcondition. We then run vc with new annotated code, obtain new formulas, and feed them to the theorem prover. This process repeats until we find good preconditions, postconditions, and loop invariants.

The aim of the current research is to mainly build systems which reduce the role of the slow and expensive human expert to a minimum. This can be achieved by the following:

- Reducing the number and complexity of the annotations required
- Increasing the power of the theorem prover

From the implementation of vc, it is now clear that the success of Hoare logic to formal verification relies on (i) obtaining a good loop invariant for each while loop in a program and (ii) having a powerful theorem prover which can prove the validity of the formulas derived by vc. The first issue will be addressed in the next section. The second issue has been addressed in the previous chapters of this book and will be discussed in Chap. 12. Since proofs of correctness of programs are typically very long and boring, it is thus useful to check proofs mechanically or automatically, even if they can only be carried out with human assistance.

9.4 Obtaining Good Loop Invariants

The concept of loop invariant is one of the foundational ideas of software construction. We have seen several examples of loop invariants. We hope that these examples establish the claim that the invariant is the key to every loop: to devise a new loop so that it is correct requires summoning the proper invariant and to understand an existing loop requires understanding its invariant.

By far, finding an invariant for a loop is traditionally the responsibility of a human: either the person performing the verification or the programmer writing the loop in the first place. A better solution, when applicable, is *program synthesis*, a constructive approach to programming advocated by Dijkstra and others [3]. Automated generation of loop invariants is still an ongoing research topic.

For the triple $\{I\}$ **while** B **do** S **od** $\{Q\}$, I is a correct invariant for the loop if it satisfies the following conditions:

1. I is true in the state preceding the loop execution.
2. Every execution of the loop body S, started in any state in which both I and B are true, will yield a state in which I holds again.

If these properties hold, then any terminating execution of the loop will yield a state in which both I and $\neg B$ hold; the latter is called the *exit condition*.

We may look at the notion of loop invariant from the constructive perspective of a programmer directing his or her program to reach a state satisfying a certain desired property, the postcondition. In this view, program construction is a form of problem-solving, and the various control structures are problem-solving techniques; a loop solves a problem through successive approximation. This idea has been strongly advocated in Furia et al.'s survey article [4] on loop invariants. In the following, our presentation is based on their survey.

- Generalize the postcondition (the characterization of possible solutions) into a broader condition: the invariant.
- As a result, the postcondition can be defined as the conjunction of the invariant and another condition, i.e., the exit condition.
- Find a way to reach the invariant from the previous state of the computation: the initialization.
- Find a way, given a state that satisfies the invariant, to get to another state, still satisfying the invariant but closer, in some appropriate sense, to the exit condition: the body.

The importance of this presentation of the loop process is that it highlights the nature of the invariant: it is a generalized form of the desired postcondition, which in a special case (represented by the exit condition) will give us that postcondition.

9.4 Obtaining Good Loop Invariants

This view of the invariant, as a particular way of generalizing the desired goal of the loop computation, explains why the loop invariant is such an important property of loops; one can argue that understanding a loop means understanding its invariant.

9.4.1 Invariants from Generalizing Postconditions

Many programmers still find it hard to come up with invariants. Even though the basic idea is often clear, coming up with a good invariant is an arduous task. We need methods to generate loop invariants automatically. To illustrate the idea of generalizing the postcondition to obtain an invariant, we will use the thousands-year-old Euclid's algorithm for computing the greatest common divisor (gcd) of two positive integers.

Example 9.4.1 The postcondition of the gcd algorithm is

$$y \doteq gcd(a, b),$$

where the precondition is $a > 0 \wedge b > 0$. The first step of the generalization is to replace this condition by

$$x \doteq 0 \wedge gcd(x, y) \doteq gcd(a, b)$$

with a new variable x, taking advantage of the mathematical property that, for every y, $gcd(0, y) \doteq y$. The second conjunct, i.e., $gcd(x, y) \doteq gcd(a, b)$, a generalization of the postcondition, will serve as the invariant; the first conjunct will serve as the exit condition. To obtain the loop body, we take advantage of another mathematical property: for every x and y, $gcd(x, y) \doteq gcd(y, x)$ and $gcd(x, y) \doteq gcd(x - y, y)$ if $x > y$, yielding the well-known algorithm:

```
x := a; y := b
{I : x ≥ 0 ∧ y > 0 ∧ gcd(x, y) ≐ gcd(a, b)}
while x ≠ 0 do
    if x < y then z := x; x := y; y := z fi
    x := x - y
od
```

The proof of correctness follows directly from the mathematical property of *gcd*. This form of Euclid's algorithm uses subtraction; we may replace the subtraction by the remainder function, i.e., replace $x - y$ by $x\%y$ (% is the remainder function), resulting in a more efficient algorithm. In this case, we need the property that $gcd(x, y) \doteq gcd(x\%y, y)$. □

Using the classification criteria proposed in Furia et al.'s survey article, the invariant for the gcd algorithm, i.e., $I : x \geq 0 \wedge y > 0 \wedge gcd(x, y) \doteq gcd(a, b)$, is divided into two categories:

- The last conjunct of the invariant is an *essential* invariant, representing a weakening of the postcondition. The first two conjuncts are *bounding* invariants, indicating that the state remains within certain general boundaries and ensuring that the "essential" part is well defined.
- The strategy that leads to the essential invariant is *uncoupling*, which replaces a property of one variable (y in this case), used in the postcondition, by a property of two variables (y and x), used in the invariant.

As illustrated by the above example, the essential invariant is a mutation (often, a weakening) of the loop's postcondition. The following mutation techniques are particularly common.

- **Constant Relaxation**
  Constant relaxation replaces a constant n (more generally, an expression which does not change during the execution of the algorithm) by a variable i and use $i \doteq n$ as part or all of the exit condition. In Example 9.2.7, constant relaxation is used to obtain the loop invariant, i.e., $s \doteq sum(i, A)$, from the postcondition $s \doteq sum(n, A)$. This condition is trivial to establish initially for $i \doteq 0$ and $s \doteq 0$ and easy to extend to an incremented i, yielding the postcondition when i reaches n.
- **Uncoupling**
  Uncoupling is to replace a variable by two, using their equality as part or all of the exit condition. Uncoupling is used in the gcd algorithm where we replace $y \doteq gcd(a, b)$ by $x \doteq 0 \wedge gcd(x, y) \doteq gcd(a, b)$. Uncoupling is also used in Example 9.3.7 where we replace $y \doteq n!$ by $x \doteq 0 \wedge x!y \doteq n!$.
- **Term dropping**
  Term dropping removes a subformula (typically a conjunct), which gives a straightforward weakening. Term dropping often applies after uncoupling. $x \doteq 0$ is dropped from $x \doteq 0 \wedge gcd(x, y) \doteq gcd(a, b)$ in the gcd algorithm and also from $x \doteq 0 \wedge x!y \doteq n!$ in the factorial algorithm, to obtain the essential invariant. The dropped term often goes into the exit condition of the loop.
- **Aging**
  Aging is to replace a variable (more generally, an expression) by an expression that represents the value the variable had at previous iterations of the loop. Aging typically accommodates "off-by-one" discrepancies between when a variable is evaluated in the invariant and when it is updated in the loop body. Typically, aging is used for generating bounding invariants. For example, if $i > 0$ appears in the loop condition and $i := i - 1$ appears in the loop body, aging will generate $i + 1 > 0$ (equivalently $i \geq 0$) as a bounding invariant. On the other hand, if $j < n$ appears in the loop condition and $j := j + 1$ appears in the body, then aging will generate $j - 1 < n$ (equivalently $j \leq n$) as a bounding invariant.

9.4 Obtaining Good Loop Invariants 343

The above generalization techniques can be implemented by heuristic methods and thus generating invariants automatically. The amount of research work on the automated generation of invariants is substantial and spread over more than three decades; this reflects the cardinal role that invariants play in the formal analysis and verification of programs.

9.4.2 Program Synthesis from Invariants

Program synthesis is the task of constructing a program that provably satisfies the postcondition. In contrast to program verification, the program is to be constructed rather than given. Both fields make use of formal proof techniques, and both comprise approaches of different degrees of automation. In contrast to automatic programming techniques, postconditions in program synthesis are usually non-algorithmic statements in an appropriate logical language.

If we are capable of generating loop invariants from the postcondition automatically, then it is relatively easy to construct the code that satisfies the postcondition. Since different algorithms for solving the same problem require different invariants, generating different invariants leads us to different algorithms.

Suppose we like to sort an array A of n elements. Let A_0 denote the original array A. The postcondition of a sorting algorithm is $permu(A, A_0) \land sorted(A, 0, n)$, where $permu(A, A_0)$ is true iff A is a permutation of A_0 and $sorted(A, 0, n)$ is true iff the elements of A from 0 to $n - 1$ are sorted. Suppose the only way to change A is by $swap(A, i, j)$, which swaps the elements $A[i]$ and $A[j]$, then $permu(A, A_0)$ will remain true through the algorithm. That is, the triple

$$\{permu(A, A_0)\}\, swap(A, i, j)\, \{permu(A, A_0)\}$$

is always true ($swap(A, i, j)$ represents the code "$z := A[i]; A[i] := A[j]; A[j] := z$"). Thus, for simplicity, we assume $swap(A, i, j)$ is the only way to modify A and drop $permu(A, A_0)$ from our discussion, focusing on $sorted(A, 0, n)$.

Intuitively, $sorted(A, i, j)$ is true iff $A[i] \leq A[i + 1] \leq \cdots \leq A[j - 1]$. For verification purpose, the program library does not need an implementation of $sorted(A, i, j)$. On the other hand, the programmer and theorem prover need to know the meaning of $sorted(A, i, j)$, which can be defined recursively as follows:

$$sorted(A, i, j) \doteq true \textbf{ if } i \geq j - 1$$
$$sorted(A, i, j + 1) \doteq sorted(A, i, j) \land A[j - 1] \leq A[j] \textbf{ if } i < j$$

Example 9.4.2 If we wish to generate an invariant for the insertion sort algorithm, the first technique to apply is constant relaxation: replace $sorted(A, 0, n)$ by $k \doteq n \land sorted(A, 0, k)$. Taking $k \doteq n$ as the exit condition, we obtain the invariant

$k > 0 \land sorted(A, 0, k)$, where $k > 0$ comes from the predcondition $n > 0$, and the outer loop looks as follows:

```
    {n > 0}
1   k := 1
    {I : k > 0 ∧ sorted(A, 0, k)}
2   while k ≠ n do
        ...
8       k := k + 1
9   od
```

It is easy to see that the exit condition of the outer loop, i.e., $k \doteq n$, will give us the postcondition $sorted(A, 0, n)$.

In order to generate an loop invariant for the inner loop, we apply uncoupling and aging to $sorted(A, 0, k)$ by introducing a new variable j, which breaks the first $k + 1$ elements of A into two sorted lists: $A[0], \ldots, A[j-1]$, and $A[j], \ldots, A[k]$. We obtain invariant I' for the inter loop:

$$I' : 0 \leq j \land j \leq k \land sorted(A, 0, j) \land sorted(A, j, k+1)$$

which has four conjuncts. The first two conjuncts are the bounding invariants. If we choose $\neg(j > 0 \land A[j-1] > A[j])$ as the exit condition of the inner loop, together with the last two conjuncts of I', we can show that $sorted(A, 0, k+1)$ is true when the exit condition is true:

```
    {n > 0}
1   k := 1
    {I : k > 0 ∧ sorted(A, 0, k)}
2   while k ≠ n do
3       j := k
        {I' : 0 ≤ j ∧ j ≤ k ∧ sorted(A, 0, j) ∧ sorted(A, j, k + 1)}
4       while (j > 0 ∧ A[j − 1] > A[j]) do
5           swap(A, j, j − 1)
6           j := j − 1
7       od
8       k := k + 1
9   od
```

It is easy to see that $sorted(A, 0, j)$ and $sorted(A, j, k+1)$ are true when $j \doteq k$, because $sorted(A, 0, k)$ comes from I and $sorted(A, k, k+1)$ from the definition of $sorted$. To show that $0 \leq j \land j \leq k$ is true when $j \doteq k$, we need to know $k \geq 0$. That is why we have already added $k > 0$ as a bounding invariant into I of the outer loop. If this were not done, we would have done it now. So, I' can be shown to be true before entering the inner loop.

To see that I' is an invariant for the inner loop, inside the second loop, $0 \leq j - 1$ is true because the loop condition $j > 0$; $sorted(A, 0, j-1)$ is true because of

9.4 Obtaining Good Loop Invariants

$sorted(A, 0, j)$, $sorted(A, j - 1, k + 1)$ is true because $sorted(A, j, k + 1)$ and $A[j - 1] \leq A[j]$ (after swapping). Thus, I' is indeed a loop invariant.

The exit condition of the inner loop is $j \leq 0 \vee A[j - 1] \leq A[j]$. If $j \leq 0$ is true, from $0 \leq j$, we know $j \doteq 0$; hence, $sorted(A, 0, k + 1)$ is true from $sorted(A, j, k + 1)$. If $A[j - 1] \leq A[j]$ is true, from $sorted(A, 0, j)$, $A[j - 1] \leq A[j]$, and $sorted(A, j, k + 1)$, we conclude also that $sorted(A, 0, k + 1)$ is true. After the execution of line 8, $k := k + 1$, we arrive at the invariant of the outer loop, i.e., $k > 0 \wedge sorted(A, 0, k)$. □

Choosing I' as the loop invariant of the inner loop with the exit condition $j \leq 0 \vee A[j - 1] \leq A[j]$ leads the insertion sort algorithm as presented in the above example. If we choose a different loop invariant with a proper exit condition, we may obtain the algorithm of *selection sort* (an exercise). These algorithms are regarded as examples of "program synthesis," if the invariants can be generated automatically.

Example 9.4.3 Suppose we are asked to write a program that "store in y the maximum of x and y." The postcondition is naturally $y \doteq max(x, y)$. Assume the precondition is $\top$, what will be the code C in $\{\top\}\ C\ \{y \doteq max(x, y)\}$. In fact, C is not unique. For example, some choices are the following:

- if $x > y$ then $y := x$ fi
- if $x > y$ then $x := y$ fi
- $y := x$
- $x := y$

What will be the "correct" code? With respect to the postcondition, the above four codes are all correct. If the intended meaning is that y contains the maximum of the original values of x and y, in this case, we need auxiliary variables, such as x_0 and y_0, to denote the original values of x and y. The correct postcondition is $y \doteq max(x_0, y_0)$, and the first of the above codes is correct with respect to this postcondition. If we ask further that $y \doteq max(x, y)$, x and y hold the original values of x and y, then the correct postcondition is $y \doteq max(x_0, y_0) \wedge ((x \doteq x_0 \wedge y \doteq y_0) \vee (x \doteq y_0 \wedge y \doteq x_0))$, and none of the above codes is correct. □

The above simple example illustrates that there exist many codes to achieve the same postcondition; postconditions should be carefully chosen to reflect precisely the requirement. When loop invariants are needed, the complexity of program synthesis grows exponentially. Since there are too many choices to create proper loop invariants, thus leading to various algorithms, program synthesis remains a very challenging task.

9.4.3 Choosing A Good Language for Assertions

Informal assertions are intuitive, but being able to express them formally is even better. Reaching this goal and, more generally, continuing to make assertions ever more mainstream in software development requires convenient, clear, and expressive notations for expressing loop invariants and other program assertions.

What will be a good underlying language for assertions? By far, assertions are written in a first-order language. Do we need a language of higher-order logic? We must make sure that any convention for assertions is compatible with the language used for preconditions and postconditions. This basic language decision leaves open the question of the level of expressiveness of assertions.

For first-order language, there are many ways to specify assertions, because first-order languages are so expressive, and we can use various predicates and functions to specify the same property. There are at least two possibilities:

- Allow assertions, in particular postconditions and loop invariants, to use functions and predicates defined using some appropriate mechanism (often, the programming language's function declaration construct) to express high-level properties based on a domain theory covering specifics of the application area. We call this the *domain theory* approach.
- Disallow the previous possibility, requiring assertions always to be expressed in terms of the constructs of the assertion language, without functions. We call this the *atomic assertion* approach.

Example 9.4.4 Example 9.2.8 is a good example of the domain theory approach. In the example, we defined a function $arrMax(n, A)$, which returns the maximal element among the first n elements in array A, in the domain of arrays. The postcondition is briefly described by $m \doteq arrMax(n, A)$. Using the atomic assertion approach, the same postcondition would be the following:

$$(\exists x\, 0 \leq x < n \wedge m \doteq A[x]) \wedge \forall y\, 0 \leq y < n \rightarrow m \geq A[y]$$

where the first part says m is an element of A and the second part says no elements of A are greater than m. Similar assertions can be written for the loop invariant. □

Even for such a simple example, the limitations of the atomic assertion approach are clear: the assertions become long and complicated; they contain quantifiers which are hard for theorem provers. In general, this approach requires going back to basic logical constructs every time, it does not scale. Past experiences have shown that the atomic assertion approach is futile.

The domain theory approach means that, before any attempt to reason about an algorithm, we should develop an appropriate model of the underlying domain by defining appropriate concepts such as greatest common divisor for algorithms on integers (Example 9.4.1) and establishing the relevant theorems (e.g., $x > y \rightarrow gcd(x, y) \doteq gcd(x - y, y)$). These concepts and theorems need only be developed once for every application domain of interest, not anew for every program over that domain. The programs can then use the corresponding functions in their assertions.

The domain theory approach takes advantage of the standard abstraction mechanism of mathematics. Its only practical disadvantage, for assertions embedded in a programming language, is that the functions over a domain (such as gcd) must come from some library and, if written in the programming language, must satisfy strict limitations.

The assertions without quantifiers are more amenable to automated reasoning. However, the domain theory approach has its own limitations: it needs some functions (e.g., $arrMax$) which are not required by the program. Some properties of these functions may be needed in the proof of the assertions, and the proofs of these properties may need other reasoning tools (e.g., inductive reasoning). For instance, in Example 9.2.8, we may need the property that $arrMax(n, A)$ is an element of A when $n > 0$.

Progress in the domain theory approach, both theoretical and on the engineering side, remains essential. Much remains to be done to bring the tools to a level where they can be integrated in a standard development environment and routinely suggest assertions, including invariants, conveniently and correctly, whenever a programmer writes a postcondition.

9.4.4 Verification Tools for Conventional Languages

Formal verification has long been an active field of research in both academy and industry. Many formal verification systems have been developed with notable success. Here, we select a few of them for conventional programming languages:

- **Verified Software Toolchain**: This tool is developed at Princeton University (website: vst.cs.princeton.edu) with worldwide participants. The tool provides a verifiable C language and assures with machine-checked proofs that the claimed assertions really hold in the C program. The tool includes static analyzers to check assertions, optimize compilers to translate C programs to machine language, and supply verified C libraries.
- **KeY**: The KeY tool is developed originally at Germany's Karlsruhe Institute of Technology and jointly by several European institutions (website: www.key-project.org). The tool includes a collection of software tools that support verified Java programming in the Eclipse IDE (integrated development environment). The tool accepts specifications written in the Java Modeling Language (JML) to Java programs. These specifications are transformed into theorems of dynamic logic and then proved either interactively (i.e., with a user) or automatically. Failed proof attempts can be used for a more efficient debugging or verification-based testing. There have been several extensions of KeY to the verification of C programs or hybrid systems.

Exercises

1. For $i = 1, 2, \ldots, 5$ in the following programs, please generate the precondition P_i and middle conditions by the Hoare rules from Q_i and show the detail of each step of the generation:

   (a) $\{P_1\}$ **if** $(x > y)$ $z := x$ **else** $z := y$ $\{Q_1 : z \doteq max(x, y)\}$
   (b) $\{P_2\}$ $x := x + 1; y := y + x$ $\{Q_2 : y \doteq x(x + 1)/2\}$
   (c) $\{P_3\}$ $y := y + x; x := x - 1$ $\{Q_3 : c - y \doteq x(x + 1)/2\}$
   (d) $\{P_4\}$ $y := y + 1; z := x * z$ $\{Q_4 : z \doteq x^y\}$
   (e) $\{P_5\}$ **if** $(y\%2 = 1) y := y - 1; z := x * z$ **else** $x := x * x; y := y/2$ $\{Q_5 : x^y z \doteq c\}$

2. Prove formally the following triple is true by choosing a proper loop invariant:

   $$\{x \doteq 0\} \textbf{ while } (x < 100) \textbf{ do } x := x + 1 \textbf{ od } \{x \doteq 100\}$$

3. Given the following annotated two programs, provide the loop invariant and prove its total correctness by Hoare logic. For every simplification step used in the proof, please state clearly which arithmetic/algebraic/logic law is used for simplification:

   $\{A : a \geq 0\}$
   $int\ x := 0; int\ y := 0$
   while $x < a$ **do**
       $x := x + 1$
       $y := y + x$
   $\{B : y \doteq a(a + 1)/2\}$

   $\{A : a \geq 0\}$
   $int\ x := a; int\ y := 0$
   while $x \neq 0$ **do**
       $y := y + x$
       $x := x - 1$
   $\{B : y \doteq a(a + 1)/2\}$

4. Repeat the previous exercise for the following two programs for computing the *power* function:

   $\{A : int\ b \geq 0\}$
   $int\ x := a; int\ y := 0; int\ z := 1$
   while $y \neq b$ **do**
       $y := y + 1$
       $z := x * z$
   $\{B : z \doteq a^b\}$

   $\{A : int\ b \geq 0\}$
   $int\ x := a; int\ y := b; int\ z := 1$

```
while y ≠ 0 do
    if (y%2 = 1) // i.e., y is odd
        y := y − 1; z := x ∗ z
    else x := x ∗ x; y := y/2
{B : z ≐ a^b}
```

5. For the four programs in the previous two exercises, please run the algorithm VC with proper loop invariants and show the returned verification conditions from the postconditions.
6. Given an array A of n integers, the problem of Max-Sum is to find the maximal sum of elements in each consecutive non-empty subarray of A. For example, if $A = [2, -5, 8, -4, 6, -3, -1, 2]$, then the solution is 10, which is the sum of the subarray $[8, -4, 6]$. Please provide a formal postcondition for the following program and prove formally that the program can solve the Max-Sum problem.
7. Repeat the previous exercises using the Prolog implementation of algorithm VC, and show the Prolog code and returned results:

```
{A[1..n] : array of integers, n ≥ 1}
int x := 2; int y := A[1]; int z := A[1]
while x ≤ n do
    y := max(y + A[x], A[x])
    z := max(y, z)
    x := x + 1
{z is the solution of Max-Sum }
```

8. Choose a different loop invariant with proper exit condition for the inner loop in Example 9.4.2, so that the resulting algorithm is "selection sort."

References

1. Krzysztof R. Apt and Ernst-Rüdiger Olderog, *Fifty years of Hoare's logic. Formal Aspects of Computing*. vol.31 (6): pp 751-807. 2019.
2. Robert S. Boyer and J. Strother Moore, *A computational logic*, Academic Press, New York, 1979.
3. Reiner Hähnle, *Dijkstra's Legacy on Program Verification*. ACM Books. 2022.
4. Carlo A. Furia, Bertrand Meyer, and Sergey Velder, *Loop invariants: analysis, classification, and examples*, ACM Computing Surveys 46, 3, Article 34 (January 2014)

Chapter 10
Temporal Logic

In the previous chapter, we have shown that Hoare logic is a framework for software verification: it can show that a program can achieve the desired property after its execution. However, when a program involves multiple concurrent processes, its verification becomes much more complex, and the first-order logic is inadequate.

For example, when you deposit cash into an automated teller machine (ATM), the program running inside the ATM will read the balance of your bank account and add the cash amount into your account. At the same time, another program running inside the bank may use the same account to pay off your credit card. The two programs cannot be run at the same time; otherwise, the balance of your account may go wrong. Modifying the same data at the same time by multiple processes is called "race condition" in concurrent programming. To avoid race conditions, a locking mechanism is needed so that an account can be accessed when it is free (unlocked). Due to locking mechanisms, two programs may go to a "deadlock" state. For instance, when both you and your friend send money to each other at the same time, one program locked account A and asks for account B and the other program locked account B and asks for account A. In such cases, we cannot prove the termination of the program by counting steps. To verify a program works correctly without race condition or deadlock, we need a formalism to express time, such as "two processes cannot access critical data at the same time," or "the program will terminate eventually."

First-order logic is an excellent logic for many applications as we have demonstrated in Chaps. 5–9. The so-called *temperature paradox* shows that it is not convenient to express time in first-order logic in a simple way. Let t be a variable for the current temperature, such as $t = 30$ to indicate that the current temperature is $30°$. Let $p(t)$ denote "temperature t is rising," then $p(30)$ is a well-formed formula. However, the interpretation of $p(30)$ is not meaningful: "temperature 30 is rising."

To express that the temperature changes over time in first-order logic, we may define the temperature as a function $t(x)$, where x is time variable. To express "temperature t is rising," we have $(s > 0) \rightarrow t(x + s) > t(x)$. In propositional logic, let *p* denote "temperature is mild" and *q* be "temperature is hot." To express the statement that "the temperature changes from mild to hot," we may use $p \wedge \circ(q)$, along with some general properties of *p* and *q*. Here, ∘ indicates "what will be true next" and is one of the temporal operators that we study in this chapter.

In logic, the term *temporal logic* is a system of rules and symbolism for representing and reasoning about time and temporal information, as well as their formal representation. Application of temporal logic includes its use as a formalism for clarifying philosophical issues about time, as a framework for defining the semantics of temporal expressions in natural language, and as a language for encoding temporal knowledge in artificial intelligence.

Temporal logic plays an important role in formal verification of concurrent programs, where it is used to specify requirements of computer programs and to conduct formal analysis and verification of program executions. For instance, one may wish to say that whenever a request is made, access to a resource is eventually granted, but it is never granted simultaneously to two parties who make requests. Such a statement can conveniently be expressed in temporal logic.

Consider the statement "I possess account A." Though its meaning does not change in time, the statement's truth value can vary in time. Sometimes it is true, and sometimes false, but never simultaneously true and false. In temporal logic, a statement can have a truth value that varies in time. This treatment of truth values of the same formula over time differentiates temporal logic from classical logic.

10.1 An Approach from Modal Logic

A modal is an expression (like "necessarily" or "possibly") that is used to qualify the truth of a statement. Modal logic is the study of the deductive behavior of the expressions containing the phrases like "it is necessary that" or "it is possible that." Modal logic has been extended to include logic for belief, for tense, and temporal expressions where "necessarily" (modal) is interpreted as "always" (temporal) and "possibly" as "eventually." An understanding of modal logic and its extensions is particularly valuable in the formal analysis of philosophical argument, where expressions from the modal family are both common and confusing.

Like most formal logical systems, modal logic as well as temporal logic can be defined as an extension to classical logic. Here, for simplicity, we present a temporal logic (TL) by adding two modal operators, □ and ◇, to propositional logic.

10.1.1 Modal Operators

The first modal operator, □, denotes "always" and takes a formula as its argument. For example, if *p* denotes "it rains today," then □*p* is a formula of TL, and its intuitive meaning is "it will rain everyday."

10.1 An Approach from Modal Logic

The second operator, ◇, denotes "eventually" and also takes a formula as its argument. Thus, if p denotes "it rains today," ◇p is a formula of TL, and its intuitive meaning is "it will rain someday."

Informally, □ states a universal property in the future, while ◇ states an existential property in the future. For example, to specify the statement "if a request is made to print a file, the file will be printed eventually," the statement can be denoted by □(r → ◇p), where r stands for "request is made" and p for "the file is printed."

The two operators, □ and ◇, appear almost in every modal logic, where □ is usually read as "necessarily" and ◇ as "possibly." We borrow them from classical modal logic for TL and give them the temporal meaning, i.e., "always" and "eventually."

Using the BNF grammar, the formulas of TL can be formally defined as follows:

Definition 10.1.1 Let op be the binary logical operators of propositional logic and V_P be the propositional variables used in the current application, then the *formulas* of TL can be defined by the following BNF grammar:

$$\langle op \rangle ::= \wedge \mid \vee \mid \rightarrow \mid \oplus \mid \leftrightarrow$$
$$\langle V_P \rangle ::= p \mid q \mid r \mid s \mid t$$
$$\langle Formulas \rangle ::= \top \mid \bot \mid \langle V_P \rangle \mid \neg \langle Formulas \rangle \mid (\langle Formulas \rangle \langle op \rangle \langle Formulas \rangle) \mid$$
$$\square \langle Formulas \rangle \mid \diamond \langle Formulas \rangle$$

Here are some examples of TL formulas:

$$(p \wedge \square \neg q), \quad \diamond(r \rightarrow (\diamond p \wedge \neg q)), \quad \square((p \wedge \neg q) \vee \diamond r), \quad (\square(p \wedge \neg q) \vee \diamond r)$$

Note the difference of the last two formulas, where □ applies to the whole formula in the former and only to the first argument of ∨ in the latter.

10.1.2 Kripke Semantics

The conventional formal semantics of modal logic is called *Kripke semantics*, also known as *Kripke frames*, *relational semantics*, or *frame semantics*. Kripke semantics is a formal semantics for modal logic systems created in the late 1950s by Saul Kripke and Andre Joyal. It was adapted to temporal logic and other nonclassical systems. The development of Kripke semantics was a breakthrough in the theory of nonclassical logics, because the model theory of such logics was almost nonexistent before Kripke.

As we have seen from Chap. 2, an interpretation of a propositional formula is a mapping of propositional variables to truth values. Given n propositional variables, we have exactly 2^n different interpretations. For example, if we have two propositional variables, say p and q, then there are four interpretations, $I_1 = pq$ (it means $I_1(p) = 1, I_1(q) = 1$), $I_2 = p\overline{q}$, $I_3 = \overline{p}q$, and $I_4 = \overline{pq}$. Using the notation

Fig. 10.1 A graphical display of a Kripke frame on two variables

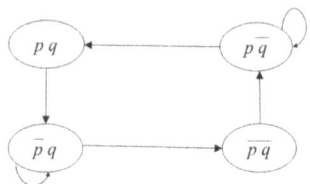

from Chap. 2, $V_P = \{p, q\}$ and $All_P = \{pq, p\overline{q}, \overline{p}q, \overline{pq}\}$. Kripke semantics are based on binary relations over All_P. Note that every binary relation R over S can be represented by a directed graph $G = (S, R)$.

Definition 10.1.2 Given a set of propositional variables V_P, a *Kripke frame* is a directed graph $K = (S, R)$, where S, called *states*, is a non-empty subset of All_P, the set of all interpretations over V_P, and R is a binary relation over S.

Example 10.1.3 Let $V_P = \{p, q\}$, $S = All_P = \{pq, p\overline{q}, \overline{p}q, \overline{pq}\}$, and $R = \{\langle pq, \overline{pq}\rangle, \langle p\overline{q}, pq\rangle, \langle p\overline{q}, p\overline{q}\rangle, \langle \overline{p}q, \overline{pq}\rangle, \langle \overline{pq}, \overline{pq}\rangle, \langle \overline{pq}, p\overline{q}\rangle\}$. Then, $G = (S, R)$, displayed in Fig. 10.1, is a Kripke frame. □

It should be understood that in an application, every TL formula uses the propositional variables from V_P and every state w in a Kripke frame $K = (S, R)$ is an interpretation of V_P. If $\langle w, w'\rangle \in R$, we say w' is a *successor* of w in K.

Definition 10.1.4 Given a Kripke frame $K = (S, R)$ and a state $s \in S$, a TL formula C is *true* in s of K, denoted by $K, s \models C$, if one of the following conditions is true:

1. C is $\top$ or 1.
2. $C \in V_P$, $s(C) = 1$.
3. C is $\neg A$, $K, s \not\models A$.
4. C is $A \wedge B$, $K, s \models A$ and $K, s \models B$.
5. C is $A \vee B$, $K, s \models A$ or $K, s \models B$.
6. C is $A \rightarrow B$, $K, s \models \neg A \vee B$.
7. C is $\Box A$, for every successor s' of s, $K, s' \models A$.
8. C is $\Diamond A$, for some successor s' of s, $K, s' \models A$.

Otherwise, we say C is false in s of K.

Apparently, the above definition is a recursive one, and we may write a recursive algorithm to compute the value of $K, s \models A$.

Example 10.1.5 Consider the Kripke frame in Fig. 10.1 and let A be $\Box \neg p$. Then, A is true in $s = pq$ because the only successor of s is $s' = \overline{pq}$ and $\neg p$ is true in s'.

Let B be $p \wedge q \rightarrow \Box \neg p$. B is true in $s = \overline{p}q$ or $s = \overline{pq}$, because p is false in s. B is true in $s = p\overline{q}$ because q is false in s. B is also true in $s = pq$ because $A = \Box \neg p$ is true in pq. That is, B is true in every state of the frame.

10.1 An Approach from Modal Logic

Using the same analysis, we can show that $p \wedge q \to \Box\Box\neg p$ is true in every state. However, $p \wedge q \to \Box\Box\Box\neg p$ is false in $s = pq$. □

In modal logic, a Kripke frame specifies the transitions of states (also called *worlds*). *Necessarily* means "in all successive states (successors)" and *possibly* means "in one of the successors." Similarly, in TL, a formula $\Box A$ is true in the current state if A is true in every successor; $\Diamond A$ is true in the current state if A is true in one of the successors. Note that if the current state does not have any successor, then $\Box A$ is true and $\Diamond A$ is false in the current state for any formula A.

Definition 10.1.6 Given a Kripke frame $K = (S, R)$, a TL formula A is *true* in K, written $K \models A$, if for every state $s \in S$, $K, s \models A$, and we say K is a *Kripke model* of A.

A is *satisfiable* in TL if A has a Kripke model. That is, there exists a Kripke frame K such that $K \models A$.

A is *valid* in TL, written $\models A$, if A is true in every Kripke frame. That is, every Kripke frame is a model of A.

Example 10.1.5 shows that both $p \wedge q \to \Box\neg p$ and $p \wedge q \to \Box\Box\neg p$ are satisfiable as they are true in the Kripke frame in Fig. 10.1. However, this frame is not a model of $p \wedge q \to \Box\Box\Box\neg p$.

Given a TL formula A, let $\mathcal{M}(A)$ denote the set of all Kripke models of A.

Definition 10.1.7 Given two TL formulas A and B, A and B are *equivalent*, written $A \equiv B$, if $\mathcal{M}(A) = \mathcal{M}(B)$. We say A entails B, or B is a *logical consequence* of A, written $A \models B$, if $\mathcal{M}(A) \subseteq \mathcal{M}(B)$.

Given a set V_P of n propositional variables, the number of choices for S in a Kripke frame is enormous ($2^{2^n} - 1$) and the number of choices for R is also huge ($2^{|S|^2}$ for each S). Hence, it is infeasible to check if a TL formula is valid by enumerating all frames. However, all tautologies in propositional logic are valid in TL, and all major results from propositional logic are still useful in TL.

Theorem 10.1.8

(a) **(Substitution)** *Any instance of a valid TL formula is valid in TL. That is, if p is a propositional variable in A, then $A[p \leftarrow B]$ is valid whenever A is valid, where B is any TL formula.*
(b) **(Substitution of equivalence)** *For any TL formulas A, B, and C, where B is a subformula of A, and $B \equiv C$, then $A \equiv A[B \leftarrow C]$.*
(c) **(Logical Equivalence)** *Given two TL formulas A and B, $\models A \leftrightarrow B$ iff $A \equiv B$.*
(d) **(Entailment)** *Given two TL formulas A and B, $\models A \to B$ iff $A \models B$.*

The proofs of the above theorem are similar to those for propositional logic and omitted here.

We show below some examples of proof in TL. Later, we will provide a proof method based on semantic tableau.

Theorem 10.1.9 (Duality) $\neg \Box p \equiv \Diamond \neg p$.

Proof For any Kripke frame $K = (S, R)$, if K is a model of $\neg \Box p$, then for every state $s \in S$, $\neg \Box p$ is true in s, or equivalently, $\Box p$ is false in s. That means there exists a successor s' of s such that p is false in s', or equivalently, $\neg p$ is true in s'. Hence, $\Diamond \neg p$ is true in s by the semantics of $\Diamond$. Since s is arbitrary, $\Diamond \neg p$ is true in every state of K. In other words, K is a model of $\Diamond \neg p$.

On the other hand, if K is not a model of $\neg \Box p$, then there exists a state $s \in S$ such that $\neg \Box p$ is false (thus $\Box p$ is true) in s. By definition, either s has no successors, or for every successor s' of s, p must be true in s'. Hence, if s has no successor, then $\Diamond \neg p$ is false by definition. If s has, since $\neg p$ is false in every successor of s, $\Diamond \neg p$ must be false in s. So, K cannot be a model of $\Diamond \neg p$.

Since K is arbitrary, $\neg \Box p$ and $\Diamond \neg p$ must share the same set of models. □

Applying substitution and equivalence, we can easily deduce from the above theorem that $\Box \neg p \equiv \neg \Diamond p$ and $\Box \neg p \leftrightarrow \neg \Diamond p$ is valid.

Example 10.1.10 We can prove that $(\Box p \to p) \equiv (\neg p \to \Diamond \neg p)$ by transforming both sides of $\equiv$ to the same formula using the known equivalence relations:

$$\Box p \to p \quad \equiv \neg \Box p \vee p$$
$$\equiv \Diamond \neg p \vee p$$
$$\neg p \to \Diamond \neg p \equiv \neg \neg p \vee \Diamond \neg p$$
$$\equiv \Diamond \neg p \vee p$$

The formula $p \vee \Diamond \neg p$ is a negation normal form (NNF) of TL, and we will discuss NNF later. □

Theorem 10.1.11 (Distribution) $\models \Box(p \to q) \to (\Box p \to \Box q)$.

Proof $\Box(p \to q) \to (\Box p \to \Box q)$ is logically equivalent to

$$(1) \qquad \neg \Box(\neg p \vee q) \vee \neg \Box p \vee \Box q.$$

Applying the duality theorem to (1), we obtain another equivalent formula:

$$(2) \qquad \Diamond(p \wedge \neg q) \vee \Diamond \neg p \vee \Box q.$$

For any Kripke frame $K = (S, R)$ and any $s \in S$, if (2) is false in s, that is, $\Diamond(p \wedge \neg q)$, $\Diamond \neg p$, and $\Box q$ are all false in s, then s must have some successors (otherwise, $\Box q$ is true in s). Because $\Box q$ is false in s, there must exist a successor s' of s such that q is false in s'. p must be true in s' because $\Diamond \neg p$ is false in s. Hence, $p \wedge \neg q$ is true in s' and $\Diamond(p \wedge \neg q)$ is true in s, a contradiction. Thus, (2) must be true in s and thus true in every Kripke frame. □

10.1.3 Restrictions and Limitations

Let r denote "it rains," then the intuitive meaning of $\Box r$ is "it always rains." Given a Kripke frame $K = (S, R)$, if $\Box r$ is true in a state $s \in S$, we know r will be true in any successor of s. What is the truth value of r in the state s? If $\Box$ means "always," should r be true now and then in every state, including s? That is, we may ask that $\Box A \to A$ be an axiom (a formula assumed to be true) of TL.

We say a Kripke frame $K = (S, R)$ is *reflexive* if every state $s \in S$ is a successor of itself, that is, $\langle s, s \rangle \in R$ (called *loop* in graph theory).

Theorem 10.1.12 (Reflexivity) *A Kripke frame K is reflexive iff K is a model of $\Box A \to A$ for any formula A.*

Proof Let $K = (S, R)$ be reflexive. If $\Box A \to A$ is false in K, then there exists $s \in S$ such that $\Box A$ is true but A is false in s. Since $\Box A$ is true, A is true in every successor of s. However, s is also a successor of s because K is reflexive. So, A is true in s, too. This is a contradiction to $\Box A \to A$ being false in K.

On the other hand, suppose $K = (S, R)$ is not reflexive. Pick a state $s \in S$ such that s is not a successor of itself. Let I_s be the propositional interpretation associated with s. If we view I_s as a conjunction of literals each of which is true in s, then $\neg(I_s)$ is a clause which is false in s. This clause will be true in every interpretation (including all successors of s) other than s. Thus, $\Box\neg(I_s)$ is true in s. Let A be $\neg(I_s)$, then $\Box A \to A$ is false in s and K is not a model of $\Box A \to A$.

Since K is arbitrary, the theorem holds. □

The above theorem plays a dual role: by restricting Kripke frames to reflexive ones, we obtain the validity of $\Box A \to A$ in these frames; on the other hand, if $\Box A \to A$ is given as an axiom, we exclude all nonreflexive Kripke frames as model candidates.

We know from Kripke frames that if $\Box p$ is true in the current state s, then p will be true in all the successors of s. What about the truth value of p in the successors of the successors of s? If $\Box$ means "always," should p be true in them, too? Essentially, we ask $\Box A \to \Box\Box A$ to be true in every state for every formula A. Let us say a Kripke frame $K = (S, R)$ is *transitive* if R is transitive: for any states u, v, w, $\langle u, v \rangle \in R$ and $\langle v, w \rangle \in R$ imply that $\langle u, w \rangle \in R$. The following theorem tells us about the relationship between the transitivity and the formula $\Box A \to \Box\Box A$.

Theorem 10.1.13 (Transitivity) *A Kripke frame K is transitive iff K is a model of $\Box A \to \Box\Box A$ for any formula A.*

Proof Let $K = (S, R)$ be transitive. If $\Box A \to \Box\Box A$ is false in K, then there exists $s \in S$ such that $\Box A$ is true but $\Box\Box A$ is false in s. The latter implies that there must exist a successor s' of s and a successor s'' of s' such that A is false in s''. Since $\Box A$ is true, A is true in every successor of s. However, s'' is also a successor of s because K is transitive. So, A is true in s'', too. This is a contradiction to $\Box A \to \Box\Box A$ being false in K.

On the other hand, suppose $K = (S, R)$ is not transitive. There must exist three states $s_1, s_2, s_3 \in S$ such that s_2 is a successor of s_1, s_3 is a successor of s_2 but not a successor of s_1. Let I_3 be the propositional interpretation associated with s_3. If we view I_3 as a conjunction of literals each of which is true in s_3, then $\neg(I_3)$ is a clause which is only false in s_3 but true in every interpretation (including all successors of s_1) other than s_3. Let A be $\neg(I_3)$, then A is false in s_3 and true in every other state, and $\Box A$ is false in s_2 and true in s_1; $\Box\Box A$ is false in s_1. Hence, $\Box A \to \Box\Box A$ is false in s_1 and K is not a model of $\Box A \to \Box\Box A$. □

Combing the above two theorems, we can conclude that if a Kripke frame is both reflexive and transitive, then $\Box A \leftrightarrow \Box\Box A$ will be true in this frame.

The *necessitation rule* of modal logic states that if A is true, so is $\Box A$. That is, if A can be proved to be true now, the same proof can be used everywhere. The axiom for expressing the necessitation rule is the formula $A \to \Box A$. The Kripke models of $A \to \Box A$ are those Kripke frames $K = (S, R)$ in which the only successor (if exists) of each state is itself, i.e., if $\langle w, v \rangle \in R$, then $w = v$. Such frames are called *discrete*. Obviously, the only edges allowed in a discrete frame are loops (i.e., $\langle w, w \rangle$), and such frames are not very useful. As some loops may be missing in a discrete frame, a discrete frame is not always reflexive.

Besides reflexivity, transitivity, and discreteness, there are many other restrictions on the relations in a Kripke frame, such as symmetry, density, etc. There are axioms corresponding to these restrictions.

Semantics is useful for logic only if the semantic consequence reflects its syntactical counterpart, i.e., the result of the proof system for the logic. It is vital to know which modal logics are sound and complete with respect to a set of Kripke frames. Let C be the set of formulas specifying satisfying some properties. For instance, C can be the reflexive (or transitive) property. We define $\mathcal{T}(C)$ to be the set of all formulas that are entailed by C. A proof system in a modal logic is *sound* with respect to C if every formula proved to be true belongs to $\mathcal{T}(C)$; it is *complete* with respect to C if every formula in $\mathcal{T}(C)$ can be proved to be true. There are over a dozen proof systems for modal logic, and the interested reader may refer to a textbook on modal logic.

10.2 Linear Temporal Logic

Typically, temporal logic has the ability to reason about a timeline, which is linear by nature. The Kripke semantics use directed graphs $K = (S, R)$ and are suitable for reasoning about multiple timelines, as each state allows multiple successors. We may put restrictions on R so that R acts like a linear relation: each state has at most one successor. This restriction can work with reflexivity by allowing each state to have at most one successor other than itself. However, this restriction does not work well with transitivity because transitivity wants a state to include successors of successors as its own successors. Moreover, the number of states in a Kripke frame is bounded by the number of propositional interpretations. We cannot model every

10.2 Linear Temporal Logic

timeline of infinite length by Kripke frames, which provide only cycles in the graph. Hence, we need semantics different from Kripke frames for linear temporal logic. Temporal logic based on Kripke frames can be viewed as a branching temporal logic.

Linear temporal logic (LTL) or linear-time temporal logic is a modal (temporal) logic with modalities referring to a unique timeline. LTL was first proposed for the formal verification of computer programs by Amir Pnueli in 1977. In LTL, one can encode formulas about the future of paths, e.g., a condition will eventually be true, or a condition will be true until another event becomes true, etc. For simplicity, our introduction is limited to an extension of propositional logic.

As in TL, in LTL, □ is interpreted as "always" and ◇ as "eventually." In addition to these two common modal operators, we have the third operator ○ which is interpreted as "next" tick (or step). Like □ and ◇, ○ takes a formula as argument: if r means "it rains today," then $\circ r$ means "it will rain tomorrow," if we interpret that a tick is a day. The operator ○ is an important part of LTL, which has been invented for the formal verification of concurrent programs. However, ○ is rarely used in the specification of concurrent programs, because not much is said about the execution of programs in the *next* step. Furthermore, we want a correctness statement about a concurrent program to hold regardless of how the interleaving processes select a *next* operation. Therefore, properties are almost invariably expressed in terms of □ and ◇, not in terms of ○.

Let V_P be the set of all propositional variables used in an application of LTL. The formal definition of the LTL formulas is given as the following BNF grammar:

$$\langle V \rangle ::= p \textbf{ if } p \in V_P$$
$$\langle op \rangle ::= \wedge \mid \vee \mid \rightarrow \mid \oplus \mid \leftrightarrow$$
$$\langle Formulas \rangle ::= \top \mid \bot \mid \langle V \rangle \mid \neg \langle Formulas \rangle \mid (\langle Formulas \rangle \langle op \rangle \langle Formulas \rangle) \mid$$
$$\Box \langle Formulas \rangle \mid \diamond \langle Formulas \rangle \mid \circ \langle Formulas \rangle$$

Here are some examples of LTL formulas:

$$(\circ p \wedge \Box \neg q), \quad \diamond (\circ r \rightarrow (\diamond p \wedge \neg q)), \quad \Box((p \wedge \neg q) \vee \circ \diamond r)$$

10.2.1 Timeline as Interpretation Sequence

The semantics of LTL will be defined by an infinite sequence of propositional interpretations so that a propositional variable has a fixed truth value at any position (called state) in the sequence. Formally, an interpretation sequence π is a mapping from the set $\mathcal{N}$ of natural numbers to the set of all propositional interpretations All_P:

$$\pi : \mathcal{N} \mapsto All_P$$

For any $i \in \mathcal{N}$, let $\pi(i) = s_i$. Then, π can be written as a sequence of interpretations:

$$\pi = s_0 s_1 s_2 \cdots$$

where $s_i = \pi(i) \in All_P$ for any $i \in \mathcal{N}$. i is called a *state* of π and s_i is the interpretation at state i. Now, we see that the concept of *state* is changed from an interpretation in TL to a natural number in LTL. Thus, we have an infinite number of states in LTL but only a finite of states in TL, and two states of LTL may have the same interpretation, i.e., $\pi(i) = \pi(j)$ for $i \neq j$.

For $i \in \mathcal{N}$, we use $\pi_{\geq i}$ to denote the *suffix* of π by removing the first i interpretations from π:

$$\pi_{\geq i} = t_0 t_1 t_2 \cdots$$

where $t_0 = s_i, t_1 = s_{i+1}, t_2 = s_{i+2}, \ldots$. Obviously, $\pi = \pi_{\geq 0} = \pi(0)\pi_{\geq 1}$, where $\pi(0) = s_0$. We say $\pi_{\geq 1}$ is the *immediate suffix* of π.

Definition 10.2.1 Given an interpretation sequence π, an LTL formula C is said to be *true* in π recursively, denoted by $\pi \models C$, if one of the following conditions is true:

1. C is $\top$ or 1.
2. C is $\neg A$ and $\pi \not\models A$.
3. C is $A \wedge B$, $\pi \models A$, and $\pi \models B$.
4. C is $A \vee B$ and either $\pi \models A$ or $\pi \models B$.
5. C is $A \to B$ and $\pi \models \neg A \vee B$.
6. C is $\circ A$ and $\pi_{\geq 1} \models A$.
7. C is $\Box A$ and for all $i \geq 0$, $\pi_{\geq i} \models A$.
8. C is $\Diamond A$ and there exists $i \geq 0$, $\pi_{\geq i} \models A$.

Otherwise, we say C is *false* in π.

From the above recursive definition, we see that a formula $\circ A$ is true in an interpretation sequence π iff A is true in $\pi_{\geq 1}$, the sequence obtained by removing the first item from π. A formula $\Box A$ is true in π iff A is *always* true in every suffix of π; $\Diamond A$ is true in π iff A is *eventually* true in one suffix of π. In contrast, in the Kripke semantics for branching time, $\Box A$ is true at the current state iff A is true in every *successor* state. Note that the above definition of truthfulness can be extended to sequences of finite length. If π is an empty sequence, for any formula LTL A, $\Box A$ is true in π but $\circ A$ and $\Diamond A$ are false in π.

For $\Diamond \Box p$ to be true in π, we need to find a state in π such that p is true in every interpretation of π from that state. That means, only a prefix of π may contain a finite number of interpretations which falsify p; p must be true in every interpretation not in the prefix. That is, there exists $i \geq 0$, for all $j \geq i$, $\pi(j)(p) = 1$.

For $\Box \Diamond p$ to be true in π, we just need p to be true in one of the interpretations of any suffix of π: for all $i \geq 0$, there exists $j \geq i$, $\pi(j)(p) = 1$. Thus, in first-order language, π for $\Diamond\Box p$ satisfies $\exists i \forall j R(i, j)$, while π for $\Box \Diamond p$ satisfies $\forall i \exists j R(i, j)$, where $R(i, j)$ denotes the formula "$i \geq 0 \land j \geq i \land \pi(j)(p) = 1$." From the view of the first-order formulas, $\Box \Diamond p$ is not equivalent to $\Diamond \Box p$.

Example 10.2.2 For the formula $\Box \Diamond p \land \Box \Diamond \neg p$, we choose a sequence π where the truth value of p is alternating. $\Box \Diamond p$ is true in π because $\Diamond p$ is true in every suffix of π. Note that $\Diamond \Box p$ is not true in π, another evidence that $\Box \Diamond p$ and $\Diamond \Box p$ are not equivalent. $\Box \Diamond \neg p$ is also true in π because $\Diamond \neg p$ is true in every suffix of π. Thus, π is a model of $\Box \Diamond p \land \Box \Diamond \neg p$. Neither $\Box \Diamond p$ nor $\Box \Diamond \neg p$ is valid because there exist sequences in which one formula is false and the other is true. □

Example 10.2.3 Formula $\Diamond \Box p \land \Diamond \Box \neg p$ is unsatisfiable because it is not true in any sequence π. That is, $\Diamond \Box p$ requires p be true everywhere in one suffix of π. $\Diamond \Box \neg p$ requires p be false everywhere in one suffix of π. The two conditions cannot be met at the same time. □

Definition 10.2.4 An LTL formula A is *satisfiable* if there exists an interpretation sequence π such that $\pi \models A$ and we say π is a *model* of A. A is *valid* in LTL, written $\models A$, if every interpretation sequence is a model of A.

Both $\Diamond \Box p$ and $\Box \Diamond p$ are satisfiable but not valid. From Example 10.2.2, we know $\Box \Diamond p \land \Box \Diamond \neg p$ is satisfiable but not valid. From Example 10.2.3, $\Diamond \Box p \land \Diamond \Box \neg p$ is unsatisfiable, and its negation is valid.

Theorem 10.2.5 *For any LTL formula A*

(a) (**reflexivity**) $\models \Box A \to A$;
(b) (**transitivity**) $\models \Box A \to \Box\Box A$.

Proof (a): To show $\Box A \to A$ is valid in LTL, consider any interpretation sequence π for A. If $\Box A \to A$ is false in π, then $\Box A$ is true and A is false in π. $\Box A$ means A is true in $\pi_{\geq i}$ for every $i \in \mathcal{N}$, including $i = 0$. That is, A is true in $\pi_{\geq 0} = \pi$. That is a contradiction to A being false in π. Hence, $\Box A \to A$ is true in π.

The proof of (b) is left as an exercise. □

Note that neither $\Box A \to A$ nor $\Box A \to \Box\Box A$ is valid in TL. Theorem 10.1.12 states that $\Box A \to A$ is true in every reflexive Kripke frame, and Theorem 10.1.13 states that $\Box A \to \Box\Box A$ is true in every transitive Kripke frame. The semantics of interpretation sequences is not a special case of Kripke semantics as Kripke frames have no way to specify the unique path explicitly defined by an interpretation sequence. On the other hand, interpretation sequences are not expressive enough to specify various relations of propositional interpretations provided by Kripke frames. In Sect. 10.2.3, we will see that Kripke frames are still very useful to represent possible models of LTL and can serve as a base for model checking with LTL.

Given an LTL formula A, let $\mathcal{M}(A)$ denote the set of all models of A.

Definition 10.2.6 Given two LTL formulas A and B, A and B are *equivalent*, written $A \equiv B$, if $\mathcal{M}(A) = \mathcal{M}(B)$.

We say *A entails B*, or *B is a logical consequence* of *A*, written $A \models B$, if $\mathcal{M}(A) \subseteq \mathcal{M}(B)$.

Theorem 10.2.7 *For any LTL formula A*

(a) $\Box A \models \circ A$;
(b) $\circ A \models \Diamond A$;
(c) $\Box A \models \Diamond A$.

The proofs of (a) and (b) come from the semantic definitions of these modal operators; (c) comes from the transitivity of implication.

Like TL, all tautologies in propositional logic are valid in LTL and all major results from propositional logic are still useful here.

Theorem 10.2.8

(a) (**Substitution**) *Any instance of a valid LTL formula is valid in LTL. That is, if p is a propositional variable in A and B is any formula, then $A[p \leftarrow B]$ is valid whenever A is valid.*
(b) (**Substitution of equivalence**) *For any LTL formulas A, B, and C, where B is a subformula of A, and $B \equiv C$, then $A \equiv A[B \leftarrow C]$.*
(c) (**Logical Equivalence**) *Given two LTL formulas A and B, $\models A \leftrightarrow B$ iff $A \equiv B$.*
(d) (**Entailment**) *Given two LTL formulas A and B, $\models A \rightarrow B$ iff $A \models B$.*

The proofs of the above theorem are analog to those for propositional logic and omitted here.

10.2.2 Properties of LTL

The proofs of the following theorems are examples of proving a formula is valid or equivalent to another formula. In the next section, we will provide some proof techniques based on semantic tableau. The operator $\circ$ commutes with $\Box$, $\Diamond$, and $\neg$, but $\Box$ and $\Diamond$ do not commute with each other, as shown in Example 10.2.3.

Theorem 10.2.9 (Commutativity)

(a) $\Box \circ p \equiv \circ \Box p$;
(b) $\Diamond \circ p \equiv \circ \Diamond p$;
(c) $\neg \circ p \equiv \circ \neg p$.

Proof The proofs of (a) and (b) are similar to that of (c): given any interpretation sequence π, π is a model of $\neg \circ p$ iff $\circ p$ is false in π; $\circ p$ is false in π iff p is false in $\pi_{\geq 1}$, or equivalently, $\neg p$ is true in $\pi_{\geq 1}$; $\neg p$ is true in $\pi_{\geq 1}$ iff $\circ \neg p$ is true in π. □

10.2 Linear Temporal Logic

Theorem 10.2.10 (Duality)

$$(a) \quad \neg \Box p \equiv \Diamond \neg p;$$
$$(b) \quad \neg \Diamond p \equiv \Box \neg p.$$

Proof (a) Given any interpretation sequence π, $\neg \Box p$ is true in π iff $\Box p$ is false in π; $\Box p$ is false in π iff there exists $i \geq 0$, $\pi(i)(p) = 0$, or equivalent, $\pi(i)(\neg p) = 1$. There exists $i \geq 0$, $\pi(i)(\neg p) = 1$ iff $\Diamond \neg p$ is true in π.

(b) Replace p by $\neg q$ in (a), and apply $\neg \neg q \equiv q$, we have $\neg \Box \neg q \equiv \Diamond q$, or $\neg \Diamond q \equiv \Box \neg q$. □

In Chap. 2, we introduced a concept called "negation normal form," which is a propositional formula where every argument of $\neg$ is a propositional variable. This concept applies to formulas of LTL as well as TL. Applying the two theorems above, we have the following theorem.

Theorem 10.2.11 *For every LTL formula A, there exists an LTL formula B such that $A \equiv B$ and B is in negation normal form.*

Proof The additional rules needed for LTL are Theorems 10.2.9(c) and 10.2.10. □

We will use negation normal form as a proving method, as shown by the next example.

Example 10.2.12 To show that $\Diamond \Box p \rightarrow \Box \Diamond p$ is valid, we transform $\neg(\Diamond \Box p \rightarrow \Box \Diamond p)$ into negation normal form:

$$\neg(\Diamond \Box p \rightarrow \Box \Diamond p) \equiv \Diamond \Box p \wedge \neg \Box \Diamond p$$
$$\equiv \Diamond \Box p \wedge \Diamond \neg \Diamond p$$
$$\equiv \Diamond \Box p \wedge \Diamond \Box \neg p$$

It is shown in Example 10.2.3 that $\Diamond \Box p \wedge \Diamond \Box \neg p$ is unsatisfiable. Thus, $\Diamond \Box p \rightarrow \Box \Diamond p$ is valid.

On the other hand, a negation normal form of $\neg(\Box \Diamond p \rightarrow \Diamond \Box p)$ is $\Box \Diamond p \wedge \Box \Diamond \neg p$, which is satisfiable (Example 10.2.2). Thus, $\Box \Diamond p \rightarrow \Diamond \Box p$ is not valid. □

Formulas in propositional logic can be transformed into conjunctive normal form (CNF) or disjunctive normal form (DNF). However, this is possible neither in LTL nor in TL. Like $\forall$ in first-order logic, $\Box$ does not distribute over $\vee$; like $\exists$, $\Diamond$ does not distribute over $\wedge$.

Theorem 10.2.13 (Distributivity)

$$(a) \quad \Box(p \wedge q) \equiv \Box p \wedge \Box q;$$
$$(b) \quad \Diamond(p \vee q) \equiv \Diamond p \vee \Diamond q;$$
$$(c) \quad \circ(p \wedge q) \equiv \circ p \wedge \circ q;$$
$$(d) \quad \circ(p \vee q) \equiv \circ p \vee \circ q.$$

Proof
(a) For any interpretation sequence π, $\Box(p \wedge q)$ is true in π iff for any $i \geq 0$, $\pi(i)(p \wedge q) = 1$. That is, $\pi(i)(p) = 1$ and $\pi(i)(q) = 1$ for any $i \geq 0$. This is the same as $\Box p$ and $\Box q$, respectively.
(b) can be obtained from *(a)* by duality. The proofs of *(c)* and *(d)* are left as exercises.
□

Example 10.2.14 To show that $\Box(p \vee q)$ is not equivalent to $\Box p \vee \Box q$, consider an interpretation sequence π where p is true only at each odd-numbered state of π and q is true only at each even-numbered state of π. Then, $p \vee q$ is true at every state of π; hence, $\Box(p \vee q)$ is true in π. It is easy to see that neither $\Box p$ nor $\Box q$ is true in π. Note that $\Box p \vee \Box q \models \Box(p \vee q)$. □

Theorem 10.2.15 (Absorption)

$$\begin{aligned}(a) \quad & \Box\Box p \equiv \Box p \\ (b) \quad & \Diamond\Diamond p \equiv \Diamond p \\ (c) \quad & \Box\Diamond\Box p \equiv \Diamond\Box p \\ (d) \quad & \Diamond\Box\Diamond p \equiv \Box\Diamond p\end{aligned}$$

Proof *(a)* comes from Theorem 10.2.5. *(b)* is a dual of *(a)*. *(c)* can be obtained from *(d)* by duality. Here is the proof of *(d)*: for any interpretation sequence π, $\Diamond\Box\Diamond p$ is true in π iff there exists $i \geq 0$, $\Box\Diamond p$ is true in $\pi_{\geq i}$. $\Box\Diamond p$ is true in $\pi_{\geq i}$ iff for all $j \geq i$, $\Diamond p$ is true in $\pi_{\geq j}$. $\Diamond p$ is true in $\pi_{\geq j}$ for $j \geq i$ iff $\Diamond p$ is true in $\pi_{\geq k}$ for any $k \geq 0$, which is equivalent to $\Box\Diamond p$ is true in π. □

The above theorem allows us to compress series of $\Box$ and $\Diamond$ operators to no more than two. They are useful in the simplification of LTL formulas.

10.2.3 Model Checking with Kripke Frames

In many applications of LTL, a relation among interpretations of propositional variables can be fixed, and this relation can be represented by a directed graph $K = (S, R)$, where S is a set of interpretations and R is a binary relation over S. K is called a Kripke frame for branching time logic, and we will use K for LTL, too. Any path of infinite length of K gives us an interpretation sequence, and we may check if this sequence is a model of LTL or not. This idea is called "model checking" in the community of formal methods.

Example 10.2.16 Suppose we would like to build a model for operating an electric oven and expect that once the power is "*on*," the oven will be eventually "*hot*." In LTL, this is expressed as $\Box(on \rightarrow \Diamond hot)$. We also request that the oven won't be *hot* if the oven door is "*open*." In LTL, we specify it as $\Box(open \rightarrow \Diamond\neg hot)$. The relation between the three propositional variables is given by the Kripke frame in

10.2 Linear Temporal Logic

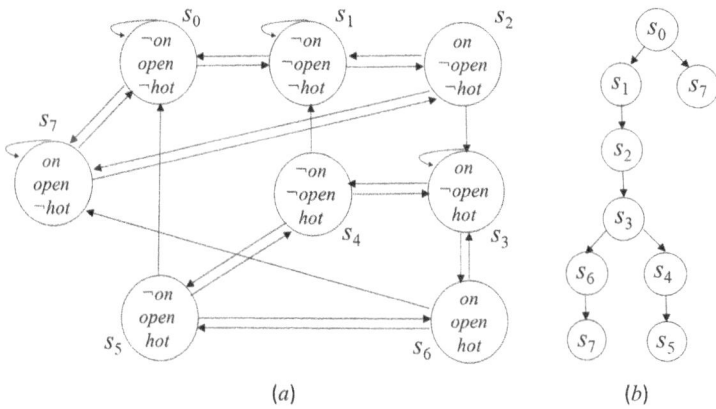

Fig. 10.2 (a) A Kripke frame for the oven model, where a link exists between two interpretations that differ by one truth value; (b) a partial search tree of the frame

Fig. 10.2a, showing the results of "the oven power is on," "open or close the oven door," and usual oven functions. Of course, this is an oversimplified oven model for the purpose of illustration. □

A *path* (repeated nodes are allowed) in a Kripke frame $K = (S, R)$ is a sequence of states $s_1, s_2, s_3, \cdots, s_i, \cdots$ over S, such that $\langle s_i, s_{i+1} \rangle \in R$ for all $i \geq 1$. Since K is finite, an infinite path of the frame is conveniently expressed as a sequence of states ending with a subsequence which can repeat any number of times. For instance, s_0^*, $s_0 s_1^*$, $(s_0 s_1)^*$, $s_0 (s_1 s_2 s_3 s_4)^*$, etc. are such paths in the oven example, where x^* indicates that we repeat x any number of times.

If an infinite sequence π of interpretations is represented by $s_0 s_1 s_2 \cdots s_k (s_{k+1} s_{k+2} \cdots s_{k+m})^*$, we say it is *regular* (called *regular expression* in theory of computation).

It is relatively easy to check if $s_0 s_1 \cdots s_k (s_{k+1} s_{k+2} \cdots s_{k+m})^*$ is a model of LTL formula C by following Definition 10.2.1, where condition $i \geq 0$ in the last two cases can be replaced by $0 \leq i \leq k + m$:

7. C is $\Box A$ and for all $0 \leq i \leq k + m$, $\pi_{\geq i} \models A$.
8. C is $\Diamond A$ and there exists $0 \leq i \leq k + m$, $\pi_{\geq i} \models A$.

Example 10.2.17 Let's continue with Example 10.2.16 and let A be $\Box(on \to \Diamond hot)$. We show that A is true in $s_0 s_1 s_2 s_3^*$, a regular sequence in Fig. 10.2. For A to be true, $on \to \Diamond hot$ needs to be true in all suffixes of $s_0 s_1 s_2 s_3^*$: They are $s_0 s_1 s_2 s_3^*$, $s_1 s_2 s_3^*$, $s_2 s_3^*$, and s_3^*. Indeed, $on \to \Diamond hot$ is true in $s_0 s_1 s_2 s_3^*$ and $s_1 s_2 s_3^*$ because on is false in s_0 and s_1. $on \to \Diamond hot$ is also true in $s_2 s_3^*$ and s_3^* because $\Diamond hot$ is true in $s_2 s_3^*$ and s_3^* (because hot is true in s_3). Thus, A is satisfiable. □

The above example shows a typical working process of model checking:

1. Build a model of the system abstracting out irrelevant details.

2. Write a formal specification about the desired properties, such as no deadlocks, reachability issues, system invariant, etc.
3. Run the model checker that performs some variation of reachability analysis on the model with answers that the specification is satisfied or a counterexample is found.

For the time being, we consider Kripke frames as our models. Kripke frames can be extended to finite state machines so that more complex systems can be modeled. How to control the search complexity of finding counterexamples is crucial for the success of model checking, and efficient strategies are needed to avoid state space explosion. For instance, $s_0 s_1 s_2 \cdots s_k (s_{k+1} s_{k+2} \cdots s_{k+m})^*$ is equivalent to

$$s_0 s_1 s_2 \cdots s_k s_{k+1} s_{k+2} \cdots s_{k+m} (s_{k+1} s_{k+2} \cdots s_{k+m})^*.$$

We may use this equivalence to reduce the number of regular sequences in model checking.

For a Kripke frame $K = (S, R)$, model checking will examine all regular sequences of K as follows: choose a node in S as the *root* of a tree T and expand repeatedly each leaf node of T according to R. A stopping criterion of the expansion is that the path from the root to a leaf contains no more than one copy of any node. This way, the tree will be finite for K, and all nonequivalent regular sequences represented by T will be also finite. For each regular sequence represented by T, model checking will examine if it is a model of the given formula.

For the Kripke frame given in Fig. 10.2a, a search tree is shown in (b). The rightmost path, i.e., $s_0 s_7$, represents the following regular sequences: s_0^*, $s_0 s_7^*$, and $(s_0 s_7)^*$. To show that "once the power is *on*, the oven will be eventually *hot*," we have to show that $\neg A$, where A is $\Box(on \rightarrow \diamond hot)$, has no models. However, $s_0 s_7^*$ is a model of $\neg A$, which is equivalent to $\diamond(on \wedge \Box \neg hot)$; thus, A is not a theorem of the oven model. We need to modify either A or the oven model if we want A to be a theorem.

10.3 Semantic Tableaux for LTL

The method of semantic tableaux introduced in Chap. 3 is a decision procedure for satisfiability in propositional logic. In this section, we extend this method for LTL by adding rules for the modal operators. The extension for TL is similar.

An LTL model is more complex than a propositional model because a propositional model is a single propositional interpretation; an LTL model is a sequence of propositional interpretations. We need to find an interpretation for each state in the sequence. Therefore, we need to group nodes in the semantic tableau by states. For example, the β-rule for $\vee$ will generate two children for the node containing $p \vee q$, and these two children belong to the group of their parent. On the other hand, $\circ p$ will generate a child node which belongs to a different group associated with the next state of the current state.

10.3 Semantic Tableaux for LTL

Like propositional logic, the extended tableau method takes an input formula A, creates the initial node containing A, and then applies the rules repeatedly to any leaf node to generate its successors. The tableau rules for LTL consist of the rules for propositional logic shown in Sect. 3.1.2, plus a new α-rule for $\Box$, a new β-rule for $\Diamond$, a new type of rule called X-rule (the neXt rule) for $\circ$. To simplify the discussion, we will apply some simplification rules to the initial formula. The simplification rules include the following rewrite rules and some rules involving the constants $\top$ and $\bot$ (such as $\bot \vee A \rightarrow A$ and $\Box\top \rightarrow \top$):

$$
\begin{aligned}
&(1) &&\neg\neg A \rightarrow A \\
&(2) &&\neg(A \wedge B) \rightarrow \neg A \vee \neg B \\
&(3) &&\neg(A \vee B) \rightarrow \neg A \wedge \neg B \\
&(4) &&\neg(A \rightarrow B) \rightarrow A \wedge \neg B \\
&(5) &&\neg(A \leftrightarrow B) \rightarrow (A \wedge \neg B) \vee (\neg A \wedge B) \\
&(6) &&\neg\Box A \rightarrow \Diamond \neg A \\
&(7) &&\neg\Diamond A \rightarrow \Box \neg A \\
&(8) &&\neg\circ A \rightarrow \circ \neg A \\
&(9) &&\Box \circ A \rightarrow \circ \Box A \\
&(10) &&\Diamond \circ A \rightarrow \circ \Diamond A
\end{aligned}
$$

The last three rules are the commutativity rules which move $\circ$ outward; the other rules move $\neg$ inward. These simplification rules are based on logical equivalences and preserve the equivalence relation between the two formulas before and after the application of a rewrite rule. The rules will transform the initial formula to negation normal forms where $\neg, \vee, \wedge, \circ, \Box, \Diamond$ are the only operators.

10.3.1 Rules for Modal Operators

In Chap. 3, we showed how a tableau is created by creating the initial node of the input formula and applying α- and β-rules to a formula in a node to create successor nodes. A α-rule creates one successor, and a β-rule creates two successors. Let n be the current node and $F(n)$ be the set of formulas contained in n. If a α-rule can replace A by $\{B, C\}$ for $A \in F(n)$, then the successor of n, denoted n', is created and $F(n') = F(n) - \{A\} \cup \{B, C\}$. If a β-rule can replace A by B and C for $A \in F(n)$, then two successors of n, denoted n' and n'', are created such that $F(n') = F(n) - \{A\} \cup \{B\}$ and $F(n'') = F(n) - \{A\} \cup \{C\}$.

As in Chap. 3, given a tableau, we say a node is *closed* if it contains a pair of complementary literals. A node is *open* if it is not closed and contains only literals. A node is *expandable* if it is a leaf node which is neither closed nor open. Expandable nodes always contain a formula which allows one of α- or β-rules to apply. For LTL, we have a new type of nodes, called X-nodes.

Definition 10.3.1 (X-**Formula,** X-**Node, and** X-**Rule**) If the root symbol of an LTL formula A is $\circ$ and no rules can be applied to A, A is called an X-formula or a next-formula.

A node of a semantic tableau is called an X-*node* if the node contains at least one X-formula, and each formula in the node is either a literal or an X-formula.

Let $F(n)$ be the formulas in an X-node n:

$$F(n) = \{A_1, A_2, \ldots, A_k, \circ B_1, \circ B_2, \ldots, \circ B_m\},$$

where $k \geq 0$, $m \geq 1$, and $\{A_1, A_2, \ldots, A_k\}$ is a set of consistent literals, then the X-rule will generate an interpretation, which makes all $A_1, A_2, \ldots, A_n$ true, for the current state and create a new node n', where $F(n') = \{B_1, B_2, \ldots, B_m\}$ for the next state.

Example 10.3.2 Assume that the initial node contains the input formula $\neg p \wedge \circ p$. Applying α-rule for $\wedge$, we obtain the second node which contains $\{\neg p, \circ p\}$. Here, $\neg p$ is a literal, $\circ p$ is an X-formula, and no rules can be applied except the X-rule. The X-rule will generate the third node containing p, and this node belongs to a different state. From the second node, we create $s_0 = \{\neg p\}$; from the third node, we create $s_1 = \{p\}$. The interpretation sequence created from this tableau is $s_0, s_1, s_1, s_1, \ldots$, which is a model of $\neg p \wedge \circ p$. In fact, any sequence whose first two states are s_0 and s_1 will be a model of this formula. □

By definition, the X-rule can be applied to X-nodes only. Neither simplification rules nor α and β rules can be applied to any formula in X-nodes. More application examples of X-rules will be given shortly, after presenting the rules for $\square$ and $\diamond$.

To construct a model for $\square A$, we need to find an interpretation sequence π such that A is true in every suffix of π. That is, A is true in $\pi(0)$ and $\square A$ must be true in all subsequent states. That is, instead of $\square A$, we will construct a model for $A \wedge \circ \square A$. Intuitively, if $\circ$ represents "tomorrow," then A is true everyday iff A is true today and true every tomorrow.

Theorem 10.3.3 (Induction)

$$(a) \ \square A \equiv A \wedge \circ \square A$$
$$(b) \ \diamond A \equiv A \vee \circ \diamond A$$

Proof
(a) For any interpretation sequence π, $A \wedge \circ \square A$ is true in π iff A is true in $\pi_{\geq 0}$ and $\square A$ is true in $\pi_{\geq 1}$. $\square A$ is true in $\pi_{\geq 1}$ iff A is true in $\pi_{\geq i}$ for every $i \geq 1$. Thus, A must be true in $\pi_{\geq i}$ for every $i \geq 0$, which is exactly "$\square A$ being true in π" by definition.

(b) is the dual case of (a). □

The above theorem serves as an inductive tool for constructing a model of $\square A$ and $\diamond A$: from (a), A is the base case and $\circ \square A$ is the inductive case. Once both cases are finished, we obtain a model for $\square A$. From (b), A is the base case; if a model of

10.3 Semantic Tableaux for LTL

A is found, we use it as a model of $\diamond A$. Otherwise, from the inductive case, we try to find a model of $\circ \diamond A$.

Definition 10.3.4 The α-rule for $\Box A$ and the β-rule for $\diamond A$ are given below:

α	α_1, α_2
$\Box A$	$A, \circ \Box A$

β	β_1	β_2
$\diamond A$	A	$\circ \diamond A$

The soundness of these two rules is ensured by Theorem 10.3.3.

The construction of semantic tableaux for LTL is similar to that of propositional logic by repeating α-, β-, and X-rules to expandable nodes, until no rules can be applied. However, there are one minor and one major differences. The minor difference is that we apply the simplification rules first to the input formula to get a negation normal form before the start of tableau construction. This difference can be removed if we treat the simplification rules as α-rules.

The major difference is that we need to check if a new node contains the same set of formulas as an existing node. If yes, instead of creating a new node, we create a link to the existing node. Because of these links, the created tableau is still a directed graph but no longer a tree. This graph may contain cycles and each node has at most two successors (i.e., two outgoing links). The termination of the tableau construction depends on this major difference as a tableau may contain paths of infinite length.

Example 10.3.5 The tableau for $\Box \diamond p$ is shown below, where the X-nodes are boxed.

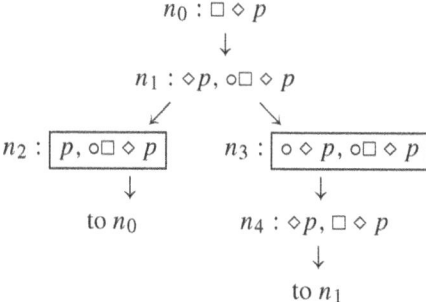

This tableau has no leaf nodes (neither open nor closed nodes). The boxed nodes are X-nodes. We will show later how to derive a model from this tableau. □

Given a tableau as directed graph, we still call a node without outgoing links a *leaf*, which is either closed, open, or expandable. Given two nodes n and n' of a directed graph, n' is said to be *reachable* from n if there exists a path from n to n' in the graph.

For any node n of a tableau, let $F(n)$ denote the set of formulas appearing in n. As in Chap. 3, we assume that $F(n) \equiv \bigwedge_{A \in F(n)} A$, that is, a set of formulas is equivalent to the conjunction of the formulas in the set.

Theorem 10.3.6

(a) The construction of a tableau for an LTL formula always terminates.
(b) If node n' is derived from n by an α-rule, then $F(n) \equiv F(n')$; if n' and n'' are derived from n by a β-rule, then $F(n) \equiv F(n') \vee F(n'')$.
(c) If node n' is derived from n by the X-rule, then an interpretation sequence π' is a model of $F(n')$ iff $\pi = s_0\pi'$ is a model of $F(n)$, where $s_0 \in All_P$ is an interpretation which makes every literal $A \in F(n)$ true.

Proof
(a) The construction stops when no expandable nodes are available. If the construction does not stop, then there will be an infinite path of nodes in the tableau such that each node contains a different set of formulas. Hence, the set of formulas from these nodes is infinite. However, the rules used in the construction can only generate subformulas of the original formula or add at most one ○ operator to them. In other words, only a finite number of formulas can be generated. This is a contradiction, so the tableau cannot have an infinite path other than cycles.
(b) All the α and β rules used in the construction preserve the equivalence relation.
(c) When the X-rule is applied to n to generate n', where $F(n) = \{A_1, A_2, \ldots, A_k, \circ B_1, \circ B_2, \ldots \circ B_m\}$, $F(n') = \{B_1, B_2, \ldots B_m\}$, where B_i is any formula. By the assumption, $F(n) \equiv (\bigwedge_{1 \leq i \leq k} A_i) \wedge \circ(\bigwedge_{1 \leq j \leq m} B_j)$. Let π' be any interpretation sequence and s_0 be an interpretation such that $s_0(A_i) = 1$ for $1 \leq i \leq k$. Then, π' is a model of $F(n')$ iff $s\pi'$ is a model of $\circ(\bigwedge_{1 \leq j \leq m} B_j)$ for any $s \in All_P$. Take $s = s_0$, then $\pi = s_0\pi'$ is a model of $(\bigwedge_{1 \leq i \leq k} A_i) \wedge \circ(\bigwedge_{1 \leq j \leq m} B_j)$, since s_0 is a model of $\bigwedge_{1 \leq i \leq k} A_i$. That is, $\pi = s_0\pi'$ is a model of $F(n)$. □

Part *(a)* of the above theorem tells that the construction of a tableau for an LTL formula will terminate such that every leaf is either closed or open, without expandable nodes. If the tableau has an open node, parts *(b)* and *(c)* tell us how to construct a model (for the input formula in the initial node) from this open node bottom-up, thus have shown that the original formula is satisfiable. What happens when a tableau has no open nodes? We will discuss this case next.

10.3.2 Deciding Satisfiability by Tableaux

We have seen how a tableau can be constructed for every LTL formula. For propositional logic, a formula is satisfiable iff its tableau has an open node. For LTL, at least one side is true.

Theorem 10.3.7 *If the tableau for an LTL formula A has an open node, then A is satisfiable.*

10.3 Semantic Tableaux for LTL

Fig. 10.3 A tableau for $\circ p \wedge \diamond \neg p$

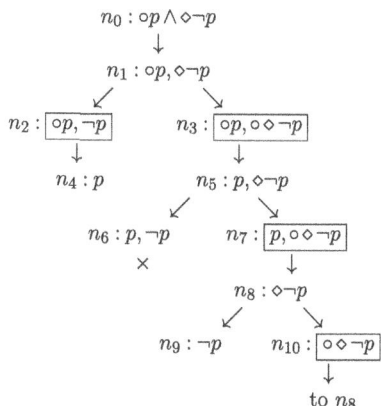

Proof Let $(n_0, n_1, \ldots, n_m)$ be a simple path from the initial node n_0 to the open node n_m in the tableau. We show that $F(n_j)$ has a model π_j for $j = m, m-1, \ldots, 1, 0$ by induction.

For the base case when $j = m$, $F(n_m)$ is a set of consistent literals. Let s_m be the interpretation which makes every literal of $F(n_m)$ true and $\pi_m = s_m^+$, where s_m^+ means we repeat s_m forever, then π_m is a model of $F(n_m)$.

As induction hypothesis, we assume that π_j is a model of $F(n_j)$, $0 < j \leq m$. If n_j is a successor of n_{j-1} by either α-rule or β-rule, by Theorem 10.3.6 (b), $\pi_{j-1} = \pi_j$ is also a model of $F(n_{j-1})$. If n_j is the successor of n_{j-1} by the X-rule, by Theorem 10.3.6 (c), $\pi_{j-1} = s_{j-1}\pi_j$ is a model of $F(n_{j-1})$, where s_{j-1} is an interpretation made from the literals $A_i \in F(n_{j-1})$.

Finally, when $j = 0$, π_0 is a model of $F(n_0) = \{A\}$. Thus, A is satisfiable. □

For LTL, a tableau may not have an open node, as seen in Example 10.3.5. That tableau does not have closed nodes, either. How can we tell if the input formula of the tableau is satisfiable? Let us look at another example.

Example 10.3.8 To show that $\circ p \wedge \diamond \neg p$ is satisfiable, we may construct a model by semantic tableau. Using the tableau rules, we obtain the tableau shown in Fig. 10.3, where X-nodes are boxed.

A model is a sequence of interpretations and an interpretation can be created either from an open node or X-node. For instance, consider the path (n_0, n_1, n_2, n_4): from n_2, we create $s_0 = \{\neg p\}$; from n_4, we create $s_1 = \{p\}$. The first two interpretations of our model are s_0 and s_1, respectively. By repeating s_1, we obtain an infinite interpretation sequence denoted by $s_0 s_1^+$, where s_1^+ means we repeat s_1 one or more times. It is easy to check that $s_0 s_1^+$ is a model of $\circ p \wedge \diamond \neg p$. Since the interpretations after s_1 do not affect the truth value of p, let s_2 denote an interpretation where p takes any value, then a general representation of the models from this path is $s_0 s_1 s_2^*$, where s_2^* means we can repeat s_2 any number of times.

Consider another path ending with an open node:

$$(n_0, n_1, n_3, n_5, n_7, n_8, n_9)$$

where n_3 and n_7 are X-nodes and n_9 is open. n_3 does not specify the truth value of p, and we name the interpretation as s_2 where p can take any truth value. The interpretations created from n_7 and n_9 are $s_1 = \{p\}$ and $s_0 = \{\neg p\}$, respectively. The model created from this path is $s_2 s_1 s_0 s_2^*$.

Consider the following path:

$$(n_0, n_1, n_3, n_5, n_7, n_8, n_{10})$$

where n_3, n_7, and n_{10} are X-nodes. If the interpretations created from them always make p true, then no model can be found from this path. On the other hand, if the path is

$$(n_0, n_1, n_3, n_5, n_7, n_8, n_{10}, n_8, n_9),$$

the model from this path is $s_2 s_1 s_2 s_0 s_2^*$. If we repeat n_8 and n_{10} a couple of times, then all of the models from this path can be represented by $s_2 s_1 s_2^* s_0 s_2^*$, where p can take any value in s_2.

In fact, all the models of $\circ p \wedge \diamond \neg p$ can be denoted by $s_0 s_1 s_2^* \cup s_2 s_1 s_2^* s_0 s_2^*$. Each model corresponds to some path in the tableau. □

The above example tells us that when there is an open node, we can construct a model for the input formula. When the path ends with a cycle, a model may or may not be constructed.

Example 10.3.9 Look back at the tableau in Example 10.3.5: it has no open nodes. If we pick the cycle (n_0, n_1, n_2), let $s_1 = \{p\}$ be the interpretation derived from the X-node n_2, then s_1^+ is a model of the input formula $\square \diamond p$.

Let us look at the path (n_0, n_1, n_3, n_4), where the last three nodes are in a cycle. The interpretation created from n_3 does not care about the truth value of p and we denote it by s_2, then s_2^+ is not a model.

On the other hand, consider the path:

$$(n_0, n_1, n_2, n_0, n_1, n_3, n_4).$$

The sequence $(s_1 s_2)^+$ is a model of $\square \diamond p$, while the sequence $s_1 s_2^+$ is not a model. In other words, we need to repeat altogether all the interpretations generated from the nodes in a cycle.

Note that all the models of $\square \diamond p$ can be denoted by the sequence $(s_2^* s_1)^+$. □

The above examples show that some paths ending with a cycle generate models and some don't. Our purpose of using tableaux is to decide if the input formula is satisfiable or not. A model is just evidence that the input formula is satisfiable. If a

10.3 Semantic Tableaux for LTL

tableau has an open node, we know the input formula is satisfiable because a model can be constructed from the path which connects the initial node and the open node. If the tableau has no open nodes, we define below a marking mechanism to check effectively if the input formula is satisfiable or not.

Definition 10.3.10 Given a tableau which has no expandable nodes. A node n is marked *dead* if one of the following conditions is true:

1. n is closed.
2. All successors of n are marked dead.
3. There exists a formula $\diamond A \in F(n)$ such that there is no node n' reachable from n and $A \in F(n')$.

In the third case, the reachable condition is checked among the nodes which are not marked dead. That is, if a node is marked dead, it cannot be used in any path. Node n is marked dead because it contains an *unfulfilled* $\diamond A$: $\diamond A$ is true in a sequence π iff A is true in a suffix of π. If A does not appear in any node n' reachable from n, then it is impossible to construct such a suffix of π such that π is a model of $F(n)$.

Example 10.3.11 To show that $\Box(p \wedge q) \rightarrow \Box p$ is valid, we show that its negation is unsatisfiable by semantic tableau. The negation normal form of $\neg(\Box(p \wedge q) \rightarrow \Box p)$ is $\Box(p \wedge q) \wedge \diamond \neg p$, whose tableau is given below:

$$n_0 : \Box(p \wedge q) \wedge \diamond \neg p$$
$$\downarrow$$
$$n_1 : \Box(p \wedge q), \diamond \neg p$$
$$\downarrow$$
$$n_2 : p \wedge q, \circ\Box(p \wedge q), \diamond \neg p$$
$$\downarrow$$
$$n_3 : p, q, \circ\Box(p \wedge q), \diamond \neg p$$
$$\swarrow \qquad \searrow$$
$$n_4 : p, q, \circ\Box(p \wedge q), \neg p \qquad n_5 : \boxed{p, q, \circ\Box(p \wedge q), \circ \diamond \neg p}$$
$$\times \qquad\qquad \downarrow$$
$$\text{to } n_1$$

For $\diamond \neg p \in F(n_1)$, there is no node n' reachable from n_1 and $\neg p \in F(n')$. Thus, n_1 is marked dead. n_0 is marked dead because its successor is marked dead. In fact, every node in this tableau can be marked dead. □

Example 10.3.12 For the tableau given in Example 10.3.5, $\diamond p$ appears in n_4. Since n_2 is reachable from n_4 through the path (n_4, n_1, n_2) and $p \in F(n_2)$, n_4 cannot be marked dead. □

Algorithm 10.3.13 The algorithm *satisfiabilityByTableau* takes an LTL formula A as input and returns "satisfiable" if A is satisfiable, otherwise returns "unsatisfiable."

proc *satisfiabilityByTableau(A)*
1. Create a tableau for A:
    1.1 Create the initial node that contains the negation normal form of A.
    1.2 Apply the α-, β-, and X-rules until no expandable nodes exist.
2. Decide if A is satisfiable:
    2.1 If the tableau has an open node, return "satisfiable."
    2.2 Apply Definition 10.3.10 until no nodes can be marked dead.
    2.3 If the initial node is marked dead, return "unsatisfiable."
    2.4 Otherwise return "satisfiable."

Although the algorithm returns only "satisfiable" or "unsatisfiable," we can still construct a model from a path of the tableau if the algorithm returns "satisfiable," as illustrated by the following examples.

Example 10.3.14 To show that $\Diamond p \wedge \Diamond \neg p$ is satisfiable, we construct its semantic tableau.

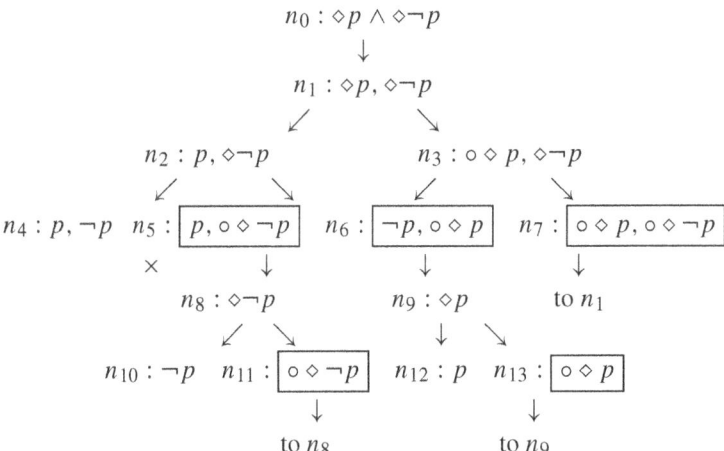

For $\Diamond p \in F(n_1)$, either n_5 or n_{12} contains p; for $\Diamond \neg p \in F(n_1)$, either n_6 or n_{10} contains $\neg p$. All the four nodes are reachable from n_1. n_0 cannot be marked dead, so the algorithm will return "satisfiable."

Various models can be constructed from this tableau, depending on which path we choose in the tableau. Let $s_0 = \{\neg p\}$, $s_1 = \{p\}$, and s_2 denotes an interpretation where we do not care about the truth value of p.

1. $(n_0, n_1, n_2, n_5, n_8, n_{10})$: The derived model is $s_1 s_0 s_2^*$, where s_1 is derived from n_5 and s_0 from n_{10}. s_2^* is added at the end because n_{10} is open.
2. $(n_0, n_1, n_3, n_6, n_9, n_{12})$: The derived model is $s_0 s_1 s_2^*$, where s_0 is derived from n_6 and s_1 from n_{12}.
3. $(n_0, n_1, n_2, n_5, n_8, n_{11})$: The last two nodes are in a cycle and no model can be found from this path. To fulfill $\Diamond \neg p \in F(n_1)$, we need n_6 or n_{10}.

10.3 Semantic Tableaux for LTL

4. $(n_0, n_1, n_3, n_6, n_9, n_{13})$: The last two nodes are in a cycle and no model can be found from this path.
5. (n_0, n_1, n_3, n_7): The last three nodes are in a cycle and no model can be found from this path.

All models of $\diamond p \wedge \diamond \neg p$ can be denoted by $s_2^* s_0 s_2^* s_1 s_2^* \cup s_2^* s_1 s_2^* s_0 s_2^*$ and each model corresponds to a path in the tableau. □

Example 10.3.15 To show that $\square(\diamond p \wedge \diamond \neg p)$ is satisfiable, we construct a semantic tableau where we replace $\wedge$ by ",", and A stands for $(\diamond p, \diamond \neg p)$.

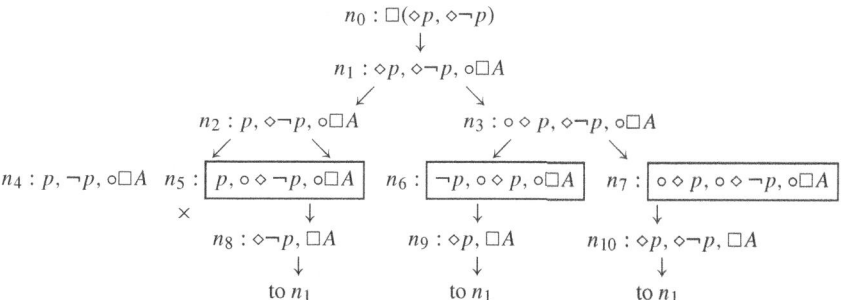

This tableau has one closed node, i.e., n_4, and no open nodes. n_4 is the only node marked dead. $\diamond p$ and $\diamond \neg p$ in n_1 are fulfilled by n_5 and n_6, respectively. n_1 cannot be marked dead, neither does n_0. So, the algorithm will return "satisfiable."

If we wish to construct a model from the tableau, we need to consider a path starting from n_0 and containing both n_5 and n_6. Let $s_0 = \{p\}$ and $s_1 = \{\neg p\}$ be the interpretations derived from X-nodes n_5 and n_6, respectively. Then, $(s_0 s_1)^+$ is a model of the input formula. Any model requires that we repeat these two interpretations together; repeating a single interpretation infinitely does not produce a model. Let s_2 be any interpretation where we do not care the value of p. All the models of $\square(\diamond p \wedge \diamond \neg p)$ can be represented by $(s_2^* s_0 s_2^* s_1 s_2^* \cup s_2^* s_1 s_2^* s_0 s_2^*)^+$, and each model corresponds to some path in the tableau. □

The following theorem provides the correctness of Algorithm 10.3.13.

Theorem 10.3.16 *An LTL formula A is satisfiable iff Algorithm 10.3.13 returns "satisfiable."*

Proof At first, the algorithm will terminate because (1) the construction of the tableau will terminate (Theorem 10.3.6 (a)), and (2) checking if the initial node is marked dead takes a finite number of steps.

If the tableau has an open node, A is satisfiable by Theorem 10.3.7.

If the initial node is not marked dead, we show that there is a model for A. This model is constructed from a path that starts with the initial node and contains all the nodes required by the *fulfilling* condition.

The path is selected among the nodes not marked dead as follows: (1) the path starts with the initial node; (2) if the last node of the current path has only one unmarked successor, add the successor to the path; (3) if the current path contains a node n and $\diamond A$ in n has not fulfilled yet, find node n' reachable from n and containing A. If n' does not appear in the path, append the path from n to n' to the current path. If n' already appears in the path, add only necessary nodes so that all the nodes of the cycle containing n and n' are present in the path. That is, the last portion of the path contains a list of nodes from the same cycle, and they are reachable to each other.

Note that if both $\diamond A$ and $\diamond B$ appear in $F(n)$, we will work on them one at a time: when $\diamond A$ is fulfilled in n', either $\diamond B$ has been fulfilled in the path up to n' or $\diamond B$ is still present in $F(n')$. In the latter case, there is another node n'' reachable from n' and containing B (if not, n' would have been marked dead).

By repeating the selection, one obtains a path ending with a cycle and all formulas of form $\diamond C$ are fulfilled. Let $n_{k_0}, n_{k_1}, \ldots, n_{k_m}$ be the X-nodes in the path and $s_0 s_1 \cdots s_m$ be the interpretations from these X-nodes (keeping them in the same order), we define interpretation sequence π as follows:

$$\pi = s_0 s_1 s_2 \cdots (s_j \cdots s_m)^+,$$

where $s_j, \ldots, s_m$ are the interpretations generated from the X-nodes in the cycle. It is ready to check that π is a model of the initial formula by a structural induction on the formulas in each node.

We claim that $\pi_{\geq i}$ is a model of $F(n_{k_i})$ for $0 \leq i \leq m$. For any node n other than X-nodes, if n_{k_i} is the first X-node following n in the path for some i, then $F(n) \equiv F(n_{k_i})$ and $\pi_{\geq i}$ is also a model of $F(n)$. For any formula $A \in F(n)$, the following statements are true:

- If A does not have any temporal operator, then A is true in $\pi_{\geq i}$.
- If A is $\circ B$, then A is true in $\pi_{\geq i}$ because B appears in the node n' following n_{k_i} by the X-rule and B is true in the model of n' by induction hypotheses (because B is smaller than $\circ B$ and this model is the immediate suffix of $\pi_{\geq i}$).
- If A is $\diamond B$, then A is true in $\pi_{\geq i}$ because B appears in a reachable node n' from n, B is true in the model for n' by induction hypotheses (B is smaller than A), and this model is a suffix of $\pi_{\geq i}$.
- If A is $\square B$, then A is true in $\pi_{\geq i}$ because B is true in $\pi_{\geq i}$ (by the α-rule for $\square$) and $\square B$ is also true in the immediate suffix of $\pi_{\geq i}$ (by the X-rule).

On the other hand, if the initial node is marked dead, then A is unsatisfiable because for any node n which is marked dead, $F(n)$ is unsatisfiable. This statement can be proved by induction according to the recursive definition of that "n is marked dead." There are three cases to consider:

1. *n is closed*: $F(n)$ contains a pair of complementary literals and cannot be satisfiable.
2. *all successors of n are marked dead*: Using induction hypotheses, the formulas in the successors are unsatisfiable. $F(n)$ is unsatisfiable by Theorem 10.3.6 (b) and (c).

3. $\diamond A \in F(n)$ *and there is no node reachable from n and containing A*: $\diamond A$ cannot be true in any interpretation sequence π, since any model corresponds to some path in the tableau. If A does not appear in any path starting from n, A cannot be true in any suffix of π. Hence, $\diamond A$ is false in any π.

□

The above theorem ensures that Algorithm 10.3.13 is a decision procedure for LTL. In literature, there is another decision procedure based on Büchi automaton (BA), which is an extension of finite-state automaton for input strings of infinite length. For LTL, these input strings represent interpretation sequences. Given a tableau, a Büchi automaton can be easily constructed from the tableau by keeping only X-nodes and open nodes (they are not marked dead) as *states*. A transition from one state to another is added if there is a path from the corresponding node to another in the tableau. The language recognized by BA should be the set of models of an LTL formula. There are algorithms that translate an LTL formula to a Büchi automaton without tableaux. These algorithms differ in their construction strategies, but they all have a common underlying principle, i.e., each state in the constructed automaton represents a set of LTL formulas that are expected to be satisfied by the remaining input string after occurrence of the state during an execution of BA. Büchi automata have important application in finite state model checking for formal verification of concurrent programs, and its presentation is out of the score of this book.

10.4 Binary Temporal Operators

In temporal logic, we sometimes want to express a relationship of two statements in terms of time. For example, after *starting*, a computer will keep *running* **until** it's *halted*. To express this sentence, we write

$$start \rightarrow \circ(running \; \mathcal{U} \; halted)$$

where the binary temporal operation $\mathcal{U}$ has the meaning of **until**.

10.4.1 *The* **until** *and* **release** *Operators*

The intuitive meaning of $(A \; \mathcal{U} \; B)$ is that A has to be true at least until B becomes true and B must hold at the current or a future state.

The operator $\mathcal{U}$ has a dual operator called *release*, denoted by $\mathcal{R}$. The intuitive meaning of $A \; \mathcal{R} \; B$ is that B has to be true until (and including) the state where A first becomes true; if A never becomes true, B must remain true forever.

We will extend LTL by adding $\mathcal{U}$ and $\mathcal{R}$ to the definition of LTL formulas such that if A and B are LTL formulas, so are $(A \mathcal{U} B)$ and $(A \mathcal{R} B)$.

We also need to extend Definition 10.2.1 with the following condition:

Given an interpretation sequence π, an LTL formula C is true in π if

- C is $A \mathcal{U} B$ and there exists $i \geq 0$, $\pi_{\geq i} \models B$ and for all $0 \leq j < i$, $\pi_{\geq j} \models A$.
- C is $A \mathcal{R} B$ and either $\Box B$ or there exists $i \geq 0$, $\pi_{\geq i} \models A$ and for all $0 \leq j \leq i$, $\pi_{\geq j} \models B$.

In the above definition, if π is a regular sequence, i.e., π is $s_0 s_1 s_2 \cdots s_k (s_{k+1} s_{k+2} \cdots s_{k+m})^*$, then we can check if π is a model of C as follows:

- C is $A \mathcal{U} B$ and there exists $0 \leq i \leq k+m$, $\pi_{\geq i} \models B$ and for all $0 \leq j < i$, $\pi_{\geq j} \models A$.
- C is $A \mathcal{R} B$ and either $\Box B$ or there exists $0 \leq i \leq k+m$, $\pi_{\geq i} \models A$ and for all $0 \leq j \leq i$, $\pi_{\geq j} \models B$.

The above rules are useful in the approach of model checking.

Example 10.4.1 By definition, $p \mathcal{U} q$ is true in π means q will be eventually true and p must be true before q becomes true. Let $s_0 = \{p, q\}$, $s_1 = \{p, \neg q\}$, $s_2 = \{\neg p, q\}$, and $s_3 = \{\neg p, \neg q\}$. Then

$p \mathcal{U} q$ is true in s_0^+, $s_1 s_0^+$, $s_1^k s_0^+$ for $k \geq 0$, $s_1^k s_2^+$ for $k \geq 0$, and $s_2 s_3^+$; for model checking, k can choose various small values to ensure the reliability of the output.

$p \mathcal{U} q$ is false in s_1^+, s_3^+, and $s_1^k s_3^+$ for $k \geq 0$. □

The following theorem displays some properties of $\mathcal{U}$.

Theorem 10.4.2 *For any LTL formulas A and B*

(a) (**idempotency**) $A \mathcal{U} A \equiv A$
(b) (**abbreviation**) $\diamond A \equiv \top \mathcal{U} A$
(c) (**absorption**) $A \mathcal{U} (A \mathcal{U} B) \equiv A \mathcal{U} B$
(d) (**absorption**) $(A \mathcal{U} B) \mathcal{U} B \equiv A \mathcal{U} B$
(e) (**distributivity**) $\circ (A \mathcal{U} B) \equiv (\circ A) \mathcal{U} (\circ B)$
(f) (**induction**) $A \mathcal{U} B \equiv B \vee (A \wedge \circ (A \mathcal{U} B))$

Proof (a) For any interpretation sequence π, $\pi_{\geq 0} = \pi$. If A is true in $\pi_{\geq 0}$, then $i = 0$, and for $0 \leq j < i$, A is true in $\pi_{\geq j}$ vacuously. So, $A \mathcal{U} A$ is true in $\pi_{\geq 0}$. If A is false in $\pi_{\geq 0}$, then $A \mathcal{U} A$ is false in $\pi_{\geq 0}$ by definition.

(b) Here, B is $\top$ and C is A. The truthfulness of $B \mathcal{U} C$ is reduced to the first condition that there exists $i \geq 0$, $\pi_{\geq i} \models C$, identical to that of $\diamond C$.

(c) For any interpretation sequence π, if $\diamond B$ is false, then both $A \mathcal{U} (A \mathcal{U} B)$ and $A \mathcal{U} B$ are false in π. If $\diamond B$ is true, let k be the state such that B is true in $\pi_{\geq k}$ but false in $\pi_{\geq j}$ for $0 \leq j < k$. Since B is true in $\pi_{\geq k}$, $A \mathcal{U} B$ is trivially true in $\pi_{\geq k}$. $A \mathcal{U} B$ is true in π iff A is true in $\pi_{\geq j}$ for $0 \leq j < k$. $A \mathcal{U} (A \mathcal{U} B)$ is true in π iff A is true in $\pi_{\geq j}$ for $0 \leq j < k$, too.

The proofs of *(d)* and *(e)* are left as exercises.

10.4 Binary Temporal Operators

(f) For $A \,u\, B$ to be true, either B is true now, or we put off to the next tick the requirement to satisfy $A \,u\, B$ while requiring that A be true now. □

Example 10.4.3 By definition, $p \,\mathcal{R}\, q$ is true in π means either q is true forever, or after p becomes true, the condition of q being true is released. Let $s_0 = \{p, q\}$, $s_1 = \{p, \neg q\}$, $s_2 = \{\neg p, q\}$, and $s_3 = \{\neg p, \neg q\}$. Then
$p \,\mathcal{R}\, q$ is true in s_0^+, s_2^+, $s_2 s_0^+$, and $s_2 s_0 s_3^+$.
$p \,\mathcal{R}\, q$ is false in s_1^+, $s_1^k s_0^+$ for $k \geq 1$, $s_1^k s_2^+$ for $k \geq 1$, $s_2 s_1^+$, $s_2 s_3^+$, and s_3^+. □

The following theorem displays some properties of $\mathcal{R}$.

Theorem 10.4.4 *For any LTL formulas A and B*

    (a) (**idempotency**) $A \,\mathcal{R}\, A \equiv A$
    (b) (**abbreviation**) $\bot \,\mathcal{R}\, A \equiv \Box A$
    (c) (**absorption**) $A \,\mathcal{R}\, (A \,\mathcal{R}\, B) \equiv A \,\mathcal{R}\, B$
    (d) (**absorption**) $(A \,\mathcal{R}\, B) \,\mathcal{R}\, B \equiv A \,\mathcal{R}\, B$
    (e) (**distributivity**) $\circ(A \,\mathcal{R}\, B) \equiv (\circ A) \,\mathcal{R}\, (\circ B)$
    (f) (**induction**) $A \,\mathcal{R}\, B \equiv B \wedge (A \vee \circ(A \,\mathcal{R}\, B))$
    (g) (**duality**) $A \,\mathcal{R}\, B \equiv \neg(\neg A \,u\, \neg B)$

Proof (a)–(f) can be deduced from (g) and Theorem 10.4.2. To simplify the proof of (g), we prove $p \,\mathcal{R}\, q \equiv \neg(\neg p \,u\, \neg q)$, where p and q are propositional variables. Replacing $\pi(i)$ by $\pi_{\geq i}$ in the following proof would give us the proof of $A \,\mathcal{R}\, B \equiv \neg(\neg A \,u\, \neg B)$ by the substitution theorem.

Assume $p \,\mathcal{R}\, q$ is true in π. If q is always true, i.e., $\Box q$ is true, then $\diamond \neg q$ and $\neg p \,u\, \neg q$ are false. Thus, $\neg(\neg p \,u\, \neg q)$ is true in π. If q is not always true, let k be minimal such that q is false in $\pi(k)$. Since $p \,\mathcal{R}\, q$ is true in π and q is false in $\pi(k)$, by the semantics of $\mathcal{R}$, there must exist $i < k$ such that p is true in $\pi(i)$ and for all $0 \leq j \leq i$, q is true in $\pi(j)$. In this case, let us check the truth value of $\neg p \,u\, \neg q$: since $\pi(k)$ is the first interpretation in which $\neg q$ is true and $\neg p$ is false in $\pi(i)$ for $i < k$, $\neg p \,u\, \neg q$ must be false in π by the semantics of u. Hence, $\neg(\neg p \,u\, \neg q)$ must be true in π.

Now, assume $p \,\mathcal{R}\, q$ is false in π. By the semantics of $\mathcal{R}$, $\Box q$ cannot be true and let k be minimal such that q is false in $\pi(k)$. Hence, q is true in $\pi(j)$ for $0 \leq j < k$. Because $p \,\mathcal{R}\, q$ is false in π, by the semantics of $\mathcal{R}$, p must be false in $\pi(j)$ for $0 \leq j < k$. Equivalently, $\neg p$ must be true in $\pi(j)$ for $0 \leq j < k$. Hence, $\neg p \,u\, \neg q$ is true in π by the semantics of u. So, we conclude that $\neg(\neg p \,u\, \neg q)$ is false in π.

The above two cases complete the proof that $p \,\mathcal{R}\, q$ is true in π iff $\neg(\neg p \,u\, \neg q)$ is true in π for arbitrary π. □

10.4.2 The weak until and strong release Operators

Some researchers also define a *weak until* binary operator, denoted by w, with semantics similar to that of the until operator, but the stop condition is not required to occur (similar to *release*). The *strong release* binary operator, denoted s, is the dual of *weak until*. It is defined similar to the *until* operator, so that the release condition has to be true at some point. Therefore, it is stronger than the *release* operator.

Definition 10.4.5 The additional temporal operators are defined as follows:

- **Weak until** $A \, w \, B$: A has to be true at least until B becomes true; if B never becomes true, A must remain true forever.
- **Strong release** $A \, s \, B$: B has to be true until and including the state where A first becomes true, and A must be true at the current or a future state.

Instead of extending Definition 10.2.1 to include the formal meanings of $A \, w \, B$ and $A \, s \, B$, we provide below the equivalent relations as an alternative definition:

Theorem 10.4.6 *For any LTL formulas A and B*

$$(a) \; (\textbf{weak until}) \quad A \, w \, B \equiv \Box A \lor (A \, u \, B)$$
$$(b) \; (\textbf{strong release}) \; A \, s \, B \equiv \Diamond A \land (A \, \mathcal{R} \, B)$$

The above theorem shows that w and s can be defined in terms of u and $\mathcal{R}$, just like $\rightarrow$ and $\leftrightarrow$ can be defined in terms of $\land$ and $\lor$. We know that $\mathcal{R}$ can be defined in terms of u by duality. In fact, any one of the four binary temporal operators can be used to define the other three operators. We have seen in Theorems 10.4.2 and 10.4.4 that $\Diamond$ and $\Box$ can be defined in terms of u and $\mathcal{R}$. Thus, the minimal set of LTL needs only two temporal operators: o and one of the four binary operators.

Theorem 10.4.7 *For any LTL formulas A and B*

$$(a) \quad A \, u \, B \equiv \Diamond B \land (A \, w \, B)$$
$$(b) \quad A \, \mathcal{R} \, B \equiv \Box B \lor (A \, s \, B)$$
$$(c) \quad A \, u \, B \equiv B \, s \, (A \lor B)$$
$$(d) \quad A \, \mathcal{R} \, B \equiv B \, w \, (A \land B)$$
$$(e) \quad A \, w \, B \equiv B \, \mathcal{R} \, (A \lor B)$$
$$(f) \quad A \, s \, B \equiv B \, u \, (A \land B)$$
$$(g) \quad A \, w \, B \equiv A \, u \, (\Box A \lor B)$$
$$(h) \quad A \, s \, B \equiv A \, \mathcal{R} \, (\Diamond A \land B)$$
$$(i) \quad \Box A \equiv A \, w \, \bot$$
$$(j) \quad \Diamond A \equiv A \, s \, \top$$
$$(k) \quad \neg (A \, s \, B) \equiv \neg A \, w \, \neg B.$$

The last equivalence is the duality of w and s. The proof of this theorem is left as an exercise.

10.4.3 Extension of Semantic Tableau

Using the duality and the rewrite rules $A \, w \, B \rightarrow \Box A \lor (A \, u \, B)$ and $A \, s \, B \rightarrow \Diamond A \land (A \, \mathcal{R} \, B)$, the negation can be pushed downward and all the formulas of LTL can be transformed into negation normal form, where all negations appear only in front of the propositional variables, and only the temporal operators $\Box, \Diamond, \circ, u$, and $\mathcal{R}$ can appear. Note that the transformation to the negation normal form does not blow up the size of the formula.

Constructing a semantic tableau for a formula that uses the operators u and $\mathcal{R}$ is straightforward. The β-rule for $A \, u \, B$ and α-rule for $A \, \mathcal{R} \, B$ are given below:

α	α_1, α_2
$A \, \mathcal{R} \, B$	$B, A \lor \circ(A \, \mathcal{R} \, B)$

β	β_1	β_2
$A \, u \, B$	B	$A, \circ(A \, u \, B)$

The α-rule is based on Theorem 10.4.4 (f), and the β-rule is based on Theorem 10.4.2 (f).

After a tableau is constructed using these rules, we need to decide if a model can be generated from the tableau. In this case, the definition of dead nodes (Definition 10.3.10) needs to be extended to consider the occurrence of $(A \, u \, B)$ in a node of the tableau: a node n is marked *dead* if there exists a formula $(A \, u \, B) \in F(n)$ such that there is no node n' reachable from n and $B \in F(n')$.

10.5 Verification of Concurrent Programs

This chapter begins with an example of how to access a bank account concurrently and correctly. This example serves as a motivation for the application of temporal logic for formal verification of concurrent programs. We will end this chapter with an example of formal verification, which is taken from a tutorial of the software system STeP (Stanford Temporal Prover). STeP is a system developed by Zohar Manna et al. at Stanford University to support the computer-aided formal verification of concurrent and reactive systems based on Hoare logic and temporal logic. At Carnegie Mellon University, Edmund Clarke et al. developed multiple tools based on model checking (MC) for the same purpose. Efficient automated reasoning techniques such as simplification methods, decision procedures, temporal logic, etc. are used in STeP and some model checkers. Because of space limitation, we will introduce model checking briefly and leave STeP to the interested readers to explore.

10.5.1 Model Checking with Finite State Machines

Model checking is a general formal verification paradigm for both imperative and functional programs. In model checking, hardware/software/algorithm designs are modeled by *finite state machines*, properties are written in *propositional temporal logic*, and verification procedure is an exhaustive search in the state space of the design. To avoid exponential space explosion, bounds are used to limit the path lengths in finite state machines, hence the name *bounded model checking*. Model checking aims at finding diagnostic counterexamples, such as unexpected behavior of a circuit or possible violation of a safety property. When no counterexamples are found, the verification is considered "complete."

When LTL (linear temporal logic) is used with model checking, a *finite state machine* M is a Kripke frame $K = (S, R)$, where S is a set of states, each of which is an interpretation of propositional variables in LTL, and R is a binary relation over S, plus the following information:

- A nonempty subset I of S is identified as *the initial states* of M.
- Optionally, a subset F of S is identified as *the final states* of M.
- Optionally, *labels* are assigned to each transition in R.

Example 10.5.1 Suppose you are asked to implement an online ticket booth to sell tickets for an event. You come up with the following procedure, where k is the total number of tickets to be sold and *reserved* is a Boolean variable for "a reservation request is being processed."

```
proc ticketing()
       precondition: nTickets = k > 0 ∧ reserved = false
l₁     while (true) do
l₂        if (nTickets > 0 ∧ ¬ reserved)
l₃           reserved := true
l₄        if (nTickets > 0 ∧ reserved)
l₅           reserved := false
l₆           nTickets := nTickets − 1 // critical area
       od
```

To describe the above program in model checking, we introduce two propositional variables: a for "$nTickets > 0$," that is, a is true iff a ticket is available and r for *reserved*. Thus, for the Kripke frame, $S = \{ar, a\bar{r}, \bar{a}r, \overline{ar}\}$. The relation R over S is dictated by the execution of the procedure.

- With the given precondition, the only initial state is $a\bar{r}$.
- After "*reserved := true*" at line l_3, the state changes to ar.
- After "*reserved := false*" at line l_5, the state goes back to $a\bar{r}$.
- After "*nTickets := nTickets − 1*" at line l_6, the new state has two choices: either stay at $a\bar{r}$ if $nTickets > 0$ is still true or go to $\overline{ar}$ if $nTickets > 0$ is false.

10.5 Verification of Concurrent Programs

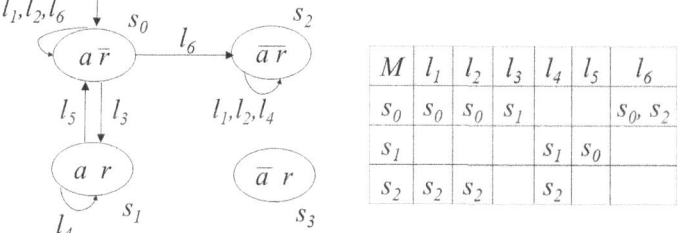

Fig. 10.4 Finite state machine M for *ticketing*()

- Other lines of the code do not change the states, and state $\bar{a}r$ is not accessible from the initial state $a\bar{r}$.

Based on the above analysis, we create a finite-state machine M shown in Fig. 10.4 for procedure *ticketing*. This machine has three useful states: s_0, s_1, and s_2, and s_0 is the initial state. The code labels are attached to each transition to help understand how the state changes during the execution. □

For the reader who knows well finite-state machines, the machine M in Fig. 10.4 is a nondeterministic one, because from state s_0, the transition may go to either s_0 or s_2 with the same label l_6. This is clear from the tabular representation of M to the right side of the figure.

In Sect. 10.2.3, we have shown that an infinite path in a Kripke frame can be conveniently represented by a regular sequence of states, that is, $s_0 s_1 \cdots s_k (s_{k+1} s_{k+2} \cdots s_{k+m})^*$. We will use them in a finite state machine as model candidates of LTL formulas.

If we concatenate the label assigned to every transition $\langle s_i, s_{i+1} \rangle$ of a given path, we obtain a string of labels, which will be called the *trace* of the execution path. For instance, the following is a path of M in Fig. 10.5.1 with explicit labels on each transition.

$$s_0 \xrightarrow{l_1} s_0 \xrightarrow{l_2} s_0 \xrightarrow{l_3} s_1 \xrightarrow{l_4} s_1 \xrightarrow{l_5} s_0 \xrightarrow{l_6} s_0 \xrightarrow{l_1} s_0 \xrightarrow{l_2} s_0 \xrightarrow{l_3} s_1 \xrightarrow{l_4} s_1 \xrightarrow{l_5} s_0 \xrightarrow{l_6} s_2$$

The trace of this path is $l_1 l_2 l_3 l_4 l_5 l_6 l_1 l_2 l_3 l_4 l_5 l_6$, which represents an execution of procedure *ticketing*() with $k = 2$. In this case, we say the path is a *run* of *ticketing*(). Not all paths are runs because not all traces show the program execution correctly. For sequential programs, a trace must follow the sequential order of the commands. For concurrent programs, a trace must be a shuffle or interleaving of two or more sequential traces. It is easy to check if a path is a run. For *ticketing*, a run must start with l_1, followed by l_2, then followed by l_3 or l_4, etc. For verification purposes, we consider only runs of programs. For being a model candidate of LTL, a path needs to be infinite, and we will use regular sequences for this purpose. For bounded model checking, finite length paths are sufficient.

10.5.2 Properties of Concurrent Programs

Suppose procedure *ticketing* is called every time a customer places an online order. It means that multiple copies of the procedure will run concurrently. The computation of a concurrent program is viewed as the interleaving of the atomic operations of its processes, where each process is a sequential program and has a control variable to locate the code of the current execution.

Concurrent procedures create many problems not seen in sequential procedures. For instance, if two copies of *ticketing* run line l_6, i.e., *nTickets* := *nTickets* − 1 simultaneously, then the end result will be wrong. For instance, if *nTickets* = 5 before the execution, and both copies will write 4, not 3, into *nTickets*. If *reserved* is a global variable and copy 1 runs l_5 right after copy 2 runs l_3, then the customer running copy 2 will not get the ticket.

Concurrent procedures should possess the following properties:

- *Mutual exclusion*: Execution is never in the critical section (line l_6 of *ticketing*) at the same time for multiple processes.
- *Accessibility*: Once a process has expressed interest in entering the critical section, it will eventually do so. For *ticketing*, if a process resides at l_3, it will always eventually reach l_6.
- *One-bounded overtaking*: If one process wants to enter the critical section, the other process can enter the critical section at most once before the first one does.

Mutual exclusion and one-bounded overtaking belong to the class of *temporal safety properties*. Informally, such properties state that something "bad" can never happen and are falsified if a bad state is ever reached: if a safety formula A is false in a model, then there is a finite prefix of the model such that A is also false in every extension of this prefix.

Accessibility, on the other hand, is a *response property* and belongs to the larger class of *progress properties*. These properties state that something "good" is guaranteed to happen eventually. The verification of each class of properties has its particular requirements. For instance, safety properties are independent of the termination requirements of the given program, whereas the verification of response properties relies on termination.

It is a challenge to design concurrent programs with these properties. Apparently, *ticketing*() cannot be used as concurrent programs because it does not have any of the above three properties.

10.5.3 The BAKERY(2) Program

Early solutions to mutual exclusion suffered from *starvation*, where a process could forever be denied access to the critical section, and some later solutions do not satisfy one-bounded overtaking. Lamport's bakery algorithm, discovered by Leslie

10.5 Verification of Concurrent Programs

local y_1, y_2 : integer where $y_1 = 0 \land y_2 = 0$

$$
\begin{bmatrix}
\text{loop forever do} \\
\quad \begin{bmatrix} \ell_0: \text{noncritical} \\ \ell_1: y_1 := y_2 + 1 \\ \ell_2: \textbf{await } (y_2 = 0 \lor y_1 \leq y_2) \\ \ell_3: \text{critical} \\ \ell_4: y_1 := 0 \end{bmatrix}
\end{bmatrix}
\quad \| \quad
\begin{bmatrix}
\text{loop forever do} \\
\quad \begin{bmatrix} m_0: \text{noncritical} \\ m_1: y_2 := y_1 + 1 \\ m_2: \textbf{await } (y_1 = 0 \lor y_2 < y_1) \\ m_3: \text{critical} \\ m_4: y_2 := 0 \end{bmatrix}
\end{bmatrix}
$$

–P1– $\qquad\qquad\qquad\qquad\qquad$ –P2–

Fig. 10.5 Program BAKERY(2)

Lamport in 1974, is the first solution that satisfied the three above properties for the general N-process (in our example, $N = 2$).

Figure 10.5 shows a program taken from a tutorial of STeP that implements Lamport's bakery algorithm for mutual exclusion. Two processes, P1 and P2, coordinate access to a critical section, where at most one process should reside at any given time. For instance, modifying the balance of a bank account should be done in a critical section. Each process selects a "ticket" which is a number in y_1 and y_2, respectively, as the customers in a bank would, and the process with the lower ticket is allowed to enter the critical section. Initially, $y_1 = y_2 = 0$. A ticket with number 0 indicates that the process is not interested in accessing the critical section. Since there are only two processes, it can be proved that $y_1, y_2 \in \{0, 1, 2\}$. This informal description of the algorithm can be made precise if the system and its properties are modeled formally.

The meaning of "**await** $(y_2 = 0 \lor y_1 \leq y_2)$" is that the execution pauses until the condition $(y_2 = 0 \lor y_1 \leq y_2)$ is true. This is the same as

$\qquad\qquad$ **while** $\neg (y_2 = 0 \lor y_1 \leq y_2)$ /* do nothing */

To describe the execution of BAKERY(2) in LTL, we introduce three propositional variables, i.e., p_1, p_2, and q, with the meaning that p_1 is $y_1 > 0$, p_2 is $y_2 > 0$, and q is $y_1 > y_2$. We then build a finite state machine based on them. Three variables may create a maximal number of eight states. However, the meaning of the three variables rules out three states: $\overline{p_1}\overline{p_2}q$, $p_1\overline{p_2}q$, $\overline{p_1}p_2q$. Hence, our finite state machine has five states. Initially, $y_1 = y_2 = 0$; hence, p_1, p_2, q are false, representing the initial state. Following the execution of BAKERY(2) line by line, we create the transitions between states, shown in Fig. 10.6. Strictly speaking, this is not a finite state machine that we see in theory of computation, because the transitions from s_0 to s_4 and s_5 model the executions of lines l_1 and m_1 at the same time with two possible outcomes ($y_1 > y_2$ is true or false).

Following the convention from Sect. 10.5.1, we call a path in a finite state machine is a *run*, if the trace of the path represents a program execution. For BAKERY(2), the trace of a run should be a "shuffle" or "interleaving" of two strings from $(l_0^+ l_1 l_2^+ l_3 l_4)^*$ and $(m_0^+ m_1 m_2^+ m_3 m_4)^*$, where l_0^+ (resp. l_2^+, m_0^+, m_2^+) means l_0 can repeat one or more times, to model the effect of working in noncritical section or the **await** command. For instance, the sequence $l_0 l_1 l_2 l_3 m_0 m_1 l_4 m_2 m_3 m_4$ shows that

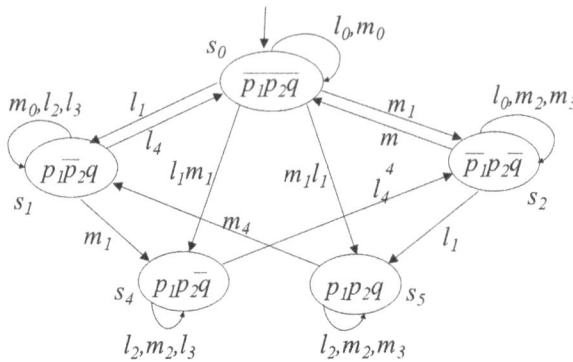

Fig. 10.6 Finite state machine M for BAKERY(2) (Figure 1 of [2])

P1 enters the **critical** section once, while P2 starts its own at that moment. The run associated with this trace is $s_0 s_0 s_1 s_1 s_1 s_1 s_4 s_2 s_2 s_2 s_0$, which can be verified below:

$$s_0 \xrightarrow{l_0} s_0 \xrightarrow{l_1} s_1 \xrightarrow{l_2} s_1 \xrightarrow{l_3} s_1 \xrightarrow{m_0} s_1 \xrightarrow{m_1} s_4 \xrightarrow{l_4} s_2 \xrightarrow{m_2} s_2 \xrightarrow{m_3} s_2 \xrightarrow{m_4} s_0$$

An advantage of finite state machines as shown in Fig. 10.6 is that program execution can be described either by traces or state transitions (to be dealt in LTL). Now, it is easy to express the three desired properties in LTL:

- *Mutual exclusion*: Lines l_3 and m_3 cannot be executed at the same time. For M in Fig. 10.6, it asks that $l_3 m_3$ or $m_3 l_3$ cannot appear in a label string of any run (this can be checked by an exhaustive graph search).

  To prove this property by LTL, we see that the only places where label l_3 appears are at states s_1 and s_4 and its existence condition is $p_1 \overline{p_2} q \lor p_1 p_2 \overline{q}$ or, equivalently, $p_1(\overline{p_2} q \lor p_2 \overline{q})$. Similarly, the only places where label m_3 appears are at states s_2 and s_5 and its existence condition is $\overline{p_1} p_2 \overline{q} \lor p_1 p_2 q$ or, equivalently, $p_2(\overline{p_1 q} \lor p_1 q)$.

  For mutual exclusion, the two existence conditions cannot be true at the same time. In LTL, we require that

$$\Box(\overline{p_1(\overline{p_2} q \lor p_2 \overline{q}) p_2(\overline{p_1 q} \lor p_1 q)})$$

  which can be easily proved to be true in LTL.

- *Accessibility*: If control resides at l_1 (resp. m_1), it will eventually reach l_3 (resp. m_3). The only obstacle for the control to go from l_1 to l_3 is l_2: **await**$(y_2 = 0 \lor y_1 \leq y_2)$. Once l_1 is executed, p_1 becomes true. To pass l_2, we need $(y_2 = 0 \lor y_1 \leq y_2)$, or equivalently, $\overline{p_2} \lor \overline{q}$, to be true. So, in LTL, the accessibility of BAKERY(2) is expressed as

$$\Box(p_1 \to \Diamond(\overline{p_2} \lor \overline{q})) \land \Box(p_2 \to \Diamond(\overline{p_1} \lor q)).$$

10.5 Verification of Concurrent Programs

We cannot prove the validity of the above formula since we do not have axioms about p_1, p_2, and q. However, we have a finite state machine in Fig. 10.6 which models the program execution. To show accessibility, we show that the negation of the above formula has no acceptable models in the finite state machine.

Let us first consider $\overline{A}$, where A is $\Box(p_1 \to \Diamond(\overline{p_2} \vee \overline{q}))$. $\overline{A}$ can simplified to $\Diamond(p_1 \wedge \Box(p_2 \wedge q))$. The only models of $\overline{A}$ in M are those regular sequences that end with s_5^*. However, they are discarded because they are not runs. Similarly, $\overline{B}$, where B is $\Box(p_2 \to \Diamond(\overline{p_1} \vee q))$, has only models which end with s_4^*, and they are discarded for the same reason.

The accessibility property rules out the case of starvation, which produces traces of form $\ldots l_0 l_1 (m_0 m_1 m_2 m_3 m_4)^*$. The next property rules out the case that m_3 appears more than once between l_1 and l_4 in any trace.

- *One-bounded overtaking*: This property is easily expressed by a property of traces of BAKERY(2): in any trace of a run, between l_1 (P1 makes request to enter the critical section) and l_4 (P1 leaves the critical section), m_3 (P2 enters the critical section) can appear at most one. The dual property is obtained by switching P1 and P2.

Since l_1 (resp. m_1) is associated with p_1 (resp. p_2) becoming true and l_4 (resp. m_4) is associated with p_1 (resp. p_2) becoming false, we may use the following formula to express the one-bounded overtaking property:

$$\Box(p_1 \to (\overline{p_2} W (p_2 W (\overline{p_2} W \overline{p_1})))).$$

This formula states that whenever control is at l_2 (p_1 is true), meaning that P1 wants to enter the critical section, the following must occur: there may be an interval in which P2 is not in the critical section (so all states in the interval satisfy $\overline{p_2}$), followed by an interval where P2 is in the critical section (states satisfying p_2), followed by an interval where P2 is again not in the critical section (states satisfying $\overline{p_2}$), followed finally by a state where P1 leaves the critical section (i.e., l_4, and $\overline{p_1}$ is false). Thus, P2 can enter the critical section (m_3) at most once before P1 leaves the critical section (p_1 becomes false).

Again, we can check that the negation of the above formula has no acceptable models for the finite state machine of BAKERY(2).

A general version of BAKERY(N), given in Fig. 10.7 [2], is an oversimplification of Lamport's bakery algorithm because when two or more processes have the same positive y-values, i.e., $y_i = y_j > 0$ for $i \neq j$ (it means they made the request at the same time), Lamport's algorithm will use $i < j$ to break the tie, whereas BAKERY(N) will allow them to finish the waiting at the same time. As a result, the *mutual exclusion* property of BAKERY(N) will not hold; the other two properties of BAKERY(2) can be carried over to BAKERY(N). As an exercise, you are asked to modify BAKERY(N) so that the *mutual exclusion* property holds for the modified BAKERY(N).

$$
\begin{array}{ll}
\text{in} & N : \text{integer where } N > 0 \\
\text{local} & y : \text{array } [1..N] \text{ of integer where } \forall i : [1..N].\, y[i] = 0 \\
\text{value} & \max : \text{array } [1..N] \text{ of (integer} \to \text{integer)}
\end{array}
$$

$$
\mathop{\|}_{i=1}^{N} :: \begin{bmatrix} \text{loop forever do} \\ \begin{bmatrix} \ell_0: \text{ noncritical} \\ \ell_1: \ y[i] := 1 + \max(y) \\ \ell_2: \ \text{await } \forall j : [1..N].(i \neq j \ \to \ y[j] = 0 \ \vee \ y[i] \leq y[j]) \\ \ell_3: \ \text{critical} \\ \ell_4: \ y[i] := 0 \end{bmatrix} \end{bmatrix}
$$

Fig. 10.7 Program BAKERY(N) (Figure 17 of [2])

There are a huge number of execution sequences when $N > 2$ in BAKERY(N), because the computation of a concurrent program is an interleaving of the atomic operations of its processes. As shown by the above example, it is very difficult by hand to formally verify concurrent programs by exhausting all interleaving sequences, to ensure that they are correct for every possible computation.

To model BAKERY(N) for $N > 2$ in LTL, we need more propositional variables. For instance, for $N = 3$, we need nine variables: p_1 (denotes $y_1 > 0$), p_2 ($y_2 > 0$), p_3 ($y_3 > 0$), q_1 ($y_1 > y_2$), q_2 ($y_2 > y_3$), q_3 ($y_3 > y_1$), r_1 ($y_1 = y_2$), r_2 ($y_2 = y_3$), and r_3 ($y_1 = y_3$). To help understand the code when $N = 3$, we replace the condition of the **await** command at line l_2 for the three processes, P1, P2, and P3, as follows:

$$
\begin{aligned}
&\text{P1: } (\overline{p_2} \vee \overline{q_1}) \wedge (\overline{p_3} \vee q_3 \vee r_3) \\
&\text{P2: } (\overline{p_1} \vee q_1 \vee r_1) \wedge (\overline{p_3} \vee \overline{q_2}) \\
&\text{P3: } (\overline{p_1} \vee \overline{q_3}) \wedge (\overline{p_2} \vee q_2 \vee r_2)
\end{aligned}
$$

It is easy to check that when $y_1 = y_2 = y_3 = 1$, all the *await* conditions in P1, P2, and P3 are true, thus entering the critical area at the same time. This bug can be fixed and this is left as an exercise.

Note that the number of interpretations over nine variables is $2^9 = 512$. A finite state machine modeling the execution of BAKERY(3) will be much more complex than that of Fig. 10.6. Obviously, software tools are indispensable for the success of formal verification. The reader should not hold a pessimistic view of formal verification because the focus of formal verification is its algorithm. Once the formal verification is complete, it can be safely used in many applications. To verify that an implementation follows correctly its algorithm is a relatively easy task, as it can be decomposed to line-by-line verification.

In short, formal verification provides a feasible tool to verify the correctness of concurrent programs. Our presentation is limited to LTL, and there are approaches based on automaton theory and model checking, which are out of the scope of this book.

Exercises

1. For the TL formula $A = p \rightarrow \Box p$, find two Kripke frames such that A is true in one and false in the other.
2. For the TL formula $A = \Box(p \vee q) \rightarrow (\Box p \vee \Box q)$, find two Kripke frames such that A is true in one and false in the other.
3. Prove that the following TL formula is valid:

$$\Box(p \rightarrow q) \rightarrow (\Diamond p \rightarrow \Diamond q).$$

4. Prove that the following equivalences are true in TL:

   (a) $\Box\Box p \equiv \Box p$
   (b) $\Diamond\Diamond p \equiv \Diamond p$.

5. Describe the following statements in LTL, using as the least number of propositional variables:

   (a) If the door is locked, it will not open until someone unlocks it.
   (b) If you press ctrl-C, you will get a command line prompt.
   (c) The saw will not run unless the safety guard is engaged.

6. For the LTL formula $A = p \rightarrow \Box p$, find two interpretation sequences such that A is true in one and false in the other.
7. For the LTL formula $A = \Box(p \vee q) \rightarrow (\Box p \vee \Box q)$, find two interpretation sequences such that A is true in one and false in the other.
8. Unlike combinational circuits, sequential circuits have "memory" and are basic building blocks of computers. The circuit (a) of Fig. 10.8 is called *SR latch*, which is one of the simplest sequential circuits. A SR latch has two Boolean inputs, S and R, and two outputs, A and B. The values of A and B are decided not only by S and R but also by the previous values of A and B. That is, $A = \overline{R \vee B'}$ and $B = \overline{S \vee A'}$, where A' and B' are the "old" values. A *state* of a SR latch is decided by the values of S, R, A, and B. Thus, there are at most 16 SR latch states. The transitions between the states are decided by the changes of S and R. Thus, a state has four successors, including itself. For instance, when $S = R = 1$, $A = B = 0$, the state is $SR\overline{A}\overline{B}$ or 1100 (this state is not allowed

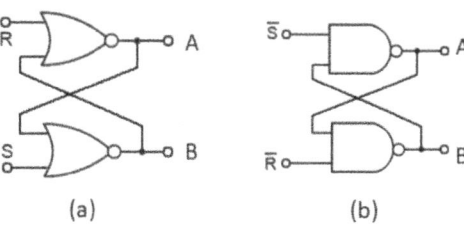

Fig. 10.8 (a) SR latch and (b) SR flip-flop

in the application of SR latches). When S changes from 1 to 0, the transition goes from 1100 to 0101. When both inputs change from 11 to 00, the output values will be indeterminate, and you need multiple transitions to all possible states. Please provide a Kripke frame to model the transitions of the SR latch in Fig. 10.8a.

9. Please repeat the previous exercise for the circuit in Fig. 10.8b, which is called SR flip-flop. Note that the inputs S and R are negated before feeding to the circuit, $A = \overline{S} \wedge B'$, and $B = \overline{R} \wedge A'$, where A' and B' are the "old" values.

10. Write a Prolog program $\mathtt{nnf}(X, Y)$, which takes LTL formula X as input and output Y as a negation normal form of X.

11. Prove that the following formulas are valid in LTL:

    (a) $\Box p \rightarrow \Box \Box p$
    (b) $\Diamond \Diamond p \rightarrow \Diamond p$
    (c) $\Box \Diamond \Box p \rightarrow \Diamond \Box p$
    (d) $\Box(p \rightarrow \Diamond q) \wedge \Box(q \rightarrow \Diamond r) \rightarrow \Box(p \rightarrow \Diamond r)$

12. Prove that the following equivalences are true in LTL:

    (a) $\circ(p \wedge q) \equiv \circ p \wedge \circ q$
    (b) $\circ(p \vee q) \equiv \circ p \vee \circ q$.

13. Prove by semantic tableau that the LTL formulas are valid:

    (a) $(\Box p \vee \Box q) \rightarrow (\Box p \vee \Box)$
    (b) $\Diamond \Diamond p \rightarrow \Diamond p$
    (c) $\Box \Diamond \Box p \rightarrow \Diamond \Box p$
    (d) $\Diamond \Box \Diamond p \rightarrow \Box \Diamond p$.

14. Find an equivalent negation formal form for the following formulas:

    (a) $(A \, u \, B) \leftrightarrow \Diamond B \wedge (A \, w \, B)$
    (b) $(A \, \mathcal{R} \, B) \leftrightarrow \Box B \vee (A \, s \, B)$
    (c) $(A \, u \, B) \leftrightarrow B \, s \, (A \vee B)$
    (d) $(A \, \mathcal{R} \, B) \leftrightarrow B \, w \, (A \wedge B)$
    (e) $(A \, w \, B) \leftrightarrow B \, \mathcal{R} \, (A \vee B)$
    (f) $(A \, s \, B) \leftrightarrow B \, u \, (A \wedge B)$
    (g) $(A \, w \, B) \leftrightarrow A \, u \, (\Box A \vee B)$
    (h) $(A \, s \, B) \leftrightarrow A \, \mathcal{R} \, (\Diamond A \wedge B)$
    (i) $\Box A \leftrightarrow A \, w \, \bot$
    (j) $\Diamond A \leftrightarrow A \, s \, \top$
    (k) $\neg (A \, s \, B) \leftrightarrow \neg A \, w \, \neg B$

15. Prove by semantic tableau that the LTL formulas are valid:

$$
\begin{aligned}
&(a) & p \mathrel{\mathcal{R}} p &\to p \\
&(b) & \bot \mathrel{\mathcal{R}} p &\to \Box p \\
&(c) & p \mathrel{\mathcal{R}} (p \mathrel{\mathcal{R}} q) &\to p \mathrel{\mathcal{R}} q \\
&(d) & (p \mathrel{\mathcal{R}} q) \mathrel{\mathcal{R}} q &\to p \mathrel{\mathcal{R}} q \\
&(e) & p \mathrel{\mathcal{R}} q &\to q \wedge (p \vee \circ(p \mathrel{\mathcal{R}} q)) \\
&(f) & p \mathrel{\mathcal{R}} q &\to \neg(\neg p \mathrel{\mathcal{U}} \neg q)
\end{aligned}
$$

16. Prove by semantic tableau that the LTL formulas are valid:

$$
\begin{aligned}
(a) &\ p \mathrel{\mathcal{U}} q \to \Diamond q \wedge (p \mathrel{\mathcal{W}} q) \\
(b) &\ p \mathrel{\mathcal{R}} q \to \Box q \vee (p \mathrel{\mathcal{S}} q) \\
(c) &\ p \mathrel{\mathcal{U}} q \to q \mathrel{\mathcal{S}} (p \vee q) \\
(d) &\ p \mathrel{\mathcal{R}} q \to q \mathrel{\mathcal{W}} (p \wedge q) \\
(e) &\ p \mathrel{\mathcal{W}} q \to q \mathrel{\mathcal{R}} (p \vee q) \\
(f) &\ p \mathrel{\mathcal{S}} q \to q \mathrel{\mathcal{U}} (p \wedge q) \\
(g) &\ p \mathrel{\mathcal{W}} q \to p \mathrel{\mathcal{U}} (\Box p \vee q) \\
(h) &\ p \mathrel{\mathcal{S}} q \to p \mathrel{\mathcal{R}} (\Diamond p \wedge q) \\
(i) &\ \Box p \to p \mathrel{\mathcal{W}} \bot \\
(j) &\ \Diamond p \to p \mathrel{\mathcal{S}} \top \\
(k) &\ \neg(p \mathrel{\mathcal{S}} q) \to \neg p \mathrel{\mathcal{W}} \neg q
\end{aligned}
$$

17. Prove the following statements are true in LTL: (a) $(A \mathrel{\mathcal{U}} B) \mathrel{\mathcal{U}} B \equiv A \mathrel{\mathcal{U}} B$ and (b) $\circ (A \mathrel{\mathcal{U}} B) \equiv (\circ A) \mathrel{\mathcal{U}} (\circ B)$.

18. Decide how many propositional variables are needed for modeling BAKERY(N) (Fig. 10.7) with $N = 4$. Please express the conditions of line l_2 for the four processes in terms of these variables.

19. (a) Modify the bakery algorithm BAKERY(N) in Fig. 10.7 to provide a correct description of Lamport's bakery algorithm.
    (b) For $N = 3$, express the condition of the *await* command in the modified BAKERY(N) as a Boolean expression of a programming language in P1, P2, and P3, respectively.
    (c) For $N = 3$, we define the following propositional variables: p_1 (denotes $y_1 > 0$), p_2 ($y_2 > 0$), p_3 ($y_3 > 0$), q_1 ($y_1 > y_2$), q_2 ($y_2 > y_3$), q_3 ($y_3 > y_1$), q_1 ($y_1 = y_2$), r_2 ($y_2 = y_3$), and r_3 ($y_1 = y_3$). Do we need all the nine variables for the modified BAKERY(N)? Please answer this question by expressing the *await* conditions of P1, P2, and P3 as a formula of these nine variables.

20. Prove the mutual exclusion and accessibility properties of the program given below:

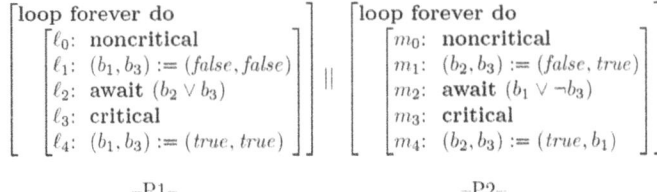

$$\text{local} \quad b_1, b_2, b_3 \;:\; \text{boolean where } b_1, b_2, b_3$$

$$\begin{bmatrix} \text{loop forever do} \\ \begin{bmatrix} \ell_0: & \text{noncritical} \\ \ell_1: & (b_1, b_3) := (\mathit{false}, \mathit{false}) \\ \ell_2: & \textbf{await } (b_2 \lor b_3) \\ \ell_3: & \textbf{critical} \\ \ell_4: & (b_1, b_3) := (\mathit{true}, \mathit{true}) \end{bmatrix} \end{bmatrix} \;\|\; \begin{bmatrix} \text{loop forever do} \\ \begin{bmatrix} m_0: & \text{noncritical} \\ m_1: & (b_2, b_3) := (\mathit{false}, \mathit{true}) \\ m_2: & \textbf{await } (b_1 \lor \neg b_3) \\ m_3: & \textbf{critical} \\ m_4: & (b_2, b_3) := (\mathit{true}, b_1) \end{bmatrix} \end{bmatrix}$$

—P1— —P2—

References

1. Yde Venema, *Temporal Logic*, in Lou Goble (ed.) The Blackwell Guide to Philosophical Logic. Blackwell, 2001
2. Nikolaj S. Bjorner, Anca Browne, Michael A. Colon, Bernd Finkbeiner, Zohar Manna, Henny B. Sipma, Tomas E. Uribe, "Verifying temporal properties of reactive systems: A STeP tutorial", *Formal Methods in System Design*, 16, 1–45, 2000
3. Edmund M. Clarke Jr., Orna Grumberg, Daniel Kroening, Doron Peled, and Helmut Veith, *Model Checking*, 2nd Ed, MIT Press, 2018. ISBN: 978-0262038836

Part IV
Logic of Computability

Chapter 11
Decidable and Undecidable Problems

In computer science, undecidability theory studies the problems which are beyond the power of computers and is a part of computability theory. In logic, undecidability concerns about what problems can be computable or uncomputable or which formulas can be proved, disproved, or unprovable. For example, all the problems of propositional logic discussed in this book are computable. On the other hand, some problems in first-order logic are uncomputable, and we will see some of them through the chapter. The objective of this chapter is to introduce the formal concepts of decidable and undecidable decision problems, as well as computable and uncomputable functions, and show that some problems in first-order logic are undecidable or uncomputable. We leave the discussion of decidable problems in first-order logic to Chap. 12, where we will present several decidable fragments of first-order logic.

11.1 A Bit of History

11.1.1 Gödel's Incompleteness Theorems

The cornerstone of undecidability theory comes from Kurt Gödel's two incompleteness theorems:

1. In any consistent formal system S which can perform common arithmetic operations on natural numbers, some formulas of S cannot be proved to be true or false in S.
2. If S is consistent, then we cannot prove in S that S itself is consistent.

A formal system is similar to a proof theory introduced in Chap. 1 (Sect. 1.3.4). A proof theory consists of two components:

1. The system of axioms, which defines which formulas are true (i.e., theorems)
2. The proof procedures which tell us how to prove theorems

We introduced four properties for a proof procedure: consistency, soundness, completeness, and termination. A decision procedure is a proof procedure that is both sound and terminating. Gödel's incompleteness theorems essentially claim that no decision procedures exist for every decision problem of a formal system which can perform common arithmetic operations.

In 1931, at the age of 25, Gödel formally published the incompleteness theorems based on the concept of ω-consistency. The reception of Gödel's results was mixed. Some important figures in the field of logic quickly assimilated the results, but there was also a lot of misunderstanding and resistance. Gödel's incompleteness theorems mark not only a milestone in proof theory and computability theory but also have great impact in the study of set theory and modern logic (modal logic, intuitionistic logic, categorical logic, etc.). *Stanford Encyclopedia of Philosophy* has a lengthy entry on "Gödel's incompleteness theorems" with detailed historic information.

Instead of introducing Gödel's formal system, we choose to present *Turing machine*, which is a formal computing model, for the following reasons:

- A Turing machine is a remarkably simple formal computing model, and each move of a Turing machine resembles a computing step by a human being. Gödel regarded Turing's approach as superior to all other previous work and commented "that this really is the correct definition of mechanical computability was established beyond any doubt by Turing."
- Turing machines are simple to describe and easily understood by people familiar with computers. Today's computers can be viewed as an implementation of the *universal Turing machine*, a special Turing machine proposed by Turing, which takes data and program code as input, executes the code on the data, and reports its output.
- Turing machines are important in computability theory due to the *Church-Turing thesis*, which states that every computing model, invented in the past or to be invented in the future, can be simulated by a Turing machine [1].
- As a result of the *Church-Turing thesis*, Gödel's formal system can be easily simulated by a Turing machine. Gödel's first incompleteness theorem is translated into a theorem of Turing machines as follows: for any Turing machine M and any input w, no computing devices can decide if M takes w as input and stops with "success" (proved), stops with "failure" (disproved), or loops forever (unprovable). That is, the termination of M on w is undecidable, which is the well-known *halting problem*.
- While the inconsistency property is not defined for Turing machines, Gödel's second incompleteness theorem corresponds to a general theorem of Turing machines (Rice's theorem): nontrivial and language-related properties of Turing machines are undecidable.

11.1.2 Three Well-Known Computing Models

Turing machine is just one of many formal computing models ever proposed. In the 1930s, several independent attempts were made to formalize the notion of computability [1]:

- In 1931, Gödel proved the first incompleteness theorem for a class called *primitive recursive functions*. In 1934, Gödel created a more general class called *general recursive functions*, also called *partial recursive functions*. The class of general recursive functions is the smallest class of functions (possibly with more than one argument) which includes all constant functions, projections, and the successor function and is closed under function composition, recursion, and minimization. If minimization is excluded, then it becomes the class of primitive recursive functions. In 1936, Stephen Kleene introduced variants of Gödel's definition and defined a recursion theory equivalent to general recursive functions.
- In 1932, Alonzo Church created a method for defining functions called *lambda calculus*, often written as λ-calculus. Later, Church defined an encoding of the natural numbers called *Church numerals*. A function on the natural numbers is called λ-computable if the corresponding function on the Church numerals can be represented by a term of the λ-calculus.
- In 1936, before learning of Church's λ-calculus, Alan Turing created a theoretical computing model, now called *Turing machines*, which could carry out calculations from inputs by manipulating symbols on a tape. Given a suitable encoding of the natural numbers as sequences of symbols, a function on the natural numbers is called *Turing computable* if there exists a Turing machine that starts with the encoded natural numbers as input and stops with the result of the corresponding function as output.

Church and Turing proved that these three formally defined classes of computable functions coincide: a function is λ-computable iff it is Turing computable and iff it is a general recursive function. This has led mathematicians and computer scientists to believe that the concept of computability is accurately characterized by these three equivalent computing models. Other formal attempts to characterize computability, including Kleene's recursion theory and Post's canonical system, have subsequently strengthened this belief. Since we cannot exhaust all possible computing models, the Church–Turing thesis, although it has near-universal acceptance, cannot be formally proved.

Today, programming languages based on general recursive functions (or lambda calculus or Kleene's recursion theory) are called *functional* programming languages. On the other hand, programming languages based on the *random-access stored-program* (RASP) machines are called *imperative* programming languages. RASP is a computing model equivalent in power to the above three computing models and is an example of the so-called *von Neumann architecture*, which uses random-access memories and stored programs. Lisp, ML, Clojure, and Haskell are examples of the former; C, C++, Java, and Python are examples of the latter.

11.1.3 Halting Problem

In computability theory, the halting problem is the problem of determining, from a description of an arbitrary computer program and an input, whether the program will finish running or continue to run forever. Turing proved in 1936 that a general algorithm to solve the halting problem for all possible pairs $\langle M, w \rangle$, where M is a Turing machine and w is an input to M, cannot exist. This result is an echo of Gödel's first incompleteness theorem for Turing machines and is one of the first cases of undecidable decision problems. The undecidability of the halting problem is significant to practical computing efforts and defines a class of problems which no computing devices can possibly solve decidedly.

The halting problem is not just a theoretical problem. This is a practical problem for many researchers. In Example 7.3.13, the unsolved Collatz conjecture is related to whether the program below terminates for any positive integer x:

> **proc** $Collatz(x)$
>   **while** $x > 1$ **do**
>   **if** $even(x)$ **then** $x := x/2$ **else** $x := (3*x+1)/2$
>   **return** 1

The same code can be expressed as a rewrite system:

$$f(1) \to 1$$
$$f(2x) \to f(x)$$
$$f(2x+1) \to f(3x+2)$$

The termination of the above rewrite system will imply that the Collatz conjecture is solved positively.

In this chapter, we will formalize the aforementioned concepts such as computability, decidability, undecidable problems, etc., based on Turing machines, and show certain problems in logic are undecidable.

11.2 Turing Machines

A *Turing machine* is a mathematical model of computation first proposed by Alan Turing in 1936. A Turing machine manipulates symbols on a tape of symbols according to a set of rules. According to the Church-Turing thesis [1], any function on the natural numbers can be calculated by an effective method iff it is computed by

11.2 Turing Machines

a Turing machine. In other words, despite the model's simplicity, Turing machines are capable of simulating any computation on today's computers, whether it is a supercomputer or a smartphone.

Turing thought that when a human calculates with pen and paper, at any given moment of the calculation, the mind can be in only one of a finite collection of states and that in each state, given the intermediate results thus far obtained, the next calculation step is completely determined. And Turing machine catches all the essential features of a human computer.

A Turing machine operates on a tape containing an infinite number of cells and each cell contains one symbol. The machine positions its "head" over a cell and "reads" the symbol there. As per the symbol and the machine's own present state in a user-specified set of instructions, the machine operates as follows:

(a) Read the tape symbol pointed by the tape head.
(b) Get the user-specified instructions according to the symbol from *(a)* and the current state. If no instructions are available, halt with "failure." Otherwise, perform *(c)*–*(e)* according to the instructions.
(c) Write a symbol into the current cell.
(d) Move the tape head one cell either left or right.
(e) Change the current state. If the new state is "success," halt with "success."
(f) Go to *(a)*.

The tape is initially filled with an input string of symbols and blank everywhere else. The machine positions its head to the first symbol with the designated state called the "initial state." The machine moves as described above through the steps *(a)*–*(f)*. There are three possible outcomes: (1) stop with "success"; (2) stop with "reject"; or (3) loop forever.

Turing was able to answer a fundamental question in the negative: does an algorithm exist that can determine whether a Turing machine stops on a given input string? This is the so-called "halting problem" of Turing machines.

A Turing machine can do everything that a real computer can do. However, their minimalist design makes them unsuitable for computation in practice: real-world computers are based on different designs that, unlike Turing machines, use random-access memory. For people who know programming, you may view a Turing machine's tape as a very long array of symbols with a single index variable of the array (tape head). The index can increase or decrease by 1 to simulate the move of the tape head. The user's instructions can be stored in a two-dimensional array indexed by the tape symbols and the state symbols.

While a Turing machine can express arbitrary computations, nonetheless, even a Turing machine cannot solve certain problems. In a very real sense, these problems are beyond the theoretical limits of computation and show the existence of fundamental limitations on the power of mechanical computation.

11.2.1 Formal Definition of Turing Machines

A Turing machine (TM) uses a tape of infinitely many cells, which provides unlimited memory for computation. The set of symbols appearing in the tape is denoted by Γ, called the *tape alphabet*. By default, Γ is a super set of Σ, which is the set of symbols used in all input strings. The blank symbol, ⊔, is in $\Gamma - \Sigma$. Σ^* denotes the set of all input strings (see Example 1.3.9) and Γ^* denotes the set of all tape contents excluding the trailing blank symbols.

For any string $w \in \Sigma^*$, let $w^0 = \epsilon$ and $w^{i+1} = ww^i$, where $i \geq 0$. For example, $a^3 = aaa$ and $(ab)^2 = abab$.

Example 11.2.1 Suppose $\Sigma = \{a, b\}$ and we want to design a Turing machine which checks whether input string $w \in A_1 = \{a^j b^k \mid 0 \leq j \leq k\}$. To design a Turing machine is the same as to design an algorithm working on an array of symbols. If the machine reads a, we may erase it (by writing another symbol into the cell), which corresponds to $j := j - 1$, and then look for the symbol b and erase it, too, which corresponds to $k := k - 1$. The pseudo-code of the algorithm may read as follows:

> **while** $(j > 0 \wedge k > 0)$ **do**
> $\quad j := j - 1; k := k - 1$
> **if** $(j = 0)$ **return** 1 **else return** 0

Based on the above algorithm, we may write out the details of the Turing machine under design. □

The machine has a state and a tape head (a pointer of the tape), can read and write the symbol in the cell pointed by the tape head, and move back or forth on the tape, one cell at a time. Initially, the tape contains only the input string in the beginning portion of the tape and is blank everywhere else. Figure 11.1 illustrates a Turing machine with initial state q_0 and input string 001110. If the machine needs to store information, it may write this information on the tape. To read the information that it has written, the machine can move the tape head back over it. The machine continues moving until it decides to produce an output. The outputs 1 (or "accept") and 0 (or "reject") are obtained by entering the designated accepting and rejecting states. If it does not enter an accepting or a rejecting state, it will go on forever, never halting, unless its move is blocked due to no definition (this case is equivalent to "reject").

To output 1, the Turing machine uses the state q_a (or q_{accept}); to output 0, the machine uses the state q_r (or q_{reject}). Since each move is decided by the symbol

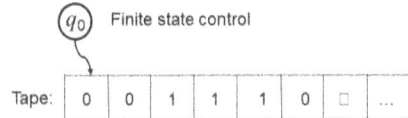

Fig. 11.1 Illustration of a Turing machine with initial state q_0 and input string 001110

11.2 Turing Machines

pointed by the head and the current state, we formally define moves as a general function δ which takes a state and a tape symbol as input and outputs (i) the next state, (ii) the symbol to be written into the current cell, and (iii) the move direction of the tape head. The definition of δ decides the feature of a Turing machine. If the user (i.e., the designer of Turing machines) does not provide the definition of δ for a state (other than q_a) and a tape symbol, it is regarded as "reject" (by blocking the next move), which is equivalent to entering q_r.

Below is the formal definition of a Turing machine.

Definition 11.2.2 A Turing machine is a 6-tuple $M = (Q, \Sigma, \Gamma, \delta, q_0, q_a)$, where Q is a finite set of states, Σ is the finite alphabet of input strings, Γ is the finite set of tape symbols, $\Sigma \subset \Gamma, \sqcup \in \Gamma - \Sigma$, the initial state $q_0 \in Q$, the accept state $q_a \in Q$, and the partial function δ.

$$\delta : Q \times \Gamma \mapsto Q \times \Gamma \times \{L, R\}$$

defines each move of M, where L and R indicate the tape head moves left or right, respectively. If the tape head points to the first symbol on the tape, it cannot move left (i.e., "move off the tape" is blocked).

Example 11.2.3 Let $A_0 = \{0^i 1^j \mid i, j > 0\}$. We want to construct $M_0 = (Q, \Sigma, \Gamma, \delta, q_0, q_a)$ to accept the strings in A_0, where $\Sigma = \{0, 1\}$, $\Gamma = \{0, 1, \sqcup\}$, and $Q = \{q_0, q_1, q_2, q_a\}$. The tasks already performed before entering a state or to be performed in this state are the following:

- q_0: check if the input starts with 0.
- q_1: the first 0 is found, skip rest 0s, and look for the first 1.
- q_2: the first 1 is found, skip rest 1s, and look for $\sqcup$, the end of the input.
- q_a: report "accept."

The δ function is defined as follows:

1. $\delta(q_0, 0) = (q_1, 0, R)$ // $i > 0$, go to q_1
2. $\delta(q_1, 0) = (q_1, 0, R)$ // skip 0 and continue
3. $\delta(q_1, 1) = (q_2, 1, R)$ // $j > 0$, go to q_2
4. $\delta(q_2, 1) = (q_2, 1, R)$ // skip 1 and continue
5. $\delta(q_2, \sqcup) = (q_a, \sqcup, R)$ // reach the end and accept

M_0 scans the tape once and does not change the tape content. Once it enters q_a, the input must contain a string of form $0^i 1^j$, where $i > 0$ and $j > 0$. For other input strings, M_0 stops at a point where $\delta(q, a)$ is undefined, and the result is "reject". Alternatively, we may enter q_r, the designated rejecting state, when one of these undefined cases happens.

□

Example 11.2.4 Continuing from Example 11.2.1, based on the algorithm in Example 11.2.1, we want to construct $M_1 = (Q, \Sigma, \Gamma, \delta, q_0, q_a)$ to accept the

(q, a)	Reason
$(q_0, \sqcup)$	The input is the empty string
$(q_0, 1)$	The input starts with 1
$(q_1, \sqcup)$	The input misses 1
$(q_2, 0)$	0 appears after 1

strings in $A_1 = \{a^j b^k \mid 0 \leq j \leq k\}$, where $\Sigma = \{a, b\}$, $\Gamma = \{a, b, x, y, \sqcup\}$, and $Q = \{q_0, q_1, q_2, q_3, q_a\}$. The symbols x and y in Γ represent erased a and b, respectively. The tasks already performed before entering a state or to be performed in this state are the following:

- q_0: at the beginning of the while loop, look for a, i.e., check $j > 0$.
- q_1: $j := j - 1$ is done, look for b, i.e., check $k > 0$.
- q_2: $k := k - 1$ is done, rewind to the last x, and go to q_0.
- q_3: no a is found, i.e., $j = 0$, check $k \geq 0$ and no a after b.

The δ function is defined as follows.

1 $\delta(q_0, \sqcup) = (q_a, \sqcup, R)$ // $j = k = 0, \epsilon \in A_1$
2 $\delta(q_0, a) = (q_1, x, R)$ // $j > 0, j := j - 1$, go to q_1
3 $\delta(q_0, b) = (q_3, b, R)$ // $j = 0, k > 0$, go to q_3
4 $\delta(q_0, y) = (q_3, y, R)$ // $j = 0$, skip y, go to q_3
5 $\delta(q_1, a) = (q_1, a, R)$ // skip a, continue right
6 $\delta(q_1, y) = (q_1, y, R)$ // skip y, continue right
7 $\delta(q_1, b) = (q_2, y, L)$ // $k := k - 1$, move back and go to q_2
8 $\delta(q_2, y) = (q_2, y, L)$ // skip y, continue left
9 $\delta(q_2, a) = (q_2, a, L)$ // skip a, continue left
10 $\delta(q_2, x) = (q_0, x, R)$ // found x, go to q_0 and start a new round
11 $\delta(q_3, y) = (q_3, y, R)$ // skip y, continue right
12 $\delta(q_3, b) = (q_3, b, R)$ // skip b, continue right
13 $\delta(q_3, \sqcup) = (q_a, \sqcup, L)$ //$j = 0, \ k \geq 0, b^k \in A_1$

From the above instructions, we see that M_1 enters q_a ("success" state) at line 1 ($\epsilon \in A_1$) and line 13 ($j = 0, k \geq 0, b^k \in A_1$, after the same number of the copies of a and b are erased). M_1 is blocked when either the state is q_1 and the current symbol is $\sqcup$ (more a symbols than b symbols in the input), or the state is q_3 and the current symbol is a (a appears after b symbols). □

Definition 11.2.5 Given an input string w to a Turing machine $M = (Q, \Sigma, \Gamma, \delta, q_0, q_a)$, let $q \in Q$ be the current state, $a \in \Gamma$ be the symbol pointed by the tape head, and α and β be the strings of tape symbols before and after the tape head, respectively. The triple $\langle \alpha, q, a\beta \rangle$, or simply $\alpha q a \beta$, is called a *configuration* of M. The initial configuration is $q_0 w$, where w is the input string. A legal move of M is a pair (C_1, C_2) of configurations, written $C_1 \vdash C_2$, such that either $C_1 = \alpha q a \beta$,

11.2 Turing Machines

$\delta(q, a) = (p, b, R)$, and $C_2 = \alpha b p \beta$ or $C_1 = \alpha c q a \beta$, $\delta(q, a) = (p, b, L)$, and $C_2 = \alpha p c b \beta$.

In the above definition, we assume that β does not contain the blank symbols after the last non-blank symbol on the tape.

Example 11.2.6 Continuing from the previous example, a sequence of moves of M_1 on the input string $aabb$ are $q_0aabb \vdash xq_1abb \vdash xaq_1bb \vdash xq_2ayb \vdash q_2xayb \vdash xq_0ayb \vdash xxq_1yb \vdash xxyq_1b \vdash xxq_2yy \vdash xq_2xyy \vdash xxq_0yy \vdash xxyq_3y \vdash xxyyq_3\sqcup \vdash xxyq_ay$. So, $aabb$ is accepted by M_1.

On the other hand, $q_0aba \vdash xq_1ba \vdash q_2xya \vdash xq_0ya \vdash xyq_3a$. Since $\delta(q_3, a)$ is not defined, the move is blocked, and the input aba is rejected by M_1. □

Definition 11.2.7 Let $M = (Q, \Sigma, \Gamma, \delta, q_0, q_a)$ be a Turing machine, and $w \in \Sigma^*$, w is said to be *accepted* by M, written $M(w) = 1$, if

$$q_0 w \vdash^* \alpha q_a \beta$$

where $\vdash^*$ denotes a sequence of zero or more successive moves (formally, $\vdash^*$ is the reflexive and transitive closure of $\vdash$), and $\alpha, \beta \in \Gamma^*$.

Definition 11.2.8 The language recognized by Turing machine M is the set $L(M)$ of all input strings accepted by M, formally

$$L(M) = \{w \mid w \in \Sigma^*, M(w) = 1\}.$$

$L \subseteq \Sigma^*$ is *recognizable* if there exists a Turing machine M such that $L = L(M)$.

It is easy to see that $a^2b^2 \in L(M_1)$ and $aba \notin L(M_1)$, as shown by Example 11.2.6. It is not an easy task to show that M_1 recognizes $A_1 = \{a^j b^k \mid 0 \leq j \leq k\}$, i.e., $L(M_1) = A_1$. To show $L(M_1) = A_1$, the interested reader may consult any textbook on theory of computation.

We would like to point out that M_1 can be also regarded as computing $k - j$, because the number of b's remaining on the tape when M_1 accepts is exactly $k - j$. This is how functions on natural numbers are computed by Turing machines. In this case, Turing machines are regarded as "computers of functions" or *procedures*, and the computed functions are called *computable*. If the computed functions are total, then the used Turing machines are called *algorithms*.

Definition 11.2.9 A set $S = \{x \mid p(x)\}$ is said to be *computable* if $p(x)$ is computable; S is *decidable* if $p(x)$ is total computable.

By definition, if L is a formal language, then L is computable iff L is recognizable. Let $\gamma(L) = \{\gamma(w) \mid w \in L\}$, where γ is the bijection from strings to natural numbers, defined in Definition 1.3.5, then $\gamma(L)$ is computable iff L is recognizable.

11.2.2 High-Level Description of Turing Machines

In general, a detailed description of a Turing machine is very tedious. We often prefer to provide a high level description such that every line in the description can be implemented by a finite number of moves of the machine.

Example 11.2.10 If we are asked to design a Turing machine M_2 to recognize $A_2 = \{b^i a^j b^k \mid i * j = k, i, j, k \geq 0\}$, we may use the following algorithm to facilitate the design of a Turing machine.

> Check if $i = 0$ or $j = 0$, then k must be 0
> **while** $(i > 0 \land k \geq j)$ **do**
> $\quad i := i - 1; k := k - j$
> **if** $(k = 0)$ **return** 1 **else return** 0

From Example 11.2.6, we know how to check $i = 0$ and how to do $i := i - 1$ and $k := k - j$. So, there is no technical difficulty to design M_2 for A_2.

M_2 = "On input $w \in \{a, b\}^*$
1. Read the first symbol, if it is a and the input contains no b's (i.e., $i = k = 0$), or it is b and the input contains no a's (i.e., $j = k = 0$), and then, **return** 1.
2. Check if w is of form $b^i a^j b^k$; if not, **return** 0, else go back to the first symbol.
3. Decrease i by one by erasing the first b.
4. Zigzag between a^j and b^k to check if $k < j$; if yes, **return** 0, else do $k := k - j$ as in M_1.
5. Go back to the first b and restore a.
6. If $i > 0$, go to 3.
7. Now, $i = 0$. Check if $k = 0$; if yes, **return** 1, else **return** 0."

How to implement each of the above steps by a sequence of Turing machine moves is left as an exercise. □

Once we know how to check $i * j = k$, we can modify the machine code to compute $i * j$, k/i, or j^k. We can also compare two natural numbers or check two strings to see if they are identical or copy one string from one place to another. This kind of operations can be assumed in a high-level description of a Turing machine. In many cases, a high-level description is the only way to design a Turing machine. However, a basic algorithm or procedure, which can be expressed by a high-level description, is a must in the design of some Turing machines.

11.2.3 Recognizable Versus Decidable

Given any Turing machine $M = (Q, \Sigma, \Gamma, \delta, q_0, q_a)$, if we feed $w \in \Sigma^*$ to M, there will be three possible outcomes: (1) M halts with "accept" ($M(w) = 1$); (2) M halts with "reject" ($M(w) = 0$); and (3) M loops on w ($M(w)$ is undefined). In other words, M partitions Σ^* into three disjoint sets:

- $L(M) = \{w \mid w \in \Sigma^*, M(w) = 1\}$.
- $reject(M) = \{w \mid w \in \Sigma^*, M(w) = 0\}$.
- $loop(M) = \{w \mid w \in \Sigma^*, M(w) \text{ is undefined}\}$.

Obviously, $L(M) \cup reject(M) \cup loop(M) = \Sigma^*$. By convention, $\overline{L(M)} = \Sigma^* - L(M) = reject(M) \cup loop(M)$.

In computability theory, any $A \subseteq \Sigma^*$ is said to be a *formal language*. A can be $\emptyset$ or Σ^*. Turing machines can be used to classify formal languages.

Definition 11.2.11 (Recognizable and Decidable) A formal language $A \subseteq \Sigma^*$ is *recognizable* if there exists a Turing machine M such that $A = L(M)$ and M is said to be a *recognizer* of L; otherwise, A is *unrecognizable*.

A is *decidable* if there exists a Turing machine M such that $A = L(M)$ and $loop(M) = \emptyset$. In this case, M is said to be a *decider* of A.

If A is decidable and $L(M) = A$ for some Turing machine M, it does not mean that $loop(M) = \emptyset$. The definition just ensures that there exists a decider M' such that $L = L(M')$ and $loop(M') = \emptyset$.

So, all formal languages are divided into three disjoint groups: (*a*) decidable, (*b*) recognizable but undecidable, and (*c*) unrecognizable. Some problems are known in one of the three groups; some problems are not known in which group, and these are so-called "open problems."

In literature, a recognizable language is also called *recursively enumerable, partially decidable, semi-decidable, partially computable,* or *computable*. A decidable language is also called *recursive, total recognizable,* or *total computable*.

For each formal language A, there exists a unique function $f_A : \Sigma^* \mapsto \{0, 1\}$: $f_A(w) = 1$ iff $w \in A$. f_A is the *characteristic function* of A. Since there exists a bijection between $\mathcal{N}$ and Σ^* (Proposition 1.3.4), f_A can be viewed as a decision function.

A decider always halts on every input with output 0 or 1, corresponding to a total decision function (Fig. 11.2a). A recognizer halts with output 1 and may loop when the output is not 1, corresponding to a partial decision function. In case we want to distinguish between "reject" and "loop," we use Fig. 11.2c instead of (*b*).

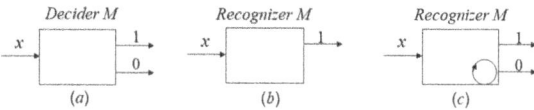

Fig. 11.2 Illustration of deciders and recognizers

11.3 Decidability of Problems

In computability theory, a problem is typically a decision problem which has a simple yes/no answer. An optimization problem like "finding the longest simple cycle in graph G" is usually converted to a question like "does graph G have a simple cycle of length at least m?" by introducing a number m, so that it becomes a decision problem. Typically, an optimization problem and its decision problem belong to the same class of complexity.

11.3.1 Encoding of Decision Problems

A common practice in computability theory is to encode every decision problem by a set of strings.

For instance, the problem that "does G have a simple cycle of length at least m?" can be encoded as a set B_1:

$$B_1 = \{\langle G, m \rangle \mid \text{graph } G \text{ has a simple cycle of length at least } m\},$$

where $\langle G, m \rangle$ is a string which represents G and m; we may think $\langle G, m \rangle = \langle G \rangle \langle m \rangle$. If G contains n vertices and $m \leq n$, then the alphabet for $\langle G, m \rangle$ can be $\{1, 2, \ldots, n, (,), \cdot\}$, where each vertex is represented by a number i, $1 \leq i \leq n$ and each edge is represented by $(i \cdot j)$, $1 \leq i, j \leq n$. The alphabet can also be $\{0, 1\}$ if every symbol in $\{1, 2, \ldots, n, (,), \cdot\}$ is encoded by a binary string of equal length, just like we use the ASCII code in a computer. Now, a graph G has a simple cycle of length at least m iff $\langle G, m \rangle \in B_1$. B_1 catches the essence of this decision problem.

"Does there exist an algorithm A which finds a simple cycle in G of length at least m?" This is also a decision problem and can be encoded as

$$B_2 = \{\langle A, G, m \rangle \mid \text{algorithm } A \text{ finds in } G \text{ a simple cycle of length at least } m\}$$

where $\langle A \rangle$ can be the code for implementing A in a programming language.

Turing machines can also be encoded as a string. For instance, the definition of M_1 in Example 11.2.1 takes less than 20 lines of this book, and each line has less than 100 characters, so a string of 2000 characters will be more than enough to encode M_1. We may also view the encoding of any Turing machine as a binary string because every file is stored in binary code in a computer. In short, there are many ways to encode a Turing machine or any input accepted by a Turing machine. Any encoding method will work for us as long as the following assumptions are true:

- Every encoding shares a common alphabet. In theory, a binary alphabet is enough for all.

11.3 Decidability of Problems

- Different objects have different codes. That is, an encoding is an injective function from objects to strings.
- The computation of encoding and decoding is efficient, taking linear time or quadratic time in the worst case. The decoding can tell if the encoded Turing machine is good (well-defined) or not.

There are many decision problems about Turing machines (TM) and the following are some of them:

- $A_{TM} = \{\langle M, w \rangle \mid \text{TM } M \text{ accepts input } w\}$
- $H_{TM} = \{\langle M, w \rangle \mid \text{TM } M \text{ halts on input } w\}$
- $D_{TM} = \{\langle M \rangle \mid \text{TM } M \text{ halts on every input }\}$
- $E_{TM} = \{\langle M \rangle \mid L(M) = \emptyset \text{ for TM } M\}$
- $One_{TM} = \{\langle M \rangle \mid M \text{ is a TM}, |L(M)| = 1\}$
- $Fin_{TM} = \{\langle M \rangle \mid L(M) \text{ is finite for TM } M\}$

Language A_{TM} encodes the *acceptance problem* of Turing machines: given any Turing machine M and any input w, does M accept w? This is one of the first problems of Turing machines shown to be undecidable, that is, there exists no algorithm to solve the acceptance problem.

Language H_{TM} encodes the well-known halting problem of Turing machines. Alan Turing proved in 1936 that a general algorithm to solve the halting problem for all possible $\langle M, w \rangle$ pairs cannot exist. We will show that both A_{TM} and H_{TM} are recognizable but not decidable.

Language D_{TM} is the encoding of all Turing machines that halt on every input. These Turing machines are called deciders or algorithms. D_{TM} encodes a problem called *the universal halting problem*, which is harder than the problem encoded by H_{TM}.

Language E_{TM} encodes the *emptiness problem* of Turing machines. Unlike A_{TM} and H_{TM}, E_{TM} is not recognizable.

One_{TM} (or Fin_{TM}) encodes a problem less popular: Turing machines that accept exactly one string (or a finite number of strings). You may create your own languages for any decision problem.

For the convenience of discussion, we assume further that $\langle M \rangle$ can be any string, including ϵ (the empty string). Some strings are the codes of "good" (i.e., well-defined) Turing machines; some are the codes of "bad" Turing machines, i.e., those under construction or with garbled characters. We assume that bad Turing machines have zero states, move zero steps on any input, and accept nothing, just like "bad" programs cannot be executed.

For understanding $x = \langle M, w \rangle$, we may assume that $x = \langle M \rangle \# w$ if M is well-defined, where # is a special character not used by M and w. If x contains more than one copy of #, then $\langle M \rangle$ is the code of a bad Turing machine and w is the suffix after the last #. If x contains no #, we assume $\langle M \rangle = \epsilon$ and $x = w$.

By the above assumptions, $\langle M \rangle$ or $\langle M, w \rangle$ can be any string of Σ^*. Now, the complement of any language in the above list makes sense. Let $\overline{L} = \Sigma^* - L$:

$$\overline{A_{TM}} = \{\langle M, w \rangle \mid \text{TM } M \text{ does not accept input } w\}$$
$$\overline{H_{TM}} = \{\langle M, w \rangle \mid \text{TM } M \text{ does not halt on input } w\}$$
$$\overline{D_{TM}} = \{\langle M \rangle \mid \text{TM } M \text{ loops on some input }\}$$
$$\overline{E_{TM}} = \{\langle M \rangle \mid L(M) \neq \emptyset \text{ for TM } M\}$$
...

By the assumption, the codes of bad Turing machines do not appear in A_{TM}, $\overline{H_{TM}}$, $\overline{E_{TM}}$, and $\overline{D_{TM}}$.

11.3.2 Decidable Problems

Since every decision problem can be encoded as a formal language $L \subseteq \Sigma^*$, we wish to design a Turing machine M which can answer the question of $w \in L$ or not, for any $w \in \Sigma^*$. This is only possible when L is decidable; if L is recognizable but not decidable, we can only answer the question when M halts on w; if L is not recognizable, then no such Turing machines can exist.

Definition 11.3.1 (Decidable and Computable Problem) A decision problem is said to be *decidable* if the language of its encoding is decidable. A decision problem is said to be *computable* if its encoding is recognizable.

The above formal definition is consistent with the definition given in Sect. 1.3.2, where we divided all decision problems into three disjoint groups: (*a*) decidable, (*b*) computable but undecidable, and (*c*) uncomputable. Recognizable languages correspond to computable decision problems. The following table summarizes the terminology used for the three domains under study: formal languages, decision problems, and general functions, along with special names of Turing machines for each domain.

Domain	Distinction	Name of Turing machines
Formal languages	Decidable	Decider
	Recognizable	Recognizer
Decision problems	Decidable	Decider
	Computable	Procedure
General functions	Total computable	Algorithm
	Computable	Procedure

All the decision problems we have met in the courses on data structures and algorithms, where their complexity is bound by a big O function, belong to the class

11.3 Decidability of Problems

of decidable problems. In computability theory, a decider for a decision problem is also called a *decision procedure*, or an *algorithm*, which coincides with the definition of *algorithm* in any course on data structures and algorithms: a finite sequence of instructions which can be executed in a finite number of steps for any input.

Recall that a *decision function* (Proposition 1.3.7) is a function that takes a natural number as input and returns 0 or 1 as output. Any decision function $f : \mathcal{N} \to \{0, 1\}$ is a decision problem about the natural numbers: does $f(n) = 1$ for input $n \in \mathcal{N}$? Apparently, a decision function is a decision problem as well as a general function. Hence, a decision function can be both decidable (by a decider) and total computable (by an algorithm). Note that our total computable function is called *computable* by Michael Sipser (Definition 5.17 of [2]). We avoid using "computable" for "total computable," because the Turing thesis applies not only to total functions but also to partial functions.

Some decision problems are decidable; some are computable but undecidable; some are uncomputable. According to Proposition 1.3.7, the set of all decision functions is uncountable. That is, there are more decision functions than the available Turing machines, which are countable. For any decision function $f : \mathcal{N} \to \{0, 1\}$, it is not always possible to have a Turing machine M_f such that M accepts i iff $f(i) = 1$ for any natural number i. There exist many decision functions which cannot be computed by any computing device in view of the Church-Turing thesis.

Here, we are interested in decision problems about Turing machines. When a Turing machine is encoded as a string or as a natural number, these decision problems are also decision functions.

Example 11.3.2 Let $S10_{TM} = \{\langle M \rangle \mid M \text{ has at least ten states}\}$. Then, $S10_{TM}$ is decidable because we can design a Turing machine which counts the number of different states in $\langle M \rangle$. It takes no more than the linear time to count, and if the count goes beyond 10, stop with "accept'; otherwise, stop with "reject." □

Example 11.3.3 Let $B10_{TM} = \{\langle M \rangle \mid M \text{ moves at least ten steps on any input}\}$.

$B10_{TM}$ is decidable because we need to consider as input only strings of length less than or equal to 10, and for each input, we need to simulate M at most ten steps. So, the total amount of effort is finite. That is, we do not need to consider strings of length more than 10 because you need at least ten steps to read the 11th symbol.

Let M_B be the decider that we will construct for $B10_{TM}$. The basic algorithm of M_B works as follows: the tape of M_B is divided into three parts separated by the special symbols # and $. Part 1 (from the first cell to #) stores the input string $\langle M \rangle$, where $M = (Q, \Sigma, \Gamma, \delta, q_0, q_a)$, and will never change. If $\langle M \rangle$ is the code of a bad Turing machine, stop with "reject."

Part 2 (between # and $) has 11 cells and stores in turn the input string $w \in \Sigma^*$ of M for every possible w such that $|w| \leq 10$. We assume that the strings in Σ^* are generated by the canonical order (Definition 1.3.5). Thus, if $\Sigma = \{0, 1\}$, then the strings are in this order: $\epsilon, 0, 1, 00, 01, 10, 11, 000, \ldots$. Assuming this order, we say the *successor* of ϵ is 0, the *successor* of 0 is 1, and so on.

Part 3 (after $) will be the space for simulating the moves of M.

M_B = "On input $\langle M \rangle$, M is a Turing machine.
1. Scan the input; if $\langle M \rangle$ is a bad code, **return** 0.
   Write # at the end of $\langle M \rangle$; move right 11 steps and write $.
   Treat the string in part 2 (between # and $, excluding the symbol ⊔)) as w.
   Initially, $w = \epsilon$, the empty string.
2. Move to the beginning of part 3 and write q_0 of M after $.
3. Copy string w in part 2 to part 3, after q_0.
   Now, part 3 contains $q_0 w$, an initial configuration of M.
4. Zigzag between part 1 and part 3, simulate M on w:
   4.1 Use the states of M_B to remember the pair (q, a) in part 3.
   4.2 Find the delta definition for (q, a) in part 1.
   4.3 If no $\delta(q, a)$ is found, **return** 0. // M hasn't moved ten times.
   4.4 Use the states of M_B to remember $\delta(q, a) = (p, b, D)$, $D \in \{L, R\}$.
   4.5 Move right to find the state of M in part 3 and update
        the configuration of M according to $\delta(q, a) = (p, b, D)$.
   4.6 If 4.5 has been done less than ten times, go to 4.1.
5. Erase the content of part 3, move to part 2, and replace w by its successor.
6. If w exceeds 10 symbols, **return** 1, else go to 2."

Lines 2–3 write the initial configuration $q_0 w$ of M in part 3, preparing for simulation of M on w. Lines 4.1–4.6 simulate the moves of M for ten moves. If this is possible, M_B continues to line 5 for a new w: M_B erases the current configuration of M in part 3 and replaces w by the successor of w in part 2 (line 5). If the move of M is blocked before moving 10 steps, M_B rejects (line 4.3). At line 6, M_B checks if $|w| > 10$. If yes, it means M has exhausted all strings x, $|x| \leq 10$, and has not found any one on which M cannot move at least ten steps. So, M_B stops with "accept"; otherwise, it starts to simulate M on new w by going to line 2 (line 6). □

We would like to point out that lines 4.1–4.5 of the above example describe how a move of M is simulated by M_B. Later, when we say "M' simulates M," we refer to this example for the details of the simulation. You cannot use the states of any Turing machine to remember the information like a^i, where i is any number. However, we may use states of M_B to remember the δ function of M because each $\delta(q, a) = (p, b, D)$ contains only five symbols and there are only a finite number of them (both the set of states and the set of tape symbols are finite).

11.3.3 Undecidable Problems

Corresponding to Gödel's first incompleteness theorem, the following theorem provides the first formal proof of an undecidable problem in this book.

Theorem 11.3.4 *The acceptance problem of Turing machines, A_{TM}, is undecidable.*

11.3 Decidability of Problems

Proof Assume that A_{TM} is decidable, then there exists a hypothetical decider H for A_{TM}. H accepts $\langle M, w \rangle$ iff M accepts w. In particular, H accepts $\langle M, \langle M \rangle \rangle$ iff M accepts $\langle M \rangle$.

Using H, we construct another decider Di, which checks if a Turing machine M does not accept its own code:

Di = "On input $\langle M \rangle$, M is a Turing machine.
1. Simulate H on $\langle M, \langle M \rangle \rangle$ until H halts.
2. **return** $1 - H(\langle M, \langle M \rangle \rangle)$."

$Di(\langle M \rangle) = 1 - H(\langle M, \langle M \rangle \rangle)$ means $Di(\langle M \rangle) = \neg H(\langle M, \langle M \rangle \rangle)$ if we view 1 as *true* and 0 as *false*. That is, if H accepts $\langle M, \langle M \rangle \rangle$, Di rejects $\langle M \rangle$; if H rejects $\langle M, \langle M \rangle \rangle$, Di accepts $\langle M \rangle$. Hence, Di accepts $\langle M \rangle$ iff H rejects $\langle M, \langle M \rangle \rangle$, iff M rejects $\langle M \rangle$, iff $M(\langle M \rangle) = 0$. Hence, Di is a decider for

$$X = \{\langle M \rangle \mid M \text{ is a Turing machine}, M(\langle M \rangle) = 0\}$$

which is decidable, because Di halts on any input.

Let us check if $\langle Di \rangle \in X$ and there are only two cases:

- $\langle Di \rangle \in X$: Every member $\langle M \rangle$ of X, including $\langle Di \rangle$, satisfies $M(\langle M \rangle) = 0$. Replacing M by Di, we have $Di(\langle Di \rangle) = 0$, i.e., Di rejects $\langle Di \rangle$. On the other hand, from $Di(\langle M \rangle) = 1 - H(\langle M, \langle M \rangle \rangle)$, we have $H(\langle Di, \langle Di \rangle \rangle) = 1 - Di(\langle Di \rangle) = 1 - 0 = 1$, i.e., Di accepts $\langle Di \rangle$, a contradiction.
- $\langle Di \rangle \notin X$: That means $Di(\langle Di \rangle) = 1$ by the definition of X; hence, $H(\langle Di, \langle Di \rangle \rangle) = 1$. From $Di(\langle M \rangle) = 1 - H(\langle M, \langle M \rangle \rangle)$, we have $Di(\langle Di \rangle) = 1 - H(\langle Di, \langle Di \rangle \rangle) = 1 - 1 = 0$, a contradiction.

In both cases, we have a contradiction, so Di cannot exist. Di is constructed from H; hence, H cannot exist. H comes from the assumption that A_{TM} is decidable. Thus, A_{TM} is undecidable. □

The above proof is analogous to that of "Barber's paradox" in Sect. 1.3.1. The Turing machine Di, which accepts the codes of those Turing machines that do not accept their own encoding, corresponds to "the Barber who shaves all those who do not shave themselves." Both proofs originated from Cantor's diagonal method, which uses a matrix. In our proof, the rows of the matrix are indexed by $\langle M_i \rangle$, the columns are indexed by w_i, and both $\langle M_i \rangle$ and w_i are members of Σ^*. By our assumption, $w_i = \langle M_i \rangle$. To see this in detail, we recommend [2]. The above proof is analogous to the proof of Gödel's first incompleteness theorem as Gödel's proof is also based on Cantor's diagonal method.

11.3.4 Reduction

Once we have found one undecidable language, other undecidable languages can be found through reduction.

Definition 11.3.5 Let X and Y be two formal languages. We say Y is *reduced* to X if we can construct a decider or recognizer B for Y, using a decider or recognizer A for X as a component of B, such that one of the following is true:

- If X is decidable, then Y is decidable.
- If X is recognizable, then Y is recognizable.

The concept of reduction can be generalized to computation problems X and Y: when we want to construct an algorithm (or procedure) B for Y, we may use the algorithm (or procedure) A for X as a component of B. Intuitively, reduction resembles the idea of "procedure call" in computer programming. As an experienced programmer, one has seen numerous procedure calls in object-oriented programming as a way of problem-solving. That is, if we have a method A for problem X, to solve problem Y, we may construct method B from A for Y. For instance, let X be "measuring the diameter of a circle" and Y be "measuring the area of a circle," then method A of X can be used for constructing method B of Y. That is exactly the meaning of reducing Y to X.

Suppose we have a decider (recognizer) A to answer the question of whether $x \in X$ or not. Using A, we then build a decider (recognizer) B for answering the question of $y \in Y$ for any y. Inside B, we may create an input x and call A on x. The output $A(x)$ is used to create the output of B. If decider (recognizer) B works correctly for Y, that is, $B(y)$ returns true iff $y \in Y$, then Y is successfully reduced to X.

There are two special cases of reduction. In the first case, decider (recognizer) B with input y uses an algorithm f which takes input y and produces $x = f(y)$. B calls A with input x and returns $A(x)$ as the output of $B(y)$, as illustrated in Fig. 11.3a. In this case, $y \in Y$ iff $x = f(y) \in X$, and the reduction is called *mapping reduction*, because f maps yes-instances of Y to yes-instances of X and no-instances of Y to no-instances of X.

Proposition 11.3.6 *If Y is reduced to X by mapping reduction, then $\overline{Y}$ is reduced to $\overline{X}$.*

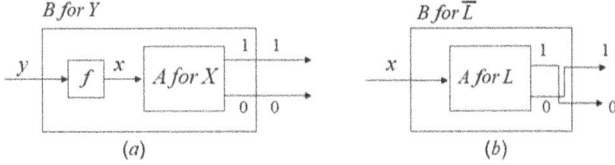

Fig. 11.3 Illustration of two special cases of reduction: (a) mapping reduction; (b) negation

11.3 Decidability of Problems

Proof The same algorithm f can be used for the reduction from $\overline{Y}$ to $\overline{X}$. □

In the second case of special reduction, the same input is used by both A and B. For instance, to show that decidable languages are closed under complement, if we have a decider A for a formal language L, we may construct B for $\overline{L} = \Sigma^* - L$ such that $B(x) = \neg A(x)$, as illustrated in Fig. 11.3b.

The application of reduction in this chapter is different and can be stated as follows: if Y is reduced to X by constructing B from A and we know B does not exist for Y, then A cannot exist for X. Formally, we have the following proposition, which comes directly from the definition of reduction.

Proposition 11.3.7 *If Y is reduced to X and Y is undecidable, then X is undecidable.*

In the following proof of the halting problem of Turing machines, X is the encoding of the halting problem and Y is A_{TM}, the first known undecidable language in this book.

Theorem 11.3.8 *The halting problem of Turing machines, H_{TM}, is undecidable.*

Proof Assume H_{TM} is decidable, then there exists a hypothetical decider H for H_{TM}. Since H is a decider, H will halt on any input, and H accepts $\langle M, w \rangle$ iff M halts on w.

Using H, we can construct another decider S for A_{TM} as follows:

$S =$ "On input $\langle M, w \rangle$, M is a Turing machine and w is an input to M.
  1. Simulate H on $\langle M, w \rangle$ until H halts.
  2. If H rejects, **return** 0.   // H rejects $\langle M, w \rangle$ if M loops on w.
  3. Simulate M on w until M halts. // M will halt because H accepts $\langle M, w \rangle$.
  4. **return** $M(w)$."

At line 2, when H rejects $\langle M, w \rangle$, it means M loops on w and justifies the rejection of w by S. At line 3, we know M will halt on w. When M halts on w, S returns the output of $M(w)$ at line 4. All four lines of S will be finished in a finite number of steps, so S is a decider.

The relationship between M and S is given below:

1. M accepts w: S accepts $\langle M, w \rangle$ (line 4).
2. M rejects w: S rejects $\langle M, w \rangle$ (line 4).
3. M loops on w: S rejects $\langle M, w \rangle$ (line 2).

Thus, M accepts w iff S accepts $\langle M, w \rangle$. If H can decide if M halts on w, then S can decide if M accepts w. Thus, S is a decider for A_{TM}. Since S cannot exist, H cannot exist, either. Thus, H_{TM} must be undecidable. □

The above proof is a reduction of the second special case, where the input strings to S and H are the same.

Reduction is a powerful tool to show that a problem X is undecidable: assume X is decidable and A is a decider for X. We use A to construct another decider B for a known undecidable problem Y. Since B cannot exist, A cannot exist, and X is thus

Fig. 11.4 Illustration of a reduction from A_{TM} to H_{TM}

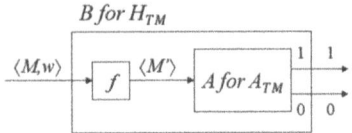

undecidable. In the proof of the above theorem, X is H_{TM}, A is H, Y is A_{TM}, and B is S. We showed that if there exists a decider for H_{TM}, then there exists a decider for A_{TM}, a contradiction.

We may also switch the roles of H_{TM} and A_{TM} by showing the following result.

Proposition 11.3.9 *If there exists a decider for A_{TM}, then there exists a decider for H_{TM}.*

Proof Suppose A_{TM} is decidable and A is its decider. We construct a decider B for H_{TM} as follows and is illustrated in Fig. 11.4.

$B = $ "On input $\langle M, w \rangle$, M is a Turing machine and w is an input to M.
  1. Call algorithm f on $\langle M \rangle$: f creates $\langle M' \rangle$ as follows:
     For any state q and tape symbol x of M, if q is q_r (the reject state), replace q_r by q_a (the accept state); if q is not q_a and $\delta(q, x)$ is not defined, add in M': $\delta'(q, x) = (q_a, x, R)$.
  2. Simulate A on $\langle M', w \rangle$ until A halts.
  3. **return** $A(\langle M', w \rangle)$."

The relationship between M and M' is given below:

1. M accepts w: M' accepts w.
2. M rejects w: M' accepts w.
3. M loops on w: M' loops on w.

Thus, M halts on w iff M' accepts w. $B(\langle M, w \rangle) = A(\langle M', w \rangle) = 1$ iff M' accepts w and M halts on w. This justifies the output of B at line 3 of B. In other words, B accepts $\langle M, w \rangle$ iff M halts on w; A accepts $\langle M', w \rangle$ iff M' accepts w. Since all the three lines of B can be done in a finite number of steps, B is a decider for H_{TM}. □

The above proof is an example of mapping reduction as we have shown that $\langle M', w \rangle \in A_{TM}$ iff $\langle M, w \rangle \in H_{TM}$.

Since A_{TM} and H_{TM} are reduced to each other, they are called *Turing equivalent*, and the class of all Turing equivalent languages (or decision problems) is called *Turing degree*, a topic out of the scope of this book.

Proposition 11.3.10 $D_{TM} = \{\langle M \rangle \mid TM\ M\ is\ a\ decider\ \}$ *is undecidable.*

Proof Suppose D_{TM} is decidable and A is its decider. We construct a decider B for A_{TM} as follows and is illustrated in Fig. 11.5

$B = $ "On input $\langle M, w \rangle$, M is a Turing machine and w is an input to M.
  1. Generate the code of Turing machine M' (see below), $\langle M' \rangle$, from $\langle M, w \rangle$.
  2. **return** $A(\langle M' \rangle)$."

11.3 Decidability of Problems

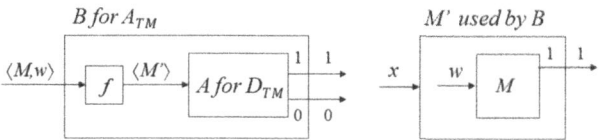

Fig. 11.5 Illustration of a reduction from A_{TM} to D_{TM}

M' = "On input x
  1. Replace x by w as new input.
  2. Simulate M on w until M halts.
  3. If $M(w) = 0$, then go to a looping state.
  4. Otherwise, **return** 1."

Note that B only creates the code of M' and does not simulate M on w. B uses the code of M to create $\langle M' \rangle$: M' will replace its input x by w on the tape of M' and then simulate M on w. If M rejects w, M' goes into an infinite loop. If M accepts w, M' accepts x. The relationship between M and M' is given as follows:

1. M accepts w: M' halts on every x.
2. M does not accept w: M' loops on every x.

Thus, M accepts w iff M' halts on every x. $B(\langle M, w \rangle) = A(\langle M' \rangle) = 1$ iff $\langle M' \rangle \in D_{TM}$ and $\langle M, w \rangle \in A_{TM}$. Hence, B is a decider for A_{TM} if A is a decider for D_{TM}. Since A_{TM} has no deciders, A cannot exist and D_{TM} must be undecidable. □

We know that both A_{TM} and H_{TM} are undecidable; are they recognizable? The answer is yes for both of them.

Proposition 11.3.11 A_{TM} *is recognizable.*

Proof We provide the following Turing machine for A_{TM}.

U = "On input $\langle M, w \rangle$, M is a Turing machine and w is an input to M.
  1. Simulate M on w until M halts.
  2. **return** $M(w)$."

It is clear that U accepts $\langle M, w \rangle$ if M accepts w; U rejects if M rejects w; U loops if M loops on w. Thus, $L(U) = A_{TM}$. □

U in the above proof is called the *universal Turing machine*, because it simulates the execution of any Turing machine on any input. If we view $\langle M, w \rangle$ as a stored program with data, U is equivalent to RASP (random-access stored-program) machines. This idea has great impact on the development of today's computers and considerable theoretical importance.

Proposition 11.3.12 H_{TM} *is recognizable.*

Proof H_{TM} can be recognized by Turing machine T.

$T=$ "On input $\langle M, w \rangle$, M is a Turing machine and w is an input to M.
1. Simulate M on w until M halts.
2. **return** 1."

T accepts $\langle M, w \rangle$ iff M halts on w, thus $L(T) = H_{TM}$. □

Proposition 11.3.13 *If both L and $\overline{L}$ are recognizable, then L is decidable.*

Proof Let A and B be recognizers of L and $\overline{L}$, respectively. We may use A and B to create a decider for L:

$T=$ "On input $\langle M, w \rangle$, M is a Turing machine and w is an input to M.
1. Simulate A on w for one move.
2. If A accepts, **return** 1.
3. Simulate B on w for one move.
4. If B accepts, **return** 0.
5. go to line 1 and continue."

For any $w \in \Sigma^*$, either $w \in L$ or $w \in \overline{L}$. So, either A accepts w or B accepts w. T will halt when one of them accepts. If $w \in L$, T will accept; otherwise, T will reject. Thus, T is a decider for L. □

The construction of T is a typical example of using recognizers as components for designing algorithms. You cannot use recognizers as deciders (like procedure calls in computer programming) because recognizers may not halt.

The above result provides an easy method for checking if a formal language is unrecognizable (see Proposition 11.4.13).

11.3.5 Rice's Theorem

We continue to hold the assumption that every string of Σ^* is the code of a Turing machine (good or bad).

Definition 11.3.14 Given a predicate $p(M)$ for all Turing machines M, let

$$L_p = \{\langle M \rangle \mid p(M) = 1, M \text{ is a Turing machine}\}$$

We say p is a *Rice property* if p satisfies the following two conditions:

1. Nontrivial: Neither $L_p = \Sigma^*$ nor $L_p = \emptyset$.
2. Language related: For any two Turing machines A and B, if $L(A) = L(B)$ and $\langle A \rangle \in L_p$, then $\langle B \rangle \in L_p$.

We say p is *positive* if $|L(M)| > 0$ whenever $p(M) = 1$; p is *nonpositive* if p is not positive.

11.3 Decidability of Problems

Obviously, if p is positive and $L(M) = \emptyset$, then $p(M) = 0$. If p is trivial, then L_p is decidable because both Σ^* and $\emptyset$ are decidable. Rice's theorem claims that L_p is undecidable if p is a Rice property.

In the beginning of this chapter, we mentioned Gödel's second incompleteness theorem: if a formal system S is consistent, then we cannot prove in S that S itself is consistent. If we regard "consistency" as a Rice property of Turing machines, then Gödel's second incompleteness theorem is generalized as Rice's theorem, which states that there are no deciders to show that a Turing machine has a Rice property or not.

Theorem 11.3.15 (Rice's Theorem) *If p is a Rice property, then L_p is undecidable.*

Proof We assume that p is positive; if this assumption fails, we take the negation of p as p in the following proof. Since the complement of a decidable language is decidable, our assumption does not invalidate the theorem.

Since L_p is nontrivial, there exists a Turing machine A such that $\langle A \rangle \in L_p$. As the first step of proof by contradiction, we assume that L_p is decidable and S is its hypothetical decider. That is, $S(\langle M \rangle) = 1$ iff $\langle M \rangle \in L_p$. Using S and A, we construct a decider T for A_{TM} as follows:

$T =$ "On input $\langle M, w \rangle$, M is a Turing machine and w is an input of M.
 1. Generate $\langle M' \rangle$, the code of M' (see below), from $\langle M, w \rangle$ and $\langle A \rangle$.
 2. **return** $S(\langle M' \rangle)$."

$M' =$ "On input x
 1. Put a special symbol # at the end of x and write w after #.
 2. Simulate M on w until M halts, and then, erase any symbol after #, including #.
 3. If M accepts w, simulate A on x until A halts and **return** $A(x)$.
 4. If M rejects w, go to a looping state."

The relationship between M and M' is given below:
1. M accepts w: $L(M') = X = L(A)$ and $p(M') = p(A) = 1$.
2. M rejects w or loops on w: M' loops, $L(M') = \emptyset$ and $p(M') = 0$.

Thus, M accepts w iff $p(M') = 1$ (or $\langle M' \rangle \in L_p$). $T(\langle M, w \rangle) = S(\langle M' \rangle) = 1$ iff $\langle M' \rangle \in L_p$ and $\langle M, w \rangle \in A_{TM}$. If S can decide $\langle M' \rangle \in L_p$, then T can decide $\langle M, w \rangle \in A_{TM}$. Since A_{TM} has no deciders, S cannot exist. The used reduction is illustrated in Fig. 11.6, where algorithm f does what in line 1 of T. □

Rice's theorem claims that any Rice property of Turing machines is undecidable. Since many properties of Turing machines are nontrivial and language related, these properties are undecidable. For instance, Rice's theorem applies to the following formal languages:

- $E_{TM} = \{\langle M \rangle \mid L(M) = \emptyset \text{ for TM } M\}$.
- $One_{TM} = \{\langle M \rangle \mid M \text{ is a TM}, |L(M)| = 1\}$.

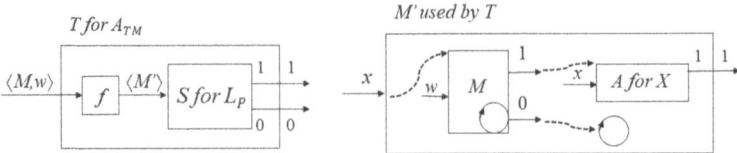

Fig. 11.6 Illustration of a reduction from A_{TM} to L_p in the proof of Rice's theorem. A dashed arrow is used to indicate a trigger for starting the simulation of a Turing machine

- $Fin_{TM} = \{\langle M \rangle \mid L(M) \text{ is finite for TM } M\}$.
- $A10_{TM} = \{\langle M \rangle \mid M \text{ accepts at least 10 strings}\}$.

The above languages are the encoding of some Rice properties about $L(M)$. For instance, E_{TM} is the encoding of property "$L(M) = \emptyset$," which is not positive. However, "$L(M) \neq \emptyset$" is a positive Rice property. By Rice's theorem, $\overline{E_{TM}}$ is undecidable. E_{TM} cannot be decidable, because decidable languages are closed under complement.

If a property is not language related, then Rice's theorem cannot be applied. For instance, Rice's theorem cannot be used to show that D_{TM} is undecidable, because there exist two Turing machines, M_1 and M_2, such that $L(M_1) = L(M_2) = \emptyset$, $\langle M_1 \rangle \in D_{TM}$ and $\langle M_2 \rangle \notin D_{TM}$, violating the "language-related" condition.

Looking back to Gödel's two incompleteness theorems introduced in Sect. 11.1.1, a formal system S which can perform common arithmetic operations on natural numbers is Turing complete because a move of any Turing machine can be simulated by arithmetic operations. If S is uncomputable, then the two incompleteness theorems are trivially true. If S is computable, by the Church-Turing thesis, S can be simulated by a Turing machine. In this case, Gödel's first incompleteness theorem is equivalent to the undecidability of A_{TM} (Theorem 11.3.4) which is the encoding of the acceptance problem of Turing machines. If the inconsistency of S is denoted by "$0 = 1$" and w is the encoding of "$1 = 0$" by Turing machine M which simulates S, then S is consistent iff $w \notin L(M)$. Let "$w \in L(M)$" be a Rice property, then this property is undecidable by Rice's theorem, an alternative proof to Gödel's second incompleteness theorem.

11.4 Computability of Counting Bijections

The concept of countable set is attributed to Georg Cantor, who made the distinction between countable and uncountable sets in 1874. The concept of computable set is younger and arose in the study of computing models in 1930s by the founders of computer science, including Gödel, Church, and Turing. In the following, we will see what will happen if we enforce that bijections are computable.

11.4.1 Properties of Counting Bijections

In combinatorics, given two sets, say X and Y, of combinatorial objects, a *bijective proof* is a technique of finding a bijection $f : X \mapsto Y$ to show that X and Y have the same size. This technique is old and useful and is mostly applied to finite sets as a way of finding a formula for the size of X if the size of Y is easier to count. Cantor used this popular technique to define the countability of an infinite set.

Example 11.4.1 The number of total functions from A to B, where A and B are finite, is known to be $|B|^{|A|} = n^m$, where $|A| = m$ and $|B| = n$. Let $F = \{f : A \mapsto B\}$ is the set of all functions from A to B, total or partial. Let $F' = \{f : A \mapsto (B \cup \{u\}) \mid f \text{ is total }\}$, where $u \notin B$. For every $f \in F$, define $f'(a) = f(a)$ if $f(a)$ is defined and $f'(a) = u$, otherwise. That is, the meaning of "$f'(a) = u$" is "$f(a)$ is undefined for $f \in F$," then the function $g : F \mapsto F'$ such that $g(f) = f'$ is bijective. So, $|F| = |F'| = |B \cup \{u\}|^{|A|} = (n+1)^m$.

Without the bijection g, to compute $|F|$, we may use the following formula:

$$|F| = \sum_{S \subseteq A} |B|^{|S|} = \sum_{k=0}^{m} \binom{m}{k} n^k = (n+1)^m$$

where $\binom{m}{k} = m!/(k!(m-k)!)$ is a binomial coefficient. $\square$

Every bijection must be total, injective, and surjective. If it is not total, its inverse is not surjective. If $f : X \mapsto Y$ is a bijection, the following properties of f are desired:

- *Counting*. Y is $\mathcal{N}$ (the set of natural numbers), and we say $f : \mathcal{N} \mapsto Y$ is a counting bijection of Y.
- *Computable*. For every $x \in X$, we compute $y = f(x) \in Y$, and for every $y \in Y$, we compute $x = f^{-1}(y) \in X$.
- *Increasing*. Let $f : \mathcal{N} \mapsto Y$ be a counting bijection of Y and $\succ$ be a well-order of Y. For any $i \in \mathcal{N}$, $f(i+1) \succ f(i)$.

By Definition 1.3.2, the inverse of a rank function is a counting bijection, which occurs in the study of countability. If you want to show an infinite set is countable, you must find a counting bijection for this set. Here are some examples of bijective proofs:

- Example 11.4.1 is one of many examples in combinatorics to show two finite sets have the same size. The bijection in this example is computable.
- In Sect. 1.3.1, we used bijective proofs to show some infinite sets are countable (Proposition 1.3.4). All bijections used in the proof of Proposition 1.3.4 are computable counting bijections. They are also increasing if we choose a well-order properly for these sets.

- We also used bijective proofs to show some uncountable sets have the same size (Proposition 1.3.7). These bijections are neither computable nor counting bijections.

The above examples suggest that bijections between countable sets are computable. In the days of Cantor, the notion of computable was not formally defined. It is reasonable to guess that Cantor used bijections as other people, assuming that counting bijections can be computed by effective methods. This assumption is deceiving because we will show that if a formal language has a computable counting bijection, then this language is recognizable. If every formal language is countable, an unrecognizable language must have an uncomputable counting bijection. To facilitate the discussion, we introduce the following concept.

Definition 11.4.2 A set S is said to be *computably countable* if either S is finite or S has a computable counting bijection.

In the above definition, if the word "computable" is dropped, we obtain the definition of *countable* (Sect. 1.3.1). From the definition, the relationship between countable and computably countable is given by the following proposition.

Proposition 11.4.3

(a) *If S is computably countable, then S is countable.*
(b) *S is not computably countable iff either S is uncountable or every counting bijection of S is uncomputable.*

In the following, we show that a formal language is computably countable iff it is recognizable. To understand the equivalence proof of computably countable and recognizable, we need the concept of *enumerator*, which is a variant of Turing machine.

11.4.2 Equivalence of Recognizable and Recursively Enumerable

An *enumerator* is a Turing machine M that runs forever once started and has an attached printer as illustrated in Fig. 11.7. M can use that printer as an output device to print strings. The set of printed strings, denoted by $E(M)$, is the language *enumerated* by M. A formal language L is called *recursively enumerable* if there exists an enumerator M such that $L = E(M)$. Note that an enumerator does not have an input, so, the tape is purely working space. The printer can be formally defined as a write-only tape with the tape head going one direction. The following result is known for enumerators:

Theorem 11.4.4

(a) *If a formal language L is recursively enumerable, then L is recognizable.*

(b) *If L is recognizable, then L is recursively enumerable, and there exists an enumerator M such that $L = E(M)$ and every printed string by M is unique.*

Proof
(a) If L can be enumerated by an enumerator M, we may construct a recognizer S to check if $w \in L(S)$ by simulating M: whenever M prints a string x, S compares x with the input string w. If $x = w$, S accepts; otherwise, S continues the simulation. It is easy to show that $w \in L(S)$ iff $w \in E(M)$; thus, $L(S) = E(M)$ is recognizable.

(b) Since L is recognizable, let $L = L(S)$ for recognizer S. We will use S to construct an enumerator M such that $L = E(M)$ and every printed string by M is unique.

Let both $f : \mathcal{N} \mapsto \mathcal{N} \times \mathcal{N}$ and $\gamma : \Sigma^* \mapsto \mathcal{N}$ be bijections (see Proposition 1.3.4). M will work as follows:

$M =$ "1. $k := 0$.
    2. $\langle i, j \rangle := f(k)$; $w_j := \gamma^{-1}(j)$.
    3. Simulate S on w_j for i steps.
    4. If S accepts w_j on the ith step, **print** w_j.
    5. $k := k + 1$; go to line 2."

For any $x \in L(S)$ and $j = \gamma(x)$, S will take a fixed i steps to accept x. Hence, $\langle i, j \rangle$ is unique for $x \in L(S)$ and corresponds to a unique k, because $\langle i, j \rangle = f(k)$. In the above M, each $k \in \mathcal{N}$ is considered exactly once, so each $x \in L(S)$ will be printed out once by M. □

The above theorem shows the equivalence of "recognizable" and "recursively enumerable." If an enumerator can print strings increasingly in the canonical order (Definition 1.3.5), then the enumerated language is decidable [2].

Theorem 11.4.5 *A formal language L is decidable iff there exists an enumerator M for L such that the printed strings are increasing in the canonical order.*

This theorem is an exercise problem of [2] (Exercise 3.18, page 190).

11.4.3 Equivalence of Recognizable and Computably Countable

If a formal language L is finite, then it is both countable and decidable. In the following discussion, we will focus on the case when L is infinite. The

Fig. 11.7 Illustration of an enumerator which is a Turing machine with a printer

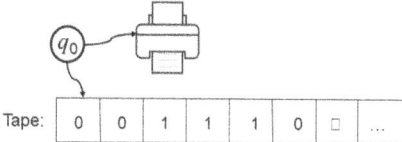

following two lemmas show that a formal language is computably countable iff it is recognizable. Thus, recognizable, recursively enumerable, and computably countable are equivalent concepts even though they have different definitions.

Lemma 11.4.6 *Every recognizable formal language is computably countable.*

Proof Suppose L is recognizable. The case when L is finite is trivial. If L is infinite, by Theorem 11.4.4(b), we have an enumerator M such that $L = E(M)$ and every printed string by M is unique, then $E(M)$ is countable because the order in which the strings are printed by M defines a computable bijection $f : \mathcal{N} \mapsto E(M)$. That is, for $n \in \mathcal{N}$, we compute $f(n)$ by algorithm $A(n)$ as follows:

> **Algorithm** $A(n)$: Let $c := 0$ and simulate M. When a string x is printed by M, check if $c = n$. If yes, return x; otherwise $c := c + 1$ and continue the simulation.

Algorithm A will terminate because $E(M)$ is infinite and c, which records the number of printed strings by M, will reach n eventually. $A(n)$ can be modified as $A(w)$ to compute $f^{-1}(w)$ for $w \in E(M)$: instead of checking $c = n$, check if $x = w$ and, if yes, return c. Since both f and f^{-1} are total functions, f is a bijection (Proposition 1.3.1). Algorithm $A(n)$ is the evidence that f is computable. □

Lemma 11.4.7 *Every computably countable formal language is recognizable.*

Proof The case when L is finite is trivial. If L is computably countable, then there exists a computable bijection $f : \mathcal{N} \mapsto L$, Thus, we may use f to design an enumerator M that prints strings $f(0), f(1), f(2)$, and so on. It is evident that $E(M) = L$. By Theorem 11.4.4(a), L is recognizable. □

The following theorem extends the above result from formal languages to general sets.

Theorem 11.4.8 *A set is computably countable iff it is computable.*

Proof Let S be any set. The case when S is finite is trivial. If S is computably countable, then there exists a computable bijection $f : \mathcal{N} \mapsto S$. Hence, S can be represented by the formal language $\{\langle f(0)\rangle, \langle f(1)\rangle, \ldots\}$, where $\langle f(i)\rangle$ is a string representing $f(i)$. By Lemma 11.4.7, S is recognizable, thus, computable.

When S is not computably countable, there are two cases to consider: S can be represented by a formal language or not. If yes, then S is unrecognizable by Lemma 11.4.6, thus uncomputable. If not, then S cannot be recognized by any Turing machine, thus uncomputable. □

Proposition 11.4.9 *If X is countably infinite, then*

1. *X is decidable iff there exists a computable and increasing bijection $f : \mathcal{N} \mapsto X$.*
2. *X is computable iff there exists a computable bijection $f : \mathcal{N} \mapsto X$.*
3. *X is uncomputable iff every bijection $f : \mathcal{N} \mapsto X$ is uncomputable.*

Proof By Theorem 11.4.8, X is computable iff X is computably countable. (3) comes from Proposition 11.4.3, although the existence of an uncomputable counting bijection is unknown. (2) is logically equivalent to (3). For (1), X is decidable iff

11.4 Computability of Counting Bijections

the elements of X can be enumerated in increasing order. This enumeration defines an increasing and computable bijection as in the proof of Lemma 11.4.6, and an increasing and computable bijection will produce an enumerator that enumerates elements in increasing order as in the proof of Lemma 11.4.7. □

The above proposition shows that the properties of counting bijection $f : \mathcal{N} \mapsto X$ are crucial to dictate if X is decidable or computable. In particular, if X is computable but undecidable, then there must exist a computable but nonincreasing counting bijection of X.

For any recognizable language L, there exists a Turing machine M such that

$$L = \{w \in \Sigma^* \mid M(w) = 1\}.$$

That is, L is recognizable or computable because $M(w)$ is computable, where M is viewed as a decision function (see Definition 11.2.9). From Theorem 11.4.8, we have the following result.

Corollary 11.4.10 *Let $p : \mathcal{N} \mapsto \{0, 1\}$ be a decision function and $S = \{x \in \mathcal{N} \mid p(x) = 1\}$.*

(a) S is total computable iff $p(x)$ is total computable.
(b) S is computably countable iff $p(x)$ is computable.
(c) S is not computably countable iff $p(x)$ is uncomputable.

The above result shows that, when $p(x)$ becomes hard to compute, S may be not computably countable. Some people take for granted that $p(x)$ is computable and arrive at the conclusion that S is computably countable. If $p(x)$ is a first-order formula, then the existence of uncomputable sets gives us uncomputable first-order formulas. In the next section, we will show the goal of this chapter: some first-order formulas (e.g., $p(x)$) are uncomputable and some decision problems in first-order logic are uncomputable.

Many formal languages are unrecognizable, and any unrecognizable language cannot be computably countable by Lemma 11.4.7. That is, Lemmas 11.4.6 and 11.4.7 provide an alternative way to show if a language is computably countable or not. Consider the following encoding of some decision problems:

- $G_{TM} = \{\langle M \rangle \mid$ Turing machine M is well-defined $\}$
- $E_{TM} = \{\langle M \rangle \mid L(M) = \emptyset$ for Turing machine $M\}$
- $N_{TM} = \{\langle M \rangle \mid L(M) \neq \emptyset$ for Turing machine $M\}$
- $All_{TM} = \{\langle M \rangle \mid L(M) = \Sigma^*$ for Turing machine $M\}$
- $D_{TM} = \{\langle M \rangle \mid$ Turing machine M is a decider $\}$
- $Al_{TM} = \{\langle M \rangle \mid$ Turing machine M is an algorithm $\}$

They are formal languages since $\langle M \rangle \in \Sigma^*$. Σ^* is decidable because we have a decider which accepts every input. G_{TM} is the encoding of all well-defined Turing machines. G_{TM} is decidable by the assumption that we have an efficient decoder to check if $\langle M \rangle$ comes from a well-defined M. By Lemma 11.4.6, G_{TM} is computably countable, hence countable.

E_{TM}, N_{TM}, and All_{TM} are the encoding of all Turing machines that accept, respectively, nothing, something, and everything. N_{TM} is the complement of E_{TM}. It is known that N_{TM} is recognizable and E_{TM} is unrecognizable (Proposition 11.4.15). By Lemma 11.4.6, N_{TM} is computably countable. By Lemma 11.4.7, E_{TM} is not computably countable. Both All_{TM} and its complement are unrecognizable (Proposition 11.4.17), hence, not computably countable.

Proposition 11.4.11 *The set of all deciders is not computably countable; hence, D_{TM} is unrecognizable.*

Proof Assume D_{TM} is computably countable with a computable bijection $h : \mathcal{N} \mapsto D_{TM}$ such that $h(i) = \langle M_i \rangle$ for each $i \in \mathcal{N}$. Using h and γ (Definition 1.3.5), we construct the following Turing machine X:

X = "On input $w \in \Sigma^*$
  1. Compute $i = \gamma(w)$ and $h(i) = \langle M_i \rangle \in D_{TM}$.
  2. Simulate M_i on w until M_i halts. // M_i is a decider.
  3. **return** $1 - M_i(w)$." // i.e., $X(w) = \neg M_i(w)$.

Since every step of X stops, $\langle X \rangle \in D_{TM}$. Let $X = M_k$ for some $k \in \mathcal{N}$. However, $X(w_k) \neq M_k(w_k)$, a contradiction to $X = M_k$. So, the assumption is wrong and D_{TM} is not computably countable. By Lemma 11.4.6, D_{TM} is unrecognizable. □

Proposition 11.4.12 *The set of all algorithms is not computably countable; hence, Al_{TM} is unrecognizable.*

Proof The proof is analogous to that of Proposition 11.4.11, where line 3 of X is changed to "**return** $1 + M_i(w)$." □

Note that $D_{TM} \subset G_{TM}$ and G_{TM} is countable. If D_{TM} is countable, then every counting bijection of D_{TM} is uncomputable (Proposition 11.4.3(b)).

11.4.4 Identifying Unrecognizable Languages

To show a language is unrecognizable, we may prove that this language is not computably countable, and D_{TM} is such an example (Proposition 11.4.11). In practice, we have several methods to show the unrecognizability of a language, and these methods can be used to show that a language is not computably countable. In the following, we examine these methods.

Proposition 11.4.13 *If a formal language L is recognizable but undecidable, then $\overline{L}$ is not recognizable.*

Proof This is a logical consequence of Proposition 11.3.13. □

Applying the above proposition to A_{TM} and H_{TM} with Propositions 11.3.11 and 11.3.12, we have the following result.

Proposition 11.4.14 $\overline{A_{TM}}$ *and* $\overline{H_{TM}}$ *are not recognizable.*

11.4 Computability of Counting Bijections

Rice's theorem is very useful for showing some languages are undecidable. If p is a Rice property, then $L_p = \{\langle M \rangle \mid p(M) = 1\}$ is undecidable. If we also know that L_p is recognizable, then $\overline{L_p}$ must be unrecognizable by Proposition 11.4.13. For instance, E_{TM} can be shown unrecognizable this way, because $\overline{E_{TM}}$ is recognizable (an exercise).

Proposition 11.4.15 E_{TM} is unrecognizable.

Analogous to Proposition 11.3.7, we have the following result from Definition 11.3.5:

Proposition 11.4.16 *If Y is reduced to X and Y is unrecognizable, then X is unrecognizable.*

The above proposition is very useful to show some formal languages are unrecognizable.

Proposition 11.4.17 *Both All_{TM} and $\overline{All_{TM}}$ are unrecognizable.*

Proof We will first work on All_{TM}. Assume All_{TM} is recognizable and A is its recognizer. We reduce $\overline{A_{TM}}$ to All_{TM} by constructing recognizer B for $\overline{A_{TM}}$ as follows, assuming the initial state of M' is not the accept state:

B = "On input $\langle M, w \rangle$, M is a Turing machine and w is an input to M.
  1. Generate the code of Turing machine M' (see below), $\langle M' \rangle$, from $\langle M, w \rangle$.
  2. **return** $A(\langle M' \rangle)$."

M' = "On input x
  1. Simulate M on w for at most $|x|$ moves.
  2. If M accepts w in line 1, go to a loop state.
  3. **return** 1."

The relationship between M and M' is given as follows:

1. M accepts w in n moves: M' loops on every x if $|x| \geq n$ and accepts x if $|x| < n$.
2. M does not accept w: M' accepts every x.

Thus, M does not accept w iff M' accepts every x. $B(\langle M, w \rangle) = A(\langle M' \rangle) = 1$ iff $\langle M' \rangle \in All_{TM}$ and $\langle M, w \rangle \in \overline{A_{TM}}$. Hence, B is a recognizer for $\overline{A_{TM}}$ if A is a recognizer for All_{TM}. Since $\overline{A_{TM}}$ has no recognizers, A cannot exist and All_{TM} must be unrecognizable.

Now on $\overline{All_{TM}}$. Assume it is recognizable and A is its recognizer. We will use the same B in the proof of Proposition 11.3.10 to reduce $\overline{A_{TM}}$ to $\overline{All_{TM}}$. The only difference is that now A is a hypothetical recognizer for $\overline{All_{TM}}$ instead of a decider for D_{TM}, and B is a recognizer for $\overline{A_{TM}}$. The relationship between M and M' is given as follows:

1. M accepts w: M' accepts every x and $L(M') = \Sigma^*$.
2. M does not accept w: M' does not accept any x and $L(M') = \emptyset$.

Thus, M does not accept w iff $L(M') \neq \Sigma^*$. $B(\langle M, w \rangle) = A(\langle M' \rangle) = 1$ iff $L(M') \neq \Sigma^*$ and $\langle M, w \rangle \in \overline{A_{TM}}$. Hence, B is indeed a recognizer for $\overline{A_{TM}}$, and $\overline{A_{TM}}$ is reduced to $\overline{All_{TM}}$. By Proposition 11.4.14, $\overline{A_{TM}}$ is unrecognizable. By Proposition 11.4.16, $\overline{All_{TM}}$ must be unrecognizable. □

Note that the two reductions used in the above proof are mapping reduction. The first reduction comes from Prof. Tian Liu of Beijing University. The same mapping of the second reduction was used to show that A_{TM} is reduced to D_{TM} (Proposition 11.3.10). Applying Proposition 11.3.6, $\overline{A_{TM}}$ is reduced to $\overline{D_{TM}}$. Applying Proposition 11.4.16, we have the following result.

Proposition 11.4.18 $\overline{D_{TM}}$ *is unrecognizable.*

All_{TM} is our first example to the statement "both L and $\overline{L}$ are unrecognizable" and D_{TM} is our second example.

In summary, we have three methods to show that a formal language X is unrecognizable:

1. If X is not computably countable, then X is unrecognizable (Lemma 11.4.6).
   Example: D_{TM}, Al_{TM}
2. If X is recognizable but undecidable, then $\overline{X}$ is unrecognizable (Proposition 11.4.13).
   Example: $E_{TM}, \overline{A_{TM}}, \overline{H_{TM}}$
3. If Y is reduced to X and Y is unrecognizable, then X is unrecognizable (Proposition 11.4.16).
   Example: $All_{TM}, \overline{All_{TM}}, \overline{D_{TM}}$

Combining the above results with Lemma 11.4.7, we have the following result.

Proposition 11.4.19 *The following formal languages are not computably countable:*

$$E_{TM}, All_{TM}, \overline{All_{TM}}, \overline{D_{TM}}, \overline{A_{TM}}, \overline{H_{TM}}.$$

11.5 Turing Completeness

According to the Church-Turing thesis, any computing model can be simulated by a Turing machine. On the other hand, it becomes easy to design a new computing model which has the same power as a Turing machine, as long as the new computing model is shown to do what a Turing machine can do. The concept of *Turing completeness* just expresses this property: if a computing model has the ability to simulate a Turing machine, it is *Turing complete*. Both Gödel's recursive functions and Church's λ-calculus are Turing complete.

Example 11.5.1 A two-stack machine is a variation of Turing machine in which the tape is replaced by two stacks: the first stack stores all the symbols from the tape

11.5 Turing Completeness

head (inclusive) to right (excluding the trailing blanks); the second stack stores all the symbols from the tape head (exclusive) to left. The state information remains the same. It is easy to show that the two-stack machine is Turing complete by simulating all the Turing moves by the operations on stacks. Initially, the input string is stored in the first stack with the first symbol on the top of the stack. The simulation of a Turing move goes as follows: a symbol a is popped off from the first stack; if the stack is empty, use the blank symbol. If $\delta(q, a) = (p, b, R)$, push b into the second stack; if $\delta(q, a) = (p, b, L)$, push b into the first stack and then move the top symbol from the second stack (if the second stack is empty, use the blank symbol) to the first stack by stack operations. □

The significance of being Turing complete is that all the undecidable properties of Turing machines will be inherited by the new computing model. Thus, the halting problem of a two-stack machine is undecidable, because if we have an algorithm to tell the termination of a two-stack machine, this algorithm can be used to show the termination of a Turing machine. Similarly, if we can show the Turing completeness of a logic, we can derive the same result.

11.5.1 Turing Completeness of Prolog

Proposition 11.5.2 *Prolog is Turing complete.*

Proof We show how to simulate a Turing machine by a Prolog program, using the idea of two-stack machines. The content of the Turing machine tape is split into two lists, L and R, which correspond to the two stacks in a two-stack machine. Initially, $L = []$ and R contains the input string. We transform L to L' and R to R' according to moves of the Turing machine, until the Turing machine halts. The transformation is illustrated by a Prolog program for the Turing machine in Example 11.2.4.

```
% M1 checks a^jb^k for j =< k. Its delta definition is defined below.
delta(q0, bl, qa, bl, right).
delta(q0, a, q1, x, right).
delta(q0, b, q3, b, right).
delta(q0, y, q3, y, right).
delta(q1, a, q1, a, right).
delta(q1, y, q1, y, right).
delta(q1, b, q2, y, left).
delta(q2, y, q2, y, left).
delta(q2, a, q2, a, left).
delta(q2, x, q0, x, right).
delta(q3, b, q3, b, right).
delta(q3, y, q3, y, right).
delta(q3, bl, qa, bl, left).

% General code for all Turing machines
turing(W, Tape) :-                   % W: input string, Tape: final tape
    write('  '), write(q0), write(W), nl,   % initial configuration
```

```
    move([], L, q0, W, R),         % ([], q0, W) |-* (rev(L), qa, R)
    reverse(L, Lr),                 % reverse L to Lr
    append(Lr, [qa|R], Tape).       % append Lr, qa, and R to Tape

% move(L1, L2, Q, R1, R2): move from (L1, Q, R1) to (L2, P, R2)
move(L, L, qa, R, R) :- !.          % halt on qa, the accept state
move(L0, L, Q, R0, R) :-            % simulate one TM move
  current(R0, A, Rest),             % load the current symbol in A
  delta(Q, A, P, B, Direct),        % delta(Q, A) = (P, B, Direct)
  result(Direct, L0, L1, B, Rest, R1),  % update L and R
  write('|- '), write(rev(L1)), write(P), write(R1), nl,
  move(L1, L, P, R1, R).            % continue to move

% current(R1, Sym, R2): pick current symbol Sym from R1
current([], bl, []).                % current symbol is the blank bl
current([Sym|Rest], Sym, Rest).     % current symbol is Sym

% result(dir, L1, L2, B, R1, R2): B is the symbol pointed by tape head
%      According to "dir", stack1 changes from L1 to L2.
%                          stack2 changes from R1 to R2,
result(left, [C|L], L, B, R, [C,B|R]). % move top symbol of L to R
result(right, L, [B|L], B, R, R).      % push B at the top of L

?- turing([a,a,b,b], Tape).
   q0[a,a,b,b]
|- rev([x])q1[a,b,b]
|- rev([a,x])q1[b,b]
|- rev([x])q2[a,y,b]
|- rev([])q2[x,a,y,b]
|- rev([x])q0[a,y,b]
|- rev([x,x])q1[y,b]
|- rev([y,x,x])q1[b]
|- rev([x,x])q2[y,y]
|- rev([x])q2[x,y,y]
|- rev([x,x])q0[y,y]
|- rev([y,x,x])q3[y]
|- rev([y,y,x,x])q3[]
|- rev([y,x,x])qa[y,bl]
Tape = [x,x,y,qa,y,bl].
```

This illustrates how a sequence of Turing machine configurations is implemented in Prolog as a relation between the configurations.

For any Turing machine, we may specify its delta function as illustrated in the above example, and the code after the definition of delta can be reused. □

The above proof shows that any Turing machine can be simulated by the above code as long as the memory used by Prolog has no limit. The Prolog code may be used as a debugging tool for designing Turing machines on the given input.

Theorem 11.5.3 *The termination of a Prolog program is undecidable.*

Proof of Sketch If we have an algorithm to decide the termination of a Prolog program on any input, then we can use the algorithm to decide the termination of a

11.5 Turing Completeness

Turing machine on an input, as the moves of the Turing machine can be simulated by a Prolog program. The termination of the Prolog program in the simulation ensures the termination of the Turing machine. □

Theorem 11.5.4 *The satisfiability of a set of Horn clauses is undecidable.*

Proof of Sketch Let H be the clauses used in the Prolog program from the proof of Proposition 11.5.2 for simulating Turing machine M, including the negative clause resulting from the query derived from $q_0 w$. It is easy to check that H is a set of Horn clauses. If we have an algorithm A to tell that H is unsatisfiable, then there exists a resolution proof of H. This resolution proof ensures that w is accepted by M. Thus, algorithm A can tell if Turing machine M accepts w or not. This is a contradiction to the fact that the acceptance problem of Turing machines is undecidable. □

Theorem 11.5.5 *The satisfiability of a first-order formula is undecidable.*

The proof is left as an exercise.

11.5.2 Turing Completeness of Rewrite Systems

In Sect. 7.2.5, we briefly introduced the concept of *string rewrite system* (SRS). Axel Thue introduced this notion hoping to solve the *word problem* for Thue systems of finite alphabets: given an SRS (Σ, R), let $=_R$ denote the symmetric closure of $\Rightarrow$, a transitive and monotonic relation induced by R over Σ^*, how to decide $s =_R t$ for any $s, t \in \Sigma^*$? Only in 1947 was the word problem shown to be undecidable—this result was obtained independently by Emil Post and Andrey Markov.

Proposition 11.5.6 *SRS is Turing complete.*

Proof Let M be any Turing machine. Assuming the states and the tape do not share symbols, any configuration of M is a string of tape and state symbols of form $\alpha q \beta$, where q is a state symbol, and α and β are strings of tape symbols. We may use an SRS R to describe all the moves of M as follows: the initial configuration $q_0 w$ is represented by the string $q_0 w \$$, where $\$$ is the end marker of all non-blank symbols on the tape of M. If $\delta(q, a) = (p, b, R)$, we have a rewrite rule $qa \to bp$; moreover, if a is ⊔ (the blank symbol), i.e., $\delta(q, ⊔) = (p, b, R)$, we add a rewrite rule $q\$ \to bp\$$. If $\delta(q, a) = (p, b, L)$, we have a set of rewrite rules: for any tape symbol c, $cqa \to pcb$; moreover, if a is ⊔, we add the following rewrite rules: $cq\$ \to pcb\$$, where c is any tape symbol. Then, it is trivial to show that $q_0 w \vdash^* \alpha q_a \beta$ in M iff $q_0 w\$ \Rightarrow^* \alpha q_a \beta \$$ in R. □

Example 11.5.7 For Turing machine M_0 given in Example 11.2.3, the rewrite rules for each delta move generated by the above proof are given below:

1	$\delta(q_0, 0) = (q_1, 0, R)$	$q_0 0 \to 0 q_1$
2	$\delta(q_1, 0) = (q_1, 0, R)$	$q_1 0 \to 0 q_1$
3	$\delta(q_1, 1) = (q_2, 1, R)$	$q_1 1 \to 1 q_2$
4	$\delta(q_2, 1) = (q_2, 1, R)$	$q_2 1 \to 1 q_2$
5	$\delta(q_2, \sqcup) = (q_a, \sqcup, R)$	$q_2 \sqcup \to \sqcup q_2, q_a \$ \to \sqcup q_a \$$

It is easy to see that $q_0 0011 \$ \Rightarrow^* 0011 \sqcup \$$ by the SRS consisting of the above rewrite rules. □

Since SRS is a special case of TRS (term rewrite system), the Turing completeness of SRS implies that of TRS.

Theorem 11.5.8 *The termination of rewrite systems is undecidable.*

Proof of Sketch If we have an algorithm to decide the termination of any rewrite system (either SRS or TRS), then we can use the algorithm to decide the termination of a Turing machine as the moves of the Turing machine can be expressed as a set of rewrite rules. The termination of the rewrite system ensures the termination of the Turing machine. □

Theorem 11.5.9 *The word problem of Thue systems is undecidable.*

Proof of Sketch If we have an algorithm to decide the word problem of Thue systems, then we can use the algorithm to decide the acceptance problem of Turing machines (encoded as A_{TM}): given any Turing machine M and input string w, we create M' from M by introducing a new *accept* state q_f such that M' simulates M on w, and once M enters the accept state q_a, M' will erase every non-blank symbol (replaced by the blank symbol) before entering q_f, so that the accepting configuration of M' is unique. Now, let $s = q_0 w \$$ (the initial configuration) and $t = q_f \$$ (the accepting configuration), then M accepts w iff $s =_R t$, where R is the SRS created from the definition of M' as we did in Example 11.5.7. □

Recall that the congruence problem of first-order logic with equality (Definition 7.1.7) is the problem of deciding if $s =_E t$ or not for any two terms s and t, where E is a given set of equations. The following is a consequence of the word problem.

Corollary 11.5.10 *The congruence problem of first-order logic with equality is undecidable.*

In this chapter, we have introduced the concept of Turing machines and proved formally that the well-known halting problem of Turing machines is undecidable. Since a Prolog program or a rewrite system can simulate the moves of any Turing machine, the termination of Prolog programs or rewrite systems is undecidable, too. The acceptance problem of Turing machines is also undecidable. Since this problem can be specified by a set of Horn clauses, it implies that the satisfiability of Horn clauses is undecidable. As a consequence, the satisfiability of first-order logic is undecidable. The satisfiability of any logic containing first-order logic,

Exercises 431

such as higher order logic, is also undecidable. These theoretical results help us to understand the logic thoroughly and avoid unnecessary efforts to find an algorithm for these undecidable problems.

Exercises

1. For the Turing machine M_1 of Example 11.2.1, please provide the sequence of moves of M_1 for each of the input strings: *(a)* a^2b^3; *(b)* a^2b; *(c)* ab^2a. Each sequence must end with M_1 accepting or rejecting the input.
2. Provide a complete design of Turing machine M_2 in Example 11.2.10 for recognizing $A_2 = \{b^i a^j b^k \mid i * j = k, i, j \geq 0\}$. That is, provide the detailed δ definition of M_2 and group the moves of your definition according to the high-level description of M_2. Provide the sequence of moves on the input string $w = b^2 a^1 b^2$.
3. Provide a high-level description of a Turing machine M_3 for recognizing $A_3 = \{a^i b^j c^k \mid k = \lfloor i/j \rfloor, i, j \geq 0\}$. Each step of the description uses either known Turing machines or is expressed (maybe informally) by a sequence of Turing machine moves.
4. Provide a high-level description of a Turing machine M_4 for recognizing $A_4 = \{a^i b^j \mid i, j \geq 0\}$, and the tape contains exactly a^{ij} when M_4 stops with q_a. Each step of the description uses either known Turing machines or is expressed by a sequence of Turing machine moves.
5. Suppose A is a formal language. Show that A is decidable if *(a)* $|A| = 1$; *(b)* A is finite.
6. Let $C10_{TM} = \{\langle M \rangle \mid \text{TM } M \text{ moves at most ten steps for any input string}\}$. Show that $C10_{TM}$ is decidable by designing a decider.
7. Show the following languages are decidable:

    (a) A_1 is the set of all binary numbers divisible by 4.
    (b) A_2 is the set of all binary numbers divisible by 3.
    (c) $A_3 = \{0^i 1^j 0^k \mid i + j = k\}$.

8. Show that decidable languages are closed under *(a)* union, *(b)* intersection, and *(c)* complement. That is, suppose A and B are decidable, the following languages are decidable: *(a)* $A \cup B$; *(b)* $A \cap B$; *(c)* $\overline{A} = \Sigma^* - A$.
9. Show that recognizable languages are closed under *(a)* union and *(b)* intersection. That is, suppose A and B are recognizable, the following languages are recognizable: *(a)* $A \cup B$; *(b)* $A \cap B$.
10. Provide a mapping reduction from A_{TM} to H_{TM}.
11. Prove formally that the partial correctness of Hoare triples is undecidable.
12. Prove by reduction from A_{TM} that the following properties of Turing machines are undecidable (Rice's theorem cannot be used):

    (a) $One_{TM} = \{\langle M \rangle \mid L(M) = \{0\}\}$.

(b) $Two_{TM} = \{\langle M \rangle \mid L(M) = \{0, 1\}\}$.
(c) $Fin_{TM} = \{\langle M \rangle \mid L(M) \text{ is finite}\}$.
(d) $A10_{TM} = \{\langle M \rangle \mid M \text{ accepts at least ten strings}\}$.

13. (Theorem 11.4.5) Prove that a formal language L is decidable iff there exists an enumerator M for L such that the printed strings are increasing in the canonical order.
14. Prove that $\overline{E_{TM}}$ is recognizable, where $E_{TM} = \{\langle M \rangle \mid L(M) = \emptyset \text{ for TM } M\}$.
15. Prove that if both A and B are computably countable, then $A \cup B$ and $A \cap B$ are computably countable.
16. Prove that the set of decision functions (total or partial), $D_F = \{f : \mathcal{N} \mapsto \{0, 1\}\}$, is not computably countable. (Hint: extend Example 11.4.1 to $A = \mathcal{N}$ and $B = \{0, 1\}$).
17. Let $One_{TM} = \{\langle M \rangle \mid |L(M)| = 1\}$. Prove that neither One_{TM} nor $\overline{One_{TM}}$ is recognizable.
18. Let $Fin_{TM} = \{\langle M \rangle \mid L(M) \text{ is finite}\}$. Prove that neither Fin_{TM} nor $\overline{Fin_{TM}}$ is recognizable.
19. Let $De_{TM} = \{\langle M \rangle \mid L(M) \text{ is decidable for TM } M\}$. Prove that neither De_{TM} nor $\overline{De_{TM}}$ is recognizable.
20. Prove that D_{TM} is unrecognizable without Proposition 11.4.11.
21. Some Turing machines allow the tape head to stay still after a move (only changing state or the current symbol). How to simulate this action in the Prolog program of Sect. 11.5.1?
22. Some Turing machines allow the tape to have infinite symbols on the left side of the head tape (so there is no "move off the tape"). How to simulate this action in the Prolog program of Sect. 11.5.1? Please provide a detailed solution.
23. Prove that the satisfiability of a first-order formula is undecidable.

References

1. Jack Copeland, "The Church-Turing Thesis", The Stanford Encyclopedia of Philosophy, Edward N. Zalta (ed.), Summer 2020
2. Michael Sipser, *Introduction to the Theory of Computation*, 3rd ed., January 1, 2014

Chapter 12
Decision Procedures

The design of computer systems, either software or hardware, is error-prone, and formal verification of such designs requires powerful automated reasoning tools. Decision procedures are necessary components of these tools. Recall that in Chap. 1 (Sect. 1.3), a *decision procedure* is defined as an algorithm which always halts with a correct answer to a yes/no question. For propositional logic, decision questions regarding satisfiability, validity, or entailment are decidable. For linear temporal logic introduced in Chap. 10, these questions are also decidable. For first-order logic, these questions are undecidable in general (see Chap. 11). However, if we restrict ourselves to some segments of first-order logic, these questions may become decidable. For instance, the following segments of the first-order logic are decidable.

- **First-order formulas with models of given size**
  It is decidable to check if a set of first-order formulas has models of a given size, because these formulas can be converted into a set of ground formulas whose satisfiability is decidable. In Sect. 7.5.1, we introduced Mace4 as a tool for models of finite size. In Sect. 7.5.2, we showed how ground formulas are converted into equivalent propositional formulas.
- **Monadic predicate calculus**
  Monadic predicate calculus is a special case of first-order logic (also called predicate calculus) where the arity of all predicate symbols is one (e.g., they take only one argument and are called monadic or unary) and the arity of all function symbols is 0. In Sect. 5.2.4, we showed that a monadic predicate defines uniquely a set and is an interpretation of *type* in many-sorted logic. Monadic predicate calculus is useful in the study of relations between sets. It is decidable whether a monadic first-order formula with equality is valid or satisfiable. However, we are unaware of any automated tool that takes advantage of this property. If we add only one binary predicate (other than the equality) or one binary function, then the resulting logic is undecidable.

- **Classical decision problems**
  There is a complete classification of decidable classes of first-order formulas based on the form of the quantifiers of the prenex normal form (see Sect. 5.4.1) without function symbols of arity > 0: if the quantifier prefix is of form: *(a)* $\exists^*\forall^*$; *(b)* $\exists^*\forall\exists^*$; or *(c)* $\exists^*\forall^2\exists^*$, where $\exists^*$ means you can have 0 or more of $\exists$, then the satisfiability of these prenex formulas is decidable [1]. On the other hand, if the quantifier prefix is of form *(d)* $\forall^3\exists$ or *(e)* $\forall\exists\forall$, then there exists such a prenex formula whose satisfiability is undecidable. Note that every prenex normal form falls into one of *(a)–(e)*. Monadic formulas can be converted into equivalent prenex formulas of form $\exists^*\forall^* F$ (class *(a)*). Since there are no reasoning tools based on the decidability of these classical problems, we skip this topic here.

In the recent trend on decision procedures, the selection of decidable segments of first-order logic is motivated by formal verification of programs as discussed in Chap. 9. In the practice of formal verification, we need to deal with the following three difficulties of complex formulas in automated reasoning:

1. **Formulas with quantifiers**
   In Chap. 5, we showed that quantifiers can be removed by a process called Skolemization. At the end of Chap. 9, we also showed how to avoid quantifiers in loop invariants, by taking the domain theory approach and using recursively defined functions to replace the formulas with quantifiers. This approach defines a set of functions with special properties and needs decision procedures to handle these functions.
2. **Formulas with one or more special domains**
   A typical programming language provides predefined data types, such as Boolean values, 32-bit or 64-bit integers, floating numbers, arrays, etc. The semantics of operations over such data types create new problems, such as overflow or unbound, and no longer matches mathematical theory of unbounded types, for example, the natural numbers. In general, we use *domain* or *theory* (standing for *theory of the domain*) to refer to a special data type and related functions. To formally verify programs written in such a programming language, we need decision procedures which handle efficiently these theories. This approach is termed as *satisfiability modulo theories* (SMT) in the community of automated reasoning.
3. **Formulas with arbitrary Boolean combinations**
   To avoid handling arbitrary Boolean combinations in different theories, Nieuwenhuis, Oliveras, and Tinelli advocated a framework called DPLL(T) [2], which is an extension of the well-known DPLL algorithm for propositional logic with theories T. DPLL(T) handles Boolean combinations in the same way as the DPLL algorithm and lets the decision procedures for various theories T to handle only a conjunction of literals. This feature reduces greatly the complexity of decision procedures for T.

Decision procedures for satisfiability modulo theories (SMT) are of continuing interest for many verification applications. Solvers based on the SMT approach check the satisfiability of first-order formulas containing functions from various domains such as Boolean values, bit vectors, integer arithmetic, and arrays. A verification problem is often translated into an SMT formula, which is a restricted first-order formula acceptable by SMT solvers. SMT solvers take these formulas and attempt to report satisfiability or unsatisfiability of the formula.

A major advantage of SMT solvers over pure SAT solvers is the higher level of abstraction which avoids the exponential explosion of propositional encoding of the same problem. By using directly formulas in theories like arithmetic, arrays, and uninterpreted functions, SMT solvers have the promise to provide higher performance than pure SAT solvers. In fact, SMT solvers have a wide range of applications in hardware and software verification, extended static checking, constraint solving, planning, scheduling, test case generation, and computer security.

12.1 DPLL(T): Extend DPLL with Theories T

Most SMT solvers use the DPLL(T) framework [2], where T stands for theories, including equality, linear arithmetic, arrays, etc. DPLL(T) is expressed as an extension of the DPLL algorithm. At high level, DPLL(T) works by transforming an SMT formula into a SAT formula where atomic formulas are replaced by propositional variables. DPLL(T) repeatedly finds a model for the SAT formula and consults a theory solver to check consistency under the domain-specific theory, until either a theory-consistent model is found, or the SAT formula runs out of models.

12.1.1 Propositional Abstraction

In Example 9.2.8, we need to prove the validity of (2.1.1.2), which is relabeled as X here:

$$X: \ (I \wedge i < n \wedge \neg(m < A[i])) \rightarrow I[i \leftarrow i+1],$$

where I is the loop invariant $(m \doteq arrMax(i, A)) \wedge (i \leq n)$. Besides logical operators, this formula involves arithmetic (e.g., $+$, $<$, and $\leq$), arrays, and function $arrMax$.

The first step of SMT is called *propositional abstraction*, which replaces atomic formulas of a first-order formula by propositional variables. For formula X, we may introduce the following propositional variables p_1–p_6 for the atomic formulas in X:

$$\begin{aligned}
&p_1 : m \doteq arrMax(i, A) &\quad &p_2 : i \leq n \\
&p_3 : i < n &\quad &p_4 : m < A[i] \\
&p_5 : m \doteq arrMax(i+1, A) &\quad &p_6 : i+1 \leq n
\end{aligned}$$

so that the abstraction of X is a SAT formula Y:

$$Y : (p_1 \wedge p_2 \wedge p_3 \wedge \neg p_4) \rightarrow (p_5 \wedge p_6).$$

We say formula Y is an *abstract* of X in propositional logic, denoted by $Y = \gamma(X)$, if every distinct atomic formula of X is replaced by a new propositional variable and the result is Y, which is unique up to the renaming of propositional variables.

In general, if B is an abstract of formula A, i.e., $B = \gamma(A)$, what will be the relationship between A and B? First, A and B do not share predicate or function symbols; hence, the two formulas do not share models. Thus, there are no equivalence or entailment relationships between them, as the concepts of equivalence and entailment are based on their model sets.

Next, we claim that A and B are not equisatisfiable, that is, $A \approx B$ is false. For instance, let A be $a \doteq b \wedge b \doteq c \wedge a \neq c$, which is unsatisfiable by equality axioms. However, $\gamma(A)$ is $p_1 \wedge p_2 \wedge \neg p_3$, which is satisfiable.

In first-order logic, the *quantifier-free theory* (QF) denotes the set of first-order formulas without quantifiers. These formulas are the union of ground formulas and the formulas containing free variables. The meaning of these free variables is application dependent: either all universally quantified or all existentially quantified. Hence, there is no need to write quantifiers explicitly in these formulas. For instance, in Skolemized formulas, the meaning of the variables is universally quantified. For simplicity, we say these variables are universally quantified. In this case, a QF formula $F(x_1, ..., x_n)$ is equivalent to

$$\forall x_1, ..., x_n \; F(x_1, ..., x_n),$$

where F does not contain any quantifier and $free(F) = \{x_1, ..., x_n\}$. On the other hand, in the unification problem or in middle school algebra, the variables in a set of equations are existentially quantified.

In the abstraction process, each free variable is regarded as "ground": these variables have the same quantification in p and $\neg p$. For instance, if p stands for $q(x)$, then $\neg p$ stands for $\neg q(x)$. Had we given the quantification of x in p as $p \equiv \forall x \, q(x)$, we would have $\neg p \equiv \exists x \, \neg q(x)$; hence, the quantification of x would be different in p and $\neg p$.

Proposition 12.1.1 *For any QF formula A, if A is satisfiable, so is $\gamma(A)$.*

Proof Let $B = \gamma(A)$, the abstract of A. For any atomic formula A_i of A, let $p_i = \gamma(A_i)$ be the propositional variable which replaces A_i in the abstraction. If A is satisfiable, then A has a first-order model I such that $I(A) = 1$. There are three cases to consider:

- A is ground: We define a propositional interpretation σ for B as follows: $\sigma(p_i) = I(A_i)$, where $p_i = \gamma(A_i)$. It is easy to check that σ is a model of B by comparing the formula trees of A and B under σ and σ', respectively.

- Free variables of A are existentially quantified: Let $free(A) = \{x_1, \ldots, x_n\}$, then $I(A) = 1$ means $I(A, \theta) = 1$, for some assignment $\theta = \{x_i \leftarrow d_i : 1 \leq i \leq n\}$, where d_i is a domain element of I. In this case, we define a propositional interpretation σ for B as follows: $\sigma(p_i) = I(A_i, \theta)$, where $p_i = \gamma(A_i)$. It is easy to check that σ is a model of B as in the case when A is ground.
- Free variables of A are universally quantified: Let $free(A) = \{x_1, \ldots, x_n\}$, then $I(A) = 1$ means $I(A, \theta) = 1$ for every assignment $\theta = \{x_i \leftarrow d_i : 1 \leq i \leq n\}$, where d_i is any domain element of I. We just need one of the assignments to define a propositional model σ for B as in the previous case.

Since we have exhausted all the cases of A, the proposition must be true. □

There is a simple reason for the necessary condition that A is quantifier-free in the above proposition: if A contains quantifiers, then these quantifiers in $\gamma(A)$ become meaningless. The proposition can be extended to prenex formulas, where all quantifiers are on top of the other logical operators, and we may store the quantifiers for the variables before abstraction.

The above proposition still holds when only a part of atomic formulas of A are replaced by propositional variables. The proposition also implies that, if $\neg A$ is satisfiable, so is $\neg B$. This is another evidence that A does not entail B: if $A \models B$, we would have $\neg B \models \neg A$. From $\neg A \models \neg B$, we would have concluded that $A \equiv B$, a contradiction.

Corollary 12.1.2 *Let B be an abstract of A, i.e., $B = \gamma(A)$, then*

(a) *If B is unsatisfiable, so is A.*
(b) *If B is valid and A is ground, then A is valid.*

Proof
(a) From Proposition 12.1.1, if A is satisfiable, then B is satisfiable. So, if B is not satisfiable, then A is not satisfiable (contrapositive of implication).
(b) Let $X = \neg A$ and $Y = \neg B$, then it is easy to see that $Y = \gamma(X)$. By *(a)*, if Y is unsatisfiable, so is X. However, Y is unsatisfiable iff B is valid, and X is unsatisfiable iff ground A is valid. Thus, if B is valid, so A is. □

Typically, in formal verification, we want to show that A is valid by showing $\neg A$ is unsatisfiable. If $\neg A$ is satisfiable, then a model of $\neg A$ is a *counterexample* of the validity of A. For this purpose, the above corollary is rarely useful, because it is rare for the abstract of $\neg A$ to be unsatisfiable. The common usage of the abstraction is as follows: let X be the QF formula derived from $\neg A$. In order to show that X is unsatisfiable, let Y be the abstract of X. If Y is unsatisfiable, we are done; if not, then for any model of Y, we have to check that the conjunction of literals from X corresponding to the model of Y is indeed a model of X. If yes, then X is satisfiable and a counterexample of A is found; if X has no such model, then X is unsatisfiable and A is valid.

12.1.2 Examples of DPLL(T)

For the DPLL(T) framework, abstraction $\gamma(A)$ plays two roles: (1) As a filter of models of A so that we waste less effort to check if A has models. (2) We need only to check if a conjunction of literals from A has a model or not.

Example 12.1.3 Let X be $(a \doteq b \vee b \doteq a) \wedge b \doteq c \wedge a \neq c$ and $Y = \gamma(X) = (p_1 \vee p_2) \wedge p_3 \wedge \neg p_4$. Y has three models: $\{p_1, p_2, p_3, \neg p_4\}$, $\{p_1, \neg p_2, p_3, \neg p_4\}$, and $\{\neg p_1, p_2, p_3, \neg p_4\}$. The conjunction of atomic formulas corresponding to $\{p_1, p_2, p_3, \neg p_4\}$ is $a \doteq b \wedge b \doteq a \wedge b \doteq c \wedge a \neq c$, which is unsatisfiable. Similarly, the SMT formulas corresponding to the other two models of Y are also unsatisfiable. Thus, X is unsatisfiable. □

Now, back to the verification example in the beginning of this section, let A be $\neg X$ and B be $\neg Y$ in the example. That is, B is $\neg((p_1 \wedge p_2 \wedge p_3 \wedge \neg p_4) \rightarrow (p_5 \wedge p_6))$, where

$$p_1 : m \doteq arrMax(i, A) \qquad p_2 : i \leq n$$
$$p_3 : i < n \qquad p_4 : m < A[i]$$
$$p_5 : m \doteq arrMax(i+1, A) \qquad p_6 : i+1 \leq n$$

If B is unsatisfiable, then A is unsatisfiable (i.e., X is valid) by Prop. 12.1.1. If B is satisfiable, we must show that every model of B leads to a contradiction of A in the theories of arithmetic, arrays, and $arrMax$. In other words, $\neg X$ has no models; thus, X must be valid.

Now, we transform B into a set S of five clauses:

$$S = \{p_1, p_2, p_3, \overline{p_4}, (\overline{p_5} \mid \overline{p_6})\}$$

S has three models in which p_1, p_2, and p_3 are true, p_4 is false, and one of p_5 and p_6 must be false. In integer arithmetic, p_3 and p_6, i.e, $i < n$ and $i + 1 \leq n$, are equivalent. Thus, p_6 must be true when p_3 is true. Moreover, $p_1 \wedge \neg p_4 \rightarrow p_5$ is true by the definitions of $arrMax$ and max:

$$arrMax(i+1, A) = max(arrMax(i, A), A[i]) \text{ definition of } arrMax$$
$$= max(m, A[i]) \qquad p_1 : m \doteq arrMax(i, A)$$
$$= m \qquad \neg p_4 : \neg(m < A[i]),$$
$$\qquad \text{definition of } max$$

Thus, p_5 must be true when p_1 is true and p_4 is false. In other words, no models of S are acceptable for A, so, $A = \neg X$ is unsatisfiable.

In Chap. 4, we have presented the *DPLL* algorithm in detail and described several techniques for the efficient implementation of *DPLL*, which is an enumeration-based search method for propositional satisfiability. *DPLL* tries to find a model for a set

12.1 DPLL(T): Extend DPLL with Theories T

of propositional clauses. Here, a model from *DPLL* can be viewed as a set of true literals in the model.

Example 12.1.4 Consider the problem of deciding the satisfiability of the following first-order formula X in conjunctive normal form:

$$X: \ g(a) \doteq c \land (f(g(a)) \neq f(c) \lor g(a) \doteq d) \land c \neq d$$

The first step of DPLL(T) is *abstraction*, where T represents equality: use propositional variables for each atomic formula and obtain a set of propositional clauses for SAT solvers. That is, with p_1 for $g(a) \doteq c$, p_2 for $f(g(a)) \doteq f(c)$, p_3 for $g(a) \doteq c$, and p_4 for $c \doteq d$, we obtain the following propositional clauses:

$$\{p_1, \neg p_2 \lor p_3, \neg p_4\}$$

DPLL(T) will feed these clauses to a SAT solver which returns $\{p_1, \neg p_2, \neg p_4\}$ as its first model. The first-order formula corresponding to this model is

$$g(a) \doteq c \land f(g(a)) \neq f(c) \land c \neq d$$

which can be proved to be unsatisfiable by a decision procedure for equality. DPLL(T) may add the negation of the model as a new clause $(\neg p_1 \lor p_2 \lor p_4)$ into the SAT solver to ask for the next model. The SAT solver will return $\{p_1, p_3, \neg p_4\}$ as its second model. The first-order formula corresponding to this model is

$$g(a) \doteq c \land g(a) \doteq d \land c \neq d$$

which can be proved to be unsatisfiable, too. DPLL(T) then adds $(\neg p_1 \lor \neg p_3 \lor p_4)$ to the SAT solver, which will return "unsatisfiable." Finally, DPLL(T) claims that the original formula X is unsatisfiable as the SAT solver runs out of models. □

The above example illustrates a great advantage of the DPLL(T) framework: the basic *DPLL* algorithm handles the logic operators, and the decision procedures for theories T handle a conjunction of literals, thus reducing the complexity of decision procedures required by the theories. In fact, many theories of interest have efficient decision procedures for the satisfiability of sets (or conjunctions) of literals.

Let us call the decision procedures used in DPLL(T) *theory solver* for T. These theory solvers use various algorithms for each domain appearing in the input. The above example also illustrates several interesting aspects of DPLL(T).

1. Atomic formulas in the input formulas are abstracted to propositional variables. This abstraction is done once for all as a preprocessing step.
2. The SAT solver is called multiple times, each time with an increment of input clauses to avoid repeated models, Thus, an incremental SAT solver is desired to support DPLL(T).

3. The theory solver is called multiple times, each with a conjunction of literals as input. Each input corresponds to a propositional model found by the SAT solver. The theory solver does not need to be incremental because the input changes unpredictably. That is, the interaction between a SAT solver and a theory solver goes back and forth in DPLL(T).
4. This approach of solving theory satisfiability via propositional satisfiability is said to be *lazy*: domain information is used lazily when checking consistency of propositional models in theory T. In contrast, the *eager* approach converts the original theory satisfiability problem into an equisatisfiable propositional instance.
5. In this lazy approach, every tool does what it is good at:
   - SAT solver takes care of Boolean combinations.
   - Theory solver takes care of conjunction of literals in theories.

12.1.3 DPLLT(X): An Algorithm of DPLL(T)

The DPLL(T) framework can be easily implemented if we have implementations of the DPLL algorithm and decision procedures for checking consistency of a set of literals in a theory. The following algorithm *DPLLT* is a minor modification of Algo. 4.2.5, i.e., *DPLL(C)* with CDCL (conflict-driven clause learning) for SAT.

Algorithm 12.1.5 *DPLLT(X)* takes as input a set X of first-order clauses and returns *true* if X is satisfiable and returns *false*, otherwise.

proc *DPLLT(X)*: Boolean
0   $C := abstraction(X)$ // C is an abstract of X.
1   $U := initialize(C)$ // initialize data structure for C U: unit clauses.
2   $level := 0; \sigma := \{\}$ // σ is a partial model of C.
3   **while** (true) **do**
4      $res := BCP(U)$ // $BCP(U)$ extends σ.
5   **if** $(res \neq \text{"SAT"})$ // res contains a conflicting clause.
6   **if** $(level = 0)$ **return** *false* // C as well as X is unsatisfiable
7         $U := insertNewClause(conflictAnalysis(res))$ // new level is set
8         $undo()$ // undo to the new level
9   **else**
10        $A := pickLiteral()$ // pick an unassigned literal as decision literal.
11   **if** $(A = nil)$ // a model of C is found.
12   **if** $isTheoryConsistent(\sigma)$ **return** $\top$ // X is satisfiable.
13   **else** $U := insertNewClause(negateDecisions(\sigma))$
14   **else** $U := \{A\}; level := level + 1$

12.1 DPLL(T): Extend DPLL with Theories T

The only differences between *DPLLT* and Algo. 4.2.5 are the following:

- At line 0: *abstraction(X)* computes the abstract of *X*.
- At line 11–13: Instead of returning *true* after ($A = nil$) (every literal has a truth value), we call *isTheoryConsistent(σ)* to check if the corresponding SMT formula to the propositional model σ of *C* is consistent in theories of *X*. If yes, return *true*; otherwise, create a new clause from the negation of the decision literals in σ: *negateDecisions(σ)* returns the negation of the conjunction of decision literals in σ. Then, *insertNewClause* is called on the newly created clause (see line 7).

The implementation of *abstraction(X)* is straightforward. The purpose of line 13 is to support CDCL (conflict-directed clause learning).

The algorithm *DPLLT(X)* just illustrates how easily we can implement the DPLL(*T*) framework if we have a SAT solver and an implementation of *isTheoryConsistent(σ)*. The latter depends on the theories appearing in *X* and is the focus of the remaining sections.

Example 12.1.6 Suppose *T* is equality and *X* contains the following clauses:

$$(b \doteq c), \ (h(a) \doteq h(c)) \lor p, \ (a \doteq b) \lor \neg p \lor (a \doteq d), \ (a \doteq b) \lor (a \neq d)$$

The abstraction procedure will convert *X* into *C* which contains the following clauses:

$$(s), \ (t \lor p), \ (q \lor \neg p \lor r), \ (q \lor \neg r)$$

where *q* stands for $a \doteq b$, *r* for $a \doteq d$, *s* for $b \doteq c$, and *t* for $h(a) \doteq h(c)$.

After line 1, $U = \{s\}$. At line 4, *BCP(U)* will make *s* to be true, and $\sigma = \{s\}$.

Suppose *pickLiteral* chooses $\neg t$ to be true and $U = \{\neg t\}$. *BCP(U)* will make *p* to be true due to $(t \lor p)$. Now, $\sigma = \{s, \neg t, p\}$.

Next, suppose *pickLiteral* chooses $\neg r$ to be true and $U = \{\neg r\}$. *BCP(U)* will make *q* to be true due to $(q \lor \neg p \lor r)$. Now $\sigma = \{p, q, \neg r, s, \neg t\}$, which is a model of *C* and *isTheoryConsistent(σ)* is called. The corresponding SMT formula to σ is

$$p \land a \doteq b \land a \neq d \land b \doteq c \land h(a) \neq h(c)$$

From $a \doteq b$ and $b \doteq c$, $a \doteq c$ is true and $h(a) \neq h(c)$ is false by the equality axioms. Thus, *isTheoryConsistent(σ)* returns $\bot$, and the new clause from the negation of the conjunction of the two decision literals is $(t \lor r)$.

The procedure *insertNewClause($t \lor r$)* will add this clause and backtrack to level 1 with $U = \{r\}$. *BCP(U)* will make *r* and *q* to be true with the second model $\sigma = \{s, \neg t, p, r, q\}$, whose corresponding SMT formula is

$$p \land a \doteq b \land a \doteq d \land b \doteq c \land h(a) \neq h(c)$$

Again, *isTheoryConsistent*(σ) found this formula inconsistent and returns false. The new clause from the negation of the decision literal is (t).

Now, *DPLLT* will backtrack to level 0 and find model $\sigma = \{p, q, r, s, t\}$ (or $\sigma = \{p, q, \neg r, s, t\}, \ldots$) and return *true* at line 12 as *isTheoryConsistent*(σ) will return *true*. □

Naturally, the minimal modification of *DPLL* into *DPLLT* does not necessarily gives us an ideal implementation of DPLL(T). At the end of this chapter, we will discuss some issues to make *DPLLT* more practical and efficient.

Next, we will present the decision procedures for the following theories, which occur often in program verification and can be integrated in *DPLLT*:

- Equality with uninterpreted functions
- Linear arithmetic

12.2 Equality with Uninterpreted Functions

In some applications, it is convenient and efficient to keep in mind the definitions of some functions. For example, in program verification, $1 + 3$ can be replaced by 4 because $+$ is the addition of integers. In equational logic, the equality predicate "$\doteq$" is assumed to satisfy the axioms of equality (Sect. 7.1.1). Functions like $+$ and $\doteq$ with their intended meanings are called "interpreted functions." On the other hand, functions which can take any interpretation are called "uninterpreted functions."

12.2.1 Uninterpreted Functions

In first-order logic, a function symbol is assumed *uninterpreted* and can be interpreted as any function of the same arity. In other words, by default, every function symbol is an uninterpreted function.

Replacing interpreted functions by uninterpreted functions in a formula is a way of generalization (or abstraction) and can be useful in obtaining stronger results or simplifying proofs.

Example 12.2.1 We wish to prove the equivalence of the programs $P1$ and $P2$:

P_1

int $a[1..3]$
int $i := 2, x := a[3]$
while ($i > 0$)
 $\quad x := a[i] + x; i := i - 1$
return x

P_2

int $a[1..3]$
int y

$y := a[1] + (a[2] + a[3])$
return y

Since the body of the **while** loop in $P1$ is executed twice, we may unroll the loop by introducing auxiliary variables x_0, x_1, x_2 to record the values of x before and after $x := a[i] + x$. The relationship between them can be expressed by

$$x_0 \doteq a[3] \wedge x_1 \doteq a[2] + x_0 \wedge x_2 \doteq a[1] + x_1$$

Together with $y \doteq a[1] + (a[2] + a[3])$, we would like to prove $x_2 \doteq y$.

If $+$ is an interpreted function, it will be difficult to decide the value of $a[2]+a[3]$, x_2, and y. If we replace $+$ by an *uninterpreted function* f, the verification becomes to prove the following theorem:

$(x_0 \doteq a[3] \wedge x_1 \doteq f(a[2], x_0) \wedge x_2 \doteq f(a[1], x_1) \wedge y \doteq f(a[1], f(a[2], a[3])))$
$\rightarrow x_2 \doteq y.$

The above can be shown to be true by rewriting, viewing each equality as a rewrite rule from left to right. Once the theorem is proved, the equivalence of P_1 and P_2 holds for any binary operation f on any type of $a[1..3]$ (e.g., double or matrix).

On the other hand, if the generalized formula cannot be proved, it does not mean that the equivalence of the programs does not hold. For instance, in P_2, if we have $y := (a[1] + (a[2]) + a[3]$, then the two programs are still equivalent because $+$ is associative. However, $f(a[1], f(a[2] + a[3])) \doteq f(f(a[1], a[2]), a[3])$ is no longer true, unless f is known to be associative. □

We would like to point out that x and y used in P_1 and P_2 are variables of the code; they are not *variables* of first-order logic. The formula for the equivalence of P_1 and P_2 is indeed ground (i.e., variable-free) because x and y are 0-arity functions in first-order logic. To avoid confusion, we call these 0-arity functions *identifiers*.

12.2.2 The Congruence Closure Problem

Recall that the axioms of equality introduced in Chap. 7 are the following five formulas, where the variables are assumed universally quantified:

reflexivity : $x \doteq x$;
symmetry : $(x \doteq y) \rightarrow (y \doteq x)$;
transitivity : $(x \doteq y) \wedge (y \doteq z) \rightarrow (x \doteq z)$;
functionmonotonicity : For $1 \leq i \leq k$, $(x_i \doteq y_i) \rightarrow f(x_1, \ldots, x_k)$
$\doteq f(y_1, \ldots, y_k)$;
predicatemonotonicity : For $1 \leq i \leq k$, $x_i \doteq y_i \wedge p(x_1, \ldots, x_k)$
$\rightarrow p(y_1, \ldots, y_k).$

A set A of literals is called *ground equality* if the only predicate in A is "$\doteq$" and no literals contain variables. We often write $\neg(s \doteq t)$ as $s \neq t$. A positive literal $s \doteq t$ is called *equation* and a negative literal $s \neq t$ is called *inequality*. If constants c_1 and c_2 appear in A, we assume that $(c_1 \neq c_2) \in A$, so that we do not distinguish identifiers (i.e., functions of zero arity) and constants. If predicates other than $\doteq$ are needed, we treat them as functions: replace positive literal $p(x)$ by $p(x) \doteq \top$ and negative literal $\neg p(x)$ by $p(x) \doteq \bot$, and add $\top \neq \bot$ into A.

Given a set A of ground equality, is A satisfiable with the axioms of equality? For instance, in Example 12.2.1, let A be

$$\{x_0 \doteq a[3], x_1 \doteq f(a[2], x_0), x_2 \doteq f(a[1], x_1),$$
$$y \doteq f(a[1], f(a[2], a[3])), x_2 \neq y\},$$

then A is unsatisfiable iff $x_2 \doteq y$ is a logical consequence of the first four equations.

The above decision problem is traditionally called *the congruence closure of equality with uninterpreted functions*. Since a function is uninterpreted by default in first-order logic, we prefer to call it *the congruence closure of ground equality*. Note that $\doteq$ is the only interpreted function and only the axioms of equality contain universally quantified variables. Some people also called it *the congruence closure of quantifier-free equality*. The set of quantifier-free formulas is the union of ground formulas and formulas with free variables. For the congruence closure problem under discussed, no free variables are presented. Thus, "ground" is a better label than "quantifier-free" for the congruence closure problem.

The conventional approach, proposed by Shostak, Nelson and Oppen, consists of the following steps:

1. Divide a set A of ground equality into equations E (positive literals) and inequalities D (negative literals).
2. Compute the congruence closure $=_E$, which is the congruence closure generated by E (Definition 7.1.7). It is the minimum equivalence relation satisfying the monotonicity axioms.
3. If there exists $(s \neq t) \in D$ such that $s =_E t$, then return "unsatisfiable"; otherwise, return "satisfiable."

The soundness of the above approach is stated as the following theorem:

Theorem 12.2.2 *Given a set A of ground equality, including both equations and inequalities, let $E \subseteq A$ be the set of all equations in A. If there exists $(s \neq t) \in A$ such that $s =_E t$, then A is unsatisfiable; otherwise, A is satisfiable.*

The proof can be done by Herbrand models (Proposition 7.1.3) and is left as an exercise. Now, we are ready for the proof of the following theorem:

Theorem 12.2.3 *The congruence closure of ground equality is decidable.*

Proof Given a set A of ground equality, including both equations and inequalities, let $E \subseteq A$ be the set of all equations in A. According to Theorem 7.2.22, there is a decision procedure for $=_E$. For each $s \neq t \in A$, we check if $s =_E t$: if yes, return "unsatisfiable." If every check says "no," return "satisfiable." Theorem 12.2.2 ensures the correctness of the answer. □

Example 7.2.21 shows how $A = \{f(a) \neq a, f^3 a \doteq a, f^5 a \doteq a\}$, where $f^3 a$ is $f(f(f(a)))$ and $f^5 a$ is $f(f(f(f(f(a)))))$, is unsatisfiable.

12.2.3 The NOS Algorithm

Congruence is an equivalence relation satisfying the monotonicity axioms. To store an equivalence relation in computer, the best data structure is *sets*: equivalent elements are stored in the same set so that reflexive, symmetric, and transitive properties are easily maintained. Each set represents an equivalent class, and the collection of equivalent classes is a partition of all the elements.

The congruence closure algorithm is based on this idea: to compute $=_E$ for a set E of ground equations

1. Place each subterm of E into its own congruence class.
2. For each $s \doteq t \in E$, if s and t are not in the same class, the *merge* operation will (i) union the two classes containing s and t and (ii) propagate $s \doteq t$ by the monotonicity axioms. That is, we try to merge the terms like $f(s, x)$ and $f(t, x)$ into one class if they appear in the input.

This idea was credited to Nelson and Oppen as well as to Shostak; both teams published independently a similar algorithm around the same time. We will call the algorithm based on this idea the NOS algorithm, where NOS comes from the first letter of Nelson, Oppen, and Shostak.

The propagation of $s \doteq t$ through the monotonicity axioms is done as follows: if s appears in $f(\ldots, s, \ldots)$ and t appears in $f(\ldots, t, \ldots)$, then the NOS algorithm will ensure that $f(\ldots, s, \ldots)$ and $f(\ldots, t, \ldots)$ should be in the same class by checking if their arguments are pairwise equivalent.

The NOS algorithm uses term graphs and the Union-Find data structure for sets, in a way very similar to the unification algorithm (Alg. 6.1.19).

The algorithm $NOS(E)$ takes a set E of ground equations as input and creates the term graph $G = (V, E)$ from all the terms in E, sharing subterms (at least sharing a identifier node for all occurrences of the identifier). If a set D of inequalities is provided, the terms in D are also added into G. $G = (V, E)$ has the following properties:

1. G is a directed acyclic graph (DAG) with possibly multiple edges between two nodes.
2. Each node $v \in V$ has a label, $label(v)$, which is a function symbol of E.
3. If $label(v) = f$ and $arity(f) = k$, then v has k ordered successors in G. The ith successor can be accessed by $child(v, i)$. Hence, if f is an identifier ($arity(f) = 0$), v has no successors.

Fig. 12.1 DAG $G = (V, E)$ for $A = \{f(a, b) \doteq a, f(f(a, b), b) \neq a\}$, where
$V = \{1, 2, 3, 4\}$,
$use(1) = \emptyset, use(2) = \{1\}$,
$use(3) = \{2\}$,
$use(4) = \{1, 2\}$

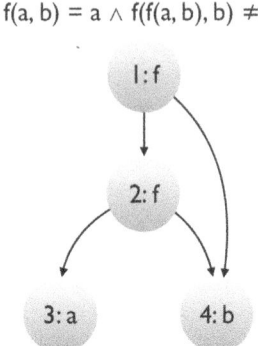

$f(a, b) = a \wedge f(f(a, b), b) \neq a$

4. Each subterm t of E is represented by a unique node $v_t \in V$ such that *(a)* t is an identifier and $label(v_t) = t$; or *(b)* if $t = f(t_1, \ldots, t_k)$, then $label(v_t) = f$ and v_t has k successors $v_{t_1}, \ldots, v_{t_k}$, such that each v_{t_i} represents t_i for $1 \leq i \leq k$.
5. To implement the Union-Find algorithm, we add the following data structures to each node $v \in V$. As a result, v has dual meanings: v is both a node of the term graph G and a node of the trees used by the Union-Find algorithm for construction of equivalent classes:

   - $up(v)$ defines the parent of v in a tree, which also represents an equivalence relation among the nodes. Initially, $up(v) = v$. That is, v is a root, and each node is initially in its own equivalence class.
   - If v is a root, $w(v)$, called the *weight* of the tree, records the number of nodes in the tree. Initially, $w(v) = 1$, that is, the tree is a single node.
   - To implement the congruence propagation, we add a list to each node $v \in V$: $use(v)$ records the list of all predecessors (also called parents) of v in G; this should not be confused with $up(v)$, which is the parent of v in the trees.

Example 12.2.4 Let $A = \{f(a, b) \doteq a, f(f(a, b), b) \neq a\}$. The DAG for the terms of A is shown in Fig. 12.1. □

Once DAG G is created, for each equation $s \doteq t$, we will call $merge(v_s, v_t)$, where v_s and v_t are the nodes of G representing s and t.

proc $merge(v_1, v_2)$
1   $s_1 := Find(v_1); s_2 := Find(v_2)$
2   **if** $s_1 = s_2$ **return** // already in the same class, exit
3   $P_1 := use(s_1); P_2 := use(s_2)$
4   $Union(s_1, s_2)$ // Union two classes into one
5   **for** $(t_1, t_2) \in P_1 \times P_2$ **do** // Congruence propagation
6   **if** $Find(t_1) \neq Find(t_2) \wedge congruent(t_1, t_2)$
7           $merge(t_1, t_2)$

The pseudo-codes for *congruent*, *Union*, and *Find* are given below.

12.2 Equality with Uninterpreted Functions

proc $congruent(t_1, t_2)$: Boolean
1   **if** $label(t_1) \neq label(t_2)$ **return** *false*
2   **for** $i := 1$ **to** $arity(label(t_1))$ **do**
3   **if** $Find(child(t_1, i)) \neq Find(child(t_2, i))$ **return** *false*
4   **return** *true*

proc $Union(s_1, s_2)$
    // Initially, for any node s, $w(s) = 1$, the weight of a singleton tree.
1   **if** $w(s_1) < w(s_2)$ $x := s_1$; $s_1 := s_2$; $s_2 := x$ // switch s_1 and s_2
2    $up(s_2) := s_1$ // s_2's representative is s_1
3    $use(s_1) := use(s_1) \cup use(s_2)$; $use(s_2) = \emptyset$
4    $w(s_1) := w(s_1) + w(s_2)$

proc $Find(v)$
1    $r := up(v)$ // look for the root of the tree containing v
2   **while** $(r \neq up(r))$ $r := up(r)$ // r is a root iff $r = up(r)$
3   **if** $(r \neq v)$ // compress the path from v to r
4   **while** $(r \neq v)$ $up(v) := r$; $v := up(v)$
5   **return** r

Algorithm 12.2.5 The algorithm $NOS(A)$ takes a set A of ground equality, which is a mix of equations or inequalities, as input, create the term graph $G = (V, E)$ from all the terms in A, construct the congruence classes for equations in A, and check the satisfiability of A.

proc $NOS(A)$: Boolean
1    $G := createGraph(A)$ // $up(v) = v$ for each v in G
2   **for** $(s \doteq t) \in A$ **do** $merge(v_s, v_t)$ // v_s, v_t represent s, t in G
3   **for** $(s \neq t) \in A$ **do**
4   **if** $Find(v_s) = Find(v_t)$ **return** $\bot$ // unsatisfiable
5   **return** $\top$ // satisfiable

Example 12.2.6 Continue from Example 12.1: to handle $f(a, b) \doteq a$, $merge(2, 3)$ is called. The first line of *merge* calls *Find* twice with the result $s_1 = Find(2) = 2$ and $s_2 = Find(3) = 3$. Since $s_1 \neq s_2$, we get $P_1 = \{1\}$ and $P_2 = \{2\}$, preparing for congruence propagation.

At line 4 of *merge*, $Union(2, 3)$ is called. Because $w(2) = w(3) = 1$, the second part of line 1 is skipped. At line 2, $up(3)$ is changed to 2, meaning node 3 uses node 2 as its class name. At line 3, $w(2)$ is changed from 1 to 2, as the class tree increases its weight. At line 4, the parents of node 3 move to those of node 2.

Now, back to line 5 of *merge*: since $P_1 \times P_2 = \{\langle 1, 2\rangle\}$, $t_1 = 1$ and $t_2 = 2$ are the only case for t_1 and t_2. Since $Find(1) = 1$ and $Find(2) = 2$, $congruent(1, 2)$ is called and returns true, because $Find(child(1, 1)) = Find(2) = 2 = Find(3) = Find(child(2, 1))$ and $Find(child(1, 2)) = Find(child(2, 2)) = 4$. Now, $merge(1, 2)$ is recursively called at line 7 and it will merge node 1 into $\{2, 3\}$ (i.e., $up(1) = 2$). The final congruence classes are $\{1, 2, 3\}$ and $\{4\}$.

For the inequality $f(f(a, b), b) \not\doteq a$ of A, since $f(f(a, b), b)$ (represented by node 1) and a (node 3) are in the same class, $f(f(a, b), b) \not\doteq a$ is false and A is unsatisfiable. □

Nelson and Oppen proved that the time complexity of $NOS(A)$ is $O(n^2)$, where n is the number of symbols of A. Various improvements and extensions of $NOS(A)$ have been proposed, and the best-known complexity is $O(n \log(n))$. Instead of presenting the complexity proof of the NOS algorithm, we will introduce Kapur's approach using the Knuth-Bendix completion procedure with detailed complexity analysis.

12.2.4 Ground Congruence by Completion

According to Theorem 12.2.3, the Knuth-Bendix completion procedure for ground equations can be used as a decision procedure for the congruence closure of ground equality. To make the completion procedure efficient, Deepak Kapur suggested two preprocessing steps which make equations or rewrite rules "flat" and "singleton" [3].

Definition 12.2.7 A term t is said to be *flat* if either t is an identifier or $t = f(a_1, \ldots, a_k)$ and $a_1, \ldots, a_k$ are identifiers. An equation $s \doteq t$ (or a rewrite rule $s \to t$) is *flat* if both s and t are flat. $s \doteq t$ (or $s \to t$) is *singleton* if t is an identifier. $s \doteq t$ (or $s \to t$) is *unit* if both s and t are identifiers.

Kapur proposed two preprocessing steps: *flattening* and *slicing* over E of ground equations. Flattening will flatten terms if the terms in an equation are not flat by introducing new identifiers for non-identifier subterms. Slicing will convert a non-singleton equation into two singleton equations by introducing a new identifier. Once the input equations are flat and singleton, so are the generated rewrite rules. Here is the formal definition of the preprocessing operations:

- **flattening**: If t in E is not flat, then t is $f(\ldots, g(s), \ldots)$ and we replace t by $f(\ldots, c, \ldots)$ and add $g(s) \doteq c$ into E, where c is a new identifier.
- **slicing**: If $(s \doteq t) \in E$, neither s nor t is an identifier, then we replace $s \doteq t$ by $s \doteq c$ and $t \doteq c$ in E, where c is a new identifier.
- **switching**: If $(s \doteq t) \in E$, s is an identifier and t is not, then we replace $s \doteq t$ by $t \doteq s$ in E.

Let $E_1 = \mathit{flattenSlicing}(E_0)$ denote the procedure which applies the above operations to E_0 until E_1 is flat and singleton.

Proposition 12.2.8 *If $E_1 = \mathit{flattenSlicing}(E_0)$, then $|E_1| = O(|E_0|)$, where $|E_i|$ denotes the total of function symbols (including identifiers) appearing in E_i, $i = 0, 1$.*

12.2 Equality with Uninterpreted Functions

Proof Let n_i be the total of function symbols (including identifiers) appearing in E_i for $i = 0, 1$. Then, $n_0 \le n_1 \le 3n_0$, because for the **flattening** operation, two copies of a new constant are added into E for each non-identifier subterm position other than the root position. Similarly, for the **slicing** operation, a new identifier is introduced for the root position of a non-identifier term. In other words, the total number of new identifier occurrences introduced by $flattenSlicing(E_0)$ cannot be more than twice of the total number of subterm positions in E_0. □

Example 12.2.9 Let $E_0 = \{f(a, g(a)) \doteq g(g(a))\}$ be the input to the modified Knuth-Bendix completion procedure. Then, $flattenSlicing(E_0)$ will return E_1:

$$E_1 = \{g(a) \doteq c_1, f(a, c_1) \doteq c_3, g(a) \doteq c_2, g(c_2) \doteq c_3\}$$

where new identifiers c_1 and c_2 are introduced by flattening and c_3 by slicing. If we feed E_1 to the Knuth-Bendix completion procedure, the procedure ends with the following rewrite rules:

(1) $g(a) \to c_1$, (2) $f(a, c_1) \to c_3$, (3) $g(c_1) \to c_3$, (4) $c_2 \to c_1$

This rewrite system is canonical (confluent and terminating), flat, and singleton. It can serve as a decision procedure for $=_{E_0}$. □

Let U be the set of all unit rules returned by the Knuth-Bendix completion procedure. Any rule in U whose left side is an identifier introduced by flattening and slicing can be removed, because this identifier occurs nowhere other than in this rule; thus, there is no chance to use this rule. For the above example, the last rule, i.e., $c_2 \to c_1$, can be deleted.

Example 12.2.10 Let $E_0 = \{f^3 a \doteq a, f^5 a \doteq a\}$. Then, $E_1 = flattenSlicing(E_0)$, where E_1 contains the following equations:

$$\{f(a) \doteq c_1, f(c_1) \doteq c_2, f(c_2) \doteq a, f(c_2) \doteq c_3, f(c_3) \doteq c_4, f(c_4) \doteq a\}$$

The first three equations come from $f^3 a \doteq a$; all the equations excluding $f(c_2) \doteq a$ come from $f^5 a \doteq a$, as we reuse the first two equations.

Feeding E_1 to the Knuth-Bendix procedure, the first three equations will produce three rewrite rules:

$$\{(1)\ f(a) \to c_1,\ (2)\ f(c_1) \to c_2,\ (3)\ f(c_2) \to a\}$$

The fourth equation, $f(c_2) \doteq c_3$, is rewritten to $a \doteq c_3$, from which we obtain

(4) $c_3 \to a$.

The fifth equation, $f(c_3) \doteq c_4$, is rewritten by (4) and (1) to $c_1 \doteq c_4$, from which we obtain

$$(5) \quad c_4 \to c_1.$$

The last equation, $f(c_4) \doteq a$, is rewritten by (5) and (2) to $c_2 \doteq a$, from which we obtain

$$(6) \quad c_2 \to a,$$

which rewrites (2) to $f(c_1) \to a$. Furthermore, (6) reduces (3) to $f(a) \doteq a$, then by (1) to $c_1 \doteq a$, from which we obtain

$$(7) \quad c_1 \to a.$$

(1) becomes $f(a) \to a$ by (7). (7) also reduces (2) to $f(a) \doteq a$, then by (1) to $a \doteq a$.

The final rewrite system is

$$\{(1)\ f(a) \to a,\ (4)\ c_3 \to a,\ (5)\ c_4 \to a,\ (6)\ c_2 \to a,\ (7)\ c_1 \to a\}$$

If we remove those rules in U whose left side is a new identifier, then only the first rule, i.e., $f(a) \to a$, is needed for the decision procedure of $=_{E_0}$. □

During the execution of the Knuth-Bendix completion procedure (Proc. 7.2.15), let E be the set of remaining equations and R be the current set of rewrite rules, both E and R are ground, flat, and singleton. R can be partitioned into U and N, i.e., $R = U \cup N$, where U are unit rules and N are non-unit rules. The above two examples illustrate some properties of (E, R) summarized by the following lemma.

Lemma 12.2.11 *Suppose the input E to the Knuth-Bendix completion procedure is a set of ground equations and R is the current set of rewrite rules. When a new rewrite rule $l \to r$ is made, for any $x \to y$ of R, x is not rewritable by $l \to r$ at the root of x. Moreover, if x is a constant, then $x \to y$ will stay in R forever.*

Proof If $l \to r$ can rewrite x at its root, then $l = x$, because they are ground. So, l can be rewritten to y by $x \to y$, a contradiction to the assumption that l is in normal form by R. If x is a constant, then the only way to rewrite x is at its root and that is not possible. □

We may observe the following properties when a rewrite system is kept flat and singleton in the Knuth-Bendix completion procedure:

1. By Lemma 12.2.11, for any $u \to v \in U$, i.e., it is a unit rule, $u \to v$ will never be deleted by any other rewrite rule during the execution of Proc. 7.2.15.
2. When a new unit rule $l \to r$ is made, this unit rule can rewrite some rules of N where l appears in the rules.

12.2 Equality with Uninterpreted Functions 451

3. When a new non-unit rule $l \to r$ is made, where r is an identifier and l is not, this non-unit rule cannot rewrite any rule of R, because $l \to r$ cannot rewrite the left side of any rule of R at its root (Lemma 12.2.11); the subterms at other positions of the rules are all identifiers, because R is flat and singleton.

Based on the above properties and using some ideas of Kapur's algorithm as well as Nieuwenhuis and Oliveras' algorithm [3], the Knuth-Bendix procedure is modified by using the following three data structures for each identifier c:

- $nf(c)$: the normal form of c by U; initially, $nf(c) = c$.
- $class(c)$: the set of all identifiers whose normal form is c, i.e., $class(c) = \{b : nf(b) = c\}$; initially, $class(c) = \{c\}$.
- $use(c)$: the set of rules in N which uses c in its left side, i.e., $use(c) = \{l \to r \in N : l \text{ contains } c\}$; initially $use(c) = \emptyset$.

Procedure 12.2.12 The procedure $KBG(E, \succ)$ takes a set E of ground, flat, and singleton equations and a total order $\succ$ where $f \succ c$ for any function symbol f and identifier c and generates a canonical rewrite system from E.

proc $KBG(E)$
1    $U := N := \emptyset$; initialize $nf()$, $class()$, and $use()$
2   **while** $(E \neq \emptyset)$ **do**
3      $(s \doteq t) := pickEquation(E); E := E - \{s \doteq t\}$
4      $s := NF(s); t := NF(t)$ //normalize s and t by $R = U \cup N$
5   **if** $(s = t)$ **continue** // identity is discarded
6   **else if** $identifier?(s)$ // if s is also an identifier
7   **if** $(|class(t)| < |class(s)|)$ $x := s; s := t; t := x$ // switch s and t
8        $U := U \cup \{s \to t\}$ // $s \to t$ is a unit rule
9   **for** $c \in class(s)$ **do** $nf(c) := t$
10       $class(t) := class(t) \cup class(s); class(s) := \emptyset$
11   **for** $(l \to r) \in use(s)$ **do** // inter-reduction
12         $N := N - \{l \to r\}; E := E \cup \{l \doteq r\}$
13   **else** // $s \to t$ is non-unit
14       $N := N \cup \{s \to t\}$ // suppose $s = f(a_1, \ldots, a_k)$
15   **for** $i := 1$ to k **do** $use(a_i) := use(a_i) \cup \{s \to t\}$
16   **return** $U \cup N$

proc $NF(t)$
17   **if** $identifier?(t)$ **return** $nf(t)$
18   **else** assume $t = f(t_1, \ldots, t_k)$
19     $s := f(NF(a_1), \ldots, NF(a_k))$
20   **if** s is flat and $(s \to b) \in N$ **return** $nf(b)$
21   **else return** s

Like the Knuth-Bendix completion procedure (Proc. 7.2.15), the above procedure takes one equation from E at each iteration of the main loop, normalizes the equation by $R = U \cup N$, tries to make a rewrite rule from the normalized equation, and adds the rule into R while keeping R inter-reduced. Some rules of R go back to E as the result of inter-reduction. The main loop terminates when E becomes empty. However, *KBG* has special features which make it efficient than the general completion procedure:

- Critical pair computation is omitted because inter-reduction is sufficient for ground equations.
- The normalization procedure $NF(t)$ uses $nf(c)$ instead of U for each identifier c; the rewrite rules of N can be used at most once because t is flat. In fact, $NF(t)$ can be done in $O(|t|)$ time since we can use a hash table or a trie structure to store N. Note that, since t is flat, $|t|$ is a constant if the arity of any function is a constant.
- When a unit rule $s \to t$ is created, we update $nf(c)$ by t for any $c \in class(s)$ (line 9) and merge $class(s)$ into $class(t)$ (line 10). Recall that $nf(c)$ is the normal form of c by U. Due to the existence of $nf(c)$, rewrite rules in U are superficial: they are neither deleted nor simplified.
- The data structure $use(s)$ records the set of non-unit rules which use identifier s in its left side (line 14). This data structure is useful at line 9 to remove rules $l \to r$ from N when l is rewritable by a new rule $s \to t$.
- When a non-unit rule is added into N (line 14), N as well as $R = U \cup N$ remains inter-reduced, so there is no need to rewrite N by the new rule.

The correctness of the above procedure comes directly from Theorem 7.2.22. The time complexity of the procedure is given explicitly by the following theorem.

Theorem 12.2.13 *If E is a set of ground, flat, and singleton equations, and n is the total number of symbols in E, then procedure* **KBG** *runs in $O(n \log(n))$.*

Proof The main loop (lines 3–15) handles $s \doteq t$ of E and each line of the main loop takes various times to execute. Line 4 takes $O(|s| + |t|)$ or $O(1)$ time if we consider the arity of any symbol is a constant. Line 9 takes $O(|class(s)|)$ time; lines 11–12 take $O(|use(s)|)$ time. All other lines take constant time. If each equation of E is processed only once, then the total time would be $O(n)$, because the sums of $|s| + |t|$, $|class(s)|$, or $|use(s)|$ for all equations $s \doteq t$ and identifies c are bound by $O(n)$. If no equations are processed more than m times, then the total time of *KBG* would be $O(nm)$.

How many times an equation can go back from N to E (line 12)? Note that the root cause of moving an equation from N to E at line 12 is that we have a new unit rule $s \to t$ (line 8) and s appears in the moved equations. When making a rewrite rule $s \to t$ from $s \doteq t$, we ensure that $|class(s)| \leq |class(t)|$ (line 7) and then merge $class(s)$ into $class(t)$ (line 10). In other words, an equation moves from N to E only when (the left side of) this equation contains an identifier whose class will be merged into another class. Since a smaller class cannot be merged into a larger class more than $log(n)$ times, no equation can be moved from N to E more than $log(n)$ times. So, the total complexity of *KBG* is bound by $O(n \log(n))$. □

Theorem 12.2.14 *The congruence closure of ground equality can be decided in $O(n \log(n))$, where n is the input size, i.e., the total number of symbols appearing in the input.*

Proof Given a set A of *ground equality*, let $E \subseteq A$ be all the equations of A. At first, we call *KBG(flattenSlicing(E))* to obtain a decision procedure R for $=_E$. Next, for each inequality $(s \neq t) \in A$, we check if $NF(s) = NF(t)$ using R; if yes, return "unsatisfiable." The whole process takes $O(n \log(n))$ time (Theorem 12.2.13) as $NF(s)$ and $NF(t)$ take $O(|s| + |t|)$ time. Note that s and t are not necessarily flat.
□

While Kapur's ideas of flattening and slicing are used in the above procedure, flattening is optional as the same result can be achieved by slicing alone (assuming each hash function takes constant time); the proof is left as an exercise. Flattening and slicing have been used successfully by Kapur to handle function symbols which are commutative and associative. Nieuwenhuis and Oliveras' algorithm can also produce automated proofs of $s =_E t$.

12.3 Linear Arithmetic

Linear arithmetic is a theory of first-order logic where the formulas are quantifier-free and contain existentially quantified variables over a domain D, where D can be of type rational or integer and support usual arithmetic operators such as $+$, $-$, $\cdot$, $=$, $\leq$, $\geq$, etc., in addition to the conventional Boolean operators. The formulas are *linear* as the multiplication, i.e., $\cdot$, is not allowed over two variables. An atomic linear arithmetic formula looks like

$$a_1 \cdot x_1 + \cdots + a_n \cdot x_n \leq b, \quad \text{or} \quad a_1 x_1 + \cdots + a_n x_n \leq b,$$

where $\leq$ can be replaced by other relation operators, variables $x_1, \ldots, x_n$ are of type D, and constants $a_1, \ldots, a_n, b \in D$. Complex linear arithmetic formulas can be obtained from atomic formulas by Boolean operators as in first-order logic. The theory of linear arithmetic also defines the interpretation of linear arithmetic formulas: all arithmetic functions, predicates, and constants are interpreted by their well-known mathematical definitions, and all Boolean operators are interpreted as we are used to from first-order logic.

In integer linear arithmetic, the constraints using predicates other than $\leq$ can be easily converted into a standard form, as shown by the following table.

input	output	input	output
$A \geq B$	$B \leq A$	$\neg(A > B)$	$A \leq B$
$A = B$	$A \leq B \wedge B \leq A$	$\neg(A < B)$	$B \leq A$
$A > B$	$B + 1 \leq A$	$\neg(A \geq B)$	$A + 1 \leq B$
$A < B$	$A + 1 \leq B$	$\neg(A \leq B)$	$B + 1 \leq A$

Example 12.3.1 Consider the constraint $\neg(x_1 - 2x_2 < 3)$. Apply $\neg(A < B) \equiv B \leq A$, $3 \leq x_1 - 2x_2$. Moving 3 to the right and $x_1 - 2x_2$ to the left of $\leq$, we have $-x_1 + 2x_2 \leq -3$. □

The top two rows of the above table can be also used for preprocessing linear programming problems. Constraints like $x \geq a$ for $a \neq 0$ can be replaced by $y \geq 0$ with $x = y + a$.

Linear arithmetic is very expressive as numerous problems in scientific and industrial computation can be expressed in linear arithmetic. Linear arithmetic is relevant to automated reasoning because most programs use arithmetic variables (e.g., integers) and perform arithmetic operations on those variables. Therefore, software verification has to deal with arithmetic. Linear arithmetic has been thoroughly investigated in at least two directions: (i) optimization via linear programming (LP), integer programming (ILP), and mixed real integer programming (MILP) and (ii) first-order quantifier elimination. In this chapter, we will focus on (i) only. In particular, we are interested in the satisfiability of linear arithmetic constraint problems in the context of the combination of theories, as they occur, e.g., in SMT (satisfiability modulo theories) solving.

12.3.1 Simplex Method by Example

Dantzig's simplex method (or simplex algorithm), invented in 1947 [4], is a popular method for linear equational systems. Each equational system defines an optimization problem over the domain of reals by an objective function and a list of linear arithmetic equations. The simplex algorithm is regarded by many as one of the ten algorithms with the greatest influence on the development and practice of science and engineering in the twentieth century.

Let us consider the following example of a linear equational system:

$$\text{Maximize } 4x_1 + 3x_2 \qquad (12.1)$$
$$\text{subject to}$$
$$x_1 + 2x_2 + x_3 \qquad\qquad = 16$$
$$x_1 + x_2 + x_4 \qquad = 9$$
$$3x_1 + 2x_2 + x_5 = 24$$
$$x_1, x_2, x_3, x_4, x_5 \geq 0$$

12.3 Linear Arithmetic

For the objective function $4x_1 + 3x_2$, we may introduce a new variable x_0 and write it in standard form as $x_0 - 4x_1 - 3x_2 = 0$, then (12.1) becomes (12.2):

Maximize x_0 where (12.2)

$$
\begin{aligned}
(a) \quad & x_0 - 4x_1 - 3x_2 && = 0 \\
(b) \quad & x_1 + 2x_2 + x_3 && = 16 \\
(c) \quad & x_1 + x_2 + x_4 && = 9 \\
(d) \quad & 3x_1 + 2x_2 + x_5 && = 24 \\
& x_1, \ x_2, \ x_3, \ x_4, \ x_5 \geq 0
\end{aligned}
$$

Note that (12.2) has a simple matrix representation, which is used for the implementation of the simplex method.

x_0	x_1	x_2	x_3	x_4	x_5	rhs
1	−4	−3	0	0	0	0
0	1	2	1	0	0	16
0	1	1	0	1	0	9
0	3	2	0	0	1	24

Given a linear equational system, a *basic variable* is a variable which appears only once in the system. In its standard form, the coefficient of a basic variable is 1. For the present example, x_0, x_3, x_4, and x_5 are basic variables; x_1 and x_2 are nonbasic variables. It is easy to spot the basic variables in the matrix where the column corresponding to the variable has only one nonzero entry. The objective function x_0 is always basic. In a linear equational system, if the number of variables is more than the number of equations, it is easy to locate a basic variable for each equation by arithmetic operations.

In the simplex method, a basic variable and a nonbasic variable can exchange their status if this exchange increases the objective function value. The simplex method repeats this kind of exchange until an optimal value is obtained.

In the present example, x_0 is a function of x_1, x_2, x_3, x_4, x_5. The value of $(x_1, x_2, x_3, x_4, x_5)$ defines a point in the search space. The points considered by the simplex method are those points where nonbasic variables are zero and the values of the basic variables are decided by the equations. In the present example, we set nonbasic variables $x_1 = x_2 = 0$. From the four equations of (12.2), we have $x_0 = 0$, $x_3 = 16$, $x_4 = 9$, and $x_5 = 24$. So, the starting point considered by the simplex method is $(0, 0, 16, 9, 24)$.

The simplex method searches for an optimal solution by repeatedly exchanging a basic variable and a nonbasic variable. Next, let us see how to choose basic and nonbasic variables for exchange.

From (12.2)(a), $x_0 = 4x_1 + 3x_2$. In order to increase the value of x_0, we need to increase x_1 or x_2, one variable at a time. If x_1 is increased by 1, x_0 gains 4. If x_2 is increased by 1, x_0 gains 3. Obviously, x_1 is preferred for the faster increment of x_0. Hence, x_1 is chosen for exchange.

Once the nonbasic variable x_1 is chosen, what basic variable, x_3, x_4, or x_5, should we replace? To answer this question, consider (12.2)(b) – (d) under the condition of $x_2 = 0$:

$$\begin{array}{ll} x_1 + x_3 = 16 & x_3 = 16 - x_1 \geq 0 \\ x_1 + x_4 = 9 \quad \text{which give us} & x_4 = 9 - x_1 \geq 0 \\ 3x_1 + x_5 = 24 & x_5 = 24 - 3x_1 \geq 0 \end{array}$$

because $x_3, x_4, x_5 \geq 0$. From $16 - x_1 \geq 0$, $x_1 \leq 16$; from $9 - x_1 \geq 0$, $x_1 \leq 9$; from $24 - 3x_1 \geq 0$, $x_1 \leq 8$. Since x_1 must satisfy all the three inequalities, we choose $x_1 = min\{16, 9, 8\} = 8$. Since this value comes from $x_5 = 24 - 3x_1 \geq 0$, x_5 is chosen for becoming non-basic. The equation $3x_1 + 2x_2 + x_5 = 24$ is called the *pivot equation* and x_5 is called the *pivot variable*; the process of choosing a pivot equation is called *pivoting*.

From $x_1 = 8$ and $x_1 + x_3 = 16$, x_3 has a new value of 8. From $x_1 + x_4 = 9$, $x_4 = 1$. So, the next search point considered by the simplex method is $(x_1, x_2, x_3, x_4, x_5) = (8, 0, 8, 1, 0)$, where x_2 and x_5 are non-basic.

Now, we need to normalize (12.2) so that basic variables appear only once and x_0 is expressed as a function of nonbasic variables. This can be done as follows: from the pivot equation (12.2)(c), we have

$$x_1 + (2/3)x_2 + (1/3)x_5 = 8 \qquad (12.3)$$

Adding (12.3) 4 times to (12.2)(a), we obtain

$$x_0 - (1/3)x_2 + (4/3)x_5 = 32$$

Subtracting (12.3) from (12.2)(b), we obtain

$$(4/3)x_2 + x_3 - (1/3)x_5 = 8$$

Subtracting (12.3) from (12.2)(c), we obtain

$$(1/3)x_2 + x_4 - (1/3)x_5 = 1$$

Putting these equations together, we obtain a new equational system in standard form:

$$\begin{array}{ll} \text{Maximize } x_0 \text{ where} & \qquad (12.4) \\ (a) \; x_0 \quad - (1/3)x_2 \qquad\qquad + (4/3)x_5 = 32 & \\ (b) \qquad\qquad (4/3)x_2 + x_3 \quad - (1/3)x_5 = 8 & \\ (c) \qquad\qquad (1/3)x_2 \qquad + x_4 - (1/3)x_5 = 1 & \\ (d) \quad x_1 + (2/3)x_2 \qquad\qquad + (1/3)x_5 = 8 & \\ x_1, \quad x_2, \; x_3, \; x_4, \qquad x_5 \geq 0 & \end{array}$$

12.3 Linear Arithmetic

The corresponding matrix is given below:

$$
\begin{array}{c c}
\begin{matrix} x_0 & x_1 & x_2 & x_3 & x_4 & x_5 & rhs \end{matrix} \\
\begin{bmatrix}
1 & 0 & -1/3 & 0 & 0 & 4/3 & 32 \\
0 & 0 & 4/3 & 1 & 0 & -1/3 & 8 \\
0 & 0 & 1/3 & 0 & 1 & -1/3 & 1 \\
0 & 1 & 2/3 & 0 & 0 & 1/3 & 8
\end{bmatrix}
\end{array}
$$

Note that the above matrix can be obtained from the matrix representation of (12.2) by the corresponding arithmetic operations on (12.2). For example, (12.3) is the same as "dividing the fourth row of the matrix for (12.2) by 3." "Adding (12.3) 4 times to (12.2)(a)" can be done by "adding the multiplication of the fourth row by 4 to the first row of the matrix." "Subtracting (12.3) from (12.2)(b)" can be done by "subtracting the fourth row of the matrix from the second row of the matrix." "Subtracting (12.3) from (12.2)(c)" can be done by "subtracting the fourth row of the matrix from the third row of the matrix." The result is exactly the matrix for (12.4). This is how the simplex method is implemented.

At the point $(8, 0, 8, 1, 0)$, the value of x_0 is 32, increasing from its old value 0. Is this the optimal value for x_0? From (12.4)(a), we have

$$x_0 = (1/3)x_2 - (4/3)x_5 + 32$$

Now, $x_0 = 32$ when $x_2 = x_5 = 0$. If x_2 is increased by 1, x_0 gains $1/3$. That is, there is a space for increment of x_0 by choosing x_2 as a new basic variable.

The process of selecting an existing basic variable for exchange is the same as before: finding the pivot variable in a pivot equation. Assuming $x_5 = 0$, from (12.4)(b) – (d), we have

$$
\begin{aligned}
(4/3)x_2 + x_3 &= 8 & x_3 &= 8 - (4/3)x_2 \geq 0 \\
(1/3)x_2 + x_4 &= 1 \quad \text{which give us} \quad & x_4 &= 1 - (1/3)x_2 \geq 0 \\
x_1 + (2/3)x_2 &= 8 & x_1 &= 8 - (2/3)x_2 \geq 0
\end{aligned}
$$

From $8 - (4/3)x_2 \geq 0$, $x_2 \leq 6$; from $1 - (1/3)x_2 \geq 0$, $x_2 \leq 3$; from $8 - (2/3)x_2 \geq 0$, $x_2 \leq 12$. Hence, we choose $x_2 = 3$ to satisfy all the three inequalities, and the pivot equation is $(1/3)x_2 + x_4 - (1/3)x_5 = 1$ ((12.4)(c)). Now, multiply (12.4)(c) by 3, we obtain

$$x_2 + 3x_4 - x_5 = 3 \qquad (12.5)$$

Multiply (12.5) by $1/3$ and add the result to (12.4)(a), we have

$$x_0 + x_4 + x_5 = 33 \qquad (12.6)$$

Multiply (12.5) by $-4/3$ and add the result to (12.4)(b), we have

$$x_3 - 4x_4 + x_5 = 4 \qquad (12.7)$$

Multiply (12.5) by $-2/3$ and add the result to (12.4)(d), we have

$$x_1 - 2x_4 + x_5 = 6 \qquad (12.8)$$

The matrix representation of these equations (12.5)–(12.8) is given as follows:

x_0	x_1	x_2	x_3	x_4	x_5	rhs
1	0	0	0	1	1	33
0	0	0	1	-4	1	4
0	0	1	0	3	-1	3
0	1	0	0	-2	1	6

From (12.6), we have $x_0 = 33 - x_4 - x_5$. Since $x_4, x_5 \geq 0$, x_0 has the maximal value 33 when $x_4 = x_5 = 0$. The values of basic variables can be computed from the equations with $x_4 = x_5 = 0$: $x_1 = 6$, $x_2 = 3$, and $x_3 = 4$. That is, x_0 has the optimal value 33 at point $(x_1, x_2, x_3, x_4, x_5) = (6, 3, 4, 0, 0)$. The simplex method stops with this result.

From this example, we can see that the simplex method is a local search procedure which starts from $(0, 0, 16, 9, 24)$, finds its best neighboring point $(8, 0, 8, 1, 0)$ through the pivot equation, and then finds the next best point $(6, 3, 4, 0, 0)$. The search stops when it reaches a local optimum. Fortunately, since all the constraints are linear, a local optimum is the global optimum. The detailed discussion on this statement is out of the scope of the book.

12.3.2 The Simplex Algorithm

Suppose the linear equational system has m constraints and n variables, with the first $m + 1$ variables as basic variables [4]. Hence, we assume $m \leq n$ because, when $m > n$, either the linear system has no solutions or one of its constraints is redundant. Its standard form is given below:

Maximize x_0 where (12.9)

$$\begin{aligned}
x_0 + &\cdots + a_{0,m+1}x_{m+1} + \cdots + a_{0,n}x_n = b_0 \\
x_1 + &\cdots + a_{1,m+1}x_{m+1} + \cdots + a_{1,n}x_n = b_1 \\
x_2 + &\cdots + a_{2,m+1}x_{m+1} + \cdots + a_{2,n}x_n = b_2 \\
&\ddots \qquad\qquad\qquad\qquad\qquad\qquad \vdots \\
x_m + a_{m,m+1}&x_{m+1} + \cdots + a_{m,n}x_n = b_m \\
&x_1, x_2, \ldots, x_n \geq 0
\end{aligned}$$

12.3 Linear Arithmetic

Theorem 12.3.2 (Optimality Criterion) *For the linear equational system of (12.9), if $a_{0,j} \geq 0$ for $j = m+1, \ldots, n$, then the maximal value of x_0 is b_0 and is attained at the point $(b_1, b_2, \ldots, b_m, 0, \ldots, 0)$.*

Proof From the first row of (12.9), we have $x_0 = b_0 - (a_{0,m+1}x_{m+1} + \cdots + a_{0,n}x_n)$. Since $a_{0,j} \geq 0$ and $x_j \geq 0$ for $j = m+1, \ldots, n$, $a_{0,m+1}x_{m+1} + \cdots + a_{0,n}x_n \geq 0$. Hence, the maximal value of x_0 is b_0 when $x_j = 0$ for $j = m+1, \ldots, n$. In this situation, $x_i = b_i$ for $i = 1, \ldots, m$. □

Theorem 12.3.3 (Unbounded Criterion) *For the linear equational system of (12.9), if there is an index j, $m+1 \leq j \leq n$, such that $a_{0,j} < 0$ and $a_{i,j} \leq 0$ for all $i = 1, \ldots, m$ then the optimal value of x_0 is ∞.*

Proof If there is an index j, $m+1 \leq j \leq n$, $a_{0,j} < 0$, and $a_{i,j} \leq 0$ for all $i = 1, \ldots, m$, then we have infinite many points, where x_j can have any value, $x_i = b_i - a_{i,j}x_j$ for $i = 1, \ldots, m$, and $x_i = 0$ for $m+1 \leq i \leq n$ and $i \neq j$. Each of these points defines a feasible solution point for x_0, i.e., $x_0 = b_0 - a_{0,j}x_j$. Thus, x_0 goes to ∞ when x_j goes to ∞ because $a_{0,j} < 0$. □

Example 12.3.4 Suppose we wish to maximize x_0 with

$$\begin{aligned} x_0 \quad & + 2x_3 - 3x_4 = 10 \\ x_1 \quad & + 2x_3 - x_4 = 3 \\ x_2 - & x_3 - 2x_4 = 2 \end{aligned}$$

Let $x_3 = 0$, then $(x_0, x_1, x_2, x_3, x_4) = (3x_4 + 10, x_4 + 3, 2x_4 + 2, 0, x_4)$ is a feasible solution point for all the constraints. The value of x_0 goes to ∞ when x_4 does. □

The above two theorems give us two halting criteria for the simplex algorithm: one gives us the optimum solution and the other gives us an unbounded solution. Let *haltCondition(M)* be the procedure which returns true iff the conditions in the above two theorems are true. Then, the outline of the simplex algorithm can be described below, following the illustration of the previous subsection.

proc *simplex(S)*
1   Convert S in standard form and store it in matrix M.
2   Let r be the row of M with basic variable x_0, which is to be maximized.
3   **while** $\neg$ *haltCondition(M)* **do**
4   select a nonbasic variable v_1 in r with the minimal coefficient.
5   select a basic variable v_2 by pivoting with respect to v_1.
6   switch the roles of v_1 and v_2 and convert M in standard form.

The complexity of the simplex algorithm is exponential in the worst case. Since there are polynomial decision procedures for linear programming, this means that there are at least theoretically faster decision procedures. But in contrast to those polynomial decision procedures, the simplex algorithm has all properties necessary for an efficient linear programming solver: it produces minimal conflict explanations, handles backtracking efficiently, and is highly incremental. In practice, these properties are more important than the difference between polynomial and exponential worst-case complexity. It also helps that the simplex algorithm rarely reaches its worst case in practice.

12.3.3 Linear Programming

The name "linear programming" existed before the age of computers. Here "programming" means "table," as the data of a linear programming problem, or *linear program* for short, are usually stored in a table.

The standard form of a linear program is given as follows:

$$\text{Maximize } a_{0,1}x_1 + \cdots + a_{0,n}x_n \quad (12.8)$$
subject to
$$a_{1,1}x_1 + a_{1,2}x_2 + \cdots + a_{1,n}x_n \leq b_1$$
$$a_{2,1}x_1 + a_{2,2}x_2 + \cdots + a_{2,n}x_n \leq b_2$$
$$\cdots$$
$$a_{m,1}x_1 + a_{m,2}x_2 + \cdots + a_{m,n}x_n \leq b_m$$
$$x_1, x_2, \ldots, x_n \geq 0$$

where each $a_{i,1}x_1 + a_{i,2}x_2 + \cdots + a_{i,n}x_n \leq b_i$ is an *atomic linear constraint* and we are looking for values of x_i, $1 \leq i \leq n$, such that the objective function is maximized.

Example 12.3.5 Below is a linear program with $n = 2$ and $m = 3$:

$$\text{Maximize } 4x_1 + 3x_2 \quad (12.9)$$
subject to
$$x_1 + 2x_2 \leq 16$$
$$x_1 + x_2 \leq 9$$
$$3x_1 + 2x_2 \leq 24$$
$$x_1, x_2 \geq 0$$

Its solution will be discussed shortly. □

Many application problems can be specified as linear programs. To solve this problem using the simplex method, we introduce m new variables, called *slack variables*, one for each of the m constraints, to convert them into equations.

12.3 Linear Arithmetic

Example 12.3.6 Continuing from the previous example, we introduce three variables, x_3, x_4, and x_5, and convert (12.8) into a linear equational system:

$$\text{Maximize } 4x_1 + 3x_2 \tag{12.10}$$
subject to
$$\begin{aligned} x_1 + 2x_2 + x_3 &= 16 \\ x_1 + x_2 + x_4 &= 9 \\ 3x_1 + 2x_2 + x_5 &= 24 \\ x_1, x_2, x_3, x_4, x_5 &\geq 0 \end{aligned}$$

(12.10) is identical to (12.1) in Sect. 12.3.1 and we have seen how (12.1) is solved by the simplex method. Hence, the optimal solution of $4x_1 + 3x_2$ is 33 when $x_1 = 6$ and $x_2 = 3$. □

By definition, linear programming is an optimization problem. We can convert it into a decision problem by introducing a constant c, so that "Maximize $a_{0,1}x_1 + \cdots + a_{0,n}x_n$" becomes "Decide $a_{0,1}x_1 + \cdots + a_{0,n}x_n \geq c$." Another decision problem related to linear programming is to decide if the set of all constraints is consistent. That is, can we find values of $x_1, x_2, \ldots, x_n$ such that all constraints are true? This problem can be easily solved if we convert linear programming into linear equational system through slack variables.

12.3.4 Integer Programming

The standard form of an integer program is the same as that of linear programming, with the additional constraints that the variables take only integers:

$$\text{Maximize } a_{0,1}x_1 + \cdots + a_{0,n}x_n \tag{12.11}$$
subject to
$$\begin{aligned} a_{1,1}x_1 + a_{1,2}x_2 + \cdots + a_{1,n}x_n &\leq b_1 \\ a_{2,1}x_1 + a_{2,2}x_2 + \cdots + a_{2,n}x_n &\leq b_2 \\ &\cdots \\ a_{m,1}x_1 + a_{m,2}x_2 + \cdots + a_{m,n}x_n &\leq b_m \\ x_1, x_2, \ldots, x_n \text{ are integers}, \end{aligned}$$

where $a_{i,j}$, b_i are assumed to be integers.

If the variables take only 0 or 1, integer programming becomes 0-1 linear programming. While linear programming can be solved in polynomial time, 0-1 linear programming is NP-complete. In other words, the chance of finding a polynomial time algorithm for 0-1 linear programming, or integer programming in general, is slim.

Fig. 12.2 An example of integer programming

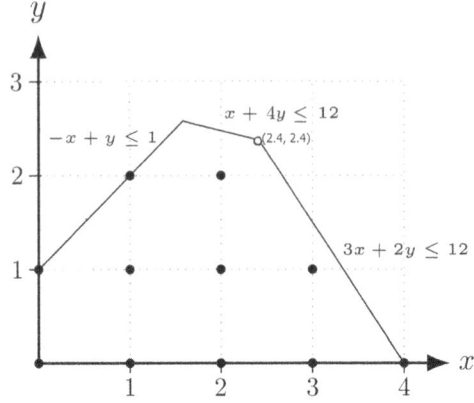

Example 12.3.7 Consider the following example of integer programming:

$$\text{Maximize } 4x + 3y \quad (12.12)$$
subject to
$$-x + y \leq 1$$
$$x + 5y \leq 12$$
$$3x + 2y \leq 12$$
x, y are non-negative integers.

If we drop the constraint that "x, y are integers," the problem becomes an example of linear programming. An optimal solution can be found by the simplex method with the optimal solution $(x, y) = (2.4, 2.4)$, i.e., $4x + 3y = 16.8$. With the constraint that x, y are integers, there are 11 feasible solutions (black dots in Fig. 12.2) and one optimal solution, i.e., $(x, y) = (4, 0)$, or $4x + 3y = 16$. □

Whereas the simplex method is effective for solving linear programs, there is no single effective technique for solving integer programs. Instead, a number of techniques have been developed, and the performance of any particular technique appears to be highly problem dependent. Methods to date can be classified broadly as one of three approaches:

- Local search techniques: Start with some promising candidate solutions, and then, choose some variables (e.g., those occurring in the violated constraints) to modify their values (e.g., add or subtract 1) in turn to find the neighbors of these candidate solutions. Select the best among these neighbors and repeat. Candidate solutions can be found by (1) relaxing integer programming to linear programming; (2) using the simplex method to find an optimal solution for the relaxed linear programming; and (3) rounding up/down the optimal solution to integer solutions.
- Enumeration techniques: The well-known enumeration technique is the branch-and-bound procedure. This procedure is very much like the branch-and-bound

12.3 Linear Arithmetic

procedure for maximum satisfiability (Sect. 4.4.4) and especially effective for 0-1 linear programming.
- Cutting-plane techniques: The basic idea of the cutting plane method is to cut off parts of the feasible region of the linear programming search space, so that the optimal integer solution becomes an extreme point and therefore can be found by the simplex method. In practice, the branch-and-bound method outperforms mostly the cutting plane method. Cutting plane is the first algorithm developed for integer programming that could be proved to converge in a finite number of steps. Even though the technique is considered not efficient, it has provided insights into integer programming that have led to other, more efficient, algorithms.

Example 12.3.8 In Example 12.3.7, $(x, y) = (2.4, 2.4)$ is an optimal solution of the relaxed linear programming problem. The round-down solution is $(x, y) = (2, 2)$, which is one of the starting points for local search. The neighbors of $(x, y) = (2, 2)$ are $\{(1, 1), (1, 2), (1, 3), (2, 1), (2, 2), (2, 3), (3, 1), (3, 2), (3, 3)\}$. From $(2, 2)$, the local search will find $(3, 1)$, then $(4, 0)$, the optimal solution.

Applying the enumeration techniques to this example, we work on y first with $y = 0$. The constraints give us $0 \leq x \leq 4$ and the optimal value is $(x, y) = (4, 0)$. When $y = 1, 2$, the optimal value is $(x, y) = (3, 1), (2, 2)$, respectively. The overall optimal value is $(x, y) = (4, 0)$.

Applying the cutting-plane techniques to this example, we may add $x < 2$ and $x \geq 2$, respectively, to the original constraints and search optimal solutions for the more constrained subproblems. The overall optimal solution comes from $x \geq 2$. □

In addition, several composite procedures have been proposed, which combine techniques using several of these approaches. In fact, there is a trend in computer systems for integer programming to include a number of approaches and possibly utilize them all when analyzing a given problem. The interested reader may consult a textbook on integer programming for the details of these techniques.

12.3.5 Difference Logic

Difference logic is a subclass of linear arithmetic where each atomic constraint contains two variables, and the coefficients of the two variables are 1 and -1. For instance, $x_1 - x_2 \leq 3$, or equivalently $x_1 \leq x_2 + 3$, is a difference constraint. The type of variables can be either real or integer, giving us variation of the same problem. A solution is an assignment of all variables with all constraints being satisfied.

These difference constraints naturally arise in many applications. For instance, the array bounds' check in a program and the timing constraints in job scheduling can be specified as difference constraints. Anther application is real-time scheduling: the time that is requested by a task can be modeled as the minimal difference between this task and the next one: each variable corresponds to a machine and a task and represents the time when this task is executed on this machine. If the

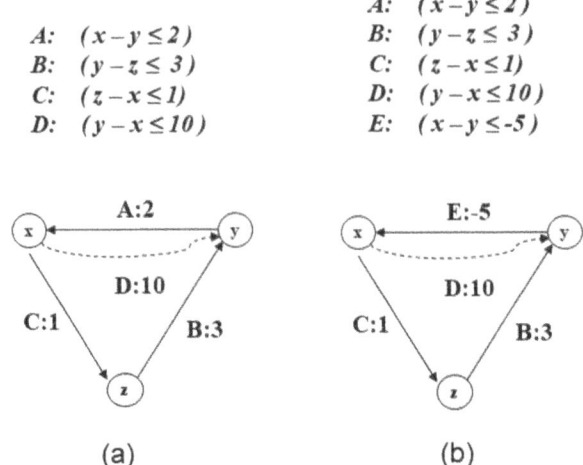

Fig. 12.3 Two examples of integer difference arithmetic

duration of each execution is fixed, one can use difference arithmetic to express constraints like earliest start and end, order of machine switches, and sequentially of the executions.

There is a well-known, efficient decision procedure for difference constraints. Given a set of difference constraints, one can reduce the problem of checking its satisfiability to the problem of detecting negative cycles in an appropriately generated graph. Then, any of the negative-cycle-detection algorithms can be used to decide the given constraints. For instance, the classic Bellman-Ford algorithm can decide m difference constraints on n variables in $O(nm)$ time and $O(n+m)$ space complexity.

Example 12.3.9 Consider the following set of integer difference constraints:

$$(A): (x - y \le 2), \quad B: (y - z \le 3), \quad (C): (z - x \le 1), \quad (D): (y - x \le 10)$$

The directed weighted graph corresponding to this set is shown in Fig. 12.3a. Obviously, there exist no negative cycles in the graph. There exist infinite assignments to x, y, and z, such that all the constraints are satisfied. For instance, $(x, y, z) = (0, 0, 0)$, or $(2, 6, 3)$.

If we add $(E): (x - y \le -5)$ to the above set of constraints, then (E) makes (A) trivially true, thus, replacing (A). The new graph is (b) of Fig. 12.3. Now, there exists a negative cycle in (b): $x \Rightarrow z \Rightarrow y \Rightarrow x$. No matter how we assign values to x, y, and z, we would have $0 \le -1$, a contradiction. □

We have now seen a basic theory solver for difference arithmetic. It determines satisfiability of a conjunction of literals by checking the absence of negative cycles in the induced weighted graph.

12.4 Making DPLL(T) Practical

The development of the DPLL(T) framework is highly motivated by the SMT problems appearing in formal verification of computer designs, including software and hardware designs. These SMT problems involve a much larger range of theories than what we have presented in the previous sections. What are these problems and decision procedures for them? To make DPLL(T) practical, we need to know these problems and, more importantly, to develop decision procedures for these problems.

Some SMT problems involve more than one theory. We may call these decision procedures one by one until a model consistent with all theories is found. Is there any efficient way to combine the involved decision procedures so that a model consistent with all theories can be quickly found?

In Sect. 12.1 of this chapter, we presented the *DPLLT* algorithm, which illustrates how DPLL(T) can be implemented with minimal modification to the *DPLL* algorithm. How to obtain an efficient implementation of DPLL(T)?

The above questions are the ongoing research topics, and we will address them briefly in the current section.

12.4.1 Making DPLL(T) Efficient

For making DPLL(T) practical, we have to implement DPLL(T) efficiently. In the first section, we showed how to implement DPLL(T) easily if we have procedure *isTheoryConsistent*(σ) which implements decision procedures for a set of SMT literals corresponding to model σ. A major shortcoming of this approach is that a theory solver may detect unsatisfiability for a subset of SMT literals corresponding to a partial propositional model. We may use this information to speed up DPLL(T), since the search time spent from this partial model to a full model can be saved. To implement this idea, we need to call *isTheoryConsistent*(σ) whenever σ is expanded.

In the following, we will present a new version of *DPLLT*(X). The following assumptions and procedures will be used:

- We will use global variables in *DPLLT*(X) for X (a set of SMT clauses), $\gamma(X)$ (the abstract of X), σ (a partial model of $\gamma(X)$), and *level* (the level number of *DPLL*, which is the number of active decision literals), and *conflictClause* (the latest conflicting clause), so that they do not appear in the parameters of procedures.
- Procedure *abstraction*(X) computes $\gamma(X)$ and stores the pair $(A, \gamma(A))$ for every atomic formula A of X.
- Procedure *initialize*(C) stores non-unit clauses of C in the watch-literal data structures and returns unit clauses of C.

- Procedure *BCP(U)* performs Boolean constraint propagation on $\gamma(X)$ with unit clauses U and σ and extends σ. It will return false iff a conflicting clause is found, and this clause will be stored in *conflictClause*.
- Procedure *isTheoryConsistent(σ)* checks the consistency of SMT literals corresponding to σ. If an inconsistency is found, it will identify a subset of σ which leads to inconsistency. The negation of this subset creates a conflicting clause, to be stored in *conflictClause*, to support conflict-driven clause learning (CDCL). This is an alternative and better approach than creating a clause from the negation of all decision literals.
- Procedure *theoryPropagate(σ)* is new and optional. This is a continuation of the procedure *isTheoryConsistent(σ)*. The theory solver will compute some new literals in theory as logical consequences of the literals derived from σ. The used literals will be collected as *reason* of the derived literals. The corresponding propositional literals will be fed to *BCP(U)* as new unit clauses with the corresponding reasons. This procedure is not needed for completion. If the propagation is expensive, it can be omitted or is called when a full propositional model is found.

Following the algorithm proposed by Cesare Tinelli, the new version of *DPLLT(X)* is given below, along with procedures *propagateGivesConflict(U)* and *analyzeConflict()*.

Algorithm 12.4.1 *DPLLT(X)* takes as input a set X of SMT clauses and returns *true* if X is satisfiable and returns *false*, otherwise.

proc *DPLLT(X)*: Boolean
    // return true iff SMT clauses X is satisfiable
1    $U := initialize(abstraction(X))$ // U is the set of unit clauses in $\gamma(X)$.
2    $level := 0; \sigma := \{\}$ // σ is a partial model of $\gamma(X)$.
3   **while** (true) **do**
4   **while** (*propagateGivesConflict(U)*)
5   **if** ($level = 0$) **return** *false* // $\gamma(X)$ as well as X is unsatisfiable
6   **else** $U := analyzeConflict()$
7   *restartIfApplicable()* // implement restart strategy
8        $A := pickLiteral()$ // pick an unassigned literal as decision literal.
9   **if** ($A = nil$) // a model of $\gamma(X)$ is found.
10   **if** *isTheoryConsistent(σ)* **return** *true* // X is satisfiable.
11   **else** $U := analyzeConflict()$
12   **else** $U := \{A\}; level := level + 1$

proc *propagateGivesConflict(U)*: Boolean
    // return true iff an inconsistency is found
1    $good := true$
2   **while** ($good$) **do**
3   **if** (not *BCP(U)*) **return** *true*
        // check theory consistency of σ

12.4 Making DPLL(T) Practical

```
4       if (not isTheoryConsistent(σ)) return true
5       good := theoryPropagate(σ)
6   return false
```

proc *analyzeConflict()*
    // return a unit clause learned from a conflicting clause
1    $c := conflictClause$ // this is the latest conflicting clause
2    $U := insertNewClause(conflictAnalysis(c))$ // new level is set
3   *undo()*; // undo to the new level
4   **return** U // new clause is unit under σ after *undo()*.

The procedures *conflictAnalysis*, *undo*, and *insertNewClause* are the same as those in Sect. 4.2 for conflict-directed clause learning.

Example 12.4.2 Suppose X contains the clauses from Example 12.1.6:

$$(b \doteq c), \ (h(a) \doteq h(c)) \vee p, \ (a \doteq b) \vee \neg p \vee (a \doteq d), \ (a \doteq b) \vee (a \neq d)$$

The abstraction procedure produces the same set of propositional clauses:

$$(s), \ (t \vee p), \ (q \vee \neg p \vee r), \ (q \vee \neg r)$$

where q stands for $a \doteq b$, r for $a \doteq d$, s for $b \doteq c$, and t for $h(a) \doteq h(c)$.

After execution of line 1 of $DPLLT(X)$, $U = \{s\}$. At line 4, *propagateGivesConflict(U)* will return *false* with $\sigma = \{s\}$ (i.e., $b \doteq c$ is true).

At line 8, suppose *pickLiteral* chooses $\neg t$ (i.e., $h(a) \neq h(c)$) to be true. Now, line 12 will be executed, $U = \{\neg t\}$ and $level = 1$.

At line 4, *propagateGivesConflict(U)* performs the following steps:

1. $BCP(U)$ will make $\neg t$ and p to be true (due to $(t \vee p)$). Hence, $\sigma = \{s, \neg t, p\}$.
2. *theoryPropagate* will make $\neg q$ (i.e., $a \neq b$) to be true, since $\neg t$ (i.e., $h(a) \neq h(c)$) and s (i.e., $b \doteq c$) are true, $a \doteq b$ cannot be true. So, $Reason(\neg q)$ is $(s \wedge \neg t) \to \neg q$, or equivalently, $(\neg q \vee \neg s \vee t)$. *theoryPropagate* returns true.
3. With $\neg q$ being added into U, $BCP(U)$ will add $\neg q$ into σ and find that $(q \vee \neg p \vee r)$ becomes unit. So, r (i.e., $a \doteq d$) is added into σ, and $BCP(U)$ will find that $(q \vee \neg r)$ becomes conflicting. $BCP(U)$ returns *false* after $(q \vee \neg r)$ is stored in *conflictClause*, and *propagateGivesConflict(U)* returns *true*.

With $(q \vee \neg r)$ being stored in *conflictClause*, at line 6 of $DPLLT(X)$, *analyzeConflict()* is executed and *conflictAnalysis* performs the following resolutions:

- Start with clause $c = (q \vee \neg r)$.
- Resolve on r with $Reason(r) = (q \vee \neg p \vee r)$ and $(q \vee \neg r)$, the resolvent is $(q \vee \neg p)$.
- Resolve on q with $Reason(\neg q) = (\neg q \vee \neg s \vee t)$ and $(q \vee \neg p)$, the resolvent is $(\neg p \vee \neg s \vee t)$.

- Resolve on p with $Reason(p) = (t \vee p)$ and $(\neg p \vee \neg s \vee t)$, the resolvent is $(\neg s \vee t)$.

The new clause generated by $conflictAnalysis()$ is $(\neg s \vee t)$. This clause becomes unit after $DPLLT(X)$ backtracks to $level = 0$. $analyzeConflict()$ returns this clause at line 6 of $DPLLT(X)$ and t will become true by $BCP(U)$. Compared to Example 12.1.6, the old version of $DPLLT(X)$ needs two decision literals to find that t must be true; the new version needs only one. □

12.4.2 Combination of Theories

In software verification, formulas like the following one arise:

$$a \doteq b + 1 \wedge (A \doteq \text{write}(B, a + 1, 4) \vee f(a - 1) \neq f(b)) \wedge \text{read}(A, b + 2) \doteq 5$$

To deal efficiently with this formula, we need decision procedures for the following theories:

- The theory of uninterpreted functions (T_{EUF})
- The theory of linear integer arithmetic (T_{LIA})
- The theory of arrays (T_{arr})

T_{EUF} and T_{LIA} have been introduced in the previous sections. For T_{arr}, we need two operators over arrays, *read* and *write*:

- $read(A, i)$: read from array A at index i. It returns $A[i]$.
- $write(A, i, d)$: write d to array A at index i. It returns A after the execution of "$A[i] := d$."

The following three axioms are assumed in T_{arr}:

$$read(write(A, i, d), i) \doteq d.$$
$$read(write(A, i, d), j) \doteq read(A, j) \text{ if } i \neq j.$$
$$(\forall i \; read(A, i) \doteq read(B, i)) \leftrightarrow (A \doteq B).$$

The last axiom is called *extension axiom*, which is needed in the current example. In the example, "$A \doteq \text{write}(B, a + 1, 4)$" asks to show that array A is equal to array B after the execution of $B[a + 1] := 4$.

Example 12.4.3 We may prove the unsatisfiability of

$$a \doteq b + 1 \wedge (A \doteq \text{write}(B, a + 1, 4) \vee f(a - 1) \neq f(b)) \wedge \text{read}(A, b + 2) \doteq 5$$

as follows: by T_{LIA}, $a - 1 \doteq b$ and $a + 1 \doteq b + 2$ are true when $a \doteq b + 1$ is true. By T_{EUF}, $f(a - 1) \neq f(b)$ is false when $a - 1 \doteq b$ is true. By T_{arr}, read(

12.4 Making DPLL(T) Practical

$write(B, a+1, 4), a+1) \doteq 4$ is true and $A \doteq write(B, a+1, 4)$ is false when $read(write(B, a+1, 4), a+1) \doteq 4$, $read(A, b+2) \doteq 5$, $a+1 \doteq b+2$, and $4 \neq 5$ are true. Since both $f(a-1) \neq f(b)$ and $A \doteq write(B, a+1, 4)$ are false, the input formula is false. □

Now, the question is: given theory solvers for the three individual theories, can we combine them to obtain one for $(T_{LIA} \cup T_{arr} \cup T_{EUF})$? It appears from the proof of the above example that we may call the solver for T_{EUF} first to obtain all the equality. However, new equality may be created from T_{LIA} (e.g., $a \doteq b$ from $a \geq b$ and $b \geq a$) or T_{arr}. It is indeed a difficult task to combine theory solvers into one.

Nelson and Oppen's Combination Algorithm

Nelson and Oppen proposed a combination algorithm that will combine theory solvers under some sufficient conditions. We will use the previous example to illustrate Nelson and Oppen's ideas.

The first step is to *purify* the input, which is a set of SMT literals, by introducing interface constants and dividing the input by theories.

For the input formula in Example 12.4.3, there are three theories and one case using the DPLL(T) framework is to check the following conjunction is satisfiable or not:

$$a \doteq b+1 \wedge f(a-1) \neq f(b) \wedge read(A, b+2) \doteq 5$$

The input is divided into three groups, F_1, F_2, and F_3, plus a common group CA over the set C of interface constants. For instance, literal $(f(a-1) \neq f(b))$ converts to $\{a - 1 \doteq e_1, f(e_1) \doteq e_2, f(b) \doteq e_3, e_2 \neq e_3\}$ by introducing three constants. Literal $(read(A, b+2) \doteq 5)$ converts into $\{ read(A, e_4) \doteq e_5, b+2 \doteq e_4, 5 \doteq e_5\}$. For the whole input, we need seven interface constants, i.e., $C = \{a, b, e_1, e_2, e_3, e_4, e_5\}$.

F_1	F_2	F_3
$b+1 \doteq a$	$f(e_1) \doteq e_2$	$read(A, e_4) \doteq e_5$
$a - 1 \doteq e_1$	$f(b) \doteq e_3$	
$b+2 \doteq e_4$		
$5 \doteq e_5$		

and $CA = \{e_2 \neq e_3\}$ initially. It is evident that $F_1 \cup CA$ belongs to T_{LIA}, $F_2 \cup CA$ to T_{EUF}, and $F_3 \cup CA$ to T_{arr}.

The second step of the combination procedure is to call each theory solver on $F_i \cup CA$ to (1) check if $F_i \cup CA$ is satisfiable and (2) generate new equations among the interface constants and add them into CA.

For the example, since from $b + 1 \doteq a$ we deduce $a - 1 \doteq b$ in T_{LIA}, we add $b \doteq e_1$ into CA (because $a-1 \doteq b$ and $a-1 \doteq e_1$). Now, $CA = \{e_2 \neq e_3, b \doteq e_1\}$.

It is easy to see that $F_2 \cup CA$ is unsatisfiable in T_{EUF} (because $b \doteq e_1$ implies $f(e_1) \doteq f(b)$, thus $e_2 \doteq e_3$). Hence, the second step ends with "unsatisfiable."

The set CA is called a *constant arrangement*. If $F_i \cup CA$ is unsatisfiable in theory T_i, then the input is unsatisfiable. Otherwise, we add a new equality into CA which is true in T_i until no new equality exists. In this case, we say CA is *saturated*. When a saturated constant arrangement is found and for every i, $F_i \cup CA$ is satisfiable in T_i, we claim that the input is satisfiable. These ideas can be described by the following procedure.

proc *NelsonOppen(S)*: Boolean
    // S is a conjunction of SMT literals.
    Purify S and split S into F_i, $1 \leq i \leq n$, and CA.
    while (*true*) **do**
        for $i := 1$ **to** n **do**
            if $(F_i \cup CA)$ is T_i-unsatisfiable **return** *false*
        $X := \emptyset$
        for $i := 1$ **to** n **do**
            if $F_i \models (x \doteq y)$ for $(x \doteq y) \notin CA$
                $X := X \cup \{x \doteq y\}$
        if $X = \emptyset$ **return** *true* // CA is saturated.
        else $CA := CA \cup X$

Nelson and Oppen showed that the above procedure works correctly under two sufficient conditions.

Definition 12.4.4 A theory T is *stably infinite* if every T-satisfiable quantifier-free formula has an infinite model.

Many interesting theories are stably infinite, including theories of an infinite structure (e.g., integer arithmetic). T_{EUF} is also known to be stably infinite. However, there are interesting theories that are not stably infinite. They are theories of a finite structure, e.g., theory of bit vectors of finite size, or arithmetic modulo n, or theory of strings of bounded length.

Definition 12.4.5 A theory T is *convex* if for any set S of literals, $S \models a_1 \doteq b_1 \vee \cdots \vee a_n \doteq b_n$, then $S \models a_i \doteq b_i$ for some $1 \leq i \leq n$.

Theorem 12.4.6 *If a conjunction S of literals can be split into signature-disjoint, stably infinite, and convex theories T_i, $1 \leq i \leq n$, the $(T_1 \cup \cdots \cup T_n)$-satisfiability of S can be checked with the Nelson-Oppen algorithm.*

Proof of the above theorem can be found in [5].

Some popular theories like integer linear arithmetic T_{LIA} are not convex. For instance, in T_{LIA}, $1 \leq x \leq 2$ is equivalent to $x \doteq 1 \vee x \doteq 2$. However, neither T_{LIA}, $1 \leq x \leq 2 \models (x \doteq 1)$ is true, nor T_{LIA}, $1 \leq x \leq 2 \models (x \doteq 2)$.

Attempts to get rid of either the stably infinite condition or the convex condition have been reported with success [5].

12.4 Making DPLL(T) Practical

Example 12.4.7 Consider input $S = \{f(1) \doteq a, f(x) \doteq b, f(2) \doteq c, 1 \leq x, x \leq 2, b \doteq a - 1, b \doteq c + 1\}$. To show S is unsatisfiable, we may add *case analysis* in Nelson and Oppen's combination algorithm by considering the two cases when $x \doteq 1$ is true or false. The detail is left as exercise. □

12.4.3 SMT-LIB: A Library of SMT Problems

To see what theories involved in the SMT problems, the best place is the SMT-LIB website:

$$\texttt{smtlib.cs.uiowa.edu/index.shtml}$$

According to the website, SMT-LIB, the satisfiability modulo theories library, is an international initiative aimed at facilitating research and development in SMT. Since its inception in 2003, the initiative has pursued these aims by focusing on the following concrete goals:

- Provide standard rigorous descriptions of background theories used in SMT systems.
- Develop and promote common input and output languages for SMT solvers.
- Connect developers, researchers, and users of SMT, and develop a community around it.
- Establish and make available to the research community a large library of benchmarks for SMT solvers.
- Collect and promote software tools useful to the SMT community.

In SMT-LIB, the problems are grouped by theories, and Fig. 12.4 shows the structure of problem groups. Each group has a name that evokes the theories

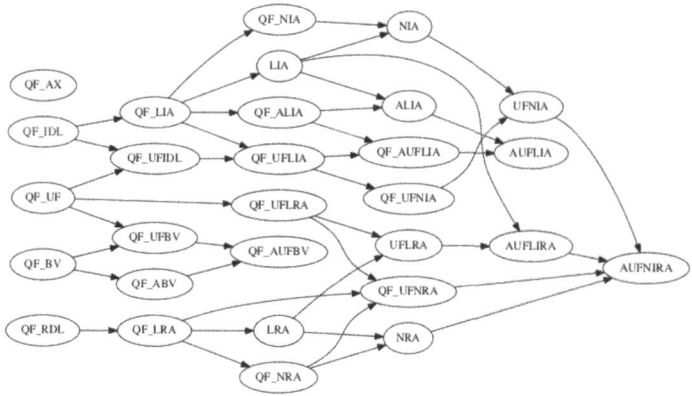

Fig. 12.4 SMT-LIB: The group structure of SMT problems by theories. The arrow shows the subgroup relation. Source: `smtlib.cs.uiowa.edu/index.shtml`

used by the groups and some major restriction of the group, with the following nomenclature:

- **QF** for the restriction to quantifier free formulas
- **A** or **AX** for the theory of arrays
- **BV** for the theory of fixed size bit vectors
- **IA** for the theory of integer arithmetic
- **RA** for the theory of real-number arithmetic
- **IRA** for the theory of mixed integer and real arithmetic
- **IDL** for the theory of integer difference logic
- **UF** for uninterpreted function symbols (other than the operations from arithmetic and equality)
- **L** before **IA**, **RA**, or **IRA** for the linear fragment of those arithmetic
- **N** before **IA**, **RA**, or **IRA** for the nonlinear fragment of those arithmetic

The above abbreviated names can be combined to form subgroups or supergroups. For instance, the supergroup is **AUFNIRA**, which is a combination of **A-UF-N-IRA** and denotes the group of problems from nonlinear (**N**) fragment of mixed integer and real arithmetic (**IRA**), with arrays (**A**) and restrictions to uninterpreted functions (**UF**).

The website notes that the reason for this nomenclature is mostly historical. For instance, the set of quantifier-free (QF) formulas is the union of ground formulas and non-ground formulas with free variables. However, the website does not have a group named "ground." When needed, it uses the word "closed" in the description of a group. Indeed, closed QF formulas are exactly ground formulas.

A great effort of SMT-LIB is to provide a standard input/output language for SMT solvers. Researchers use this language to create new problems and send them to SMT-LIB. Developers of SMT solvers use SMT-LIB to experiment and test new algorithms. SMT-LIB also provides links to SMT solvers and related tools and utilities.

12.4.4 SMT-COMP: Competition of SMT Solvers

SMT-LIB provides benchmark problems for the Satisfiability Modulo Theories Competition (SMT-COMP), which has its own website at

github.com/SMT-COMP/

According to the website, SMT-COMP arose from the SMT-LIB initiative to spur adoption of the common, community-designed SMT-LIB formats and to spark further advances in SMT, especially for verification. The first SMT-COMP was held in 2005. The experience with SMT-COMP 2005 confirmed the community's expectations that a public competition would indeed motivate researchers of SMT solvers to adopt the common SMT-LIB input format. Subsequent SMT-COMPs have provided further evidence that such a competition can stimulate improvement in solver implementations: solvers entered in each competition have improved significantly over those in previous competitions. SMT-LIB is the place where we can find SMT solvers of the state of the art.

The competition consists of several divisions by problem types and each division consists of several tracks by ways of solving. For SMT-COMP 2023, there are 24 divisions and a total of 95 tracks. Some frequent or high-performance winners of these tracks are listed below:

- CVC5: an SMT solver developed mainly at Stanford University and University of Iowa
- Yices2: an SMT solver developed and distributed by SRI International
- Bitwuzla: an SMT solver developed at Stanford University
- SMTInterpol: an SMT solver written in Java, developed at University of Freiburg
- OpenSMT: an SMT solver written in C++, developed at USI, Switzerland
- Z3: an SMT solver developed by Microsoft, and Z3++, a modification of Z3 (4.8.15), developed at Chinese Academy of Sciences
- STP: a solver for QF-BV and arrays, developed at Stanford University and University of Melbourne
- Vampire 4.8: a theorem prover for first-order logic, developed at University of Manchester and Vienna University of Technology
- iProver 3.8: a theorem prover for quantified first-order logic with support for arithmetical reasoning, developed at The University of Manchester

These SMT solvers are freely available on the internet. The interested reader may download a copy of these solvers and experiment with the problems from SMT-LIB.

There exist outstanding solvers that work well with a single theory, and some examples are listed below:

- Linear Arithmetic (LA): SPASS-SATT from Max Planck Institute (MPI) for Informatics, https://www.mpi-inf.mpg.de/departments/automation-of-logic/software/spass-workbench/spass-satt
- Nonlinear Integer Arithmetic (NIA): AProVE from RWTH Aachen University, https://aprove.informatik.rwth-aachen.de/ and BLAN from Chinese Academy of Sciences, https://github.com/fuqi-jia/BLAN
- String Constraints: OSTRICH from several universities in the UK, Germany, and China, https://github.com/uuverifiers/ostrich

This chapter will be updated in the near future to report on mature and efficient techniques for SMT problems.

Exercises

1. Please continue the execution of $DPLLT(X)$ by hand on Example 12.1.6 until a model is found and record every step of the execution.

2. In Example 12.2.1, the rewrite rules are made from

$$\{(x_0 \doteq a[3], x_1 \doteq f(a[2], x_0), x_2 \doteq f(a[1], x_1), y \doteq f(a[1], f(a[2], a[3])))\}$$

by orienting them from left to right. Please show $x_2 \doteq y$ by the rewrite rules obtained by orienting the above equations from right to left and list each step of rewriting.

3. Run the algorithm $NOS(A)$ by hand for $A = \{f^3 a \doteq a, f^5 a \doteq a\}$. Using Example 12.2.6 as model to describe the changes of $up(v)$, $h(v)$, $use(v)$ for any node v.

4. If you have a software tool to solve a linear program for maximizing an objective function, how to use this tool to solve a linear program where the goal is to minimize an objective function?

5. Given the set E of ground equations from Example 12.2.10

$$\{f(a) \doteq c_1, f(c_1) \doteq c_2, f(c_2) \doteq a, f(c_2) \doteq c_3, f(c_3) \doteq c_4, f(c_4) \doteq a\}$$

please run the Knuth-Bendix procedure on E with the ordering $f > a > c_1 > c_2 > c_3 > c_4$, and list all the rewrite rules made from the procedure in the order they were created.

6. Let $E_0 = \{f(a, g(a)) \doteq g(g(a))\}$ (Example 12.2.9). Please run $KBG(\textit{flattenSlicing}(E_0))$ by hand and show the contents of $nf(c)$, $class(c)$, and $use(c)$ for each identifier c before and after the main loop of KBG.

7. Assuming each hash function takes constant time, prove that Theorem 12.2.14 remains true if the flatten operation is omitted. In other words, the slicing operation is sufficient for Theorem 12.2.14.

8. The final matrix given in Sect. 12.3.1 is

x_0	x_1	x_2	x_3	x_4	x_5	rhs
1	0	0	0	1	1	33
0	0	0	1	−4	1	4
0	0	1	0	3	−1	3
0	1	0	0	−2	1	6

Please write the corresponding linear equational system in standard form and show step by step how the initial matrix became the final matrix through operations on the matrices (such as multiplying one row by constant or subtracting one row by another row).

9. Please continue the execution of the new version of $DPLLT(X)$ by hand on Example 12.4.2 until a model is found and record every step of the execution.

10. Please provide the details of running Nelson and Oppen's combination algorithm on the following inputs involving T_{LIA} and T_{EUF}:
    a. $S_1 = \{f(f(x-1) - f(y)) \doteq a+1, f(0) > a+2, x - y \doteq 1\}$.
    b. $S_2 = \{f(g(x-1)) \doteq x+1, f(g(y+1)) \doteq y-1, x \doteq y+2\}$.
    c. $S_3 = \{f(x) \geq 1, f(y) \doteq 0, x < y+1, x \geq y\}$.

11. Please show the execution detail of Nelson and Oppen's combination algorithm on the following input:

$$a \doteq b+1 \wedge A \doteq \text{write}(B, a+1, 4) \wedge \text{read}(A, b+2) \doteq 5$$

12. Please add *case analysis* in Nelson and Oppen's combination algorithm and show how input S is shown to be unsatisfiable (see Example 12.4.7):

$$S = \{f(1) \doteq a, f(x) \doteq b, f(2) \doteq c, 1 \leq x, x \leq 2, b \doteq a-1, b \doteq c+1\}.$$

References

1. Egon Börger, Erich Grädel, Yuri Gurevich, *The Classical Decision Problem*, Springer-Verlag, Berlin, 1997
2. Robert Nieuwenhuis, Albert Oliveras, and Cesare Tinelli, "Solving SAT and SAT Modulo Theories: From an Abstract Davis–Putnam–Logemann–Loveland Procedure to DPLL(T)", *Journal of Association for Computing Machinery*, 53 (6): 937–977, 2006
3. Deepak Kapur, "Shostak's congruence closure as completion", in Hubert Comon (ed.) *Rewriting Techniques and Applications*, 8th International Conference, RTA-97, Sitges, Spain, June 2-5, 1997, vol 1232, Lecture Notes in Computer Science, pages 23–37. Springer, 1997
4. George Dantzig, "Origins of the simplex method". In Stephen G. Nash (ed.). *A History of Scientific Computing*, Association for Computing Machinery, May 1987
5. Daniel Kroening, Ofer Strichman, *Decision Procedures: An Algorithmic Point of View*, (2nd ed.) Springer, November 2016

Index

Symbols
α-rule, 83
□, 369
 first-order logic, 185
 release, 381
β-rule, 83
◇, 369
 first-order logic, 185
 until, 381
λ-calculus, 397

A
Algebra
 Boolean, 25
 categorical, 25
 universal, 25
Algorithm, 23, 409
Alphabet, 162
Argument, 8
Aristotle, 5
Arity, 163
Assertion, 316
 middle, 324
Assignment
 rule, 320
Atom
 atomic formula, 164
Axiom, 26
 consistent, 26
 independent, 26

B
Backus, John, 21
Backus–Naur form (BNF), 21, 181
Bakery algorithm, 384
Bendix, Peter, 221, 252
Bijection, 14
 computable, 419
 counting, 419
 increasing, 419
Bijective proof, 419
Binary decision diagram (BDD), 57
Boethius, 5
Boole, George, 5, 25
Boolean, 4
 algebra, 25
 function, 14, 35
 variable, 4
Boolean constraint propagation (BCP), 106
Bounded model checking (BMC), 381
Buchberger, Bruno, 258
Buchberger's algorithm, 258

C
Chomsky's, Noam, 23
Church, Alonzo, 23, 182, 183, 397
Church-Turing thesis, 23, 24, 396
Clarke, Edmund, 381
Clause, 51
 binary, 94
 definite, 108, 288

empty, 94
fact, 108, 288
hard, 145
Horn, 108
negative, 94
positive, 94
query, 288
rule, 108, 288
soft, 145
unit, 94
Closure
congruence, 242
equivalence, 242
transitive, 242
Collatz conjecture, 268
Complete
C-complete, 260
proof procedure, 27
Turing, 24
Computable, 22, 23
function, 22
total, 405
Computably countable, 420
Congruence, 239
class, 240
Congruence closure
ground equality, 444
quantifier-free, 444
uninterpreted functions, 444
Conjecture
Collatz, 268
Conjunctive normal form (CNF), 51
2CNF, 117
Consistency, 27
Consistent, 26
C-consistent, 260
Constraint satisfaction, 173
Constructor, 259
Correctness
partial, 328
total, 328
Counterexample, 437
Curry, Haskell, 269

D
Davis, Martin, 120
Decidable, 23
Decider, 23, 405
Decision
literal, 121
procedure, 27
tree, 121
Decision function, 409

Decision procedure, 81
semi-decision procedure, 28
Definition
extensional, 180
intentional, 179
Deletion
pure, 105
subsumption, 105
tautology, 105
unit, 106
de Morgan's law, 45
Directed acyclic graph (DAG), 57
Disjunctive normal form (DNF), 53
Distributive law, 45
Domain
of discourse, 168
theory, 434

E
Einstein, Albert, 2
Entailment, 46, 175
Enumerator, 420
Equality
crossing, 262
Equality crossing
ordered, 272
Equisatisfiable, 97, 192
Equivalence, 43, 175
LTL, 361
TL, 355
Erdös, Paul, 268
Euclid, 5
Extensional definition, 180

F
Fallacy, 8
circular reasoning, 268
deductive, 8
hasty generalization, 165
logical, 8
First-order
interpretation, 170
Flattening, 278
Floyd, Robert, 316
Form
canonical, 48
normal, 48
Formal language, 405
Formula
closed, 167
dual, 50
first-order, 164

Index 479

instance, 38
model, 41
propositional, 37
satisfiable, 41
TL, 353
valid, 42
X-, 368
Fraenkel, Abraham, 20
Frege, Gottlob, 5, 88, 182
Function, 13
   bijective, 14
   Boolean, 14
   characteristic, 18, 19
   computable, 22
   decision, 18
   defined, 259
   domain, 13
   general recursive, 21, 23, 397
   image, 13
   injective, 14
   inverse, 14
   one-to-one, 14
   onto, 14
   partial, 14
   partial recursive, 21
   primitive recursive, 21, 23, 397
   range, 13
   rank, 15
   Skolem, 191
   surjective, 14
   total, 14
   total computable, 22
   uninterpreted, 442, 443, 468

G
Gentzen, Gerhard, 89
Godel, Kurt, 23, 397
Gröbner basis, 258
Ground, 164
   equality, 444
   equation, 255
Group, 241

H
Halting problem, 398, 399
Herbrand
   base, 196
   interpretation, 196
   model, 197
   universe, 196
Herbrand base
   with equality, 240
Herbrand, Jacques, 195
Hilbert system, 88
Hilbert, David, 15, 88, 182
Hoare, Tony, 316
Hoare triple, 319
Horn, Alfred, 108
Hyper-resolution, 229
   negative, 229

I
Ibn Sina, 5
Identifier, 443
   if-hen-else (ITE), 55
If-hen-else (ITE), 49
Imperative language, 316, 397
Imperative program, 316
Implication
   contrapositive, 43
   converse, 43
   rule, 321
Independent, 26
Induction
   structural rule, 261
Inequality, 444
Inference
   fair strategy, 29
   graph, 92
   rule, 28
      axiom, 86
      sound, 86
   system, 28
Inference rule
   sound, 28
Intentional definition, 179
Interpretation, 4
   first-order, 170
   full, 111
   Herbrand, 196
   partial, 111
   propositional, 39
Invariant
   bounding, 342
   essential, 342
   loop, 324
ITE normal form (INF), 55, 59

K
Kapur, Deepak, 448
Kleene, Stephen, 23, 397
Knuth, Donald, 221, 252
Knuth-Bendix completion procedure, 248, 252, 448

Knuth-Bendix order (KBO), 221, 231, 246
Kripke
 frame, 353, 354, 359, 365
 model, 355
 semantics, 353
Kripke, Saul, 353

L
Lambda calculus, 182, 397
Lambert, Johann, 5
Lamport, Leslie, 384
Language
 decidable, 405
 formal, 405
 recognizable, 405
 recursive, 405
 total recognizable, 405
 unrecognizable, 405
Leaf, 369
 closed, 367
 expandable, 367
 open, 367
Leibniz, Gottfried, 5
Lexicographical path order (LPO), 220, 231, 246
Literal, 39, 50
 decision, 121
 first-order, 188
 implied, 121
 negative, 50
 positive, 50
Local search, 147
Logic
 Boolean, 4
 first-order, 161
 formal, 3
 higher-order, 182
 Hoare, 316
 many-sorted, 169, 181
 modal, 352
 order-sorted, 180
 philosophical, 3
 probability, 4
 propositional, 4
 second-order, 182
 temporal, 352
Logical
 connective, 4
 consequence, 46, 355, 362
 equivalence, 44
 operator, 4

Loop, 320
 exit condition, 324
 invariant, 324

M
Manna, Zohar, 381
Mapping, 13
Marked dead, 373
Markov, Andrey, 429
MaxSAT
 hybrid, 145
 partial and weighted, 146
 solution, 145
 weighted, 145
Maxterm, 51
Meaning, 4
Minsky, Marvin, 23
Minterm, 53
Modal logic, 352
Model, 25
 computing, 22
 first-order, 173
 Herbrand, 197
 Kripke, 355
 LTL, 361
Model checking, 361, 366, 378, 382
Modus ponens (MP), 28, 184
Monad, 317
Monadic, 168, 179, 264, 433
Monotonic, 219
Most general unifier (mgu), 204
Multiset, 218

N
Naur, Peter, 21
Negation normal form (NNF), 49
Next
 X-formula, 368
 X-rule, 368

O
One-to-one correspondence, 14
Operator
 minimum, 59
 status, 221
 sufficient, 59
Order
 canonical, 17
 KBO, 221

Knuth-Bendix, 221
LPO, 220
minimal, 13
partial order, 13
quasi, 219
RPO, 220
simplification, 219
total, 13
well, 13
well-founded, 13

P
Paradox
Barber's, 19
the court, 7
Curry's, 269
drinker, 178
Epimenides, 268
hotel, 15
interesting number, 7
liar, 268
Pinocchio, 29
Protagoras, 7
Russell's, 19, 269
temperature, 351
Yablo's, 30
Partially computable, 405
Partially decidable, 405
Peano, Giuseppe, 5
Pivot equation, 456
Pnueli, Amir, 359
Post, Emil, 23, 429
Postcondition, 319
Precedence, 37, 164, 167, 219, 232, 246
Precondition, 319
weakest, 322
Predicate, 161
calculus, 161
Prenex normal form (PNF), 188, 190
Prior, Arthur, 269
Product
term, 53
Prolog
fact, 288
query, 288
rule, 288
Proof, 28, 86
formal, 28
procedure, 27
Property
Rice property, 416
Proposition, 3

Propositional
logic, 35
variable, 4
Protagoras, 7
Prover
refutation, 80
tautology, 80
theorem, 80

Q
Quantifier
existential, 162
universal, 162
Quantifier-free (QF), 436

R
Random-access stored-program (RASP), 397, 415
Rank, 15
Reachable, 369
Reason, 111
Recognizable, 403
Recognizer, 405
Recursive functions, 397
Recursively enumerable, 405, 420
Recursive path order (RPO), 220, 231, 246
Reduction, 412
mapping, 412
Reflexing, 271
Regular sequence, 365
Relation, 12
antisymmetric, 13
asymmetric, 13
comparable, 13
irreflexive, 12
reflexive, 12
transitive, 13
Release, 377
strong, 380
Resolution
binary, 94
core, 95
cut, 92
hyper-resolution, 229
input, 96
linear, 96
negative, 96
ordered, 96, 224
positive, 96
proof, 95
saturation, 102, 224

set-of-support, 96
unit, 96
unit-resulting (ur), 229
ur-resolution, 229
Resolvent, 94
Rice property, 416
Robinson, J. Alan, 207, 213
Rule
  assignment, 320
  consequence, 321
  implication, 321
Russell, Bertrand, 5

S

Satisfiability (SAT), 48, 119
  decoder, 138
  encoder, 138
  first-order logic, 173
  maximum, 65
  MAX-SAT, 143
  propositional, 48
Satisfiable
  equisatisfiable, 97, 192
  first-order, 173
  LTL, 361
  TL, 355
Saturation, 224
Semi-decidable, 405
Sentence, 167
Sequent, 89
Set, 11
  cardinality, 12
  characteristic function, 19
  characteristic sequence, 18
  complement, 12
  countable, 15
  countably infinite, 15
  difference, 12
  intersection, 12
  same size, 14
  union, 12
  universal, 11
Side effect, 317
Skolem
  constant, 191
  function, 191
Smullyan, Raymond, 72, 178
Soundness, 27
  weighted clauses, 151
Stable, 219
Stanford temporal prover (STeP), 381
State
  accept, 401

reject, 401
sequence, 360
Turing machine, 399
Statement
  composed, 4
  simple, 4
Status
  operator, 221
Strategy
  fair, 29, 93
Style
  deduction, 81
  enumeration, 81
  reduction, 81
  saturation, 101
Substitution, 202
  affected variable, 202
  idempotent, 202
  renaming, 202
Subsumption
  backward, 217
  forward, 217
Suffix
  immediate, 360

T

Tableau, 81
  LTL, 366
Tautology, 42
Temporal logic, 352
Term, 163
  graph, 208
Termination, 27
Theorem, 26, 46
  inductive, 259
  prover, 80
  proving, 42
Theory, 46
  convex, 470
  equality, 243
  stably infinite, 470
  type, 183
Thue, Axel, 256, 429
Total order, 13
Total recognizable, 405
Trace, 383
Truth
  table, 36
  value, 3
Tseitin
  transformation, 98
Turing
  complete, 24

computable, 397
degree, 414
equivalent, 414
Turing, Alan, 23, 397
Turing machine, 396
decider, 405
recognizer, 405
universal, 396, 415
Type, 168, 264, 433
strong typing, 181
Type theory, 183
simple, 183

U
Unifiable, 202
Unification
clash, 204
occur check, 204
Unifier, 202
mgu, 204
more general, 204
Unit deletion, 217
Unit propagation, 106
Universal halting problem, 407
Universe, 168
Herbrand, 196
of discourse, 11, 168
Until, 377
weak, 380

V
Valid, 42
first-order, 173
LTL, 361
TL, 355
Variable
basic, 455
Boolean, 4
bounded, 167
free, 167
propositional, 4
scope, 167
slack, 460
Verification condition, 331
Von Neumann, John, 23, 397

W
Weighted CNF (WCNF), 145
Well-founded
induction, 218
order, 13
Well order, 13
William of Ockham, 5
Wu, Wenjun, 258
Wu's method, 258

Z
Zermelo, Ernst, 19
Zermelo–Fraenkel set theory (ZF), 20

The manufacturer's authorised representative in the EU is Springer Nature Customer Service Centre GmbH, Europaplatz 3, 69115 Heidelberg, Germany. If you have any concerns regarding our products, please contact ProductSafety@springernature.com

Printed and bound by CPI Group (UK) Ltd, Croydon, CR0 4YY

25/03/2026

02078197-0006